Microstructure and Properties in Metals and Alloys, 2nd Volume

Microstructure and Properties in Metals and Alloys, 2nd Volume

Editors

Andrea Di Schino
Claudio Testani

Basel • Beijing • Wuhan • Barcelona • Belgrade • Novi Sad • Cluj • Manchester

Editors
Andrea Di Schino
University of Perugia
Perugia
Italy

Claudio Testani
CALEF
Rome
Italy

Editorial Office
MDPI
St. Alban-Anlage 66
4052 Basel, Switzerland

This is a reprint of articles from the Topic published online in the open access journals *Alloys* (ISSN 2674-063X), *Coatings* (ISSN 2079-6412), *Crystals* (ISSN 2073-4352), *Materials* (ISSN 1996-1944), and *Metals* (ISSN 2075-4701) (available at: https://www.mdpi.com/topics/22674A2MWM).

For citation purposes, cite each article independently as indicated on the article page online and as indicated below:

Lastname, A.A.; Lastname, B.B. Article Title. *Journal Name* **Year**, *Volume Number*, Page Range.

ISBN 978-3-7258-1387-2 (Hbk)
ISBN 978-3-7258-1388-9 (PDF)
doi.org/10.3390/books978-3-7258-1388-9

© 2024 by the authors. Articles in this book are Open Access and distributed under the Creative Commons Attribution (CC BY) license. The book as a whole is distributed by MDPI under the terms and conditions of the Creative Commons Attribution-NonCommercial-NoDerivs (CC BY-NC-ND) license.

Contents

About the Editors ix

Andrea Di Schino and Claudio Testani
Microstructure and Properties in Metals and Alloys (Volume 2)
Reprinted from: *Metals* 2024, *14*, 473, doi:10.3390/met14040473 1

Jinjian Li, Bing Hu, Liyang Zhao, Fangmin Li, Jiangli He, Qingfeng Wang and Riping Liu
Influence of Heat Input on the Microstructure and Impact Toughness in Weld Metal by High-Efficiency Submerged Arc Welding
Reprinted from: *Metals* 2023, *13*, 1217, doi:10.3390/met13071217 11

Zhenbo Zuo, Rui Hu, Xian Luo, Hongkui Tang, Zhen Zhu, Zitong Gao, et al.
Evolution Behavior of Rapidly Solidified Microstructure of a Ti-48Al-3Nb-1.5Ta Alloy Powder during Hot Isostatic Pressing
Reprinted from: *Metals* 2023, *13*, 1243, doi:10.3390/met13071243 29

Ruida Xu, Ying Li and Huichen Yu
Microstructure Evolution and Dislocation Mechanism of a Third-Generation Single-Crystal Ni-Based Superalloy during Creep at 1170 °C
Reprinted from: *Materials* 2023, *16*, 5166, doi:10.3390/ma16145166 45

Nikolay Belov, Torgom Akopyan, Kirill Tsydenov, Nikolay Letyagin and Anastasya Fortuna
Structure Evolution and Mechanical Properties of Sheet Al–2Cu–1.5Mn–1Mg–1Zn (wt.%) Alloy Designed for $Al_{20}Cu_2Mn_3$ Disperoids
Reprinted from: *Metals* 2023, *13*, 1442, doi:10.3390/met13081442 57

Bryan Ramiro Rodriguez-Vargas, Giulia Stornelli, Paolo Folgarait, Maria Rita Ridolfi, Argelia Fabiola Miranda Pérez and Andrea Di Schino
Recent Advances in Additive Manufacturing of Soft Magnetic Materials: A Review
Reprinted from: *Materials* 2023, *16*, 5610, doi:10.3390/ma16165610 77

Qian Zhao, Zhixia Qiao and Ji Dong
Effect of Tempering Time on Carbide Evolution and Mechanical Properties of a Nb-V-Ti Micro-Alloyed Steel
Reprinted from: *Metals* 2023, *13*, 1495, doi:10.3390/met13081495 110

Duo Tan, Bin Fu, Wei Guan, Yu Li, Yanhui Guo, Liqun Wei and Yi Ding
Hierarchical Multiple Precursors Induced Heterogeneous Structures in Super Austenitic Stainless Steels by Cryogenic Rolling and Annealing
Reprinted from: *Materials* 2023, *16*, 6298, doi:10.3390/ma16186298 125

Mhd Noor Ervina Efzan and Hao Jie Kong
The Properties and Microstructure of Na_2CO_3 and Al-10Sr Alloy Hybrid Modified LM6 Using Ladle Metallurgy Method
Reprinted from: *Materials* 2023, *16*, 6780, doi:10.3390/ma16206780 134

Genggen Liu, Jiao Man, Bin Yang, Qingtian Wang and Juncheng Wang
A Phase Field Study of the Influence of External Loading on the Dynamics of Martensitic Phase Transformation
Reprinted from: *Materials* 2023, *16*, 6849, doi:10.3390/ma16216849 145

Bingyang He, Juan Wang and Weipu Xu
Constitutive Models for the Strain Strengthening of Austenitic Stainless Steels at Cryogenic Temperatures with a Literature Review
Reprinted from: *Metals* 2023, *13*, 1894, doi:10.3390/met13111894 . . . . **164**

Dong-Ju Chu, Chanhee Park, Joonho Lee and Woo-Sang Jung
Effect of Ti/Al Ratio on Precipitation Behavior during Aging of Ni-Cr-Co-Based Superalloys
Reprinted from: *Metals* 2023, *13*, 1959, doi:10.3390/met13121959 . . . . **179**

Yikai Wang, Xiao Qin, Naixin Lv, Lin Gao, Changning Sun, Zhiqiang Tong and Dichen Li
Microstructure Optimization for Design of Porous Tantalum Scaffolds Based on Mechanical Properties and Permeability
Reprinted from: *Materials* 2023, *16*, 7568, doi:10.3390/ma16247568 . . . . **193**

Jorge Eduardo Hernandez-Flores, Bryan Ramiro Rodriguez-Vargas, Giulia Stornelli, Argelia Fabiola Miranda Pérez, Felipe de Jesús García-Vázquez, Josué Gómez-Casas and Andrea Di Schino
Evaluation of Austenitic Stainless Steel ER308 Coating on H13 Tool Steel by Robotic GMAW Process
Reprinted from: *Metals* 2024, *14*, 43, doi:10.3390/met14010043 . . . . **206**

Seulgee Lee, Chayanaphat Chokradjaroen, Yasuyuki Sawada, Sungmin Yoon and Nagahiro Saito
Regulated Phase Separation in Al–Ti–Cu–Co Alloys through Spark Plasma Sintering Process
Reprinted from: *Materials* 2024, *17*, 304, doi:10.3390/ma17020304 . . . . **223**

Grega Klančnik, Jaka Burja, Urška Klančnik, Barbara Šetina Batič, Luka Krajnc and Andrej Resnik
Microstructure and Properties Variation of High-Performance Grey Cast Iron via Small Boron Additions
Reprinted from: *Crystals* 2024, *14*, 103, doi:10.3390/cryst14010103 . . . . **235**

Xufeng Wang, Xufeng Gao, Yaxuan Jin, Zhenhao Zhang, Zhibo Lai, Hanyu Zhang and Yungang Li
Molecular Dynamics Simulation Research on Fe Atom Precipitation Behaviour of Cu-Fe Alloys during the Rapid Solidification Processes
Reprinted from: *Materials* 2024, *17*, 719, doi:10.3390/ma17030719 . . . . **254**

Lianfeng Yang, Huan Zhang, Xiran Zhao, Bo Liu, Xiumin Chen and Lei Zhou
Study on the Microscopic Mechanism of the Grain Refinement of Al-Ti-B Master Alloy
Reprinted from: *Metals* 2024, *14*, 197, doi:10.3390/met14020197 . . . . **268**

Xinyue Li, Kunyu Wang, Yunlong Li, Zhiqiang Wang, Yang Zhao and Jie Zhu
Mechanical and Magnetic Properties of Porous $Ni_{50}Mn_{28}Ga_{22}$ Shape Memory Alloy
Reprinted from: *Metals* 2024, *14*, 291, doi:10.3390/met14030291 . . . . **281**

Giulia Morettini, Luca Landi, Luca Burattini, Giulia Stornelli, Gianluca Foffi, Andrea Di Schino, et al.
Application of the Theory of Critical Distance (TCD) to the Breakage of Cardboard Cutting Blades in Al7075 Alloy
Reprinted from: *Metals* 2024, *14*, 301, doi:10.3390/met14030301 . . . . **296**

Ruifeng Dong, Jian Li, Zishuai Chen, Wei Zhang and Xing Zhou
Effect of Deformation Degree on Microstructure and Properties of Ni-Based Alloy Forgings
Reprinted from: *Metals* 2024, *14*, 340, doi:10.3390/met14030340 . . . . **314**

Haoxin Sun, Bo Liu and Guo Pu
Improved High-Temperature Stability and Hydrogen Penetration through a Pd/Ta Composite Membrane with a TaTiNbZr Intermediate Layer
Reprinted from: *Coatings* **2024**, *14*, 370, doi:10.3390/coatings14030370 . . . . **327**

Lucia Bajtošová, Barbora Kihoulou, Rostislav Králík, Jan Hanuš and Miroslav Cieslar
Nickel Nanoparticles: Insights into Sintering Dynamics
Reprinted from: *Crystals* **2024**, *14*, 321, doi:10.3390/cryst14040321 . . . . **338**

About the Editors

Andrea Di Schino

Andrea Di Schino holds a degree in Physics from the University of Pisa, and, in 1996, received a PhD in Materials Engineering from the University of Naples Federico II. He has previous experience as a senior scientist at Centro Sviluppo Materiali, Rome (Metallurgy Department). He is an associate professor of Metallurgy at the Engineering Department, University of Perugia, and he is also the head of the Industrial School University of Perugia-Terni branch. He is also listed as a Top Italian Scientist.

Claudio Testani

Claudio Testani holds a degree in structural aerospace engineering, and was awarded his Ph.D. in material engineering from the Tor Vergata University of Rome, Italy, in 2006. He holds a habilitation in metallurgy and is member of the Ph.D. industrial engineering scientific board of Tor Vergata University. For over 30 years, he has managed research projects and is the technical director of CALEF, an Italian public–private research consortium. In 2023, he was awarded the Prize "Materials 2023—Outstanding Reviewer Award".

Editorial

Microstructure and Properties in Metals and Alloys (Volume 2)

Andrea Di Schino [1,*] and Claudio Testani [2]

1. Dipartimento di Ingegneria, Università Degli Studi di Perugia, Via G. Duranti 93, 06125 Perugia, Italy
2. CALEF-ENEA CR Casaccia, Via Anguillarese 301, Santa Maria di Galeria, 00123 Rome, Italy; claudio.testani@consorziocalef.it
* Correspondence: andrea.dischino@unipg.it

1. Introduction and Scope

Microstructure design is key in targeting the desired material's properties. It is therefore essential to understand the relationship between these properties and the microstructure [1–10] and how to enhance them via a specific process [11–20], including, as an example, additive techniques. The following five journals participated in the current Topic: *Coatings*, *Alloys*, *Crystals*, *Materials*, and *Metals*. Contributions related to microstructure design and characterization were brought together as part of this Topic, combined with their association with the mechanical, fatigue, wear, and corrosion resistance of different types of metals and alloys. The goal of this Topic is to present contributions related to the relationship between the microstructure and properties of metals and alloys for different applications, including aeronautical and aerospace applications. Different process routes were considered (including thermo-mechanical routes and additive manufacturing) as a part of this Topic. Welding is a key issue in many applications; this is the reasoning as to why contributions related to welding are also included in this Topic.

2. Overview of the Published Articles

The present Topic includes nineteen research papers, one communication, and two review papers. Among the publications included, eleven papers were published in *Metals*, eight papers were published in *Materials*, two were published in *Crystals*, and one paper was published in *Coatings*, covering numerous aspects concerning microstructure–property relationships in the field of metals and alloys.

The contributions to the aforementioned journals are listed below:

1. Li, J.; Hu, B.; Zhao, L; Li, F.; He, J.; Wang, Q.; Liu, R. Influence of Heat Input on the microstructure and Impact Toughness in Weld Metal by High-Efficiency Submerged Arc Welding. *Metals* **2023**, *13*, 1217. https://doi.org/10.3390/met13071217.
2. Zuo, Z.; Hu. R.; Luo, X.; Tang, H.; Zhu, Z.; Gao, Z.; Li, J.; Zou, H.; Li. A.; Zhao, X.; Lai, Y., Li. S. Evolution Behavior of Rapidly Solidified Microstructure of a Ti 48Al-3Nb-1.5Ta Alloy Powder during Hot Isostatic Pressing. *Metals* **2023**, *13*, 1243. https://doi.org/10.3390/met13071243.
3. Xu, R.; Li, Y.; Yu, H. Microstructure Evolution and Dislocation Mechanism of a Third-Generation Single-Crystal Ni-Based Superalloy during Creep at 1170 °C. *Materials* **2023**, *16*, 5166. https://doi.org/10.3390/ma16145166.
4. Belov, N.; Akopyan, T.; Tsydenov, K.; Letyagin, N.; Fortuna, A. Structure Evolution and Mechanical Properties of Sheet Al–2Cu–1.5Mn–1Mg–1Zn (wt.%) Alloy Designed for $Al_{20}Cu_2Mn_3$ Disperoids. *Metals* **2023**, *13*, 1442. https://doi.org/10.3390/met13081442.
5. Rodriguez-Vargas, B.R.; Stornelli, G.; Folgarait, P.; Ridolfi, M.; Perez, A.F.M.; Di Schino, A. Recent Advances in Additive Manufacturing of Soft Magnetic Materials: A Review. *Materials* **2023**, *16*, 5610. https://doi.org/10.3390/ma16165610.

Citation: Di Schino, A.; Testani, C. Microstructure and Properties in Metals and Alloys (Volume 2). *Metals* **2024**, *14*, 473. https://doi.org/10.3390/met14040473

Received: 21 March 2024
Accepted: 8 April 2024
Published: 18 April 2024

Copyright: © 2024 by the authors. Licensee MDPI, Basel, Switzerland. This article is an open access article distributed under the terms and conditions of the Creative Commons Attribution (CC BY) license (https://creativecommons.org/licenses/by/4.0/).

6. Zhao, Q.; Qiao, Z.; Dong, J. Effect of Tempering Time on Carbide Evolution and Mechanical Properties of a Nb-V-Ti Micro-Alloyed Steel. *Metals* **2023**, *13*, 1495. https://doi.org/10.3390/met13081495.
7. Tan, D.; Fu, B.; Guan, W.; Li. Y.; Guo, Y.; Wei, L.; Ding, Y. Hierarchical Multiple Precursors Induced Heterogeneous Structures in Super Austenitic Stainless Steels by Cryogenic Rolling and Annealing. *Materials* **2023**, *16*, 6298. https://doi.org/10.3390/ma16186298.
8. Efzan, M.N.E.; Kong, J.K. The Properties and Microstructure of Na_2CO_3 and Al-10Sr Alloy Hybrid Modified LM6 Using Ladle Metallurgy Method. *Materials* **2023**, *16*, 6780. https://doi.org/10.3390/ma16206780.
9. Liu, G.; Jiao, M.; Yang, B.; Wang, Q.; Wang, J. A Phase Field Study of the Influence of External Loading on the Dynamics of Martensitic Phase Transformation. *Materials* **2023**, *16*, 6849. https://doi.org/10.3390/ma16216849.
10. He, B.; Wang, J.; Xu, W. Constitutive Models for the Strain Strengthening of Austenitic Stainless Steels at Cryogenic Temperatures with a Literature Review. *Metals* **2023**, *13*, 1894. https://doi.org/10.3390/met13111894.
11. Chu, D.J., Park, C.; Lee, J.; Jung, W.S. Effect of Ti/Al Ratio on Precipitation Behavior during Aging of Ni-Cr-Co-Based Superalloys. *Metals* **2023**, *13*, 1959. https://doi.org/10.3390/met13121959.
12. Wang, Y.; Qin, X.; Lv, N.; Gao, L.; Sun, C.; Tong, Z.; Li, D. Microstructure Optimization for Design of Porous Tantalum Scaffolds Based on Mechanical Properties and Permeability. *Materials* **2023**, *16*, 7568. https://doi.org/10.3390/ma16247568.
13. Hernandez-Flores J.E., Rodriguez-Vargas, B.R.; Stornelli, G.; Perez, A.F.M.; Gaircia-Vazquez, F.d.J.; Gomez-Casas, J.; Di Schino, A. Evaluation of Austenitic Stainless Steel ER308 Coating on H13 Tool Steel by Robotic GMAW Process. *Metals* **2024**, *14*, 43. https://doi.org/10.3390/met14010043.
14. Lee, S.; Chokradjaroen, C.; Sawada, Y.; Yoon, S.; Saito, N. Regulated Phase Separation in Al–Ti–Cu–Co Alloys through Spark Plasma Sintering Process. *Materials* **2024**, *17*, 304. https://doi.org/10.3390/ma17020304.
15. Klančnik, G.; Burja, J.; Klančnik, U.; Šetina, B.; Krajinc, L.; Resnik, A. Microstructure and Properties variation of high-performance grey cast iron via small boron additions. *Crystals* **2024**, *14*, 103. https://doi.org/10.339/cryst14010103.
16. Wang, X.; Gao, X.; Jin, Y.; Zhang, Z.; Lai, Z.; Zhang, H.; Li, Y. Molecular Dynamics Simulation research on Fe atom precipitation behavior of Cu-Fe Alloys during the rapid solidification processes. *Materials* **2024**, *17*, 719. https://doi.org/10.3390/ma17030719.
17. Yang, L.; Zhang, H.; Zhao, X.; Lui, B.; Chen, X.; Zhou, L. Study on the microscopic mechanism of grain refinement of Al-Ti-B master alloy. *Metals* **2024**, *14*, 197. https://doi.org/10.3390/met1402019.
18. Li, X.; Wang, K.; Li, Y.; Wang, Z.; Zhao, Y.; Zhu, J. Mechanical and magnetic properties of porous $Ni_{50}Mn_{28}Ga_{22}$ shape memory alloy. *Metals* **2024**, *14*, 291. https://doi.org/10.3390/met14030291.
19. Morettini, G.; Landi, L.; Burattini, L.; Stornelli, G.; Foffi, G.; Di Schino, A.; Cianetti, F.; Braccesi, C. Application of the Theory of Critical Distance (TCD) to the Breakage of Cardboard Cutting Blades in Al7075 alloy. *Metals* **2024**, *14*, 301. https://doi.org/10.3390/met14030301.
20. R. Dong; J. Li; Z. Chen; W. Zhang; X. Zhou. Efect of deformation degree on microstructure and properties of nickel base alloys for forgings. *Metals* **2024**, *14*, 340. https://doi.org/10.3390/met14030340.
21. Sun, H.; Li, J.; Du, X; Xu, F.; Yan, S.; Li, X.; Li, Z.; Liu, B.; Pu, G. Improved high temperature stability and hydrogen penetration through a Pd/Ti composite membrane with TaTiNbZr intermediate layer. *Coatings* **2024**, *14*, 370. https://doi.390/coatings14030370.

22. Bajtosova, L.; Kihoulou, B.; Kralol, R.; H., J.; Cieslar, M. Nickel nanoparticles: insights into sintering dynamics. *Crystals* **2024**, *14*, 321. https://doi.org/10.390/crust14040321.

J. Li et al. (Contribution 1) report on the influence of heat input on the microstructure and impact toughness of weld metal via high-efficiency submerged arc welding. In their paper, weld metal was obtained through the process of welding at three different high heat inputs with laboratory-developed high-efficiency submerged arc welding wire for bridges. The effect of changing different high heat inputs on the microstructure and impact toughness of high-efficiency submerged arc weld metal was systematically investigated via cutting and Charpy V-notch impact tests at $-40\ °C$ through the use of optical microscopy, scanning electron microscopy, energy-dispersive electron spectroscopy, electron backscatter diffraction, and transmission electron microscopy for characterization and analysis. With the increase in heat input from 50 kJ/cm to 100 kJ/cm, the impact absorption energy decreased significantly from 130 J to 38 J. The number of inclusions in the weld metal significantly decreased and the size increased, which led to a significant decrease in the number of inclusions, thus effectively promoting acicular ferrite nucleation, further leading to a decrease in the proportion of acicular ferrite in the weld metal. Concurrently, the microstructure of the weld metal was significantly coarsened, the percentage of high-angle grain boundaries decreased, and the size of martensite/austenite constituents was significantly increased monotonically. The crack initiation energy was reduced by the coarsened martensite/austenite constituents and inclusions, which produced larger local stress concentrations, and the process of crack propagation took place with greater ease due to the coarsened microstructure and lower critical stress for crack instability propagation. The martensite/austenite constituents and inclusions in large sizes worked together to cause premature cleavage fracture of the impact specimen, which significantly deteriorated the impact toughness. The findings of this study showed that heat input should not exceed 75 kJ/cm for high-efficiency submerged arc welding wires for bridges.

Zuo et al. (Contribution 2) report on Ti-48Al-3Nb-1.5Ta powders manufactured from cast bars via the supreme-speed plasma rotating electrode process (SS-PREP) and used to prepare hot isostatically pressed (HIPed) material at 1050–1260 $°C$ with 150 MPa for 4 h. The phase, microstructure, and mechanical performance were analyzed through the use of XRD and SEM, an electrical universal material testing machine, and other methods. The results showed that the phase constitution changed from γ phase to $\alpha 2$ phase and then to γ phase with the material changing from as-cast to powders and then to as-HIPed. Compared with the as-cast material, the grain size and element segregation were significantly reduced for both powders and as-HIPed. When the hot isostatic pressing (HIP) temperature was low, the genetic characteristics of the powder microstructure were evident. With the increase in the HIP temperature, the homogeneity of the composition and microstructure increased, and the prior particle boundaries (PPBs) gradually disappeared. The elastic moduli of the powder and as-HIPed were superior to those of th eas-cast, which increased with the increase in the HIP temperature. The hardness of as-HIPed was lower than that of the powder. The compressive strength, compressive strain, bending strength, and tensile strength of as-HIPed were higher than those of the as-cast. With the increase in the HIP temperature, the compressive strength decreased gradually, and the compressive strain was found to first decrease and then increase.

Xu et al. (Contribution 3) investigated the creep behavior and deformation mechanism of a third-generation single-crystal Ni-based superalloy at 1170 $°C$ under a range of stress levels. Scanning electron microscopes (SEMs) and transmission electron microscopes (TEMs) were employed to observe the formation of a rafted γ' phase, which exhibits a topologically close-packed (TCP) structure. The orientation relationship and elemental composition of the TCP phase and matrix were analyzed to discern their impact on the creep properties of the alloy. The primary deformation mechanism of the examined alloy was identified as dislocation slipping within the γ matrix, accompanied by the climbing of dislocations over the rafted γ' phase during the initial stage of creep. In the later stages of creep, super-dislocations with Burgers vectors of a<010> and a/2<110> were observed to shear into the γ' phase, originating from interfacial dislocation networks. Up to the fracture, the sequential activation of dislocation shearing in the primary and secondary slipping systems of the γ' phase occurs. As a consequence of this alternating dislocation shearing, a twist deformation of the rafted γ' phase ensued, ultimately contributing to the fracture mechanism observed in the alloy during creep.

Belov et al. (Contribution 4) studied the possibility of increasing the strength of non-heat-treatable sheet alloy Al2Cu1.5Mn (wt.%) through the joint addition of 1% Mg and 1% Zn. The effect of these elements on the structure and mechanical properties of the new sheet Al2Cu1.5Mn alloy designed for Al20Cu2Mn3 dispersoids was examined through calculations and the use of experimental methods. The obtained data on the phase composition, microstructure, and physical and mechanical properties of the new alloy for different processing routes (including hot rolling, cold rolling, and annealing) were compared with those for the ternary Mg- and Zn-free alloy. The authors showed that the formation of nanosized Al20Cu2Mn3 dispersoids (~7 vol.%) aids in the preservation of the non-recrystallized grain structure after annealing at up to 400 °C (3 h), while Mg and Zn have a positive effect on the strength due to the formation of alloyed aluminum solid solution. As a result, cold-rolled sheets of the Al2Cu1.5Mn1Mg1Zn model alloy showed substantially higher strength performance after annealing at 400 °C in comparison with the ternary reference alloy. Of note, the UTS was found to be ~360 vs. ~300 MPa and the YS was found to be 280 vs. 230 MPa. Regarding the Al2Cu1.5Mn1Mg1Zn model alloy, researchers have demonstrated that the system shows promise for the design of new heat-resistant alloys as a sustainable alternative to 2xxx alloys. This new alloy has an advantage over commercial alloys (particularly 2219, 2024, and 2014) not only in terms of manufacturability but also thermal stability. The sheet production cycle for the model alloy is much shorter because the stages of homogenization, solution treatment, and water quenching are excluded.

Rodriguez-Vargas et al. (Contribution 5) present a revised paper focusing on recent advances in the additive manufacturing of soft magnetic materials. They highlight the significant progress made in the field of materials science, which has enabled the development of novel materials such as high-entropy alloys (HEAs). These alloys, due to their complex chemical composition, can exhibit soft magnetic properties. The aim of the present work was to provide a critical review of the state-of-the-art SMMs manufactured through the use of different AM technologies. This review covers the influence of these technologies on microstructural changes, mechanical strength, post-processing, and magnetic parameters such as saturation magnetization (MS), coercivity (HC), remanence (Br), relative permeability (Mr), electrical resistivity (r), and thermal conductivity (k).

Zhao et al. (Contribution 6) describe the evolution of the microstructure, precipitation behavior, and mechanical performance of Nb-V-Ti micro-alloyed steel prepared under different tempering times, which were studied through the use of transmission electron microscopy (TEM), X-ray diffraction (XRD), and mechanical tests. It was found that the width of the martensite laths increased with the increasing tempering time. Several types of carbides, including M_3C, M_2C, $M_{23}C_6$, M_7C_3, and MC particles, were identified following tempering. The MC carbides were found to remain stable during tempering; however, the transformation behavior of other carbides was identified. The transformation sequence can be summarized as $M_3C \rightarrow M_2C \rightarrow M_7C_3 \rightarrow M_{23}C_6$. The strength was found to decrease and the Charpy impact toughness increased gradually with the increase in the tempering time. The ultimate strength (UTS) decreased from 1231 to 896 MPa, and the yield strength (YS) decreased from 1138 to 835 MPa. The -40 °C Charpy impact toughness was found to increase from 20 to 61 J as the tempering time increased from 10 min to 100 h. The evolution of carbides plays an important role in their mechanical performance.

Tan et al. (Contribution 7) report on a study examining deformed substructures including dislocation cells, nanotwins (NTs), and martensite, which were introduced into super austenitic stainless steels (SASSs) through the process of cryogenic rolling (Cryo-R, 77 K/22.1 mJ·m^{-2}). With the reduction increasing, a low stacking fault energy (SFE) and increased flow stress led to the activation of secondary slip and the occurrence of NTs and martensite nano-laths, while only dislocation tangles were observed under a heavy reduction via cold-rolling (Cold-R, 293 K/49.2 mJ·m^{-2}). The multiple precursors not only possess variable deformation stored energy but also experience competition between recrystallization and reverse transformation during subsequent annealing, thus contributing to the formation of a heterogeneous structure (HS). The HS, which consists of bimodal-grained austenite and retained martensite simultaneously, showed a higher yield strength (~1032 MPa) and a greater tensile elongation (~9.1%) than the annealed coarse-grained Cold-R sample. The superior strength–ductility and strain hardening observed originate from the synergistic effects of grain refinement, dislocation, and hetero-deformation-induced hardening.

Efzan et al. (Contribution 8) report on Al-10Sr alloy and Na_2CO_3 addition to LM6 (reference alloy) as hybrid modifiers through ladle metallurgy. Microstructure enhancement was analyzed through the use of an optical microscope (OM). The results were further confirmed through the use of a scanning electron microscope (SEM) and energy dispersive X-ray (EDX) spectroscopy. The results showed that Na_2CO_3 and Al-10Sr alloy successfully hybrid modified the sharp needle-like eutectic Si into fibrous eutectic Si. Soft primary Al dendrites were also discovered following hybrid modification. The formation of β-Fe flakes was suppressed, and α-Fe sludge was transformed into Chinese script morphology. A 2.13% density reduction was also recorded. A hardness test was additionally performed to examine the mechanical improvement of the hybrid modified LM6. A reduction in hardness of 2.3% was recorded in the hybrid modified LM6 through ladle metallurgy. Brittle cracks were not observed, while ductile pileups were the main features that appeared on the indentations of hybrid-modified LM6, indicating a brittle-to-ductile transformation after hybrid modification of LM6 by Na_2CO_3 and Al-10Sr alloy through ladle metallurgy.

Liu et al. (Contribution 9) adopted an elastoplastic phase field model to investigate the influence of external loading on the martensitic phase transformation kinetics in steel. The phase field model incorporates external loading and plastic deformation. During the simulation process, the authenticity of the phase field model is ensured by introducing the relevant physical parameters and comparing them with experimental data. During the calculations, loads of various magnitudes and loading conditions were considered. An analysis and discussion were conducted concerning the volume fraction and phase transition temperature during the phase transformation process. The simulation results extensively illustrate the preferential orientation of variants under different loading conditions. This model can be applied to the qualitative phase transition evolution of Fe-Ni alloys, and the crystallographic parameters adhere to the volume expansion effect. The authors of the study concluded that uniaxial loading promotes martensitic phase transformation, while triaxial compressive loading inhibits this process. From a dynamic perspective, the authors demonstrate that external uniaxial loading accelerates the kinetics of martensitic phase transformation, with uniaxial compression being more effective in accelerating the phase transformation process than uniaxial tension. When compared to experimental data, the simulation results provide evidence that, under the influence of external loading, the martensitic phase transformation is significantly influenced by the applied load, with the impact of external loading being more significant than that of plastic effects.

He et al. (Contribution 10) present a review paper examining constitutive models for the strain strengthening of austenitic stainless steels at cryogenic temperatures. In this paper, the mechanical properties and microstructure evolution of austenitic stainless steels under different temperatures, types, and strain rates are compared. The phase-transformation mechanism of austenitic stainless steels during strain at cryogenic temperatures and its influence on strength and microstructure evolution are summarized. The constitutive models of strain strengthening at cryogenic temperatures were set to calculate the volume fraction of strain-induced martensite and predict the mechanical properties of austenitic stainless steels.

Chu et al. (Contribution 11) analyzed the precipitation behaviors of Ni-Cr-Co-based superalloys with different Ti/Al ratios aged at 750, 800, and 850 °C for up to 10,000 h using scanning and transmission electron microscopy. The Ti/Al ratio did not significantly affect the diameter of the γ' phase. However, the volume fraction of the γ' phase increased with the increase in the Ti/Al ratios. The η phase was not observed in alloys with a small Ti/Al ratio; however, it was precipitated after aging at 850 °C for 1000 h in alloys with a Ti/Al ratio greater than 0.80. Higher aging temperatures and higher Ti/Al ratios led to faster η formation kinetics and accelerated the degradation of alloys. It is thought that the increase in hardness with an increase in the Ti/Al ratio is attributed to the effective inhibition of the γ' phase on dislocation movement due to the increase in the volume fraction of the γ' phase and an increase in the antiphase boundary (APB) energy.

Wang et al. (Contribution 12) report on porous tantalum (Ta) implants with important clinical application prospects due to their appropriate elastic modulus and excellent bone growth and bone conduction ability. Ta microstructure designs generally mimic titanium (Ti) implants commonly used in the clinic, and at present, there is a lack of research on the influence of such microstructures on the mechanical properties and penetration characteristics, both of which will significantly affect bone integration performance. The authors of this study explored the effects of different microstructure parameters, including the fillet radius of the middle plane and top planes, on the mechanics and permeability properties of porous Ta diamond cells through simulation and put forward an optimization design with a 0.5 mm midplane fillet radius and a 0.3 mm top-plane fillet radius in order to significantly decrease the stress concentration effect and improve permeability. On this basis, the porous Ta structures were prepared through the use of laser powder bed fusion (LPBF) technology and evaluated before and after microstructural optimization. The results showed that the elastic modulus and yield strength increased by 2.31% and 10.39%, respectively. Concurrently, the permeability of the optimized structure also increased by 8.25%. It should be noted that the optimized microstructure design of porous Ta has important medical application value.

Hernandez-Flores et al. (Contribution 13) report on austenitic stainless steel ER308 coating on H13 tool steel through the use of the robotic GMAW process. The heat input during the process was calculated to establish a relationship between the geometry obtained in the coating and its dilution percentage. Furthermore, the evolution of the microstructure of the coating, interface, and substrate was evaluated through the use of XRD and SEM techniques. Notably, the presence of martensite at the interface was observed. The mechanical behavior of the welded assembly was analyzed through Vickers microhardness, and a pin-on-disk wear test was employed to assess its wear resistance. It was found that the dilution percentage is around 18% at high heat input (0.813 kJ/mm) but then decreases to about 14% with reduced heat input. Microhardness test results revealed that at the interface, the maximum value is reached at around 625 HV due to the presence of quenched martensite. Moreover, increasing the heat input favors wear resistance.

Lee et al. (Contribution 14) studied the phase separation in Al–Ti–Cu–Co alloys through use of the spark plasma sintering process. A lightweight Al-Ti-containing multi-component alloy with excellent mechanical strength and an Al–Ti–Cu–Co alloy with a phase-separated microstructure was prepared. The granulometry of metal particles was reduced using planetary ball milling. The particle size of the metal powders decreased as the ball milling time increased from 5 to 7 and up to 15 h (i.e., 6.6 ± 6.4, 5.1 ± 4.3, and 3.2 ± 2.1 μm, respectively). The reduction in particle size and the dispersion of metal powders promoted enhanced diffusion during the spark plasma sintering process. This led to the micro-phase separation of the (Cu, Co)2AlTi (L21) phase, and the formation of a Cu-rich phase with embedded nanoscale Ti-rich (B2) precipitates. The Al–Ti–Cu–Co alloys prepared using powder metallurgy through spark plasma sintering exhibited different hardness values of 684, 710, and 791 HV while maintaining a relatively low density of 5.8–5.9 g/cm^3 (<6 g/cm^3). The mechanical properties were improved upon due to a decrease in particle size achieved through increased ball milling time, leading to a finer grain size. The L21 phase, consisting of (Cu, Co)2AlTi, is the site of basic hardness performance, and the Cu-rich phase is the mechanical buffer layer between the L21 and B2 phases. The finer network structure of the Cu-rich phase also suppresses brittle fracture. These results are important and useful for innovative applications or high process temperatures where extreme requirements are necessary for the tool material and innovative tool concept.

Klančnik et al. (Contribution 15) investigated the effects of small boron additions on the solidification and microstructure of hypo-eutectic alloyed gray cast iron. The characteristic temperatures upon crystallization of the treated metal melt were recorded with regard to small boron addition through the use of thermal analysis with the ATAS system. Additionally, a standardized wedge test was performed to observe any changes in chill performance. The microstructures of the thermal analysis samples were analyzed using a light optical microscope and field emission scanning electron microscope equipped with an energy-dispersive spectroscope, which revealed the variation in graphite count number with boron addition within the examined random and undercooled flake graphite. The effect of boron was estimated through the use of classical analytical and statistical approaches. The solidification behavior under equilibrium conditions was predicted through a thermodynamic approach using Thermo-Calc. Based on all gathered data, a response model was set with boron for a given melt quality and melt treatment using the experimentally determined data. The results of the study revealed that boron as a ferrite- and carbide-promoting element under the experimental conditions shows weak nucleation potential in synergy with other heterogenic nuclei at increased solidification rates; however, no considerable changes were observed in the TA samples solidified at slower cooling rates, indicating the loss of overall inoculation effect. The potential presence of boron nitride as an inoculator for graphite precipitation for a given melt composition and melt treatment was not confirmed in this study. From the results of the above study, it would seem that boron at increased solidification rates can contribute to overall inoculation; however, at slower cooling rates, these effects are gradually lost and in the last solidification range at increased boron content could have a carbide-forming nature, as usually expected. The results of this study suggest that boron in trace amounts could affect the microstructure and properties of hypo-eutectic alloyed grey cast iron.

Wang et al. (Contribution 16) employed molecular dynamics simulation with a cooling rate of 2×10^{10} for $Cu_{100-X}FeX$ (where X represents 1%, 3%, 5%, and 10%) alloy to explore the crystalline arrangement of the alloy and the processes involving iron (Fe) precipitation. The results showed that when the Fe content was 1%, Fe atoms consistently remained uniformly distributed as the temperature of the alloy decreased. Furthermore, there was no Fe atom aggregation phenomenon. The crystal structure was identified as FCC-based Cu crystal, and Fe atoms existed in the matrix in solid solution form. When the Fe content was 3%, Fe atoms tended to aggregate with the decrease in the temperature of the alloy. Moreover, the proportion of BCC crystal structure exhibited no obvious changes, and the crystal structure remained as FCC-based Cu crystal. When the Fe content was 5% and 10%, the Fe atoms exhibited obvious aggregation with the decreasing temperature of the alloy. At the same time, the aggregation phenomenon was found to be more significant with higher Fe content. Fe atom precipitation behavior can be delineated into three distinct stages. The initial stage involves the gradual accumulation of Fe clusters, characterized by a progressively stable cluster size. This phenomenon arises due to the interplay between atomic attraction and the thermal motion of Fe-Fe atoms. In the second stage, small Fe clusters undergo amalgamation and growth. This growth is facilitated by non-diffusive local structural rearrangements of atoms within the alloy. The third and final stage represents a phase of equilibrium where both the size and quantity of Fe clusters remain essentially constant following crystallization of the alloy.

Yang et al. (Contribution 17) studied the structure and properties of Ti_nB_n (n = 2–12) clusters and simulated the microstructure of the Al-Ti-B system through molecular dynamics to determine the grain refinement mechanism of Al-Ti-B master alloy in Al alloy. Based on the density functional theory method, the structural optimization and property calculations of Ti_nB_n (n = 2–12) clusters were carried out. The clusters at the lowest energy levels indicate that the Ti and B atoms were prone to form TiB_2 structures, and the TiB_2 structures tended to be on the surface of the clusters. The $Ti_{10}B_{10}$ cluster was determined to be the most stable structure in the range of n from 2 to 12 in relation to average binding energy and second-order difference energy. The analysis of HOMOs and LUMOs suggested that TiB_2 was the active center in the cluster and the activity of Ti was high; however, the activity of B atoms decreased as the cluster size n increased. In contrast, the prediction of reaction sites using the Fukui function, condensed Fukui function, and condensed dual descriptor showed that Ti atoms were more active than B atoms. Furthermore, the TiB_2 structures were found in the Al-Ti-B system simulated by the ab initio molecular dynamics method, and it was found that there were Al atoms growing on the Ti atoms in the TiB_2. Based on the above analysis, the results of this study suggested that TiB_2 may be a heterogeneous nucleation center of α-Al. This study helps to further understand the mechanism of Al-Ti-B-induced heterogeneous nucleation in Al alloys, which can provide theoretical guidance for related experiments.

X. Li et al. (Contribution 18) report on porous $Ni_{50}Mn_{28}Ga_{22}$ alloy produced through powder metallurgy, with NaCl serving as the pore-forming agent. The phase structure, mechanical properties, and magnetic properties of annealed bulk alloys and porous alloys with different pore sizes were analyzed. Vacuum sintering or mixed green billets in a tube furnace was employed, which facilitated the direct evaporation of NaCl, resulting in the formation of porous alloys characterized by a complete sinter neck, uniform pore distribution, and a consistent pore size. The authors of the study found that porous alloys within this size range exhibit a recoverable shape memory performance of 3.5%, as well as a notable decrease in the critical stress required for martensitic twin shear when compared to that of bulk alloys. Additionally, porous alloys demonstrated a 2% super-elastic strain when exposed to 353 K. Notably, under a 1.5 T magnetic field, the porous Ni50Mn28Ga22 alloy with a pore size ranging from 20 to 30 μm exhibited a peak saturation magnetization of 62.60 emu/g and a maximum magnetic entropy of 1.93 J/kg·K.

G. Morettini et al. (Contribution 19) present a study undertaken in response to two instances of unexpected blade breakage in the cutting blade used in a Carton Wrap machine (CW). Failure of the Al7075 alloy blade occurred at an indentation, during typical operational loading conditions. Subsequent metallographic examinations of the fractured samples confirmed that both cases were attributed to fatigue failure. The main objective of this study was to investigate potential causes of fatigue failure in the CW blade using simplified linear elastic static numerical simulations through finite element analysis (FEA). In this research, the well-established theory of critical distance (TCD) was employed and the authors provided its contextualization at an industrial level for the specific case of the cutting blade. Furthermore, the analysis focuses on a second key aspect: proposing a new blade geometry aimed at mitigating the identified issues and eliminating possible causes of failure. In this context, the actual stress concentration at the indentation is determined using the theory of critical distance with the linear method (ML). The results from the numerical simulations indicate that the new blade geometry significantly reduces the stress concentration, resulting in a risk factor reduction of approximately four when compared to the original blade design, even under non-optimal operating conditions. Overall, the proposed numerical approach, in conjunction with simple linear static FEA, provides substantial support for designers, especially in fault analysis and when comparing different industrial solutions.

R. Dong et al. (Contribution 20) report on the effect of deformation degree on the microstructure and properties of nickel-base alloys for forgings. They describe an experiment using a free-forging hammer to achieve a deformation degree ranging from 60% to 80%. The impact of the forging deformation degree on the hardness and high-temperature erosion performance was evaluated using the Rockwell hardness tester (HRC) and high-temperature erosion tester, respectively. The experimental results indicate that as the deformation degree increased, the hardness of the forged material progressively increased while the rate of high-temperature erosion gradually decreased. In order to comprehensively study the mechanism responsible for variations in forging performance, optical microscopy (OM), scanning electron microscopy (SEM), electron backscatter diffraction (EBSD), and transmission electron microscopy (TEM) were employed. The findings reveal that as the deformation degree increased, the presence of small-angle grain boundaries and an increase in grain boundary area contributed to enhanced hardness in the alloy forgings. Furthermore, it was discovered that grain boundaries with twin orientation promoted dynamic recrystallization during deformation, specifically through a discontinuous dynamic recrystallization mechanism. Additionally, the precipitated γ' phase in the alloy exhibited particle sizes ranging from 40 to 100 nm. This particle size range resulted in a higher critical shear stress value and a more pronounced strengthening effect on the alloy.

H. Sun et al. (Contribution 21) investigated the improved high temperature stability and hydrogen penetration through a Pd/Ti composite membrane with TaTiNbZr intermediate layer. In their paper, the hydrogen separation membrane, a dense TaTiNbZr amorphous layer, was prepared between Pd and Ta to form a Pd/TaTiNbZr/Ta membrane system to prevent the reaction between Pd and Ta at high temperature. The structural stability, as well as the chemical stability of the Pd/TaTiNbZr/Ta film system at high temperatures, was investigated through annealing at 600 °C for 24 h. The high-temperature hydrogen permeation properties of the Pd/TaTiNbZr/Ta film systems were investigated through the use of hydrogen permeation experiments at 600 °C after heat treatment for 6 h. The TaTiNbZr layer was significantly hydrogen permeable. With the increase in the thickness of the barrier layer, the hydrogen permeability of Pd/TaTiNbZr/Ta decreases; however, its hydrogen permeation flux was smaller than that of the highest value of Pd/Ta when it reached the steady state. The presence of the TaTiNbZr layer effectively blocks the interdiffusion between Pd and Ta to form TaPd3, thus improving the sustained working ability of the Pd/TaTiNbZr/Ta membrane system. The results of this study show that TaTiNbZr is a candidate material for the intermediate layer to improve the high-temperature stability of metal–composite hydrogen separation membranes.

L. Bajtosova et al. (Contribution 22) investigated the sintering dynamics of nickel nanoparticles (Ni NPs) through a comprehensive approach that included in situ transmission electron microscopy annealing and molecular dynamics simulations. In this study, the authors systematically examined the transformation behaviors of Ni NP agglomerates over a temperature spectrum from room temperature to 850 °C. The experimental observations, supported by molecular dynamics simulations, revealed the essential influence of rotational and translational motions of particles, especially at lower temperatures, on sintering outcomes. The effect of the orientation of particles on the sintering process was confirmed, with initial configurations markedly determining sintering efficiency and dynamics. The calculated activation energies observed in this investigation follow those reported in the literature, confirming surface diffusion as the predominant mechanism driving the sintering of Ni NPs.

Acknowledgments: We are grateful to all staff at MDPI for this valuable collaboration. We would like to express our gratitude to all of the contributing authors and reviewers; without their excellent work, it would not have been possible to publish this Topic, which we hope will be both interesting and of lasting importance in general reading and the reference literature.

Conflicts of Interest: The authors declare no conflicts of interest.

References

1. Di Schino, A.; Gaggiotti, M.; Testani, C. Heat treatment effect on microstructure evolution in a 7% Cr steel for forging. *Metals* **2020**, *10*, 808. [CrossRef]
2. Di Schino, A.; Testani, C. Corrosion behavior and mechanical properties of AISI 316 stainless steel clad Q235 plate. *Metals* **2020**, *10*, 552. [CrossRef]
3. Püttgen, W.; Pant, M.; Bleck, W.; Seidl, I.; Rabitsch, R.; Testani, C. Selection of suitable tool materials and development of tool concepts for the Thixoforging of steels. *Steel Res. Int.* **2006**, *77*, 342. [CrossRef]

4. Kim, D.W.; Yang, J.; Kim, Y.G.; Kim, W.K.; Lee, S.; Sohn, S.S. Effects of Granular Bainite and Polygonal Ferrite on Yield Strength Anisotropy in API X65 Linepipe Steel. *Mater. Sci. Eng. A* **2022**, *843*, 143151. [CrossRef]
5. Roy, S.; Romualdi, N.; Yamada, K.; Poole, W.; Militzer, M.; Collins, L. The Relationship between Microstructure and Hardness in the Heat-Affected Zone of Line Pipe Steels. *JOM* **2022**, *74*, 2395–2401. [CrossRef]
6. Fazeli, F.; Amirkhiz, B.S.; Scott, C.; Arafin, M.; Collins, L. Kinetics and Microstructural Change of Low-Carbon Bainite Due to Vanadium Microalloying. *Mater. Sci. Eng. A* **2018**, *720*, 248–256. [CrossRef]
7. Baker, T.N. Microalloyed Steels. *Ironmak. Steelmak.* **2016**, *43*, 264–307. [CrossRef]
8. Bay, Y.; Bhattacharyya, R.; Mc Cormick, M.E. Use of High Strength Steels. *Elsevier Ocean Eng. Ser.* **2001**, *3*, 353.
9. Narimani, M.; Hajjari, E.; Eskandari, M.; Szpunar, J.A. Electron Backscattered Diffraction Characterization of S900 HSLA Steel Welded Joints and Evolution of Mechanical Properties. *J. Mater. Eng. Perform.* **2022**, *31*, 3985–3997. [CrossRef]
10. Geng, R.; Li, J.; Shi, C.; Zhi, J.; Lu, B. Effect of Ce on Microstructures, Carbides and Mechanical Properties in Simulated Coarse-Grained Heat-Affected Zone of 800-MPa High-Strength Low-Alloy Steel. *Mater. Sci. Eng. A* **2022**, *840*, 142919. [CrossRef]
11. Kaščák, Ľ.; Varga, J.; Bidulská, J.; Bidulský, R.; Grande, M.A. Simulation tool fopr material behaviour prediction in additive manufacturing. *Acta Metall. Slovaca* **2023**, *19*, 113. [CrossRef]
12. Bidulský, R.; Petrousek, P.; Bidulská, J.; Hiudak, R.; Zivcak, J.; Grande, M.A. Porosity quantification of additive manufactured Ti6Al4V and CrCoW alloys produced by L-PBF. *Arch. Metall. Mater.* **2022**, *67*, 83. [CrossRef]
13. Bidulská, J.; Bidulský, R.; Petrousek, P.; Kvackaj, T.; Grande, M.A.; Radovan, H. Evaluation of materials properties of Ti and CoCr alloys prepared by laser powder bed fusion. *Mater. Sci. Forum* **2020**, *985*, 223. [CrossRef]
14. Wang, J.; Zhang, Y.; Aghda, N.H.; Pillai, A.R.; Thakkar, R.; Nokhodchi, A.; Maniruzzaman, M. Emerging 3D Printing Technologies for Drug Delivery Devices: Current Status and Future Perspective. *Adv. Drug Deliv. Rev.* **2021**, *174*, 294. [CrossRef] [PubMed]
15. Yin, H.; Qu, M.; Zhang, H.; Lim, Y. 3D Printing and Buildings: A Technology Review and Future Outlook. *Technol. Archit. Des.* **2018**, *2*, 94. [CrossRef]
16. Lee, J.-Y.; An, J.; Chua, C.K. Fundamentals and Applications of 3D Printing for Novel Materials. *Appl. Mater. Today* **2017**, *7*, 120. [CrossRef]
17. Gao, W.; Zhang, Y.; Ramanujan, D.; Ramani, K.; Chen, Y.; Williams, C.B.; Wang, C.C.L.; Shin, Y.C.; Zhang, S.; Zavattieri, P.D. The Status, Challenges, and Future of Additive Manufacturing in Engineering. *Comput. Aided Des.* **2015**, *69*, 65. [CrossRef]
18. Buchanan, C.; Gardner, L. Metal 3D Printing in Construction: A Review of Methods, Research, Applications, Opportunities and Challenges. *Eng. Struct.* **2019**, *180*, 332. [CrossRef]
19. Najmon, J.C.; Raeisi, S.; Tovar, A. 2—Review of Additive Manufacturing Technologies and Applications in the Aerospace Industry. In *Additive Manufacturing for the Aerospace Industry*; Froes, F., Boyer, R., Eds.; Elsevier: Amsterdam, The Netherlands, 2019; pp. 7–31, ISBN 978-0-12-814062-8.
20. Kustas, A.B.; Susan, D.F.; Monson, T. Emerging Opportunities in Manufacturing Bulk Soft-Magnetic Alloys for Energy Applications: A Review. *JOM* **2022**, *74*, 1306. [CrossRef]

Disclaimer/Publisher's Note: The statements, opinions and data contained in all publications are solely those of the individual author(s) and contributor(s) and not of MDPI and/or the editor(s). MDPI and/or the editor(s) disclaim responsibility for any injury to people or property resulting from any ideas, methods, instructions or products referred to in the content.

Article

Influence of Heat Input on the Microstructure and Impact Toughness in Weld Metal by High-Efficiency Submerged Arc Welding

Jinjian Li [1,2], Bing Hu [1,2], Liyang Zhao [1,2], Fangmin Li [1,3], Jiangli He [1,2], Qingfeng Wang [1,2,4,*] and Riping Liu [1,2,4]

1. State Key Laboratory of Metastable Materials Science and Technology, Yanshan University, Qinhuangdao 066004, China; ljj13315669823@163.com (J.L.); binghuysu19941631@163.com (B.H.); zly13780385805@163.com (L.Z.); lfmcrsi@163.com (F.L.); hejiangli@stumail.ysu.edu.cn (J.H.); riping@ysu.edu.cn (R.L.)
2. National Engineering Research Center for Equipment and Technology of Cold Strip Rolling, Yanshan University, Qinhuangdao 066004, China
3. China Railway Science & Industry Group Co., Ltd., Wuhan 430066, China
4. Hebei Key Lab for Optimizing Metal Product Technology and Performance, Yanshan University, Qinhuangdao 066004, China
* Correspondence: wqf67@ysu.edu.cn; Tel.: +86-335-2039067

Abstract: The development of high-efficiency multi-wire submerged arc welding technology in bridge engineering has been limited due to the high mechanical performance standards required. In this paper, weld metal was obtained by welding at three different high heat inputs with the laboratory-developed high-efficiency submerged arc welding wire for bridges. The effect of changing different high heat inputs on the microstructure and impact toughness of high efficiency submerged arc weld metal was systematically investigated by cutting and Charpy V-notch impact tests at −40 °C, using optical microscopy, scanning electron microscopy, energy-dispersive electron spectroscopy, electron backscatter diffraction, and transmission electron microscopy to characterize and analyze. With the increase in heat input from 50 kJ/cm to 100 kJ/cm, the impact absorption energy decreased significantly from 130 J to 38 J. The number of inclusions in the weld metal significantly decreased and the size increased, which led to a significant decrease in the number of inclusions that effectively promote acicular ferrite nucleation, further leading to a decrease in the proportion of acicular ferrite in the weld metal. At the same time, the microstructure of the weld metal was significantly coarsened, the percentage of high-angle grain boundaries was decreased, and the size of martensite/austenite constituents was significantly increased monotonically. The crack initiation energy was reduced by the coarsened martensite/austenite constituents and inclusions, which produced larger local stress concentrations, and the crack propagation was easier due to the coarsened microstructure and lower critical stress for crack instability propagation. The martensite/austenite constituents and inclusions in large sizes worked together to cause premature cleavage fracture of the impact specimen, which significantly deteriorated the impact toughness. The heat input should not exceed 75 kJ/cm for high-efficiency submerged arc welding wires for bridges.

Keywords: high heat input; weld metal; microstructure; M/A constituents; inclusions; impact toughness

1. Introduction

As the modern steel structure industry continues to develop in the direction of large-scale and high mechanical properties, it is increasingly important to use high-efficiency welding methods for production. For high-efficiency welding methods, researchers all over the world have carried out in-depth research and exploration. Among them, Lincoln Electric Co. (Cleveland, OH, USA) of the United States has made very important contributions in the research and development and manufacturing of high-efficiency welding materials, high-efficiency welding power sources, and high-efficiency welding robots, especially to promote the development of high-efficiency welding technology in the shipbuilding industry. The FCB submerged arc welding line developed by Ogden Engineering Co. for shipbuilding has promoted the development of shipbuilding. The company Canada Weld Process developed the T.I.M.E. welding technology, which increases the wire feeding speed with high current and is added to the special shielding gas to improve welding efficiency. The laser-arc composite welding technology, which was first developed by British scholars, has the advantages of large penetration, high efficiency, and small welding deformation and has been widely used in many countries in Europe and the United States in the manufacture of shipbuilding and marine engineering steel [1]. In the field of bridge steel structures, submerged arc welding is one of the most important welding methods. At the same time, multi-wire, high-heat input submerged arc welding is one of the most common high-efficiency welding methods.

Compared to single-wire submerged arc welding (20 kJ/cm ≤ heat input ≤ 50 kJ/cm), the heat input (E_j) of double-wire submerged arc welding (heat input ≥ 50 kJ/cm) has a large amount of deposited metal per unit time, which can significantly save production time and cost and reduce the labor intensity of workers. However, with higher E_j, severe coarsening of the microstructure occurs, which deteriorates the mechanical properties of weld metal [2,3]. The weld metal composed of a large amount of acicular ferrite (AF) has a higher impact toughness due to a higher proportion of high-angle grain boundaries (HAGBs) and a precise interlocking structure [4–6]. However, as the E_j increases, the proportion of HAGBs in the weld metal decreases, while the coarse weissite-ferrite appears, leading to a deterioration of the impact toughness [7–9]. At the same time, the volume fraction of coarse martensite/austenite (M/A) constituents increases significantly while the HAGB spacing increases, thus deteriorating the impact toughness [10,11]. When the E_j is too low, the impact toughness is reduced because of the formation of a large number of bainite and martensite in weld metal [12].

The relationship between E_j and the size or amount of inclusions in the weld metal has also been extensively investigated by researchers [7,8,13–15]. Among them, Kluke et al. [13] proposed that the inclusions diameter was proportional to the cube root of the E_j, supported by the Ostwald ripening theory. Large inclusions tend to have a stronger ability to promote ferrite nucleation than small inclusions; however, they also act as initiation points for cleavage fracture [16]. According to these two opposite effects, the size of the inclusions should neither be too large nor too small; otherwise, they will have a negative impact on the impact toughness of the weld metal.

Due to the high mechanical performance standards required in bridge engineering, single-wire submerged arc welding and gas shielded welding have been the main production welding methods, and the development of high-efficiency submerged arc welding technology has been slow in the bridge field. In this regard, to promote the application of high-efficiency welding in the field of bridges, our team developed a submerged arc welding wire by adding some micro-alloy elements based on the C-Si-Mn alloy and using a combination of mechanisms such as fine grain strengthening, phase transformation strengthening, and solid solution strengthening to ensure the weld metal can meet the mechanical properties of bridges under high-heat input. This work systematically investigated the relationship between the microstructure and impact toughness of the weld metal obtained by high-efficiency submerged arc welding wires under different high-heat inputs from the perspectives of microstructure transformation, HAGB ratio, M/A constituents, and inclusions. Thus, the applicable range of high-efficiency welding E_j was obtained by this work to satisfy high-efficiency welding on bridge steel, and a reference basis for high-efficiency welding in the field of bridges was provided by this work.

2. Experimental Procedure

The purpose of this work is to investigate the changes in the microstructure and mechanical properties of the weld metal with heat input changes and to derive the applicable heat input for the high-efficiency submerged arc welding wire. Therefore, high efficiency submerged arc welding wires were used for submerged arc welding with high heat input, followed by sampling of the weld metal and testing the mechanical properties of the weld metal at different high-heat inputs. Finally, the welded metal under different heat inputs was observed and analyzed.

2.1. Welding Tests

Three bridge steel plates were used in the sizes of 300 × 150 × 32 mm and welded using self-developed high-efficiency submerged arc welding wires for E_j = 50, 75, and 100 kJ/cm, respectively. The high-efficiency submerged arc welding steel was melted by a vacuum induction furnace and then underwent forging of the welding wire ingot, hot rolling of the wire rod, rough drawing, fine drawing to φ 5.0 mm, and copper plating. The welding parameters are shown in Table 1, and the chemical compositions of the base metal, weld wire, weld metal, and flux are shown in Table 2.

Table 1. Welding parameters.

Sample	Welding Current (I)/A	Welding Voltage (U)/V	Welding Speed (υ)/mh^{-1}	Interpass Temperature (T)/°C	Heat Input (E_j)/kJ cm^{-1}
WM-50 kJ/cm	680 (AW) 630 (PW)	30 (AW) 32 (PW)	24	150≤	50
WM-75 kJ/cm	750 (AW) 700 (PW)	32 (AW) 34 (PW)	21.5	150≤	75
WM-100 kJ/cm	800 (AW) 750 (PW)	34 (AW) 36 (PW)	21	150≤	100

AW: Anterior wire; PW: Posterior wire.

Table 2. The chemical composition of welding flux, base metal, weld wire, and weld metal.

Sample	Chemical Composition/wt%							
Weld flux	$SiO_2 + TiO_2$ 25–35	CaO + MgO 20–30	Al_2O_3 + MnO 15–30		CaF_2 15–25		S 0.06	P 0.08
Element type	C	Si	Mn	P	S	Ni	Cr	Nb + V + Ti + Al + Mo + B
Base metal	0.07	0.22	1.52	0.013	0.002	0.027	0.04	0.106
Weld wire	0.06	0.20	1.55	0.014	0.003	0.39	0.03	0.851
WM-50 kJ/cm	0.06	0.21	1.53	0.016	0.004	0.37	0.03	0.832
WM-75 kJ/cm	0.05	0.19	1.49	0.016	0.004	0.35	0.03	0.828
WM-100 kJ/cm	0.05	0.16	1.44	0.017	0.004	0.33	0.03	0.821

2.2. Mechanical Tests

Three impact specimens were taken for each kind of heat input weld metal and processed into 55 × 10 × 10 mm impact specimens in accordance with the ASTM standard E2298, as shown in Figure 1. The weld metal impact test was performed by the JBN-300B impact tester (Jinan Marxtest Technolggy Co., Ltd, Jinan, China) at a temperature of −40 °C. Three sets of impact tests were conducted on each E_j specimen and averaged to reduce the error.

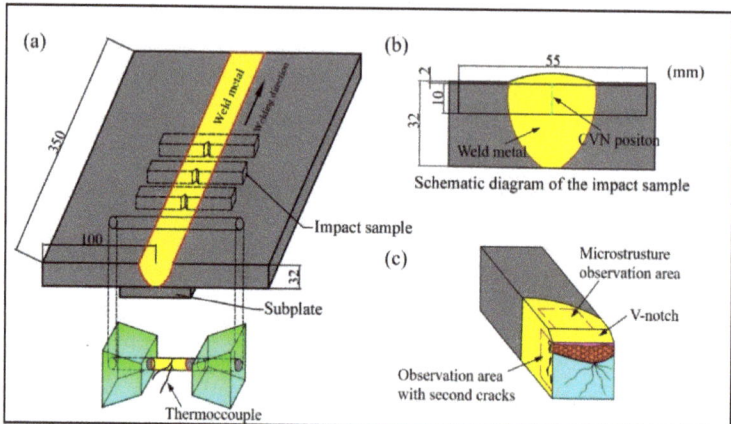

Figure 1. Schematic diagram of weld joints with welding thermal simulation and standard impact samples (**a,b**), and microstructure observation area (**c**).

2.3. Fracture Observation

The fracture surfaces of the impact samples were observed via a scanning electron microscopy (SEM SU5000, HITACHI, Tokyo, Japan). Subsequently, the samples were cut along the cross-section perpendicular to the V-notch as shown in Figure 1c, and the distribution and propagation path of the secondary cracks near the main fracture were observed under the SEM.

2.4. The Determination of Phase Transition Temperature

In order to measure the effect of E_j on the $\gamma \rightarrow \alpha$ phase transition temperature Ar3, the thermal expansion curve was recorded using a C-strain gauge, and the beginning and end temperatures of the phase transition from austenite to ferrite at the corresponding cold rate were determined using the tangent method. Figure 1a gives a schematic diagram of the sampling of the specimen for determining the phase change point of the weld metal. One round bar-type sample was taken for each heat input specimen. After the sampling was completed, the weld microstructure of the specimen was etched with a 4% nitric acid alcohol solution. Subsequently, the thermocouple wires were spot welded to the weld metal, as shown in Figure 1a. The Gleeble-3500 thermal simulation tester (Dynamic Systems Inc., New York, NY, USA) was used to conduct the welding thermal cycle from E_j = 50 kJ/cm to 100 kJ/cm on the specimens. Afterwards, the phase transition temperature Ar3 was measured for each E_j sample.

2.5. Microstructure Characterization

The remaining impact samples of three kinds of heat inputs were first cut in the cross-section perpendicular to the fracture surface. Then the specimens were sanded and polished, then etched with a 4% nitric acid alcohol solution. The microstructure was observed by an Olympus BX51M (Olympus, Aizu, Japan). The percentage of the microstructures was calculated using Image-Pro Plus software (Image-Pro ®Plus, Media Cybernetics, Bethesda, MD, USA). The specimens were repolished and etched with LePera solution for 120 s, then the M/A constituents of the specimens were observed under the metallographic microscope. The size and area fraction were statistically calculated by Image-Pro Plus software. TEM slices parallel to the metallographic specimens were obtained by cutting and sanding to 40–50 μm thickness. Small discs of φ 3 mm were pressed out of the slices and subsequently thinned using double-jet electrolytic polishing. After the thinning was completed, the typical microstructure of the weld metal was observed by a JEM-2010 high-resolution transmission electron microscopy (TEM, Japan Electronics optics Corporation, Tokyo, Japan). The metallographic specimens were polished again and electrolytically polished using perchloric acid (20%) and methanol (80%) solutions, followed by scanning using the SEM with an electron backscatter diffraction detector in 0.25 μm steps to quantify the crystal orientation of the weld metal. The reported mean equivalent diameter (MED) statistics were the average values of at least 8 electron backscatter diffraction images used. The specimens were sanded and polished again, and then the inclusions in the weld metal were observed via the SEM.

3. Results

3.1. Microstructure Observations of the Weld Metal

The microstructural metallographs and the microstructure percentage under different high-heat inputs are given in Figure 2. Under E_j = 50 kJ/cm, the microstructure was mainly composed of AF and some grain boundary ferrite (GBF), granular bainite (GB), and polygonal ferrite (PF). With increasing E_j, microstructures underwent significant coarsening, and the ferrite side plate (FSP) appeared. This kind of jagged microstructure along the GBF growth toward the interior of the prior austenite grain is very detrimental to the toughness of weld metal. At the same time, with increasing E_j, the percentages of GBF, GB, and PF increased, and the percentage of AF decreased, as shown in Figure 2d.

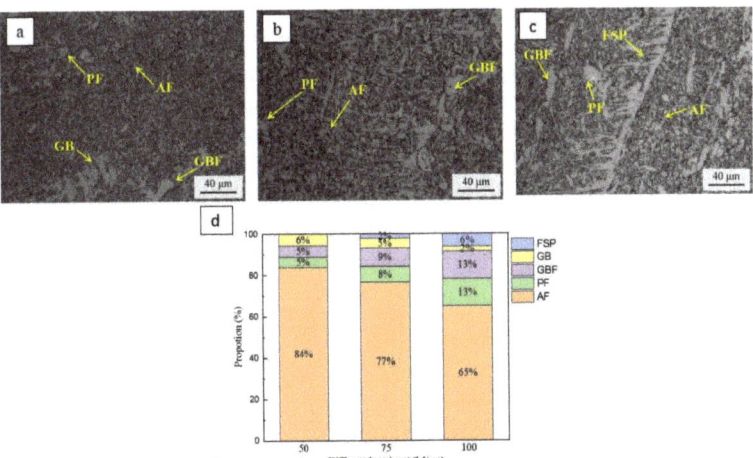

Figure 2. Microstructure morphologies of 50 kJ/cm (**a**), 75 kJ/cm (**b**), 100 kJ/cm (**c**), and a schematic diagram of the percentage of microstructures (**d**).

The M/A constituents of the weld metal are shown in Figure 3. Separately, under 10 fields of view, the size and area proportion of more than 300 M/A constituents were statistically calculated by Image Pro Plus software to minimize local statistics. The statistical results are shown in Table 3. The M/A constituents were small, with an average size of 0.92 μm, and the area proportion was 2.2% under E_j = 50 kJ/cm. However, the average size of inclusions increased significantly to 2.36 μm, and the area proportion increased to 8.3% with the E_j increasing to 100 kJ/cm. At the same time, it can be found that with the increase in E_j, the number of stripy M/A constituents increased.

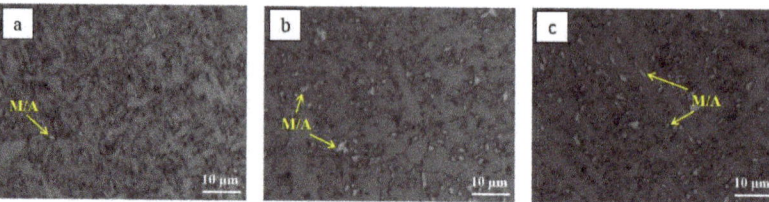

Figure 3. Four percent LePera's reagent-etched samples of 50 kJ/cm (**a**), 75 kJ/cm (**b**), and 100 kJ/cm (**c**), where the matrix is gray and the M/A constituents are white.

Table 3. Quantification results of microstructures in each sample.

Heat Input /kJ·cm^{-1}	Microstructures	$f_{M/A}$ /%	$d_{M/A}$ /μm	$f_{MTA \geq 15°}$ /%	$MED_{MTA \geq 15°}$ /μm
50	AF + PF + GBF + GB	2.2	0.92	46.3	3.1
75	AF + PF + GBF + GB + FSP	4.5	1.59	37.7	3.9
100	AF + PF + GBF + GB + FSP	8.3	2.36	24.1	5.3

The TEM images of the weld metal are presented in Figure 4. As shown in Figure 4, the main microstructure of the weld metal was AF with a little PF. As E_j increased, the size of AF increased significantly, and the number of inclusions decreased while its size increased. By comparison, it was found that not all inclusions produced effective promotion of AF nucleation, and the small-sized inclusions shown in Figure 4a (indicated by the blue arrows) were engulfed by the growth of ferrite. The larger inclusions (indicated by the yellow arrows) had a stronger tendency to stimulate the nucleation of AF. Ferrite laths nucleated on larger-sized inclusions and grew radiologically. Subsequently, neighboring ferrite laths nucleated on the previously existing laths by mutual induction, similar to the previous study [17].

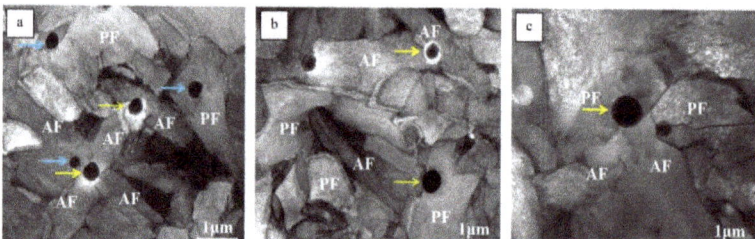

Figure 4. TEM micrographs of 50 kJ/cm (**a**), 75 kJ/cm (**b**), and 100 kJ/cm samples (**c**).

The EDS mapping image of a typical inclusion, as shown in Figure 5, revealed that the inclusions were composite oxide and sulfide composed of Al, Ti, Si, and Mn.

The distribution of inclusions in the weld metal is shown in Figure 6, where the inclusions that stimulate AF nucleation are circled in yellow. All inclusions were counted under 25 fields of scanning electron microscopy using Image-Pro Plus software to minimize errors. With E_j increasing from 50 kJ/cm to 100 kJ/cm, the total number of inclusions decreased significantly. The average size of inclusions increased from 0.59 to 1.26 μm, and the number of large-sized inclusions increased, with the maximum size reaching about 2.5 μm.

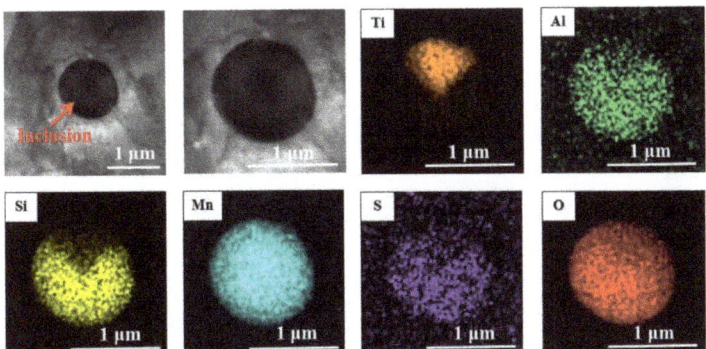

Figure 5. Element distribution by EDS of inclusions in weld metal.

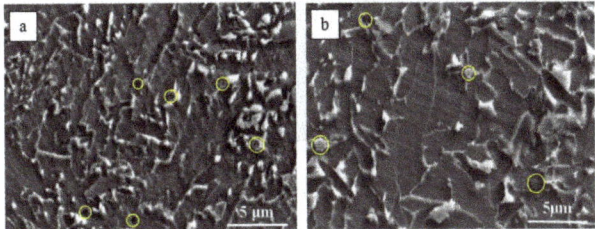

Figure 6. Observations on the inclusion of a 50 kJ/cm sample (**a**) and a 100 kJ/cm sample (**b**).

3.2. Crystallographic Characteristics of the Weld Metal

Inverse pole figures (IPF) were obtained using electron backscatter diffraction, as shown in Figure 7a–c. Previous studies [18–20] usually defined the threshold of HAGB at 15° because grain boundaries with misorientation tolerance angles (MTA) greater than 15° have higher energy to effectively force the deflection or termination of crack propagation. Therefore, in this paper, grain boundaries with an MTA less than 15° between adjacent grains were defined as low-angle grain boundaries (LAGBs), while grain boundaries were defined as high-angle grain boundaries (HAGBs) when the MTA was higher than 15°. The HAGBs were marked with black lines in Figure 7. As shown in Figure 7, HAGBs were formed between adjacent AF or PF, which could effectively hinder crack propagation [21].

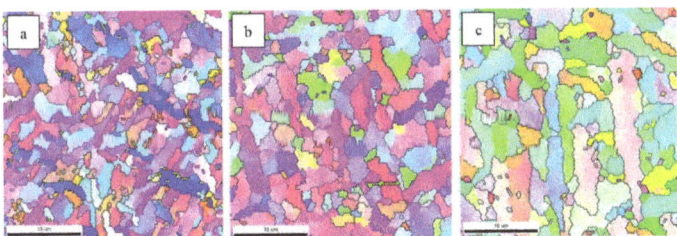

Figure 7. Inverse pole figures of 50 kJ/cm (**a**), 75 kJ/cm (**b**), and 100 kJ/cm sample (**c**).

As shown in Figure 8a, the MED defined by different MTAs increased with the increase in MTA. With the increase in E_j, the MED defined by MTA $\geq 15°$ (MED$_{MTA \geq 15°}$) increased from 3.13 μm to 5.28 μm. Figure 8b gives the percentage of HAGBs. It can be found that with the increase in E_j, the percentage of HAGBs decreased from 46.3% to 24.1%.

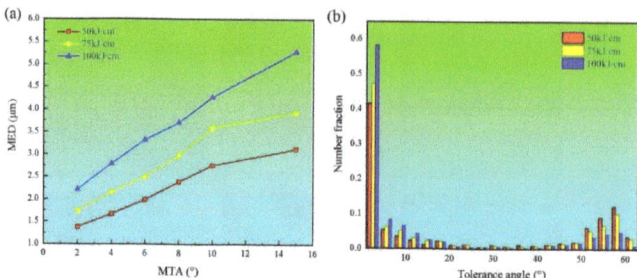

Figure 8. MED varied with MTA (**a**) and the distribution of boundary misorientation (**b**).

3.3. Impact Toughness and Fracture Behavior of the Weld Metal

Table 4 gives the impact energy of the weld metal samples, and the average value decreased from 130 J to 38 J with E_j increasing from 50 kJ/cm to 100 kJ/cm. As shown in Figure 9, the fibrous zone was more dominant under E_j = 50 kJ/cm, and this zone corresponds to the ductile crack with a higher energy consumption. Accordingly, the cleavage surfaces were found to exist in the region away from the Charpy V-notch, i.e., the cleavage fracture occurred later under E_j = 50 kJ/cm. The cleavage fracture corresponds to the stage of crack instability propagation, and the energy consumed is low. With the increase in E_j, the area of the fibrous zone was significantly reduced, and the cleavage surfaces were found close to the V-notch, i.e., the onset of cleavage fracture was significantly advanced. Meanwhile, the fracture surface was very flat, which corresponded to the lower impact energy.

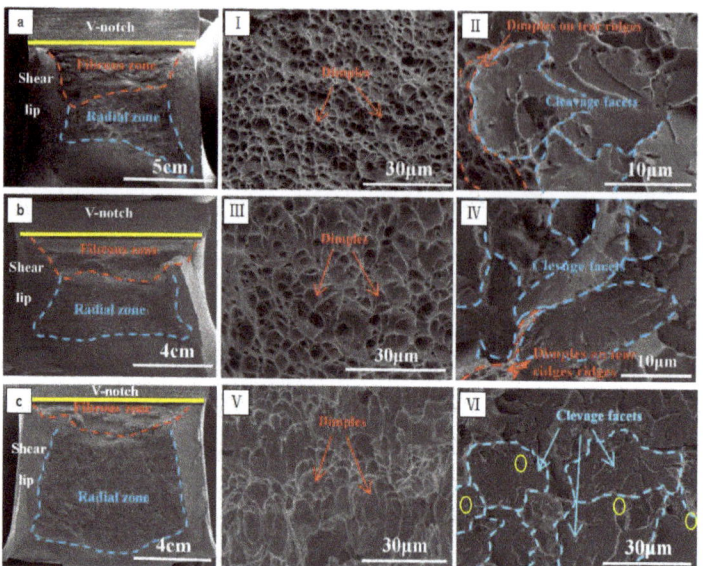

Figure 9. Fracture surfaces of the weld metal with E_j = 50 kJ/cm (**a**), 75 kJ/cm (**b**), 100 kJ/cm (**c**), and a high magnification image of the fibrous zone (**I/III/V**) and radial zone (**II/IV/VI**).

Table 4. CVN impact energy at −40 °C of weld metal under different E_j.

Heat Input kJ/cm	CVN Impact Energy at −40 °C/J			Average Value
	Sample 1	Sample 2	Sample 3	
50	139	113	137	168
75	87	75	69	119
100	32	42	40	49

The fibrous and radial zones of the three specimens were further observed separately by magnification, as shown in Figure 9I–VI. For the sample of E_j = 50 kJ/cm, the fiber zone consisted of small, uniform, and deep dimples, while the radial zone had small cleavage surfaces, and the cleavage surfaces extended from the cleavage crack to the periphery, forming a cleavage step when encountering HAGBs. Tear ridges were found at the edges of the cleavage surface, indicating a higher impediment to crack propagation. With the increase in E_j, the dimples in the fiber zone became shallower and larger and developed into a parabolic shape, which means that the degree of plastic deformation during the ductile fracture stage was reduced. At the same time, the sizes of the cleavage surfaces in the radial zone significantly increased, and the tear ridges were almost invisible, implying that crack propagation became easier during the cleavage fracture [22]. It was worth noting that when E_j increased to 100 kJ/cm, many large-size inclusions, or M/A constituents, were formed and acted as crack initiation sources for the various cleavage surfaces. Accordingly, it could be inferred that the generation of these coarse M/A constituents and inclusions was inextricably linked to the reduction in impact toughness of the specimens, which will be discussed later.

In order to further observe and analyze the initiation and propagation of cracks under different E_j, specimens were cut from the impact fracture section to observe the morphology and distribution of secondary cracks.

As shown in Figure 10a, a small number of microvoids were observed in the fracture cross-section of the 50 kJ/cm sample in the fibrous region. The growth and coalescence of the microvoids were limited by the HAGBs, such as the surrounding AF. Meanwhile, as the large M/A constituents and inclusions that promoted the nucleation of microvoids were less, the distance between microvoids was larger, and the energy required for microvoid coalescence was correspondingly higher. As a result, the size of the microvoids was relatively small under E_j = 50 kJ/cm, and the coalescence and growth of microvoids were hardly seen. For the 100 kJ/cm sample, at the fiber zone, as shown in Figure 10b, it was found that many microvoids were nucleated at the interface of M/A constituents or inclusions and matrix, and the microvoids were close to each other and appeared to grow and coalesce.

Under 50 kJ/cm, the radiation zone was relatively small, and the propagation paths of secondary cracks were short, as shown in Figure 10c. When secondary cracks propagated on the fine AF, there was blunting and deflection at the grain boundaries, indicating that large amounts of AF have a strong arresting effect on crack propagation [23]. Under 100 kJ/cm, the number and size of secondary cracks in the radial zone increased significantly. As shown in Figure 10d, the long strip crack nucleated along the interface between the ferrite matrix and M/A constituents. The secondary crack propagated straight through the ferrite grain and the elongated M/A, eventually deflecting at the grain boundary and arresting at the prior austenite grain boundary. The overall straightness of the cracks indicated that the surrounding microstructures had a weak hindering effect on the cracks propagation.

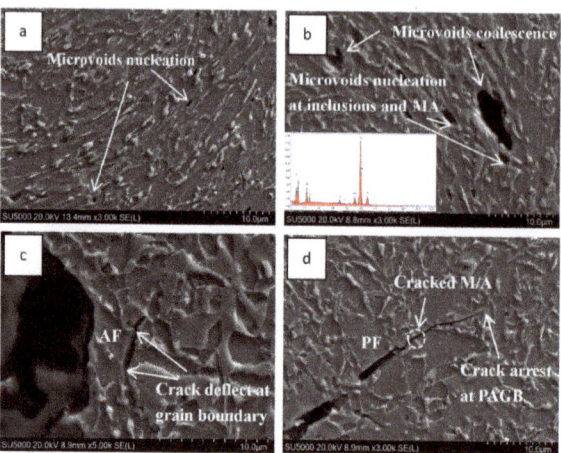

Figure 10. SEM observations of microvoids and crack morphologies at E_j = 50 kJ/cm (**a**,**c**) and 100 kJ/cm (**b**,**d**).

4. Discussion

4.1. Effect of E_j on the Inclusions of the Weld Metal

The presence of convective flow of liquid in the melt pool leads to collisions and aggregation between the inclusions, which are called gradient collisions, resulting in coarse inclusions. Therefore, with the increase in E_j, the convective flow of fluid in the weld pool was enhanced. Accordingly, the velocity gradient in the weld pool was increased, increasing the possibility of collision and aggregation of inclusions. That is the main reason why the number of inclusions in the weld metal decreased significantly, and the size coarsened obviously as E_j increased from 50 kJ/cm to 100 kJ/cm.

Ostwald ripening may occur during the final stages of cooling in the weld pool, allowing further coarsening of the inclusions [16,24,25]. Kluken et al. [24] divided the melt pool into two regions, i.e., the "hot" and "cold" parts, as shown in Figure 11. The collision and aggregation of inclusions and the floating processes mainly occurred in the hot part of the weld pool. In contrast, in the cold part, collisions and aggregation of inclusions were less likely to occur due to the low temperature gradient and the low convective intensity of the weld pool. Therefore, in the cold part, Ostwald ripening was the main way for the growth of inclusions. The growth of large inclusions was at the expense of small inclusions during Ostwald ripening, and there was a decrease in the density of inclusions. The kinetics of Oswald ripening can be calculated as shown in Equation (1).

$$d^3 = \bar{d}_{ini}^3 + \frac{64 D_O C_O V_M}{9RT} t \tag{1}$$

where d_{ini} is the average initial diameter; d is the average diameter of inclusions after a time t; V_M is the molar volume of oxide; D_O is the diffusivity of oxygen; and C_O is the nominal oxygen concentration of the liquid.

At the same time, with the increase in E_j, the cooling rate of the weld pool became slower, and the nucleation rate of the first formed oxide inclusions was very low due to the low subcooling. Since the oxides formed first grew to larger sizes at high temperatures and acted as heterogeneous sites for subsequent oxide nucleation, their reduced number further led to a lower number of inclusions. At the same time, with the increase of E_j, the growth of inclusions was also promoted at high temperatures, so that the number of inclusions decreased with the increase of E_j but the size coarsened.

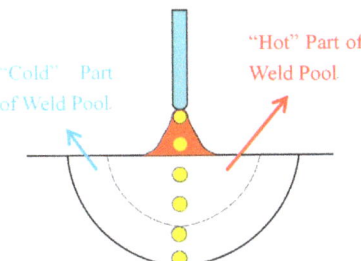

Figure 11. Schematic diagram of two main reaction zones in the weld pool.

In order to further analyze the relationship between inclusions and AF nucleation, the rates of ferrite nucleation promoted by inclusions at different sizes were calculated, as shown in Figure 12. (Inclusions smaller than 0.2 μm were neglected). The increase in the size of inclusions increased their ability to stimulate AF nucleation, which is consistent with the results of previous studies [16,26,27]. However, an interesting phenomenon is that the size of inclusions in the weld metal of a 50 kJ/cm sample could stimulate AF nucleation at almost 100% when reaching about 1 μm, while the size of inclusions needed to reach about 1.6 μm to stimulate AF nucleation at almost 100% with the E_j up to 100 kJ/cm. That means the ability to stimulate AF nucleation of inclusions in the weld metal decreased with the increase in E_j. With the increase in E_j, the increased transformation temperature from austenite to ferrite made the formation of GBF and PF in the weld metal easier than AF. At the same time, as the size of the inclusion increases, there will be relaxation of stress around the interface between large-size inclusions and austenite [28–31]. These factors all lead to a decrease in the ability of AF nucleation induced by inclusions in weld metal and an increase in E_j.

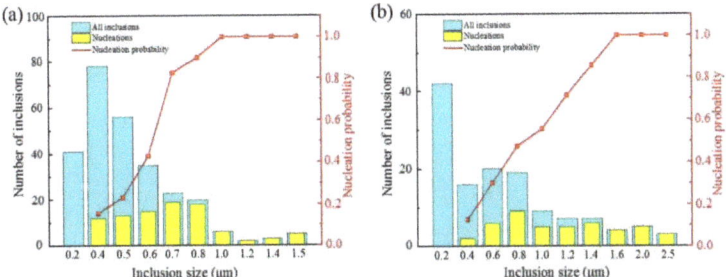

Figure 12. Size and number distribution of inclusions and statistics of nucleation rate of AF promoted by inclusions with different sizes under different E_j: (**a**) 50 kJ/cm and (**b**) 100 kJ/cm.

The inclusions formed in the weld metal can either contribute to the nucleation of AF, increasing the percentage of HAGB, thus improving the impact toughness [23], or they can act as crack sources for the cleavage fracture, which makes the cleavage fracture occur prematurely and reduce the impact toughness [16,32,33], especially when the inclusions are larger than 1 μm [34], which will be discussed further in the subsequent sections.

4.2. Effect of E_j on the Microstructures of the Weld Metal

To better understand the influence of various E_j on final microstructures, the continuous cooling expansion curves of each specimen were measured. Subsequently, the phase transition temperatures (Ar3) from austenite to ferrite were measured using the tangent method, as shown in Figure 13. As E_j increased from 50 kJ/cm to 100 kJ/cm, Ar3 decreased from 701 °C to 572 °C.

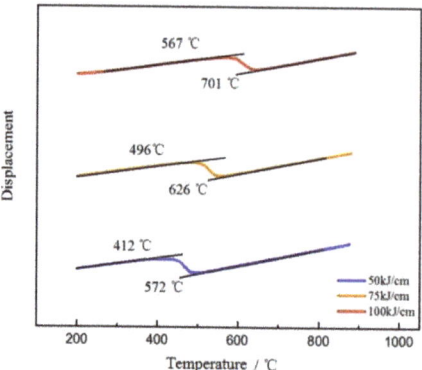

Figure 13. Dilatometric curves of the weld metal obtained at different E_j.

The phase transition from austenite to ferrite began with the formation of grain boundary ferrite (GBF) along the austenite grain boundaries at a higher transition temperature. The increase in E_j led to a coarsening of GBF, and the volume fraction of GBF in the weld metal increased [7]. Due to the high formation temperature of GBF, its C atoms were diffused more sufficiently into the surrounding environment, and the GBF was soft. Under the impact load, GBF deformed firstly resulting in stress concentration. At the same time, because GBF was continuously distributed in a network along the prior austenite grain boundary, it often promoted crack propagation [35]. So, the increase in its size and proportion deteriorates the impact toughness of the weld metal. With the increase in E_j, the ferrite sideplate (FSP) from the GBF side grew in a pickaxe shape toward the austenite interior. This kind of coarse microstructure cuts the grains apart and has a very negative effect on the toughness of the weld metal.

With further cooling, PF, AF, and GB were formed in austenite. The formation of PF is a diffusion phase transition controlled by the diffusion and migration of solute atoms near the phase interface. When welded at 50 kJ/cm, the lower transformation temperature (572 °C) hindered the diffusion of carbon atoms and inhibited the formation of PF. The higher undercooling was conducive to the transformation of AF and GB. The expression of the AF nucleation rate is shown in Equation (2) [29,36].

$$I_V = C_3 \exp = \left\{ -\frac{C_4}{RT} - \frac{C_4 \Delta G_{max}^{\gamma \to \alpha}}{C_2 RT} \right\} \qquad (2)$$

where C_3 is related to the number of AF nucleation sites, and C_2 and C_4 are constants based on experimental measurements. In this work, for the AF phase transition, C_3 is related to the number density of effective inclusions. T is the temperature in Kelvin, and $\Delta G_{max}^{\gamma \to \alpha}$ is the driving force of the transformation from austenite to ferrite. At lower E_j, the number of inclusions was large and the value of C_3 was high, which favored AF nucleation. Therefore, the AF content was higher at a lower E_j of 50 kJ/cm. As the E_j increased, phase transition temperatures (Ar3) increased, and the C atoms diffused more fully, leading to a significant increase in the PF content. At the same time, the various microstructures also underwent significant coarsening with sufficient diffusion of C atoms [37]. The microstructure of AF is small, and AF has a precise interlocking structure, forming HAGBs between each other. Therefore, the reduction of AF content and the coarsening of the microstructure reduced the hindering effect of weld metal on crack propagation.

During the cooling process of the welding thermal cycle, with the transformation from austenite to ferrite, C diffuses from AF, PF, and GBF to the residual austenite γ', thus causing the enrichment of C in γ' and improving the stability of γ' [38]. On further cooling to near room temperature, γ' transformed into M/A constituents. When E_j was low, the diffusion distance of C was shorter because of the decreased diffusion coefficient due to the faster cooling rate [37]. Accordingly, the size of the M/A constituents formed at the grain boundaries was small and mainly point M/A. With the increase of E_j, C atoms had sufficient time and dynamics to diffuse; thus, the number and size of M/A constituents increased, while more elongated M/A constituents were formed along the grain boundaries between the ferrite [39,40]. The corresponding schematic representation of the microstructural transformation is shown in Figure 14.

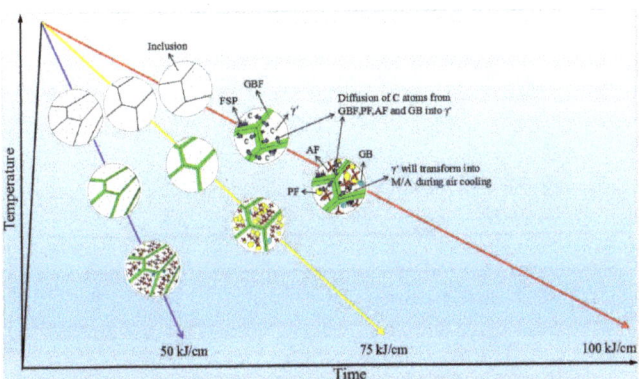

Figure 14. Schematically illustrating microstructural transformation under different E_j.

4.3. Effects of E_j on the Impact Toughness and Fracture Behaviors of the Weld Metal

Due to the obvious hardness difference between the M/A constituents, inclusions, and the ferrite matrix, when the stress was conducted, the ferrite matrix with less hardness deformed plastically first, and a large number of dislocations accumulated at the M/A constituents and inclusions or near their interfaces with the surrounding ferrite matrix, resulting in stress concentration and thus causing microcrack initiation. The kernel average misorientation map could be used to evaluate the dislocation density and estimate the stress distribution state of the material [41]. In this paper, it was mainly used to analyze the magnitude of local stress. The band contrast maps and kernel average misorientation maps are shown in Figure 15, which means that the stress concentration mainly occurred at M/A constituents or at the M/A–matrix interface in the 100 kJ/cm sample. The M/A constituents of the 50 kJ/cm sample were small, and there was rarely a high stress concentration. The degree of stress concentration increased significantly with the size of the M/A constituents. Long-strip M/A constituents produced high stress concentrations, as did large-sized blocky M/A constituents.

As shown in Figure 10, with the increase in E_j, the cleavage fracture occurred near the V-notch, i.e., the occurrence of the cleavage fracture was advanced. At the same time, it was observed that a large number of cleavage surfaces of the 100 kJ/cm sample nucleated at the large size inclusions and M/A constituents as shown in Figure 16. The bulging of the M/A constituent at the origin of the cleavage surface indicated that crack initiation should be due to the debonding of the M/A constituent at the interface with the ferrite matrix [11]. Therefore, with the increase in E_j, the coarsening of inclusions and M/A constituents significantly reduced the cleavage crack initiation energy.

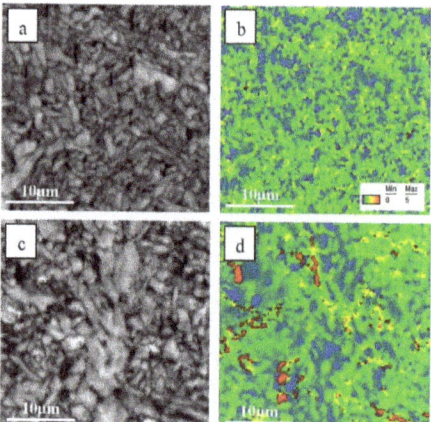

Figure 15. Band contrast map and kernel average misorientation map of the 50 kJ/cm (**a**,**b**) and 100 kJ/cm (**c**,**d**) samples.

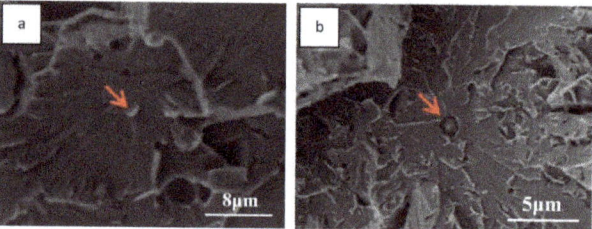

Figure 16. M/A constituents (**a**) and inclusions (**b**) acting as the nucleation site for cleavage fracture on the 100 kJ/cm sample.

According to Griffith's theory [42], the critical stress for cleavage crack instability propagation can be expressed by the following equation.

$$\sigma'_c = \left(\frac{\pi E \gamma'}{(1-v^2)d_0} \right)^{1/2} \quad (3)$$

where σ'_c is the critical stress, E is the Young's modulus, γ' is the effective surface energy of the fracture, v is the Poisson's ratio, and d is the microcrack size, where d_0 can be regarded as the maximum width of the M/A constituents or inclusions [43]. From Equation (3), it can be seen that the critical stress is mainly related to the effective surface energy of ferrite and the size of M/A constituents or inclusions. The smaller the size of M/A constituents or inclusions, the larger the critical stress for cleavage cracks and instability propagation. With the increase of E_j, the stress concentration at the large-size M/A constituents and inclusions increased significantly, and crack initiation as well as propagation became easier.

At the same time, the microstructure coarsening caused by the increase in E_j also played an important role in deteriorating the impact toughness. The research results of Cao and Martín-Meizoso A et al. [44–46] show that the cleavage fracture can be divided into the following processes: First, microcracks nucleate at the hardening phase or at the interface between the hardening phase and the matrix and then pass through the hardening phase or the interface between the hardening phase and the matrix. Second, microcracks propagate in the matrix and reach the HAGBs of the matrix. Finally, the microcracks pass through the HAGBs and coalesce, leading to the final, complete fracture.

Based on the previous studies and combined with the experimental results of this work, the schematic diagram of cleavage crack initiation and propagation of the impact samples is shown in Figure 17.

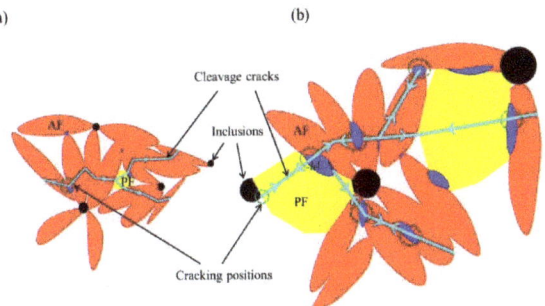

Figure 17. Schematic diagram of cleavage microcrack initiation and propagation in the weld metal under 50 kJ/cm (**a**) and 100 kJ/cm (**b**).

When the specimen was subjected to the impact load, a large number of dislocations were generated that were easier to slip in AF, which was conducive to stress release [47]. AF reduces stress concentration and crack initiation tendency due to its fine structure and superior deformation ability. The size of M/A constituents and inclusions was small under $E_j = 50$ kJ/cm, and the nucleation of microcracks was difficult. At the same time, the grain size of AF was small, and the HAGBs accounted for a high proportion. Therefore, in the subsequent process of microcrack propagation, the number of times that cracks met with the HAGBs increased, which consumed more energy.

However, with the increase of E_j to 100 kJ/cm, the proportion of AF decreased and the size of M/A constituents and inclusions increased, which makes the initiation and propagation of cleavage cracks very easy. This was crucial because it directly determined that the ductile fracture zone was very small, as shown in Figure 10, which means that the cleavage fracture occurred much earlier. At the same time, a serious weakening of the arresting effect on crack propagation occurred as the grain size of the weld metal coarsened and the proportion of HAGBs decreased. As the external force was applied, a large number of microcracks nucleated on many large hardening phases and propagated at the same time. These cracks interconnected without going through many HAGBs, leading to the catastrophic failure of the sample.

5. Conclusions

In general, the influence of E_j on the microstructure and impact toughness in the weld metal of high-efficiency submerged arc welding wires has been studied with the following conclusions:

1. The increase in E_j from 50 kJ/cm to 100 kJ/cm led to a significant reduction in the number of inclusions in the weld metal and a significant increase in their size from 0.59 to 1.26 μm. The ability to stimulate AF nucleation of inclusions was decreased with the increase of E_j due to the increased transformation temperature from austenite to ferrite and the relaxation of stress around the interface between large-size inclusions and austenite;
2. The microstructure of the weld metal welded by the high-efficiency submerged arc welding wires included GBF, FSP, PF, AF, GB, and M/A constituents. With the E_j increasing from 50 kJ/cm to 100 kJ/cm, the $MED_{MTA \geq 15°}$ of the weld metal increased from 3.1 to 5.3 μm. At the same time, AF content decreased from 84% to 65%, with an increase in PF from 5% to 13%. The average size of M/A constituents increased from 0.92 to 2.36 μm.

3. The high-efficiency submerged arc welding wires studied in this work are suitable for $E_j \leq 75$ kJ/cm. With the increase in E_j from 50 kJ/cm to 100 kJ/cm, the impact absorption energy decreased significantly from 130 J to 38 J. The fracture behavior of weld metal changed from mainly ductile fracture to mainly brittle fracture. With the increase in E_j, the local stress around the large size inclusions and M/A constituents was greatly improved, while the fraction of HAGBs decreased from 46.3% to 24.1%. These two factors led to the premature cleavage fracture of weld metal, and the impact energy decreased significantly.

Author Contributions: Conceptualization, Q.W. and J.L.; software, B.H.; validation, L.Z.; formal analysis, F.L.; investigation, J.H.; resources, Q.W.; data curation, J.L.; writing—original draft preparation, J.L.; writing—review and editing, J.L. and Q.W.; visualization, J.L.; supervision, Q.W.; project administration, Q.W.; funding acquisition, Q.W. and R.L. All authors have read and agreed to the published version of the manuscript.

Funding: This work was funded by the National Natural Science Foundation of China (52127808), the Innovation Ability Promotion Program of Hebei (22567609H), and the National Key Research and Development Program of China (Grant No. 2017YFB0304800 and Grant No. 2017YFB0304802 for the second subproject).

Institutional Review Board Statement: Not applicable.

Informed Consent Statement: Not applicable.

Data Availability Statement: Not applicable.

Conflicts of Interest: The authors declare no conflict of interest.

References

1. Ribic, B.; Palmer, T.A.; DebRoy, T. Problems and Issues in Laser-Arc Hybrid Welding. *Int. Mater. Rev.* **2009**, *54*, 223–244. [CrossRef]
2. Evans, G.M. The Effect of Heat Input on the Microstructure and Properties of C-Mn All-Weld-Metal Deposits. *Met. Constr.* **1982**, *61*, 125S–132S.
3. Qiao, M.; Fan, H.; Shi, G.; Wang, L.; Wang, Q.; Wang, Q.; Liu, R. Effect of Welding Heat Input on Microstructure and Impact Toughness in the Simulated CGHAZ of Low Carbon Mo-V-Ti-N-B Steel. *Metals* **2021**, *11*, 1997. [CrossRef]
4. Wu, H.; Xia, D.; Ma, H.; Du, Y.; Gao, C.; Gao, X.; Du, L. Study on Microstructure Characterization and Impact Toughness in the Reheated Coarse-Grained Heat Affected Zone of V-Microalloyed Steel. *J. Mater. Eng. Perform.* **2022**, *31*, 376–382. [CrossRef]
5. Stornelli, G.; Tselikova, A.; Mirabile Gattia, D.; Mortello, M.; Schmidt, R.; Sgambetterra, M.; Testani, C.; Zucca, G.; Di Schino, A. Influence of Vanadium Micro-Alloying on the Microstructure of Structural High Strength Steels Welded Joints. *Materials* **2023**, *16*, 2897. [CrossRef]
6. Narimani, M.; Hajjari, E.; Eskandari, M.; Szpunar, J.A. Electron Backscattered Diffraction Characterization of S900 HSLA Steel Welded Joints and Evolution of Mechanical Properties. *J. Mater. Eng. Perform.* **2022**, *31*, 3985–3997. [CrossRef]
7. Viano, D.M.; Ahmed, N.U.; Schumann, G.O. Influence of Heat Input and Travel Speed on Microstructure and Mechanical Properties of Double Tandem Submerged Arc High Strength Low Alloy Steel Weldments. *Sci. Technol. Weld. Join.* **2000**, *5*, 26–34. [CrossRef]
8. Wen, C.; Wang, Z.; Deng, X.; Wang, G.; Misra, R.D.K. Effect of Heat Input on the Microstructure and Mechanical Properties of Low Alloy Ultra-High Strength Structural Steel Welded Joint. *Steel Res. Int.* **2018**, *89*, 1700500. [CrossRef]
9. Li, W.; Cao, R.; Zhu, W.; Guo, X.; Jiang, Y.; Chen, J. Microstructure Evolution and Impact Toughness Variation for High Strength Steel Multi-Pass Weld Metals with Various Cooling Rates. *J. Manuf. Process.* **2021**, *65*, 245–257. [CrossRef]
10. Wang, X.L.; Tsai, Y.T.; Yang, J.R.; Wang, Z.Q.; Li, X.C.; Shang, C.J.; Misra, R.D.K. Effect of Interpass Temperature on the Microstructure and Mechanical Properties of Multi-Pass Weld Metal in a 550-MPa-Grade Offshore Engineering Steel. *Weld World* **2017**, *61*, 1155–1168. [CrossRef]
11. Lan, L.; Qiu, C.; Song, H.; Zhao, D. Correlation of Martensite–Austenite Constituent and Cleavage Crack Initiation in Welding Heat Affected Zone of Low Carbon Bainitic Steel. *Mater. Lett.* **2014**, *125*, 86–88. [CrossRef]
12. Mao, G.; Cayron, C.; Cao, R.; Logé, R.; Chen, J. The Relationship between Low-Temperature Toughness and Secondary Crack in Low-Carbon Bainitic Weld Metals. *Mater. Charact.* **2018**, *145*, 516–526. [CrossRef]
13. Kluken, A.O. Grong Mechanisms of Inclusion Formation in Al−Ti−Si−Mn Deoxidized Steel Weld Metals. *Metall. Trans. A* **1989**, *20*, 1335–1349. [CrossRef]
14. Lan, L.; Qiu, C.; Zhao, D.; Gao, X.; Du, L. Analysis of Martensite–Austenite Constituent and Its Effect on Toughness in Submerged Arc Welded Joint of Low Carbon Bainitic Steel. *J. Mater. Sci.* **2012**, *47*, 4732–4742. [CrossRef]

15. Chu, Q.; Xu, S.; Tong, X.; Li, J.; Zhang, M.; Yan, F.; Zhang, W.; Bi, Z.; Yan, C. Comparative Study of Microstructure and Mechanical Properties of X80 SAW Welds Prepared Using Different Wires and Heat Inputs. *J. Mater. Eng Perform.* **2020**, *29*, 4322–4338. [CrossRef]
16. Lan, L.; Kong, X.; Qiu, C.; Zhao, D. Influence of Microstructural Aspects on Impact Toughness of Multi-Pass Submerged Arc Welded HSLA Steel Joints. *Mater. Des.* **2016**, *90*, 488–498. [CrossRef]
17. Bhadeshia, H.K.D.H.; Christian, J.W. Bainite in Steels. *MTA* **1990**, *21*, 767–797. [CrossRef]
18. Fan, H.; Shi, G.; Peng, T.; Wang, Q.; Wang, L.; Wang, Q.; Zhang, F. N-Induced Microstructure Refinement and Toughness Improvement in the Coarse Grain Heat-Affected Zone of a Low Carbon Mo–V–Ti–B Steel Subjected to a High Heat Input Welding Thermal Cycle. *Mater. Sci. Eng. A* **2021**, *824*, 141799. [CrossRef]
19. Shi, G.; Zhao, H.; Zhang, S.; Wang, Q.; Zhang, F. Microstructural Characteristics and Impact Fracture Behaviors of Low-Carbon Vanadium-Microalloyed Steel with Different Nitrogen Contents. *Mater. Sci. Eng. A* **2020**, *769*, 138501. [CrossRef]
20. Zhang, Y.; Shi, G.; Sun, R.; Guo, K.; Zhang, C.; Wang, Q. Effect of Si Content on the Microstructures and the Impact Properties in the Coarse-Grained Heat-Affected Zone (CGHAZ) of Typical Weathering Steel. *Mater. Sci. Eng. A* **2019**, *762*, 138082. [CrossRef]
21. Zhang, J.; Xin, W.; Ge, Z.; Luo, G.; Peng, J. Effect of High Heat Input Welding on the Microstructures, Precipitates and Mechanical Properties in the Simulated Coarse Grained Heat Affected Zone of a Low Carbon Nb-V-Ti-N Microalloyed Steel. *Mater. Charact.* **2023**, *199*, 112849. [CrossRef]
22. Lan, L.; Qiu, C.; Zhao, D.; Gao, X.; Du, L. Microstructural Characteristics and Toughness of the Simulated Coarse Grained Heat Affected Zone of High Strength Low Carbon Bainitic Steel. *Mater. Sci. Eng. A* **2011**, *529*, 192–200. [CrossRef]
23. Lan, L.; Qiu, C.; Zhao, D.; Gao, X.; Du, L. Analysis of Microstructural Variation and Mechanical Behaviors in Submerged Arc Welded Joint of High Strength Low Carbon Bainitic Steel. *Mater. Sci. Eng. A* **2012**, *558*, 592–601. [CrossRef]
24. Babu, S.S.; David, S.A.; DebRoy, T. Coarsening of Oxide Inclusions in Low Alloy Steel Welds. *Sci. Technol. Weld. Join.* **1996**, *1*, 17–27. [CrossRef]
25. Lindborg, U.; Torssell, K. A Collision Model for the Growth and Separation of Deoxidation Products. *Trans. Metall. Soc. AIME* **1968**, *242*, 94–102.
26. Ricks, R.A.; Howell, P.R.; Barritte, G.S. The nature of acicular ferrite in HSLA steel weld metals. *J. Mater. Sci.* **1982**, *17*, 732–740. [CrossRef]
27. Lee, T.-K.; Kim, H.J.; Kang, B.Y.; Hwang, S.K. Effect of Inclusion Size on the Nucleation of Acicular Ferrite in Welds. *ISIJ Int.* **2000**, *40*, 1260–1268. [CrossRef]
28. Sarma, D.S.; Karasev, A.V.; Jönsson, P.G. On the Role of Non-Metallic Inclusions in the Nucleation of Acicular Ferrite in Steels. *ISIJ Int.* **2009**, *49*, 1063–1074. [CrossRef]
29. Babu, S.S. The Mechanism of Acicular Ferrite in Weld Deposits. *Curr. Opin. Solid State Mater. Sci.* **2004**, *8*, 267–278. [CrossRef]
30. Gregg, J.M.; Bhadeshia, H. Solid-State Nucleation of Acicular Ferrite on Minerals Added to Molten Steel. *Acta Mater.* **1997**, *45*, 739–748. [CrossRef]
31. Pan, T.; Yang, Z.G.; Bai, B.Z.; Fang, H.S. Study of thermal stress and strain energy in γ-fe matrix around inclusion caused by thermal coefficient difference. *Acta Met. Sin.* **2003**, *39*, 1037–1042.
32. Tweed, J.H.; Knott, J.F. Micromechanisms of failure in C Mn weld metals. *Acta Metall.* **1987**, *35*, 1401–1414. [CrossRef]
33. Stone, V.; Cox, T.B.; Low, J.R.; Psioda, J.S. Microstructural Aspects of Fracture by Dimpled Rupture. *Int. Mater. Rev.* **1985**, *30*, 157–180. [CrossRef]
34. Avazkonandeh-Gharavol, M.H.; Haddad-Sabzevar, M.; Haerian, A. Effect of Copper Content on the Microstructure and Mechanical Properties of Multipass MMA, Low Alloy Steel Weld Metal Deposits. *J. Mater. Sci.* **2009**, *30*, 1902–1912. [CrossRef]
35. Wang, X.L.; Nan, Y.R.; Xie, Z.J.; Tsai, Y.T.; Yang, J.R.; Shang, C.J. Influence of Welding Pass on Microstructure and Toughness in the Reheated Zone of Multi-Pass Weld Metal of 550 MPa Offshore Engineering Steel. *Mater. Sci. Eng. A* **2017**, *702*, 196–205. [CrossRef]
36. Rees, G.I.; Bhadeshia, H.K.D.H. Bainite Transformation Kinetics Part 1 Modified Model. *Mater. Sci. Technol.* **1992**, *8*, 985–993. [CrossRef]
37. Wang, L.; Fan, H.; Shi, G.; Wang, Q.; Wang, Q.; Zhang, F. Effect of Ferritic Morphology on Yield Strength of CGHAZ in a Low Carbon Mo-V-N-Ti-B Steel. *Metals* **2021**, *11*, 1863. [CrossRef]
38. Zhang, S.; Liu, K.; Chen, H.; Xiao, X.; Wang, Q.; Zhang, F. Effect of Increased N Content on Microstructure and Tensile Properties of Low-C V-Microalloyed Steels. *Mater. Sci. Eng. A* **2016**, *651*, 951–960. [CrossRef]
39. Wang, Z.; Shi, M.; Tang, S.; Wang, G. Effect of Heat Input and M-A Constituent on Microstructure Evolution and Mechanical Properties of Heat Affected Zone in Low Carbon Steel. *J. Wuhan Univ. Technol.-Mat. Sci. Edit.* **2017**, *32*, 1163–1170. [CrossRef]
40. Luo, X.; Chen, X.; Wang, T.; Pan, S.; Wang, Z. Effect of Morphologies of Martensite–Austenite Constituents on Impact Toughness in Intercritically Reheated Coarse-Grained Heat-Affected Zone of HSLA Steel. *Mater. Sci. Eng. A* **2018**, *710*, 192–199. [CrossRef]
41. Shi, Z.; Wang, R.; Su, H.; Chai, F.; Wang, Q.; Yang, C. Effect of Nitrogen Content on the Second Phase Particles in V–Ti Microalloyed Shipbuilding Steel during Weld Thermal Cycling. *Mater. Des.* **2016**, *96*, 241–250. [CrossRef]
42. Babu, S.S.; David, S.A.; Vitek, J.M.; Mundra, K.; DebRoy, T. Development of Macro- and Microstructures of Carbon–Manganese Low Alloy Steel Welds: Inclusion Formation. *Mater. Sci. Technol.* **1995**, *11*, 186–199. [CrossRef]
43. Zhao, J.; Hu, W.; Wang, X.; Kang, J.; Yuan, G.; Di, H.; Misra, R.D.K. Effect of Microstructure on the Crack Propagation Behavior of Microalloyed 560MPa (X80) Strip during Ultra-Fast Cooling. *Mater. Sci. Eng. A* **2016**, *666*, 214–224. [CrossRef]

44. Cao, R.; Li, J.; Liu, D.S.; Ma, J.Y.; Chen, J.H. Micromechanism of Decrease of Impact Toughness in Coarse-Grain Heat-Affected Zone of HSLA Steel with Increasing Welding Heat Input. *Metall. Mater. Trans. A* **2015**, *46*, 2999–3014. [CrossRef]
45. Martín-Meizoso, A.; Ocaña-Arizcorreta, I.; Gil-Sevillano, J.; Fuentes-Pérez, M. Modelling Cleavage Fracture of Bainitic Steels. *Acta Metall. Et Mater.* **1994**, *42*, 2057–2068. [CrossRef]
46. Lambert-Perlade, A.; Gourgues, A.F.; Besson, J.; Sturel, T.; Pineau, A. Mechanisms and Modeling of Cleavage Fracture in Simulated Heat-Affected Zone Microstructures of a High-Strength Low Alloy Steel. *Metall. Mater. Trans. A* **2004**, *35*, 1039–1053. [CrossRef]
47. Xiong, Z.; Liu, S.; Wang, X.; Shang, C.; Li, X.; Misra, R.D.K. The Contribution of Intragranular Acicular Ferrite Microstructural Constituent on Impact Toughness and Impeding Crack Initiation and Propagation in the Heat-Affected Zone (HAZ) of Low-Carbon Steels. *Mater. Sci. Eng. A* **2015**, *636*, 117–123. [CrossRef]

Disclaimer/Publisher's Note: The statements, opinions and data contained in all publications are solely those of the individual author(s) and contributor(s) and not of MDPI and/or the editor(s). MDPI and/or the editor(s) disclaim responsibility for any injury to people or property resulting from any ideas, methods, instructions or products referred to in the content.

Article

Evolution Behavior of Rapidly Solidified Microstructure of a Ti-48Al-3Nb-1.5Ta Alloy Powder during Hot Isostatic Pressing

Zhenbo Zuo [1,2], Rui Hu [1,*], Xian Luo [1], Hongkui Tang [2], Zhen Zhu [2], Zitong Gao [1], Jinguang Li [1], Hang Zou [1], An Li [2], Xiaohao Zhao [2], Yunjin Lai [2] and Shaoqiang Li [2]

[1] State Key Laboratory of Solidification Processing, Northwestern Polytechnical University, Xi'an 710072, China; zuozhenbo@mail.nwpu.edu.cn (Z.Z.)
[2] Sino-Euro Materials Technologies of Xi'an Co., Ltd., Xi'an 710018, China
* Correspondence: rhu@nwpu.edu.cn

Abstract: In this study, Ti-48Al-3Nb-1.5Ta powders were manufactured from cast bars by the supreme-speed plasma rotating electrode process (SS-PREP) and used to prepare hot isostatically pressed (HIPed) material at 1050–1260 °C with 150 MPa for 4 h. The phase, microstructure and mechanical performance were analyzed by XRD, SEM, electrical universal material testing machine and other methods. The results revealed that the phase constitution changed from γ phase to $α_2$ phase and then to γ phase with the material changing from as-cast to powders and then to as-HIPed. Compared with the as-cast material, the grain size and element segregation were significantly reduced for both powders and as-HIPed. When the hot isostatic pressing (HIP) temperature was low, the genetic characteristics of the powder microstructure were evident. With the HIP temperature increasing, the homogeneity of the composition and microstructure increased, and the prior particle boundaries (PPBs) gradually disappeared. The elastic moduli of powder and as-HIPed were superior to those of as-cast, which increased with the HIP temperature increasing. The hardness of as-HIPed was lower than that of the powder. The compressive strength, compressive strain, bending strength, and tensile strength of as-HIPed were higher than those of as-cast. With an increase in the HIP temperature, the compressive strength decreased gradually, and the compressive strain first decreased and then increased.

Keywords: Ti-48Al-3Nb-1.5Ta; HIP; powder; cast; microstructure; phase

1. Introduction

TiAl is a promising material for high-temperature applications due to its excellent properties of low density and high strength (the density is only half that of nickel-based alloys). As the best candidate material for weight reduction and efficiency [1,2], γ-TiAl is considered to be a substitute for nickel-based superalloys in the range of 800~900 °C. Blades, turbine disks, valves, and other crucial heated end components of aeroengines have been manufactured using TiAl alloy. Particularly, the GEnx engines' low-pressure turbine blades have successfully used the Ti-48Al-2Cr-2Nb (at.%) alloy, resulting in a considerable reduction in the engine weight [3]. However, the service temperature of commercial TiAl alloys such as Ti4822 is only about 650 °C and the deficiency of mechanical performance at high temperature restricts the application of TiAl alloys.

Alloying is a common method to improve the comprehensive mechanical properties of TiAl alloy. Cr, Mn, V, Mo, Ta, Nb, B, Y, C and W are usually added to TiAl alloys [1], which can improve the performance of TiAl alloys through solid solution strengthening and second phase strengthening. The high temperature strength, creep resistance and oxidation resistance can be improved by adding appropriate Nb and Ta to the TiAl alloy [1], which is expected to further improve the service temperature of the TiAl alloy. However, Nb and Ta tend to segregate in the alloy, and it is difficult to eliminate segregation, especially for Ta, which is difficult to diffuse as it is a rather heavy element [4].

The constitution segregation can be improved by powder metallurgy techniques like HIP, and HIPed TiAl alloys also have remarkable benefits such as fine grain and isotropy [2,5–7]. Research on HIPed TiAl alloys has been carried out by IMR, GKSS, Rolls-Royce and Birmingham University, and TiAl auto engine connecting rods, TiAl rotating symmetric thin-wall structure parts and high pressure compressor casings have been prepared through the HIP process [2,8–11].

According to earlier studies [12], casting Ti-48Al-3Nb-1Ta can achieve a fine lamella structure by microstructure adjustment, which results in outstanding mechanical properties. The performance at high temperatures is anticipated to be further improved by appropriately increasing the Ta concentration. With HIP's benefits, it is hoped to achieve a finer lamella structure and weaken the composition segregation, improving the overall mechanical performance of TiAl alloy.

Metal powder is the prerequisite for high-quality HIP alloys. There are many kinds of metal powder preparation processes, among which the metal powders prepared by the SS-PREP® have many advantages, such as high sphericity, outstanding flow ability, less hollow powder and oxygen increment, and the properties of powder metallurgy (P/M) parts will be ensured [1].

Although previous studies have been conducted on HIPed TiAl alloys and cast TiAl alloys with Nb and Ta additions [13,14], there is no relevant study on the evolution behavior of the rapidly solidified microstructure, which is achieved by the SS-PRE process from cast bars of a Ti-48Al-3Nb-1.5Ta alloy during HIP processes.

In this study, the microstructure evolution and properties of Ti-48Al-3Nb-1.5Ta from as-cast to powders and then to as-HIPed are investigated and clarified, providing valuable guidance for the application of TiAl alloy at higher temperatures.

2. Materials and Methods

In this study, Ti-48Al-3Nb-1.5Ta in different states, including as-cast, powders and as-HIPed, were investigated. The cast Ti-48Al-3Nb-1.5Ta was prepared by vacuum arc remelting three times, and it was also the raw material for preparing powders using SS-PREP® in Sino-Euro Materials Technologies of Xi'an Co., Ltd. (Sino-Euro, Xi'an, China) after being machined into bars. The powders were mostly perfect spherical in shape, as shown in Figure 1a. The particle distribution was determined by a laser particle analyzer (Bettersize BT-9300, Better, Dandong, China) according to ASTM B822-17, as shown in Figure 1b. Other details about Ti-48Al-3Nb-1.5Ta powders can be found in the previous study [15]. The prealloyed powders with the size of 45–250 μm were poured into 45# steel capsules and degassed at 673 K for 4 h to remove air, and then the stem on the top of the capsule was welded when the vacuum of the capsule reached 1×10^{-4} Pa. Finally, the sealed capsules were put into the HIP furnace. The capsules were heated to the specified temperatures, as shown in Figure 2, at a heating rate of 4 °C/min and pressurized up to 150 MPa for 4 h, followed by cooling at 5 °C/min to room temperature. The temperature and pressure were elevated at the same time. The HIPed billets with different temperatures were denoted as 1#–5#, as shown in Table 1.

 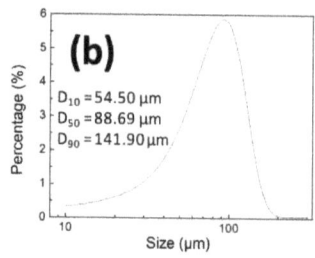

Figure 1. The typical morphology (**a**) and the particle distribution (**b**) of SS-PREP® Ti-48Al-3Nb-1.5Ta powders with the range of 45–250 μm.

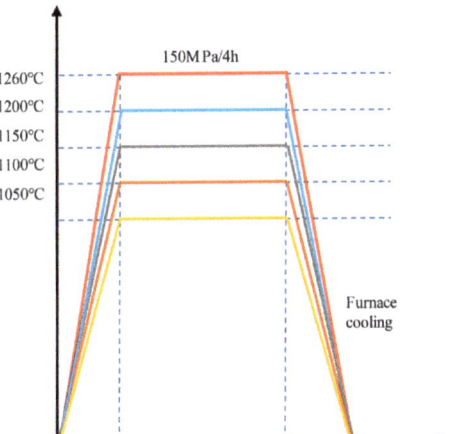

Figure 2. HIP processes.

Table 1. Numbers of HIP billets with different temperatures.

HIP Temperature/°C	1050	1100	1150	1200	1260
Billets No.	1#	2#	3#	4#	5#

The HIP process mainly includes three stages: initial compaction of powder particles, plastic creep deformation and diffusion bonding. The process of filling the powders into capsules is accompanied by vibration. The relative density of the powders is about 66%, and there are still many gaps between the particles. Due to the shielding effect of the capsule on the pressure, the thermal expansion effect of compaction is more obvious in the early stage, resulting in a small decrease in the relative density. With the increase in temperature and pressure, the compaction enters the rapid densification stage, when the powders translate and rotate, and the lap holes collapse rapidly. At this time, there is no large plastic deformation of particles, and the contact among particles is still mainly point contact, as illustrated in Figure 3. The rearrangement and viscoplastic deformation of powders is the main mechanism of rapid densification. During the stages of heat preservation and pressure preservation, viscoplastic deformation, such as diffusion and creep, is the main densification mechanism. In addition, since the shear stress of the powders exceeds the yield strength, plastic deformation will occur in the sliding mode, so obvious twinned plastic deformation occurs within the grains. During the sliding process, some powders are squeezed into adjacent pores, which reduces the porosity and increases the relative density. Therefore, plastic deformation of the powders is also one of the main mechanisms of densification. The remaining tiny pores are isolated from each other, dispersed among particles, and spheroidized under the effect of surface tension. At this time, the powder plastic flow no longer plays a role. Instead, the dislocation creep, volume diffusion, and grain diffusion are activated, slow diffusion and creep of atoms and holes occur, and atoms in the powders slowly enter the residual micropores [16]. The viscoplastic rheological effect still occurs with the unloading stage, but the significant thermal strain becomes the main factor of densification in the later stage [17].

The as-cast test specimens were cut from the cast bars and the as-HIPed test specimens were cut from the HIPed billets after the capsules were removed. For SEM microstructure analysis, the powders were sieved into 15–45, 45–106, and 106–250 μm. The same range for the powders used for HIP, 45–250 μm, was employed for other analyses. The densities of as-cast and as-HIPed samples with a size of 10 × 10 × 10 mm prepared by wire cutting, were determined by an electronic hydrometer (DH-300, DahoMeter, Shenzhen, China) through Archimedean principle and the results were obtained as arithmetic mean values from three test-sample measurements. The phase transition temperature was measured with a φ4 × 3.5 mm sample by differential scanning calorimetry (DSC, NETZSCH DSC 214, NETZSCH, Selb, Germany) with a heating rate of 10 °C/min from room temperature to 1500 °C. The phase constituent of the 10 × 10 × 10 mm samples was analyzed by X-ray diffraction (XRD, Bruker D8 DISCOVER A25, Bruker, Billerica, MA, USA) with Co Kα radiation and 2.5°/min from 20° to 90° at room temperature. The microstructures and the element distribution were analyzed by scanning electron microscopy (SEM, ZEISS Sigma 300, Oberkochen, Germany) equipped with energy dispersive X-ray spectrometry (EDX, ZEISS Sigma 300, Oberkochen, Germany) and transmission electron microscopy (TEM, FEI Talos F200X TEM, Hillsboro, OR, USA). The crystallographic feature was further analyzed by electron backscattered diffraction (EBSD, ZEISS Sigma 300, Oberkochen, Germany). The 10 × 10 × 10 mm cubic samples of as-cast and as-HIPed and powders, mounted into epoxy resin, mechanically polished with SiC abrasive papers and polishing cloth, and finally polished by a vibratory Polisher, were used for SEM and EBSD analysis. The specimens for TEM were first cut into a 10 × 10 × 0.4 mm slice, then polished to 0.05 mm thickness with SiC abrasive papers manually and were finally obtained by ion milling. Hardness and elastic modulus were analyzed by a nanomechanical testing system (Hysitron TI-950, Bruker, Billerica, MA, USA) with 10 mN force and results were obtained as arithmetic mean values from ten test point measurements approximately evenly distributed along a straight line of one 10 × 10 × 10 mm sample with the same preparation method with EBSD. The compression test samples with the size of φ5 × 20 mm, the bending test samples with the size of 50 × 8 × 2.5 mm and the tensile samples with a diameter of 3 mm and a gauge length of 15 mm were used to determine the mechanical performance by the micro-computer control electron universal testing machine (CMT5105, Sansi Yongheng, Ningbo, China) according to GB/T 7314-2017 (Chinese Standard), YB/T 5349-2014 (Chinese Ferrous Metallurgy Industry Standard) and GB/T 228.1-2021 (Chinese Standard), respectively. The mechanical performances were obtained as arithmetic mean values from two test samples measurements for each item.

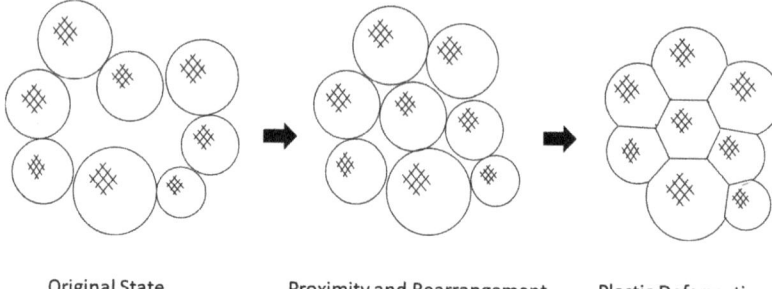

Figure 3. Schematic diagram of powder densification process by HIP.

3. Results and Discussion

3.1. Density

The density is one of the most important properties that directly determines the performance of HIPed parts. The densities of the as-cast and as-HIPed samples with different HIP temperatures were measured as shown in Figure 4. The densities of as-HIPed parts increased with increasing HIP temperatures due to the diffusion coefficient of the elements increasing and material transportation accelerating [18], and were higher than those of as-cast and the increasing rate became mild when the HIP temperature reached 1100 °C. The density ratio of the as-cast to the 5# HIPed billet was 97.95% and this means there were more voids and defects in the cast material which may cause material failure after crack initiation under service.

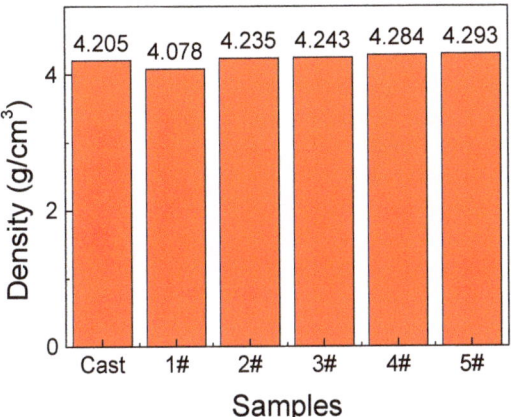

Figure 4. Densities of as-cast and as-HIPed.

3.2. Phase Identification

Since the transformation temperature has a great influence on the microstructure evolution, the phase-transition temperature was analyzed as shown in Figure 5, which showed that $T_{\gamma \to \alpha+\gamma}$ was 1247.6 °C, $T_{\alpha+\gamma \to \alpha}$ was 1342.6 °C and T_m was 1419.1 °C, according to the TiAl binary phase diagram.

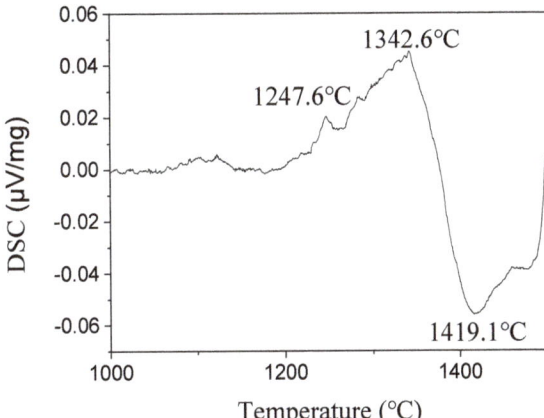

Figure 5. Simultaneous thermal analysis of HIPed Ti-48Al-3Nb-1.5Ta.

To investigate the difference in the phase constitution in different states, EBSD and XRD were conducted on as-cast, powders and as-HIPed Ti-48Al-3Nb-1.5Ta. The results are shown in Figures 6 and 7. The as-cast sample mainly consisted of γ phase, the powders mainly consisted of $α_2$ phase, due to the rapid cooling rate, and γ phase was dominant after HIP.

The rapid solidification of droplets is a non-equilibrium process in SS-PREP, and the diffusion is insufficient or even prevented. The phase transition of $α \to α_2 + γ$ is incomplete, so the volume fraction of $α_2$ in powders is higher, which is more obvious in fine particles because of the higher cooling rate [19]. Note that it is just a non-equilibrium state without a change of composition.

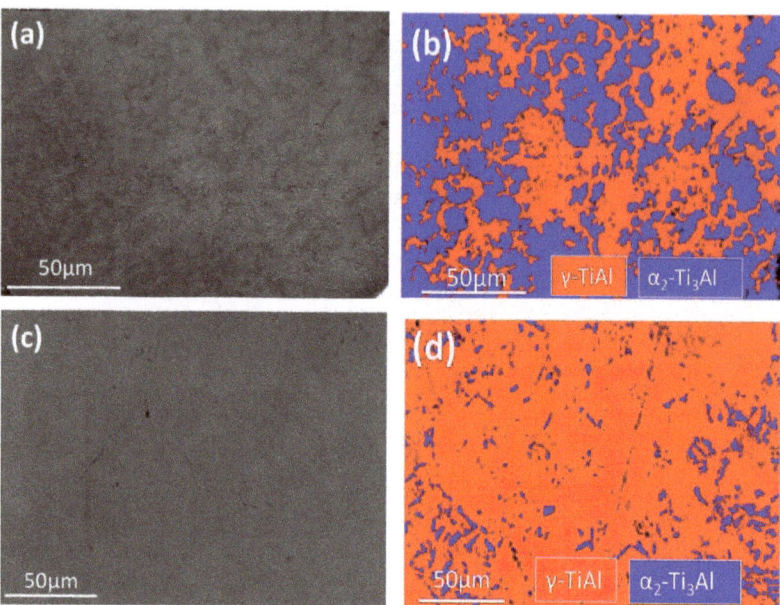

Figure 6. Original images and phase constitution of Ti-48Al-3Nb-1.5Ta: (**a,b**) powder; (**c,d**) 5# HIPed billet.

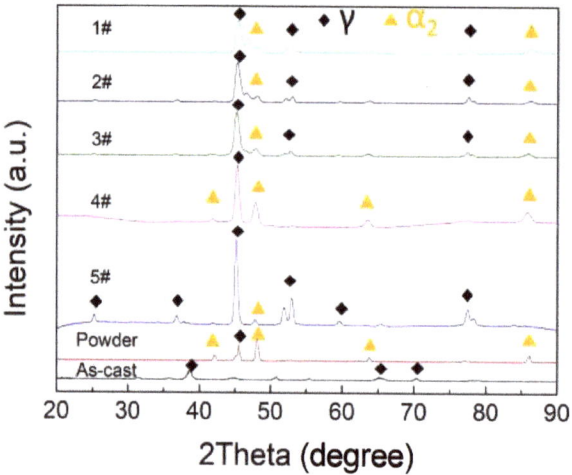

Figure 7. XRD patterns of as-cast, powders and as-HIPed.

During the HIP process, with the increase in temperature, the metastable α_2 phase in the prealloyed powders will transform into γ phase when the temperature reaches the phase transition point [20,21]. This transformation is first the crystal structure change of HCP → FCC, changing the stacking order of the local (0001) plane in the HCP matrix by decomposing the perfect dislocation $a/3\langle 11\bar{2}0\rangle$ in the HCP matrix into $a/3\langle 10\bar{1}0\rangle$ + stacking fault $a/3\langle 01\bar{1}0\rangle$. Then, atomic diffusion changes the chemical composition, which leads to the ordering transformation of the FCC structure into the $L1_0$ structure [22]. During the HIP process in the $\alpha_2+\gamma$ two-phase region, as the "atomic volume" of γ-phase is smaller than that of α_2 phase, the excess production of γ-phase is promoted under isostatic pressure, resulting in the abnormal increase in the volume fraction of γ-phase [23], which is exactly the opposite of the transformation during manufacturing powders. With the increase in HIP temperatures, the α_2 phase content decreased gradually, while the γ phase content increased gradually as the number of γ peaks and the intensity of the main γ peaks both increased in XRD patterns, which is consistent with the DSC analysis result of the powders in the previous study [15]. Namely, there is a long endothermic trend below 1261 °C for Ti-48Al-3Nb-1.5Ta powders, which is the process by which the non-equilibrium α_2 phase transforms into equilibrium γ phase. With the increase in temperature, the transformation amount of $\alpha_2 \rightarrow \gamma$ gradually increases until the state of equilibrium [9]. Additionally, the γ summit around 37° of 5# shifted to the left compared with as-cast, which is mainly caused by the slight expansion of the lattice structure [24]. Given that Ta has a larger atomic radius compared to Al, Ta atoms replace Al atoms in the lattice under conditions of enhanced diffusion at high temperatures during the HIP process, resulting in XRD peak shifting toward lower angles, while Ta is characterized by segregation in the as-cast material.

The TEM bright-field (BF) image and the selected area diffraction pattern (SADP) of the 5# HIPed billet are shown in Figure 8, demonstrating that it was composed of γ phase and α_2 phase, with a few dislocations in the γ phase, particularly close to the interface of the two phases, which is consistent with Yang's research [25]. In the initial thermal deformation stage, the dislocation movement is the main deformation mechanism, as investigated in previous studies [26,27]. As depicted in Figure 8a, numerous dislocations quickly propagated throughout the γ grain and accumulated there. While some dislocations migrate to the grain boundaries and vanish, others become jammed and entangled, obstructing further migration [28].

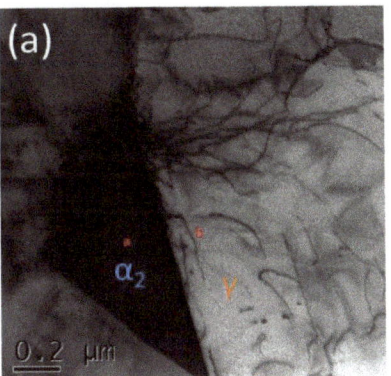 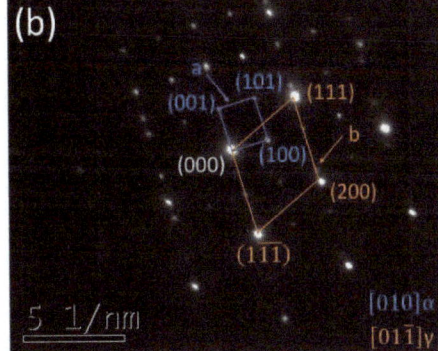

Figure 8. TEM of 5# HIPed billet: (**a**) BF image; (**b**) SADP of the area highlighted in image (**a**).

3.3. Microstructural Evolution

To investigate the microstructure evolution, the microstructures of Ti-48Al-3Nb-1.5Ta as-cast, powders and as-HIPed were analyzed, as shown in Figures 9–11. The microstructure of the as-cast is a typical dendrite structure. The microstructures of the SS-PREP® powders are directly related to the size of the droplets when they fly out of the melting pool at the end of the bar by centrifugal force during SS-PREP. With the increase in the droplet size, the specific surface region decreases, the cooling rate slows down, and the microstructure gradually changes from a featureless smooth structure to a dendritic and cellular dendritic morphology. The grain size was refined and the segregation was weakened significantly from the as-cast to powders, as we could see the particle sizes of the powders were even much smaller than a dendrite of as-cast. The microstructures of as-HIPed with different temperatures were characterized mainly as near-gamma and locally duplex, for HIP temperatures in this study located in the γ single-phase region and near $T_{\gamma \rightarrow \alpha + \gamma}$, according to the phase diagram in Figure 5.

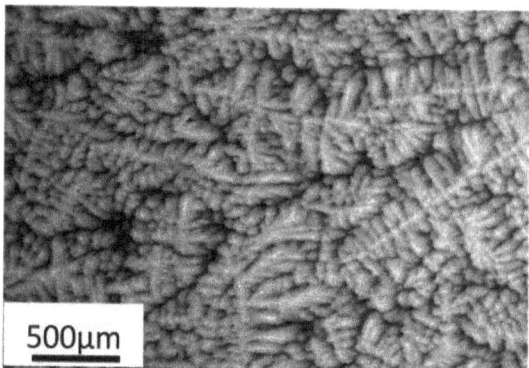

Figure 9. The microstructure of as-cast Ti-48Al-3Nb-1.5Ta.

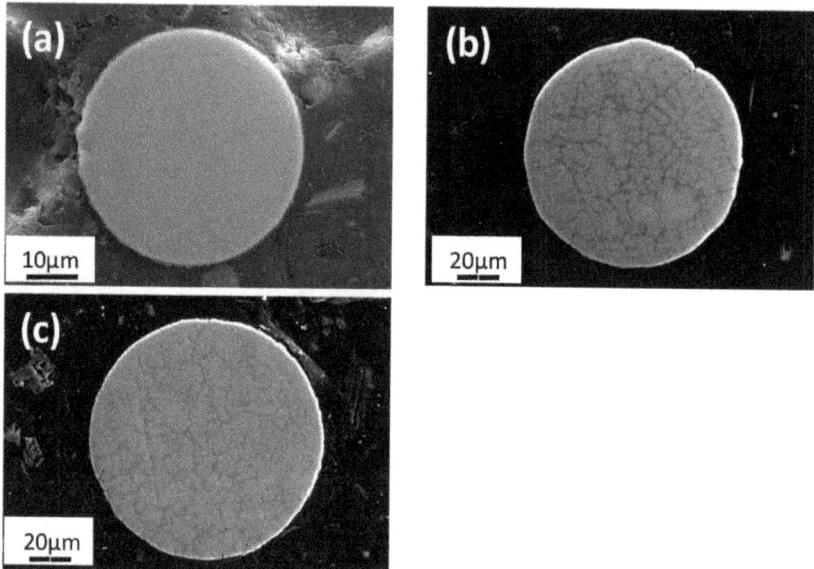

Figure 10. The microstructures of powders: (**a**) 15–45 μm, (**b**) 45–106 μm, (**c**) 106–250 μm.

During the HIP process, Ti-48Al-3Nb-1.5Ta powders soften at high temperatures with an increase in the temperature in the furnace, and the strength of the powders decreases gradually, contributing to serious deformation of the powders [29]. When the pressure is higher than the yield stress of the powders, plastic deformation occurs and the particle gaps are gradually filled. The plastic deformation is not uniform. When particles with different sizes contact each other, smaller particles will bear higher interparticle contact stress and generate larger plastic deformation, while larger particles will generate relatively smaller plastic deformation [30], as shown in the marked regions in Figure 11. For Ti-48Al-3Nb-1.5Ta powders with dendrite structures, plastic deformation begins in the Al-rich γ phase region between the dendrites. During deformation, $1/2\langle 110\rangle$ unit dislocation, $\langle 101\rangle$ super-dislocation and $1/2\langle 112\rangle$ super-dislocation are activated, where high density dislocations tangle and deformation twins appear. When the plastic deformation exceeds the recrystallization threshold value, the free energy stored by the high density defects causes dynamic recrystallization in the deformation region, refining the equiaxed grain size and greatly reducing the dislocation density, while dislocations and twins will be retained to the final structure and the grain is irregular and coarser in the region without dynamic recrystallization [31]. In the 1# HIPed billet (Figure 11a), the cellular dendrite structures of the original particles embedded in the near-spherical structure were observed. The microstructure of the different original particles was independent, and the boundary was sharp. The initial lamellar structure originates from the Al-poor dendrite stem, as shown in the spectrum, and Nb and Ta are rich in the initial lamellar structure because Nb and Ta are rich in the dendritic stem. The deformation of particles can promote dynamic recrystallization and generate equiaxed grains around PPBs [32]. Based on the size of the near-spherical structure and the absence of the equiaxed grain around the near-spherical structure boundaries, it indicated that these near-spherical structures originated from the undeformed Ti-48Al-3Nb-1.5Ta powders and the microstructure heredity of Ti-48Al-3Nb-1.5Ta powders was evident. Guo and Cai found a similar heterogeneous microstructure in Ti55 and TA15 by HIP, which are near-α titanium alloys [29,33]. The oxide layer on the surface of the prealloyed powders protects the powders from deformation during the HIP process.

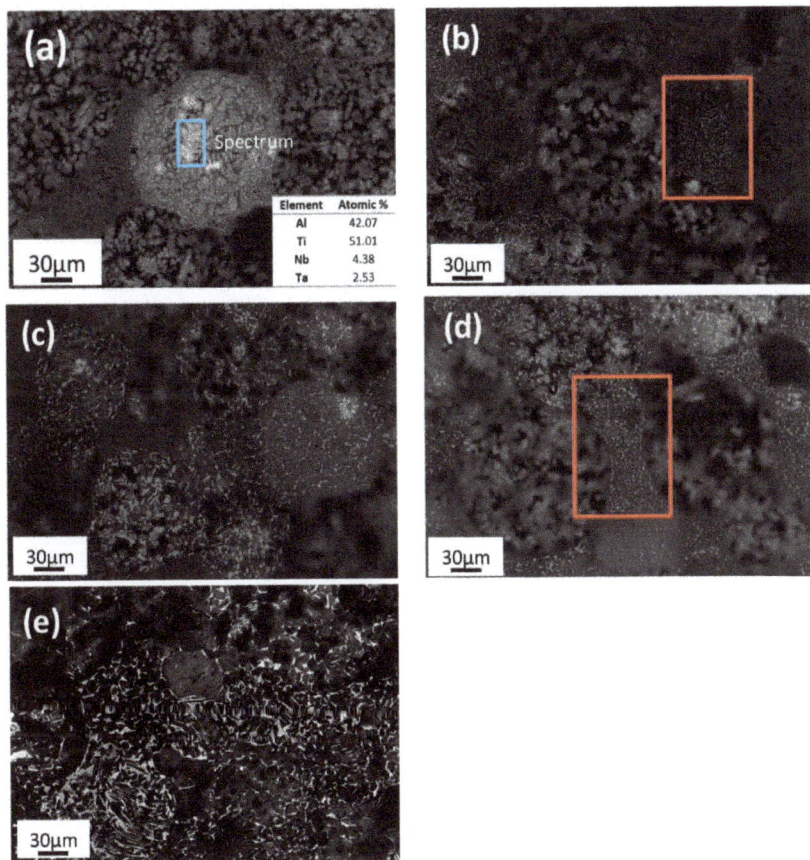

Figure 11. The microstructures of as-HIPed materials: (**a**) 1#, (**b**) 2#, (**c**) 3#, (**d**) 4#, (**e**) 5#. The red box regions show that the smaller particles will bear higher interparticle contact stress and generate larger plastic deformation.

Before plastic deformation, particles collide with each other or wedge against each other, and large plastic strain and lattice distortion only occur in the local region of the particle boundary, forming a large strain band at the boundary. Due to the accumulation of large strain energy, dynamic recovery and recrystallization occur in the large strain zone of the boundary during the preservation of heat and pressure, and many equiaxed α phases are formed, which aggregate into a spatial network at the boundary, namely PPB. Although there was a degassing process for removing foreign gas absorbed on the surface of particles with the process of 1×10^{-4} Pa and 400 °C for 4 h before HIP, TiO_2 and Al_2O_3 oxide layers were still difficult to remove and remained on the surface of the powders. The oxide layers on the surface of the powders affect the diffusion of elements and the metallurgical bonding between particles and hinder the growth of grains. Due to its poor plasticity, it cannot be coordinated with the deformation of the matrix during deforming, resulting in dislocation accumulation and stress concentration, forming cracks and holes at the interface between the PPB and the matrix, eventually weakening the binding force of the particle boundary, becoming the main crack source and propagation channel, and giving poor performance [16].

Although the PPBs could still be seen in the 2# HIPed billet (Figure 11b), element migration and microstructure transformation had taken place among the different grains, and the associated microstructures had been generated. With the increase in the HIP temperature, the element diffusion was more efficient, segregation was weakened, the homogeneity of the microstructure was improved, and the PPBs gradually disappeared because of the reduction in element migration activation energy and yield strength of the powders [34].

There is a local coarsening phenomenon after HIP at 1150 °C, although the overall microstructure is fine. The local coarsening region is randomly distributed, and the size is similar to that of coarse prealloyed powders. When the temperature rises to 1260 °C, these local coarsening regions transform into a lamellar structure. There may be two reasons for local coarsening after HIP. On the one hand, different cooling rates lead to different phase constitutions of the prealloyed powders with different sizes, and α phase transforms into γ_m phase (massive γ phase) in coarse powders due to a relatively low cooling rate [35]. During the subsequent densification process, the specific strain of γ_m phase is generated due to lattice distortion, which provides a driving force for subsequent grain growth and full lamination [36]. On the other hand, segregation during solidification leads to the reduction in or disappearance of the α phase in local regions, resulting in γ-phase coarsening without the pinning effect of the α phase during subsequent densification [37].

To analyze the crystallographic characteristics, the EBSD analysis of as-cast and as-HIPed materials was performed. The pole figures (PFs) are shown in Figure 12, indicating that all orientations were evenly distributed. It can be seen from the color distribution that random crystal orientation is an inherent characteristic of HIPed materials, independent of the HIP temperature. The filled powders experienced isotropic pressure and heat during the HIP process, which resulted in no preferential crystal orientation [29].

The average grain sizes of as-cast and as-HIPed, analyzed by EBSD, are shown in Figures 13 and 14. As expected, with the increase in the HIP temperature, the average grain size increased gradually, which is consistent with Cai's research [32]. However, the grain sizes of all HIP temperatures were much smaller than those of as-cast (Figure 13f). Twins exist in as-HIPed samples with different temperatures, as we can see from Figures 15 and 16, because dynamic recrystallization under high temperature and stress results in stacking faults [38], and the extrinsic stacking faults act as sites for twin nucleation. At the same time, Ta can reduce the stacking fault energy by replacing Al at the interface, contributing to the appearance of twin boundaries [39]. The γ lamellae present a 180° true-twin relationship with the γ matrix, according to the SADP.

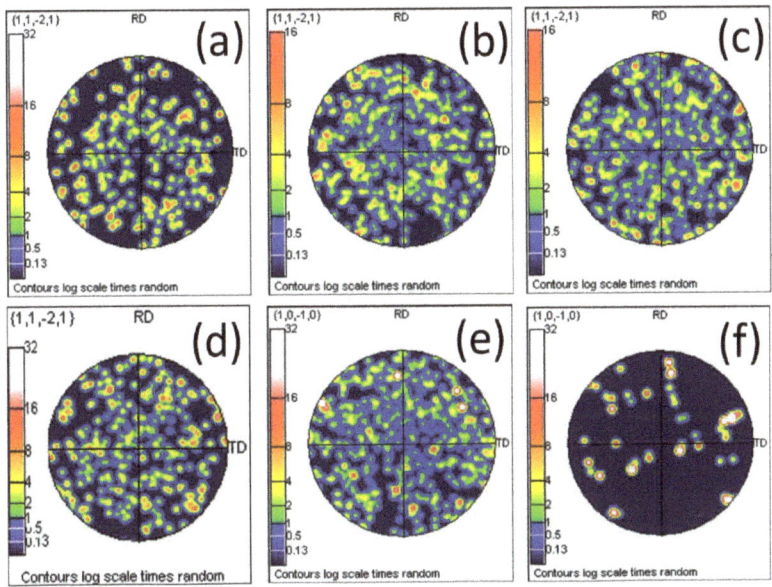

Figure 12. PF maps of as-HIPed and as-cast materials: (**a**) 1#; (**b**) 2#; (**c**) 3#; (**d**) 4#; (**e**) 5#; (**f**) as-cast.

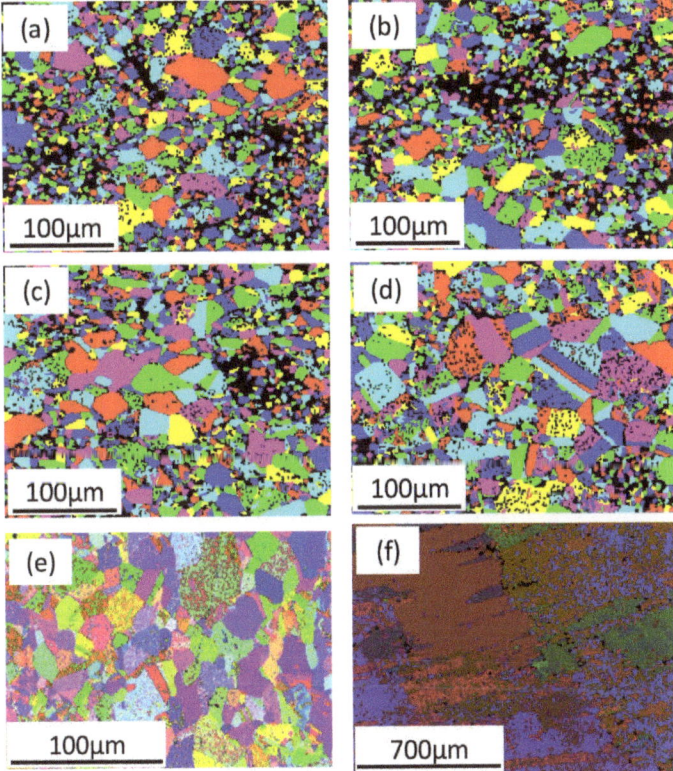

Figure 13. EBSD analysis of as-HIPed and as-cast materials: (**a**) 1#; (**b**) 2#; (**c**) 3#; (**d**) 4#; (**e**) 5#; (**f**) as-cast.

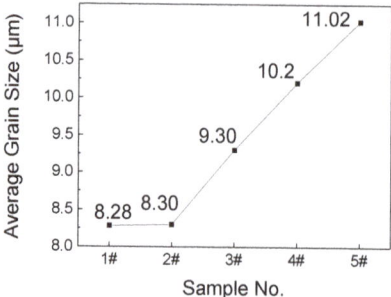

Figure 14. Average grain sizes of HIP billets with different temperatures.

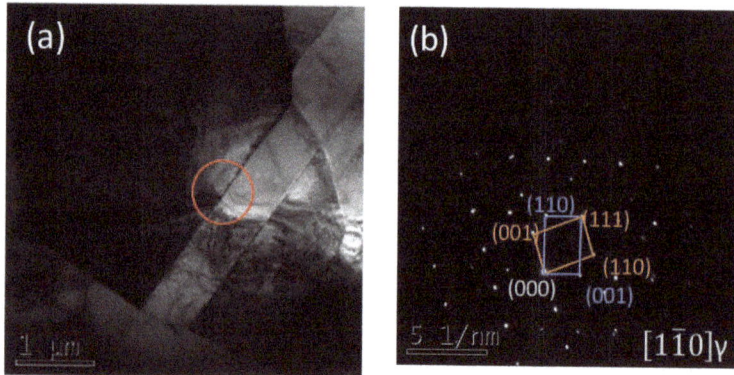

Figure 15. The twin crystal in 5# HIPed billet: (**a**) TEM BF image; (**b**) SADP of the area highlighted with a red circle in image (**a**).

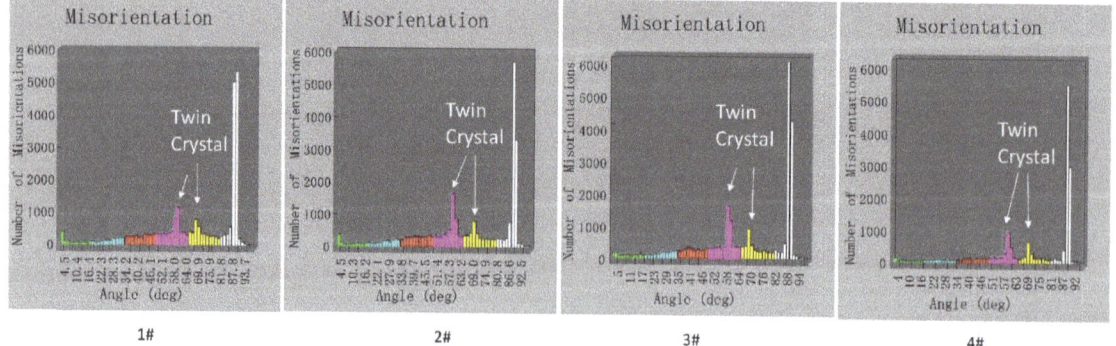

Figure 16. Examination of the grain boundaries in 1# to 4# HIPed billets.

3.4. Mechanical Properties

In order to analyze the mechanical properties of Ti-48Al-3Nb-1.5Ta from as-cast to powders and then to as-HIPed, nanoindentation analysis was carried out on the samples in different states, as shown in Figure 17, and the compression and bending performance of as-cast and as-HIPed samples were analyzed, as shown in Figures 18 and 19.

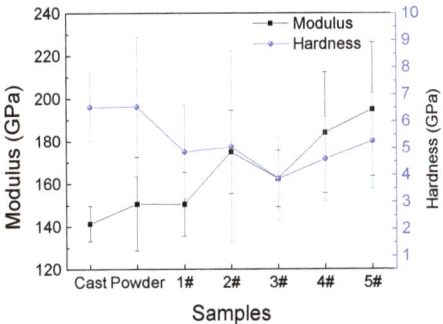

Figure 17. The modulus and hardness of as-cast, powders and as-HIPed.

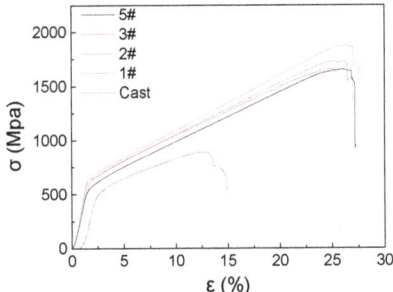

Figure 18. The compression stress–strain curve of as-cast and as-HIPed.

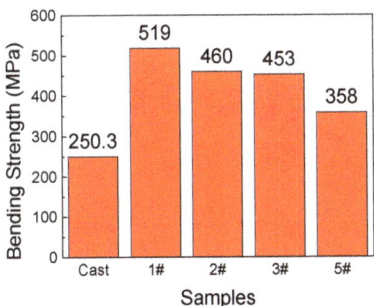

Figure 19. The bending stress of as-cast and as-HIPed.

The elastic modulus reflects the interaction force between atoms, which depends on the types of atoms and lattices [40] mainly affected by the chemical composition. The elastic moduli of the powders and as-HIPed were larger than those of the as-cast, which was mainly due to the serious constituent segregation in the cast alloy. The elastic modulus of the 1# HIPed billet was almost consistent with the powder, indicating that the composition uniformity was close, because the element diffusion was weak when HIP at a lower temperature. With the increasing HIP temperature, the elastic modulus increased, on the whole, which was mainly related to the increase in the constituent homogenization.

The hardness of as-HIPed was lower than that of the powder. It can be seen from the above that α_2 phase transforms into γ phase during the HIP process, and the hardness of γ phase is lower than that of α_2 phase [41]. The lamellar structure began to appear in local regions as the HIP temperature increased to 1150 °C, as shown in Figure 11c. In the lamellar structure, the lamellar α_2 phase reduces the interstitial atom concentration in γ phase, resulting in the weakened ability of P-N stress pinning dislocations, and the enhanced deformation ability of γ phase, reducing the hardness of the microstructure, so the hardness decreases after the lamellar structure appears [42]. With an increase in the HIP temperature, the densification increases, which is helpful for increasing hardness. Hardness first decreases and then increases with the increase in HIP temperature under comprehensive effects.

The compressive strength, compressive strain and bending strength of as-HIPed materials were higher than those of as-cast, as shown in Figures 18 and 19. In addition, the tensile strength of the 4# HIP billet was 444.7 MPa, which was higher than that of as-cast (385.9 MPa). This was mainly because of the poor mechanical properties of the cast alloy due to its coarse structure and severe constitution segregation [42]. Meanwhile, with the increase in the HIP temperature, the compressive strength and bending strength decreased gradually, and the compressive strain first decreased and then increased. On the one hand, it was affected by the grain size. According to the Hall–Petch formula, a smaller grain size means higher strength and better plasticity. The grain size increased with increasing HIP temperature. At the same time, with the HIP temperature increasing, the γ phase ratio increased and the microcracks increased, mainly caused by the phase stress and crystallographic orientation between the α_2 and γ phases [43], leading to the decrease in strength. With the increase in HIP temperature and densification, the compressive strain first decreased and then increased under the comprehensive effects. The compression, bending and tensile performance of HIPed billets were obviously superior to the as-cast material and the plasticity was more significant for applications. As-HIPed Ti-48Al-3Nb-1.5Ta is more in line with service requirements, in spite of its lower hardness, compared with as-cast.

4. Conclusions

Ti-48Al-3Nb-1.5Ta powders were manufactured from cast bars by SS-PREP® and used to prepare HIPed materials at different temperatures. The microstructure evolution and characteristics of Ti-48Al-3Nb-1.5Ta from as-cast to powders and then to as-HIPed were investigated and clarified. The main conclusions are as follows:

(1) The phase constitution of Ti-48Al-3Nb-1.5Ta changed from γ phase to α_2 phase and then to γ phase with material states from as-cast to powders and then to as-HIPed;
(2) The grain sizes and element segregation of Ti-48Al-3Nb-1.5Ta were significantly reduced for both powders and as-HIPed, compared with as-cast. When the HIP temperature was low, the genetic characteristics of the powder microstructure were evident. With increasing HIP temperature, the homogeneity of composition and microstructure increased and the PPBs gradually disappeared. The initial lamellar structure originates from the dendrite stem during the HIP process, and Nb and Ta are rich in the initial lamellar structure;
(3) The elastic moduli of the powder and as-HIPed materials were superior to those of as-cast, which increased with the increase in HIP temperature. The hardness of the as-HIPed materials was lower than that of the powder. The compressive strength, compressive strain and tensile strength of as-HIPed were higher than those of as-cast. With an increase in the HIP temperature, the compressive strength decreased gradually, and the compressive strain first decreased and then increased.

Author Contributions: Conceptualization, Z.Z. (Zhenbo Zuo), Y.L. and R.H.; methodology, Z.Z. (Zhenbo Zuo), H.Z. and S.L.; formal analysis, Z.Z. (Zhenbo Zuo) and X.Z.; investigation, Z.Z. (Zhen Zhu), H.T., J.L. and Z.G.; resources, A.L., H.Z. and X.Z.; data curation, Z.Z. (Zhenbo Zuo), A.L. and Z.Z. (Zhen Zhu); writing—original draft preparation, Z.Z. (Zhenbo Zuo); writing—review and editing, R.H. and X.L.; supervision, R.H., Y.L. and S.L. All authors have read and agreed to the published version of the manuscript.

Funding: This research was funded by Key R&D Program of Shaanxi, grant number 2022GY-388, and Xi'an Science and Technology Project, grant number 21XJZZ0077.

Data Availability Statement: All the data generated during this study are included in this article.

Acknowledgments: We would like to thank everyone supplying help.

Conflicts of Interest: The authors declare no conflict of interest.

References

1. Appel, H.F.; Paul, J.; Oehring, M. *Gamma Titanium Aluminide Alloys: Science and Technology*; John Wiley & Sons: Hoboken, NJ, USA, 2011; pp. 465–732.
2. Yang, R. Advances and challenges of TiAl base alloys. *Acta Metall. Sin.* **2015**, *51*, 129–147. [CrossRef]
3. Bewlay, B.P.; Nag, S.; Suzuki, A.; Weimer, M.J. TiAl alloys in commercial aircraft engines. *Mater. High Temp.* **2016**, *33*, 549–559. [CrossRef]
4. Zhang, R.; Liu, P.; Cui, C.; Qu, J.; Zhang, B.; Du, J.; Zhou, Y.; Sun, X. Present Research Situation and Prospect of Hot Working of Cast & Wrought Superalloys for Aero-Engine Turbine Disk in China. *Acta Metall. Sin.* **2021**, *57*, 1215–1228.
5. Li, J.; Song, B.; Nurly, H.; Xue, P.; Wen, S.; Wei, Q.; Shi, Y. Microstructure evolution and a new mechanism of B2 phase on room temperature mechanical properties of Ti-47Al-2Cr-2Nb alloy prepared by hot isostatic pressing. *Mater. Charact.* **2018**, *140*, 64–71. [CrossRef]
6. Yang, C.; Hu, D.; Wu, X.; Huang, A.; Dixon, M. Microstructures and tensile properties of hot isostatic pressed Ti4522XD powders. *Mater. Sci. Eng. A* **2012**, *534*, 268–276. [CrossRef]
7. Li, H.-Z.; Che, Y.-X.; Liang, X.-P.; Tao, H.; Zhang, Q.; Chen, F.-H.; Han, S.; Liu, B. Microstructure and high-temperature mechanical properties of near net shaped Ti−45Al−7Nb−0.3W alloy by hot isostatic pressing process. *Trans. Nonferrous Met. Soc. China* **2020**, *30*, 3006–3015. [CrossRef]
8. Clemens, H.; Mayer, S. Design, processing, microstructure, properties, and applications of advanced intermetallic TiAl alloys. *Adv. Eng. Mater.* **2013**, *15*, 191–215. [CrossRef]
9. Li, J.Z. Fundamental Study of Key Technology for TiAl Intermetallic Compound Components Fabricated by Hot Isostatic Pressing. Ph.D. Dissertation, Huazhong University of Science and Technology, Wuhan, China, 2019.
10. Wu, J.; Xu, L.; Lu, Z.G.; Guo, R.P.; Cui, Y.Y.; Yang, R. Effect of container on the microstructure and properties of powder metallurgy TiAl alloys. *Mater. Sci. Forum* **2015**, *817*, 604–609. [CrossRef]
11. Lasalmonie, A. Intermetallics: Why is it so difficult to introduce them in gas turbine engines? *Intermetallics* **2006**, *14*, 1123–1129. [CrossRef]
12. Zhang, K. Evaluation Mechanism of Solidification Microstructure and Metastable Structure of Ta-Containing Hyperperitectic TiAl Alloys. Ph.D. Dissertation, Northwestern Polytechnic University, Xi'an, China, 2020.
13. Lapin, J.; Kamyshnykova, K. Effect of Ta and W Additions on Microstructure and Mechanical Properties of Tilt-Cast Ti-45Al-5Nb-2C Alloy. *Metals* **2021**, *11*, 2052. [CrossRef]
14. Huang, X.-M.; Cai, G.-M.; Liu, H.-S. Phase equilibria and transformation in the Ti–Al–Ta system. *J. Mater. Sci.* **2022**, *57*, 2163–2179. [CrossRef]
15. Zuo, Z.; Hu, R.; Luo, X.; Wang, Q.; Li, C.; Zhu, Z.; Lan, J.; Liang, S.; Tang, H.; Zhang, K. Solidification behavior and microstructures characteristics of Ti-48Al-3Nb-1.5Ta powder produced by supreme-speed plasma rotating electrode process. *Acta Metall. Sin.* **2023**, 1–14. [CrossRef]
16. Xue, P.J. Study on Process Technology of Near Net Shape Hot Isostatic Pressing of Ti6Al4V Powders. Ph.D. Dissertation, Huazhong University of Science and Technology, Wuhan, China, 2014.
17. Liu, G.C. Metal Powders Densification under Hot Isostatic Pressing: Numerical Simulation and Experiment. Ph.D. Dissertation, Huazhong University of Science and Technology, Wuhan, China, 2011.
18. Pei, Y.; Qu, X.; Ge, Q.; Wang, T. Evolution of Microstructure and Elements Distribution of Powder Metallurgy Borated Stainless Steel during Hot Isostatic Pressing. *Metals* **2022**, *12*, 19. [CrossRef]
19. Gerling, R.; Clemens, H.; Schimansky, F. Powder metallurgical processing of intermetallic gamma titanium aluminides. *Adv. Eng. Mater.* **2004**, *6*, 23–38. [CrossRef]
20. Schaeffer, R.; Janowski, G. Phase transformation effects during hip of TiAl. *Acta Metall. Mater.* **1992**, *40*, 1645–1651. [CrossRef]
21. Choi, B.; Marschall, J.; Deng, Y.; McCullough, C.; Paden, B.; Mehrabian, R. Densification of rapidly solidified titanium aluminide powders—II. The use of a sensor to verify HIRing models. *Acta Metall. Mater.* **1990**, *38*, 2245–2252. [CrossRef]
22. Wang, G.; Xu, L.; Cui, Y.Y.; Yang, R. Densification mechanism of TiAl pre-alloy powders consolidated by hot isostatic pressing and effects of heat treatment on the microstructure of TiAl powder compacts. *Acta Metall. Sin.* **2016**, *52*, 1079–1088. [CrossRef]
23. Huang, A.; Hu, D.; Loretto, M.; Mei, J.; Wu, X. The influence of pressure on solid-state transformations in Ti-46Al-8Nb. *Scr. Mater.* **2007**, *56*, 253–256. [CrossRef]
24. Mohammadnejad, A.; Bahrami, A.; Khajavi, L.T. Microstructure and Mechanical Properties of Spark Plasma Sintered Nanocrystalline TiAl-xB Composites (0.0 < x < 1.5 at. %) Containing Carbon Nanotubes. *J. Mater. Eng. Perform.* **2021**, *30*, 4380–4392.
25. Yang, D.Y. Microstructure and Phase Structure of TiAl Alloy Prepared by Gas Atomization and Sintering. Ph.D. Dissertation, Harbin Institute of Technology, Harbin, China, 2015.

26. Singh, V.; Mondal, C.; Sarkar, R.; Bhattacharjee, P.P.; Ghosal, P. Dynamic recrystallization of a β(B2)-Stabilized γ-TiAl based Ti-45Al-8Nb-2Cr-0.2B alloy: The contributions of constituent phases and Zener-Hollomon parameter modulated recrystallization mechanisms. *J. Alloys Compd.* **2020**, *828*, 154386. [CrossRef]
27. Zhu, L.; Li, J.S.; Tang, B.; Zhao, F.T.; Hua, K.; Yan, S.P.; Kou, H.C. Dynamic recrystallization and phase transformation behavior of a wrought beta-gamma TiAl alloy during hot compression. *Prog. Nat. Sci.* **2020**, *30*, 517–525. [CrossRef]
28. Chen, X.; Tang, B.; Liu, D.; Wei, B.; Zhu, L.; Liu, R.; Kou, H.; Li, J. Dynamic recrystallization and hot processing map of Ti-48Al-2Cr-2Nb alloy during the hot deformation. *Mater. Charact.* **2021**, *179*, 111332. [CrossRef]
29. Cai, C.; Gao, X.; Teng, Q.; Kiran, R.; Liu, J.; Wei, Q.; Shi, Y. Hot isostatic pressing of a near α-Ti alloy: Temperature optimization, microstructural evolution and mechanical performance evaluation. *Mater. Sci. Eng. A* **2021**, *802*, 140426. [CrossRef]
30. Nair, S.V.; Tien, J.K. Densification mechanism maps for hot isostatic pressing (HIP) of unequal sized particles. *Metall. Trans. A* **1987**, *18*, 97–107. [CrossRef]
31. Wang, X.N.; Zhu, L.P.; Yu, W.; Ding, X.F.; Nan, H. Research progress of powder hot isostatic pressing for intermetallic titanium aluminide. *Rare Met. Mater. Eng.* **2021**, *50*, 3797–3808.
32. Cai, C.; Song, B.; Xue, P.; Wei, Q.; Yan, C.; Shi, Y. A novel near α-Ti alloy prepared by hot isostatic pressing: Microstructure evolution mechanism and high temperature tensile properties. *Mater. Des.* **2016**, *106*, 371–379. [CrossRef]
33. Guo, R.; Xu, L.; Chen, Z.; Wang, Q.; Zong, B.Y.; Yang, R. Effect of powder surface state on microstructure and tensile properties of a novel near α-Ti alloy using hot isostatic pressing. *Mater. Sci. Eng. A* **2017**, *706*, 57–63. [CrossRef]
34. Wang, G.; Zheng, Z.; Chang, L.T.; Xu, L.; Cui, Y.Y.; Yang, R. Characterization of TiAl prealloyed powder and its densification microstructure. *Acta Metall. Sin.* **2011**, *47*, 1263–1269. [CrossRef]
35. Wang, P.; Viswanathan, G.B.; Vasudevan, V.K. Observation of a massive transformation from α to γ in quenched Ti-48 At. pct al alloys. *Metall. Trans. A* **1992**, *23*, 690–697. [CrossRef]
36. Fischer, F.; Cha, L.; Dehm, G.; Clemens, H. Can local hot spots induce α2/γ lamellae during incomplete massive transformation of γ-TiAl alloys? *Intermetallics* **2010**, *18*, 972–976. [CrossRef]
37. Adams, A.; Rahaman, M.; Dutton, R. Microstructure of dense thin sheets of γ-TiAl fabricated by hot isostatic pressing of tape-cast monotapes. *Mater. Sci. Eng. A* **2008**, *477*, 137–144. [CrossRef]
38. Ge, X.X. Study on Solidification Microstructure and Diffusion Behavior of a High-W Containing Ni-Based PM Superalloy. Master's Dissertation, Northwestern Polytechnical University, Xi'an, China, 2020.
39. Singh, S.; Howe, J. Effect of Ta on twinning in TiAl. *Scr. Metall. Mater.* **1991**, *25*, 485–490. [CrossRef]
40. Guo, Z.; Wang, X.; Yang, X.; Jiang, D.; Ma, X.; Song, H.; Young, R.B. Relationships between Young's modulus, hardness and orientation of grain in polycrystalline copper. *Acta Metall. Sin.* **2008**, *44*, 901–904. [CrossRef]
41. Schloffer, M.; Iqbal, F.; Gabrisch, H.; Schwaighofer, E.; Schimansky, F.-P.; Mayer, S.; Stark, A.; Lippmann, T.; Göken, M.; Pyczak, F.; et al. Microstructure development and hardness of a powder metallurgical multi phase γ-TiAl based alloy. *Intermetallics* **2012**, *22*, 231–240. [CrossRef]
42. Zhang, H.F. Microstructure and Properties of TiAl-Based Alloy Prepared by Powder Metallurgy. Master's Dissertation, Harbin Institute of Technology, Harbin, China, 2012.
43. Li, J.; Song, B.; Wen, S.; Shi, Y. A new insight of the relationship between crystallographic orientation and micro-cracks of Ti-47Al-2Cr-2 Nb alloy. *Mater. Sci. Eng. A* **2018**, *731*, 156–160. [CrossRef]

Disclaimer/Publisher's Note: The statements, opinions and data contained in all publications are solely those of the individual author(s) and contributor(s) and not of MDPI and/or the editor(s). MDPI and/or the editor(s) disclaim responsibility for any injury to people or property resulting from any ideas, methods, instructions or products referred to in the content.

Article

Microstructure Evolution and Dislocation Mechanism of a Third-Generation Single-Crystal Ni-Based Superalloy during Creep at 1170 °C

Ruida Xu, Ying Li and Huichen Yu *

Science and Technology on Advanced High Temperature Structure Materials Laboratory, AECC Key Laboratory of Aeronautical Materials Testing and Evaluation, Beijing Key Laboratory of Aeronautical Materials Testing and Evaluation, Beijing Institute of Aeronautical Materials, Beijing 100095, China; xjxxrd@163.com (R.X.); liying_patent@163.com (Y.L.)
* Correspondence: yhcyu@126.com; Tel.: +86-6249-6718

Abstract: The present study investigates the creep behavior and deformation mechanism of a third-generation single-crystal Ni-based superalloy at 1170 °C under a range of stress levels. Scanning electron microscopes (SEM) and transmission electron microscopes (TEM) were employed to observe the formation of a rafted γ′ phase, which exhibits a topologically close-packed (TCP) structure. The orientation relationship and elemental composition of the TCP phase and matrix were analyzed to discern their impact on the creep properties of the alloy. The primary deformation mechanism of the examined alloy was identified as dislocation slipping within the γ matrix, accompanied by the climbing of dislocations over the rafted γ′ phase during the initial stage of creep. In the later stages of creep, super-dislocations with Burgers vectors of a<010> and a/2<110> were observed to shear into the γ′ phase, originating from interfacial dislocation networks. Up to the fracture, the sequential activation of dislocation shearing in the primary and secondary slipping systems of the γ′ phase occurs. As a consequence of this alternating dislocation shearing, a twist deformation of the rafted γ′ phase ensued, ultimately contributing to the fracture mechanism observed in the alloy during creep.

Keywords: single-crystal Ni-based superalloy; creep; dislocation; deformation mechanism; γ/γ′ phases; topologically close-packed phase

Citation: Xu, R.; Li, Y.; Yu, H. Microstructure Evolution and Dislocation Mechanism of a Third-Generation Single-Crystal Ni-Based Superalloy during Creep at 1170 °C. *Materials* **2023**, *16*, 5166. https://doi.org/10.3390/ma16145166

Academic Editors: Andrea Di Schino and Claudio Testani

Received: 16 June 2023
Revised: 16 July 2023
Accepted: 20 July 2023
Published: 22 July 2023

Copyright: © 2023 by the authors. Licensee MDPI, Basel, Switzerland. This article is an open access article distributed under the terms and conditions of the Creative Commons Attribution (CC BY) license (https://creativecommons.org/licenses/by/4.0/).

1. Introduction

Single-crystal (SX) nickel-based superalloys are widely regarded as the primary material choice for turbine blades in aero-engines due to their exceptional mechanical properties, particularly at high temperatures [1,2]. As one of the main fracture modes for turbine blades at high temperatures, extensive research has been conducted to investigate the creep behavior of nickel-based superalloys in high-temperature environments [3–8]. During the initial creep stage, deformation is primarily governed by dislocation slipping in the γ matrix and climbing over the γ′ phase [8,9]. Then, the dislocations pile up at the interface of the γ phase and γ′ phase to form dislocation networks. Studies have demonstrated that the density of dislocation networks is inversely proportional to the minimum creep rate observed in third- and fourth-generation SX superalloys [10,11], which is influenced by alloying elements. To serve in harsher environments with elevated temperatures, the refractory elements are incorporated in the superalloys [7,12]. Notably, the addition of rhenium (Re), which predominantly dissolves in the γ matrix, acts as a solid solution strengthener and effectively slows down diffusion processes [13–15]. However, the presence of Re within the γ matrix increases lattice strain and enhances the solution-strengthening capacity of the γ matrix, impeding dislocation movement within the matrix [16]. Re has been reported to decelerate the creep strain rate during the steady-creep stage through its interaction with interfacial dislocations [17]. In the tertiary creep stage, dislocations shear into the γ′ phase in the form of super-dislocations, partials with anti-phase boundaries (APBs), and stacking faults [18,19]. Tian et al. [6] have shown that the inclusion of Re can reduce the stacking fault energy of the alloy, thereby enhancing its resistance to creep.

Meanwhile, the creep properties of SX nickel-based superalloys exhibit a close correlation with microstructure evolution. At elevated temperatures, the shape of the γ' phase undergoes a transformation, resulting in the formation of an N-type or P-type rafted structure during the initial creep stage [20–22]. The transformation is influenced by the lattice misfit. The formation of an N-type rafted structure occurs with the negative lattice misfit between the γ phase and γ' phase, whereas the positive lattice misfit would lead to a P-type rafted structure. When the dislocations shear, bypass, and climb over the γ' phase, they are influenced by the microstructure degradation [23]. This microstructural degradation leads to reduced resistance to creep and a decreased creep lifetime [24,25]. However, the addition of rhenium (Re) can significantly impede the growth of the γ' phase and contribute to microstructural stability [26]. It is worth noting that previous investigations into the deformation mechanisms of SX superalloys during creep have mostly focused on temperatures below 1100 °C [27]. The behavior and deformation mechanisms during creep at temperatures exceeding 1100 °C have been rarely explored, despite the fact that Re addition at these temperatures promotes the precipitation of topologically close-packed (TCP) phases [28–30]. The growth of TCP phases causes the depletion of solid solution-strengthening elements and consequently leads to a deterioration in mechanical properties [28,31]. The presence of refractory elements, as a consequence of high-temperature exposure or creep, contributes to the segregation, leading to the formation of TCP phases. The dislocation movement is also influenced by the occurrence of the TCP phase [32].

Hence, the purpose of this study is to investigate the creep behavior of an SX nickel-based superalloy at 1170 °C within a certain range of applied stresses. The creep characteristics, as microstructure evolution and the movement of dislocations in γ and γ' phases, are investigated in detail. The deformation mechanism of the superalloy is discussed and determined.

2. Materials and Methods

For the present investigation, a third-generation single-crystal (SX) nickel-based superalloy was employed. The alloy composition includes a total of 20.5 wt.% of refractory elements, such as tungsten (W), molybdenum (Mo), tantalum (Ta), rhenium (Re), and niobium (Nb), where the percentage of Re is 4.5 wt.% [12]. The specific chemical compositions can be found in Table 1. The superalloy was directionally solidified to form [001]-oriented single-crystal bars in a vacuum furnace. Only single-crystal bars with a maximum deviation of 5° from the [001] orientation were selected for subsequent creep experiments, utilizing Laue-back reflection techniques. The heat treatment of the alloy involved a sequential process of 1613 K/6 h/air cooling + 1393 K/4 h/air cooling + 1143 K/32 h/air cooling. This treatment was carefully designed to control the morphology and volume of γ' precipitates.

Table 1. Chemical composition of the alloy used in this study (wt.%) [12].

Cr	Co	Mo	W	Ta	Re	Nb	Al	Hf	C	Y	Ni
3.5	7	2	6.5	7.5	4.5	0.5	5.6	0.1	0.008	0.001	Bal.

Following the heat treatment, the single-crystal bars were machined into specimens for creep testing. The schematic of the samples is illustrated in Figure 1, wherein the gage segment of the specimens possessed a length of 51.6 mm and a diameter of 10 mm. Creep tests at 1170 °C were conducted in accordance with ASTM E139-11 standards [33], applying tensile loads of 100 MPa. The creep test machine is an RD-100 Creep/Rupture Test Machine made by Changchunkexin Test Instrument Co., Ltd., Changchun, China. Additionally, three interrupted creep tests were performed under the conditions of 1170 °C/100 MPa after certain creep durations of 2 h, 10 h, and 50 h, respectively. A Quanta FEG 450 SEM microscope (FEI, Hillsboro, OR, USA) was employed for the observation of the microstructure evolution of the alloy. For transmission electron microscopy (TEM) observations, specimens were sliced from the middle portion of the gage section of the creep specimens and oriented along the [001] and [011] crystallographic directions. Before TEM analysis, these specimens underwent grinding with metallographic sandpaper to a thickness of 50 µm, followed by twin-jet electropolishing utilizing an electrolyte consisting of 10 wt.% perchloric acid and 90 wt.% absolute ethyl alcohol. TEM investigations were performed using a JEM-2010F microscope, operated at an accelerating voltage of 200 kV.

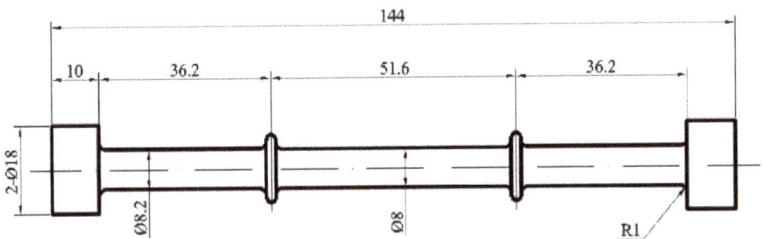

Figure 1. Schematic diagram of the creep specimen (unit: mm).

3. Results

3.1. Creep Behaviors of the Alloy

Figure 2 presents the creep curves and the strain-rate vs. time curves of the investigated alloy, while Table 2 provides a comprehensive overview of the quantitative creep characteristics exhibited by the alloy. The creep deformation behavior can be deduced from the creep curves as follows: (i) The creep curves display typical features corresponding to three creep stages: the initial creep stage, steady-creep stage, and tertiary creep stage. Notably, the creep strain rates exhibit a noticeable decrease during the onset of creep. Subsequently, the alloy enters the steady-creep stage, which constitutes the most significant portion of its overall life under a stress level of 100 MPa. When the creep enters the tertiary creep stage, the creep strain rate increases immediately; (ii) The minimum strain rate observed during the steady-creep stage under a stress level of 100 MPa measures at 0.122×10^{-7} s^{-1}.

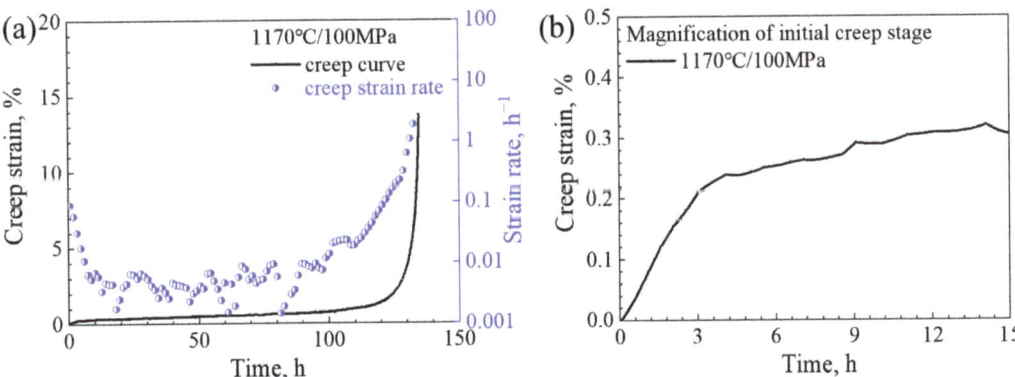

Figure 2. Creep properties of the alloy at 1170 °C. (**a**) Overall creep strain curves and strain-rate vs. time curves; (**b**) magnification of the initial creep stage.

Table 2. Quantitative creep features of the alloy at 1170 °C/100 MPa.

Stress/MPa	Creep Life/h	Duration of Steady-Creep Stage/h	Minimum Strain Rate/×10^{-7}s^{-1}
	134.9	92.9	0.122

Figure 3a displays the microstructure of the alloy after undergoing full heat treatment. Within the γ matrix, the cubic γ′ phase is uniformly distributed, with an average particle size of 410 nm and an average width of 65 nm for the γ matrix channels. The measured volume fraction of the γ′ phase amounts to approximately 68 vol.%. Subsequently, Figure 3b–d illustrate the microstructures after creeping for 2 h, 10 h, and up to fracture under conditions of 1170 °C/100 MPa. After 2 h of creep, the cubic γ′ phase undergoes a transformation into an ellipsoidal shape, while certain portions of the γ′ phase connect along the direction perpendicular to the applied stress, transforming into an N-type rafted structure. Moreover, the width of the horizontal γ matrix channel significantly increases, as evident in Figure 3b. These observations indicate the occurrence and completion of the rafting process at 1170 °C during the initial stages of creep. As the creep duration extends to 10 h (Figure 3c), the width of the horizontal γ matrix channel continues to grow, while the vertical γ matrix channel experiences degeneration and even disappearance. After fracture at 1170°C/100 MPa, a topological inversion phenomenon occurs, wherein the continuous γ channels isolating the γ′ phase develop into continuous γ′ rafts with an isolated γ phase. Furthermore, the γ′ phase exhibits a twisted morphology due to large plastic deformation. The presence of TCP phases is observed, as indicated by black arrows.

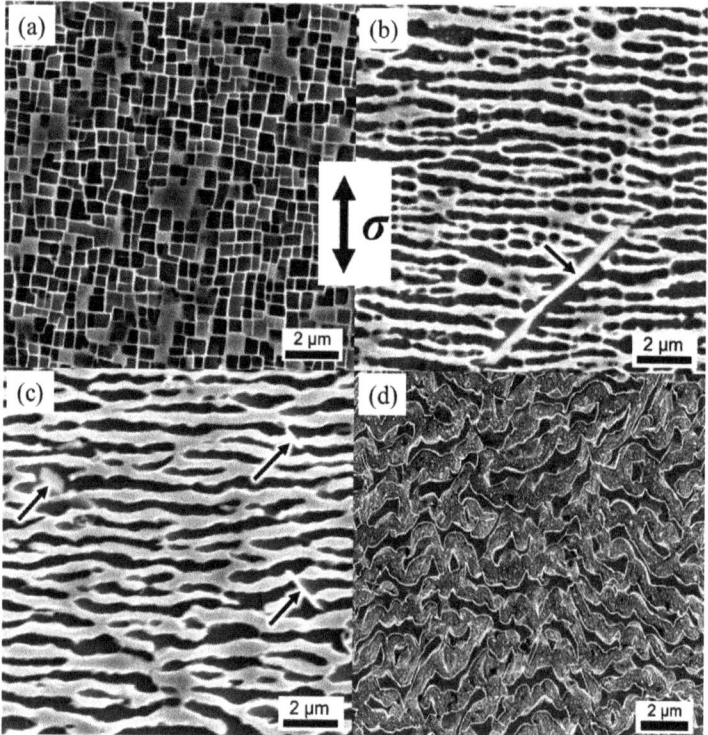

Figure 3. Microstructure evolution of the alloy during creep at 1170 °C and 100 MPa. (**a**) After full heat treatment; (**b**) creeping for 2 h; (**c**) creeping for 10 h; (**d**) failure.

3.2. Analysis of Dislocation Configuration

The TEM images presented in Figure 4 depict the microstructural evolution of the alloy subjected to various durations of creep at 1170 °C/100 MPa. The plane of observation for the specimens aligns parallel to the (101) crystallographic plane. Figure 4a displays an illustration of the alloy's morphology after a creep duration of 2 h. It reveals that the γ' phase displays a rafted structure, while interfaces between the rafted γ and γ' phases exhibit dislocation networks. The accumulation of dislocations within the γ matrix at the γ/γ' interface leads to the gradual formation of denser dislocation networks, resulting in localized stress concentration. This stress concentration mechanism promotes the dislocation shear into the γ' phase. Notably, upon subjecting the alloy to 10 h of creep at 1170 °C/100 MPa, the γ' phase exhibits complete rafting perpendicular to the applied stress axis, as depicted in Figure 4b. The presence of a super-dislocation, indicated by arrows, is observed to shear into the γ' phase. The morphology of the alloy approaching fracture is demonstrated in Figure 4c. At this critical stage, an abundance of dislocation arrays is observed within the γ' phase, leading to a significant increase in dislocation density within the γ' phase.

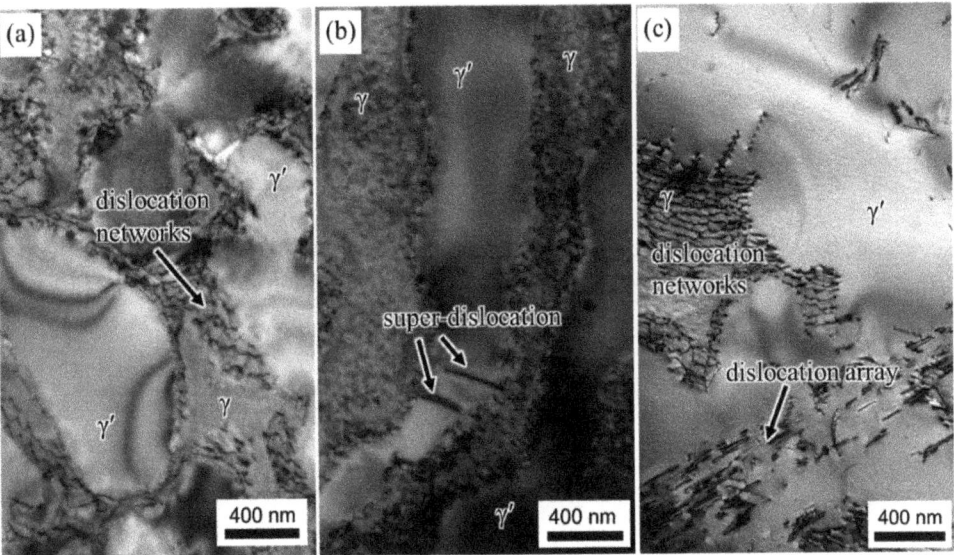

Figure 4. Microstructures after the alloy crept for different durations: (**a**) 2 h; (**b**) 10 h; (**c**) 135 h (up to fracture) at 1170 °C/100 MPa.

Figure 5 presents the dislocation configuration of the alloy after 2 h of creep. The dislocations primarily pile up at the interface between the γ and γ' phases, and a super-dislocation, referred to as dislocation A, is observed shearing into the γ' phase from the γ/γ' interface. Dislocation A exhibits contrast when the diffraction vectors **g** = [200] and **g** = [$\bar{1}1\bar{1}$] are considered, while it disappears when the diffraction vectors **g** = [02$\bar{2}$] and **g** = [11$\bar{1}$] are considered. Based on the invisible criteria **b·g** = 0 or ±2/3 () for dislocations, dislocation A is identified as having a Burgers vector $\mathbf{b}_A = \mathbf{g}_{02\bar{2}} \times \mathbf{g}_{11\bar{1}} = [011]$. The trace vector of dislocation A is [21$\bar{1}$], and the slipping plane is determined as ($\bar{1}11$) based on **b** × **μ**. Therefore, dislocation A is characterized as the [011] super-dislocation shearing into the γ' phase. The magnified view of the interfacial dislocations is shown in Figure 5c, revealing dislocations with jog features indicated by the dashed lines. This suggests that the dislocations may traverse the rafted γ' phase by climbing along the jogs.

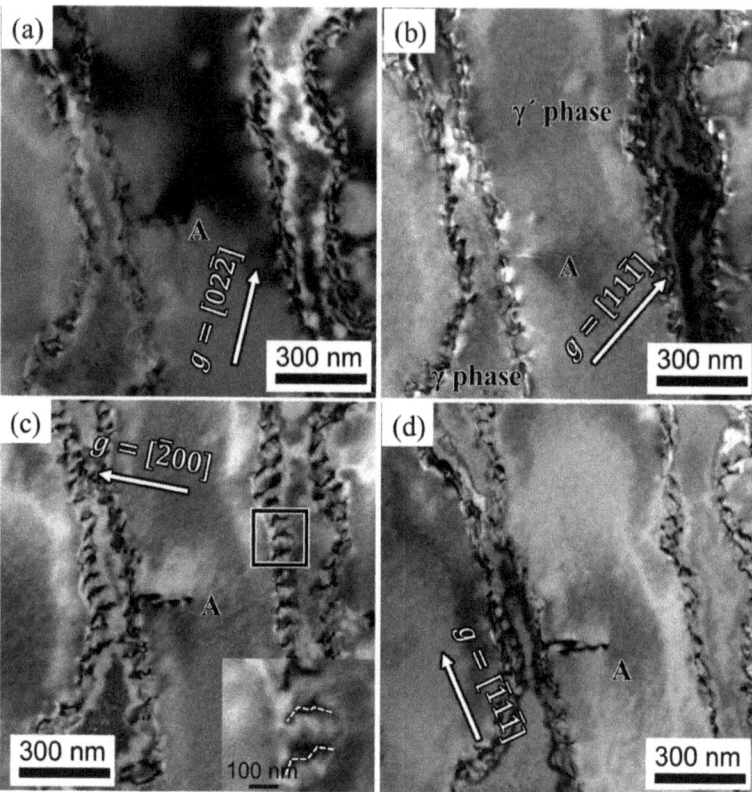

Figure 5. Dislocation configurations of the alloy after creeping for 2 h at 1170 °C/100 MPa: (a) $\mathbf{g} = [02\bar{2}]$, (b) $\mathbf{g} = [11\bar{1}]$, (c) $\mathbf{g} = [\bar{2}00]$, (d) $\mathbf{g} = [\bar{1}1\bar{1}]$.

Figure 6 presents the dislocation configuration within the γ' phase of the alloy after 10 h of creep at 1170 °C/100 MPa. Dislocation networks are distributed at the interface between the γ and γ' phases. The dislocations within the γ' phase are labeled as B, C, and D, respectively. Dislocation B displays contrast when the diffraction vectors are $\mathbf{g} = [1\bar{1}\bar{1}]$ and $\mathbf{g} = [1\bar{1}1]$, whereas its contrast disappears at $\mathbf{g} = [00\bar{2}]$ and $\mathbf{g} = [\bar{2}20]$. Therefore, dislocation B is identified as having a Burgers vector $\mathbf{b}_B = [110]$. The trace vector of dislocation B is $\mu_B = [1\bar{1}2]$, and its slipping plane is identified as $(\bar{1}1\bar{1})$ based on $\mathbf{b} \times \mu$. Dislocation C exhibits contrast when $\mathbf{g} = [00\bar{2}]$ and $\mathbf{g} = [1\bar{1}\bar{1}]$ are considered, while its contrast disappears at $\mathbf{g} = [1\bar{1}1]$ and $\mathbf{g} = [\bar{2}20]$. Consequently, dislocation C is identified as having a Burgers vector $\mathbf{b}_C = [110]$. The slipping plane of dislocation C is determined as $(1\bar{1}\bar{1})$, and its trace vector is $[1\bar{1}2]$. Dislocation D loses contrast at $\mathbf{g} = [1\bar{1}1]$ but displays contrast at $\mathbf{g} = [00\bar{2}]$, $\mathbf{g} = [\bar{2}20]$, and $\mathbf{g} = [1\bar{1}\bar{1}]$. Dislocation D is identified as having a Burgers vector $\mathbf{b}_D = [011]$. The trace vector of dislocation D is $\mu_D = \bar{2}20$, and its slipping plane is determined as $(\bar{1}1\bar{1})$.

Additionally, another dislocation with a zigzag character in the γ' phase, marked as dislocation E, is observed originating from the γ/γ' interface and may be influenced by adjacent dislocation networks. Dislocation E disappears in contrast at $\mathbf{g} = [00\bar{2}]$ but exhibits contrast at $\mathbf{g} = [1\bar{1}1]$, $\mathbf{g} = [\bar{2}20]$, and $\mathbf{g} = [1\bar{1}\bar{1}]$. Accordingly, dislocation E is identified as a[010] super-dislocation according to the invisible criteria. The trace vector of dislocation E is $[\bar{2}00]$, leading to the identification of its slipping plane as $(00\bar{1})$, based on $\mathbf{b} \times \mu$.

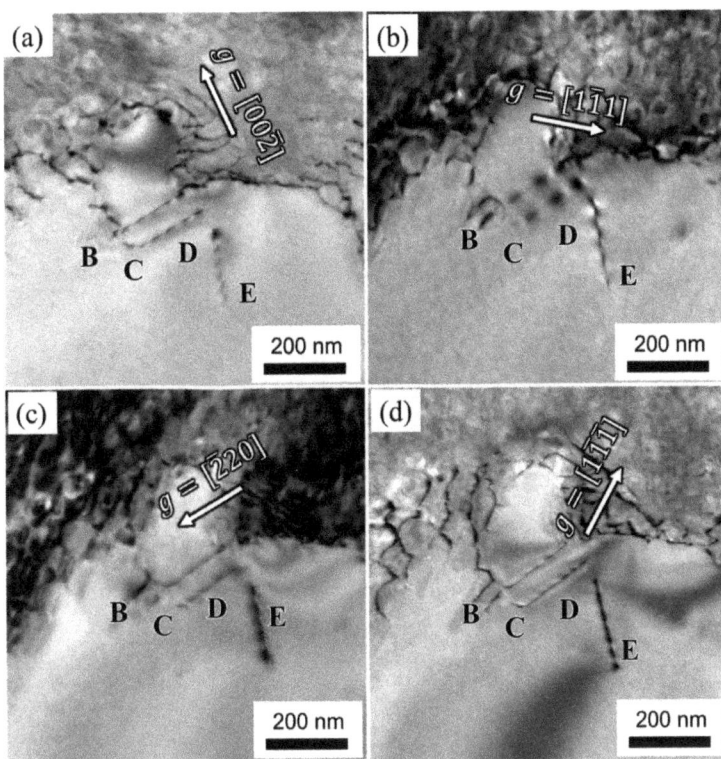

Figure 6. Dislocation configurations of the alloy after creeping for 10 h at 1170 °C/100 MPa: (a) **g** = [00$\bar{2}$], (b) **g** = [1$\bar{1}$1], (c) **g** = [$\bar{2}$20], (d) **g** = [1$\bar{1}$1].

In summary, the dominant deformation feature under the condition of 1170 °C is characterized by dislocation networks in the γ matrix and super-dislocations shearing into the γ' phase, as observed from the analysis presented above.

4. Discussion

4.1. Analysis of Microstructure Evolution during Creep

During the creep process at 1170 °C, the microstructure of the alloy undergoes significant changes, characterized by the phenomena of rafting and coarsening. These transformations are influenced by both the lattice misfit and the applied stress. Specifically, when an alloy with a negative lattice misfit is subjected to tensile stress, its microstructure can transform into an N-type rafted structure. The formation of this structure is achieved through the process of orientational diffusion, wherein elements that promote the formation of the γ phase (referred to as γ-forming elements) diffuse horizontally in the direction of the applied stress, while elements that contribute to the formation of the γ' phase (referred to as γ'-forming elements) diffuse vertically. Consequently, adjacent γ' phases become connected, resulting in the development of a rafted structure. The phenomenon of topological inversion, as observed in Figure 3d, where the γ matrix is surrounded by the γ' phase, has also been documented in Ref. [19].

Figure 7a illustrates the morphology of the TCP phase in the alloy subsequent to fracture at 1170 °C and under a stress of 100 MPa. The TCP phase appears in a needle-like form and is identified as the σ phase, which assumes a lamellar structure within the alloy. Figure 7b presents the diffraction pattern associated with the TCP phase. The crystallographic relationship is acquired as $[010]_{\gamma/\gamma'} \parallel [100]_{\sigma}$. Typically, the lamellar σ phase aligns parallel to the {111} planes of the γ matrix. The angle between the σ phase and the γ′ phase is approximately 45°, as depicted in Figure 3b,c. The reported orientation relationship between the σ phase and the γ phase is $(00\bar{1})_\sigma \parallel (1\bar{1}1)_\gamma$ [24]. The habit plane for the precipitation of the σ phase is the {111} planes of the γ matrix, coinciding with the slipping planes of dislocations within the γ matrix. The presence of the lamellar σ phase hinders dislocation movement by impeding their climb or shear into the σ phase. Consequently, dislocations pile up at the interface between the σ phase and the γ matrix, ultimately leading to crack nucleation.

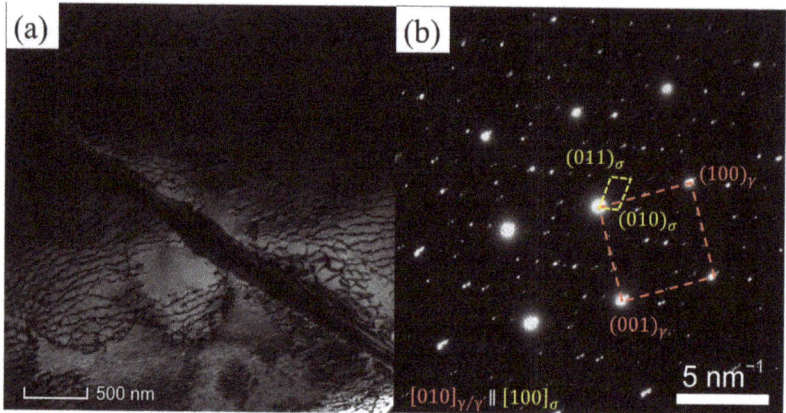

Figure 7. (a) TEM image of the TCP phase after fracturing at 1170 °C/100 MPa, (b) the diffraction pattern of the TCP phase in (a).

Figure 8 presents a high-angle annular dark-field (HAADF) image and corresponding elemental distribution maps of the region encompassing the TCP phase. The chemical compositions of the TCP phase, γ matrix, and γ′ phase in Figure 8 are measured and listed in Table 3. Notably, the TCP phase exhibits element segregation, particularly with respect to elements such as W, Cr, and Re, while elements such as Ni, Al, and Co are present in lower concentrations within the TCP phase. This phenomenon of element segregation is attributed to the presence of γ-forming elements in the TCP phase surrounding the γ′ phase. The segregation of refractory elements like Re, W, and Cr within the TCP phase serves to weaken the alloy's solution-strengthening characteristics.

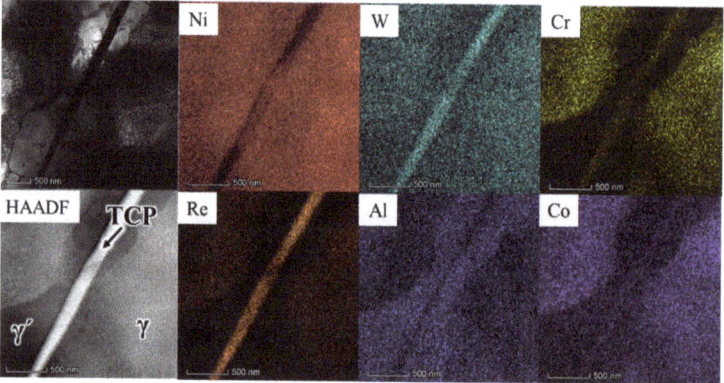

Figure 8. The bright TEM image, HAADF image, and EDS elemental distribution maps of the TCP phase.

Table 3. Chemical composition of the γ matrix, γ′ precipitates, and TCP phase in Figure 8 (atomic fraction, %).

	Al	Cr	Co	Ni	W	Re
γ matrix	6.72	8.01	9.94	65.13	3.2	3.05
γ′ phase	14.83	1.33	5.92	69.46	2.84	0.35
TCP phase	5.03	6.59	7.12	53.37	8.52	16.05

In summary, during the creep process at 1170 °C, the alloy's microstructure undergoes rafting and coarsening, influenced by both the lattice misfit and the applied stress. Tensile stress induces the formation of an N-type rafted structure through orientational diffusion, leading to the connection of adjacent γ′ phases. The presence of the TCP phase, identified as the σ phase, with its lamellar structure and its alignment along the {111} planes of the γ matrix, hampers dislocation movement, eventually resulting in crack nucleation. Furthermore, the TCP phase exhibits element segregation, particularly involving refractory elements, thereby weakening the alloy's solution-strengthening properties.

4.2. Deformation Mechanisms of Creep

During the initial creep stage, as the γ′ phase transforms into an N-type rafted structure, dislocations within the γ matrix become activated and pile up at the interface of the γ/γ′ phase. This phenomenon is clearly depicted in Figures 5 and 6, where most dislocations slip within the γ matrix and form immobile interfacial dislocation networks. However, mobile dislocations that reach the interface can interact with the dislocation network, altering their slipping direction and facilitating dislocation climbing over the rafted γ′ phase. The presence of jog features in the interfacial dislocations, as observed in Figure 5, further supports the idea that dislocations can traverse the γ′ phase by moving along jogs. A two-dimensional, simplified schematic representation of dislocation climbing is provided in Figure 9.

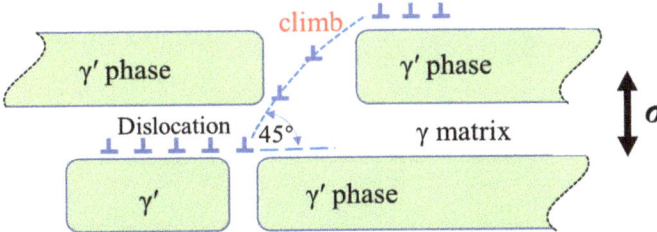

Figure 9. The simplified schematic of dislocation climbing over the rafted γ′ phase.

The process of dislocation climbing is influenced by both temperature and applied stress. The critical stress (σ_i) required for dislocation climbing at elevated temperatures can be described by Equation (1) [16]:

$$\sigma_i = \frac{G \cdot b}{[8\pi(1-\nu)h]kT} \qquad (1)$$

Here, G represents the shear modulus, b represents the magnitude of the Burgers vector, ν represents the Poisson's ratio, h represents the height of the γ′ phase, and T represents the temperature. In Equation (1), it is observed that the critical stress for the dislocation climb is influenced by the shear modulus (G), the height of the rafted γ′ phase (h), and the temperature (T). Equation (1) specifically emphasizes the inverse relationship between the critical stress (σ_i) and temperature (T). Furthermore, the shear modulus (G) and the volume fraction and size of the γ′ phase are affected by temperature. As the temperature increases, the alloy's shear modulus decreases accordingly. Similarly, the height of the γ′ phase is also influenced by temperature, manifesting primarily through the rafting and coarsening of the microstructure. For SX Ni-based alloys, the rafting and coarsening of the γ' phase typically occur above 1000 °C, leading to an increase in the size of the γ' phase. Therefore, due to the combined influence of these three factors, compared to creep at lower temperatures, the critical stress for dislocation climbing decreases with increasing temperature, making dislocation climbing more likely to occur at higher temperatures. In the creep deformation at 1170 °C, the climbing process is regarded as the dominant mechanism governing the plastic strain during creep [34].

As creep progresses, the stress concentration resulting from the dense dislocation networks within the γ matrix promotes the shearing of dislocations into the γ' phase. The dislocations within the γ' phase primarily manifest as a/2<110> super-dislocations, as depicted by dislocation A in Figure 5. These dislocations in the γ' phase may undergo decomposition, leading to a configuration comprising two a/3<112> Shockley partial dislocations separated by stacking faults [17]. Figure 10 illustrates a set of dislocation partials situated within the {111} slipping plane. The magnification of the dashed-square area in Figure 10a shows the stacking fault ribbon appearing as a parallel fringe pattern clear, as displayed in Figure 10b. This particular dislocation structure arises from the decomposition of the a/2<110> super-dislocation within the γ' phase, resulting in partial dislocations accumulating at obstacles. Furthermore, a zigzag-shaped a<010> super-dislocation is observed within the γ' phase, represented by dislocation E in Figure 6. As the a<010> super-dislocation slips along the (00$\bar{1}$) planes, its Schmid factor is determined to be zero when the applied stress aligns with the [001] direction. Consequently, the dislocation remains immobile and can only move through climbing within the γ' phase. The formation of the a<010> super-dislocation can be reasonably attributed to the following reaction at the γ/γ' interface:

$$\frac{a}{2}\langle 110 \rangle_\gamma + \frac{a}{2}\langle \bar{1}10 \rangle_\gamma \rightarrow a \langle 010 \rangle_{\gamma'} \qquad (2)$$

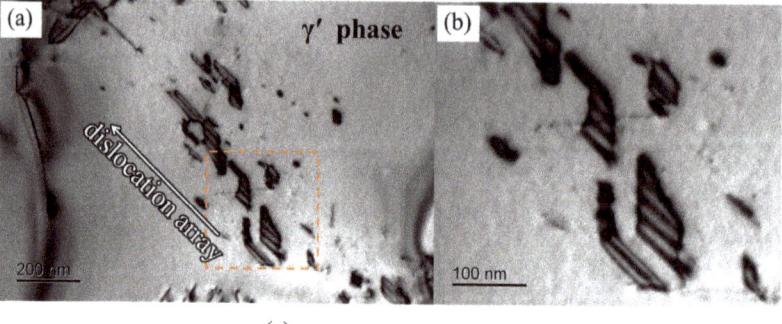

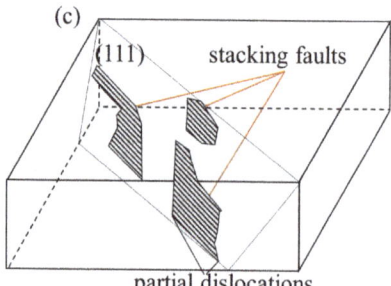

Figure 10. Dislocation arrays pile up in the γ' phase of the alloy after creeping for 50 h at 1170 °C/100 MPa: (**a**) the TEM morphology; (**b**) magnification of the dashed-square area in (**a**); (**c**) the schematic diagram of stacking faults.

In the later stages of creep, dislocations from different <111> slip systems alternately shear into the rafted γ′ phase. This alternating activation of primary and secondary slipping systems results in the twisting of the rafted γ′ phase, as observed in Figure 3d. The degree of twist in the rafted γ′ phase increases the strain experienced by the alloy during creep, ultimately leading to fracture.

5. Conclusions

In summary, the investigation clarified the creep behavior and deformation mechanism of the third-generation single-crystal Ni-based superalloy at an elevated temperature. The observation of the rafted γ′ phase and its interaction with the γ matrix, along with the identification of dislocation slipping and shearing as the principal deformation mechanisms, provides valuable insights into the alloy's mechanical properties under creep conditions.

(1) During creep, the initially cubic γ′ phases undergo a transformation into a rafted morphology through thickening and coarsening. Concurrently, the presence of the TCP phase is observed at various stages of creep at a temperature of 1170 °C. Through HAADF-EDS mapping, it is revealed that elements such as Re, W, and Cr exhibit significant segregation within the TCP phase. This selective distribution of refractory elements within the TCP phase has the detrimental effect of weakening the alloy's solution-strengthening mechanism and consequently diminishing its creep performance.

(2) In the initial creep stage at 1170 °C, the primary mechanism of deformation in the alloy is identified as dislocation slipping within the γ matrix, accompanied by the process of dislocation climbing over the rafted γ′ phase.

(3) As the creep progresses to later stages, super-dislocations characterized by a Burgers vector of a<010> and a/2<110> shear into the γ′ phase, originating from interfacial dislocation networks. These networks not only serve as a source of dislocations for the γ′ phase but also impede the movement of dislocations within the γ phase.

Author Contributions: Conceptualization, H.Y.; Validation, R.X. and Y.L.; Formal analysis, Y.L.; Investigation, R.X.; Writing—original draft preparation, R.X.; Writing—review and editing, H.Y. All authors have read and agreed to the published version of the manuscript.

Funding: This research was funded by the Ye Qisun Science Foundation of the National Natural Science Foundation of China, grant number U2141204, the Science Center for Gas Turbine Project, grant number P2021-A-IV-001-001, the National Science and Technology Major Project, grant number J2019-VI-0002-0115, and the National Key Research and Development Program of China, grant number 2017YFB0702004.

Data Availability Statement: The data presented in this study are available on request from the corresponding author.

Conflicts of Interest: The authors declare no conflict of interest. The funders had no role in the design of the study; in the collection, analyses, or interpretation of the data; in the writing of the manuscript; or in the decision to publish the results.

References

1. Reed, R.C. *The Superalloys Fundamentals and Applications*; Cambridge University Press: New York, NY, USA, 2006.
2. Pineau, A.; Antolovich, S.D. High temperature fatigue of nickel-base superalloys—A review with special emphasis on deformation modes and oxidation. *Eng. Fail. Anal.* **2009**, *16*, 2668–2697. [CrossRef]
3. Escamez, J.-M.; Strudel, J.L. Creep Damage and Failure of Several PM Nickel Base Superalloys. *Fracture* **1984**, *84*, 2231–2238.
4. Zhang, Y.; Hu, J.; Kang, L.; He, Y.; Xu, W. Creep Behavior Characterization of Nickel-Based Single-Crystal Superalloy DD6 Thin-Walled Specimens Based on a 3D-DIC Method. *Materials* **2023**, *16*, 3137. [CrossRef]
5. Tian, S.; Zeng, Z.; Fushun, L.; Zhang, C.; Liu, C. Creep behavior of a 4.5%-Re single crystal nickel-based superalloy at intermediate temperatures. *Mater. Sci. Eng. A* **2012**, *543*, 104–109. [CrossRef]
6. Tian, S.; Zhu, X.; Wu, J.; Yu, H.; Shu, D.; Qian, B. Influence of Temperature on Stacking Fault Energy and Creep Mechanism of a Single Crystal Nickel-based Superalloy. *J. Mater. Sci. Technol.* **2016**, *32*, 790–798. [CrossRef]
7. Li, Y.; Wang, L.; Zhao, S.; Zhang, G.; Lou, L. Creep anisotropy of a 3rd generation nickel-base single crystal superalloy in the vicinity of [001] orientation. *Mater. Sci. Eng. A* **2022**, *848*, 143479. [CrossRef]

8. Lv, P.; Liu, L.; Zhao, G.; Guo, S.; Zhou, Z.; Chen, R.; Zhao, Y.; Zhang, J. Creep properties and relevant deformation mechanisms of two low-cost nickel-based single crystal superalloys at elevated temperatures. *Mater. Sci. Eng. A* **2022**, *851*, 143561. [CrossRef]
9. Chang, H.-J.; Fivel, M.C.; Strudel, J.-L. Micromechanics of primary creep in Ni base superalloys. *Int. J. Plast.* **2018**, *108*, 21–39. [CrossRef]
10. Li, Y.; Wang, L.; He, Y.; Zheng, W.; Lou, L.; Zhang, J. Role of interfacial dislocation networks during secondary creep at elevated temperatures in a single crystal Ni-based superalloy. *Scr. Mater.* **2022**, *217*, 114769. [CrossRef]
11. Zhang, J.X.; Wang, J.C.; Harada, H.; Koizumi, Y. The effect of lattice misfit on the dislocation motion in superalloys during high-temperature low-stress creep. *Acta Mater.* **2005**, *53*, 4623–4633. [CrossRef]
12. Li, J.R.; Liu, S.Z.; Wang, X.G.; Shi, Z.X.; Zhao, J.Q. Development of A Low-Cost Third Generation Single Crystal Superalloy DD9. *Superalloys* **2016**, *2016*, 55–63.
13. Wang, W.Z.; Jin, T.; Liu, J.L.; Sun, X.F.; Guan, H.R.; Hu, Z.Q. Role of Re and Co on microstructures and γ′ coarsening in single crystal superalloys. *Mater. Sci. Eng. A* **2008**, *479*, 148–156. [CrossRef]
14. He, C.; Liu, L.; Huang, T.; Yang, W.; Wang, X.; Zhang, J.; Guo, M.; Fu, H. The effects of misfit and diffusivity on γ′ rafting in Re and Ru containing Nickel based single crystal superalloys—Details in thermodynamics and dynamics. *Vacuum* **2021**, *183*, 109839. [CrossRef]
15. Ji, J.Y.; Zhang, Z.; Chen, J.; Zhang, H.; Zhang, Y.Z.; Lu, H. Effect of refractory elements M,(=Re, W, Mo or Ta) on the diffusion properties of boron in nickel-based single crystal superalloys. *Vacuum* **2023**, *211*, 111923. [CrossRef]
16. Tian, S.; Su, Y.; Qian, B.; Yu, X.; Liang, F.; Li, A. Creep behavior of a single crystal nickel-based superalloy containing 4.2% Re. *Mater. Design* **2012**, *37*, 236–242. [CrossRef]
17. Shu, D.L.; Tian, S.G.; Liang, S.; Zhang, S.B. Deformation and Damage Mechanism of a 4.5%Re-containing Nickel-based Single Crystal Superalloy During Creep at 980 °C. *J. Mater. Eng.* **2017**, *45*, 93–100.
18. Tang, Y.; Huang, M.; Xiong, J.; Li, J.; Zhu, J. Evolution of superdislocation structures during tertiary creep of a nickel-based single-crystal superalloy at high temperature and low stress. *Acta Mater.* **2017**, *126*, 336–345. [CrossRef]
19. Li, Y.; Wang, L.; Zhang, G.; Zheng, W.; Lou, L.; Zhang, J. On the role of topological inversion and dislocation structures during tertiary creep at elevated temperatures for a Ni-based single crystal superalloy. *Mater. Sci. Eng. A* **2021**, *809*, 140982. [CrossRef]
20. Wang, G.; Zhang, S.; Tian, S.; Tian, N.; Zhao, G.; Yan, H. Microstructure evolution and deformation mechanism of a [111]-oriented nickel-based single-crystal superalloy during high-temperature creep. *J. Mater. Res. Technol.* **2022**, *16*, 495–504. [CrossRef]
21. Caccuri, V.; Cormier, J.; Desmorat, R. γ′-Rafting mechanisms under complex mechanical stress state in Ni-based single crystalline superalloys. *Mater. Design* **2017**, *131*, 487–497. [CrossRef]
22. Utada, S.; Despres, L.; Cormier, J. Ultra-High Temperature Creep of Ni-Based SX Superalloys at 1250 °C. *Metals* **2021**, *11*, 1610. [CrossRef]
23. Naze, L.; Strudel, J.-L. Strain rate effects and hardening mechanisms in Ni base superalloys. *Mater. Sci. Forum* **2010**, *638–642*, 53–60. [CrossRef]
24. Yang, W.; Yue, Q.; Cao, K.; Chen, F.; Zhang, J.; Zhang, R.; Liu, L. Negative influence of rafted γ′ phases on 750 °C/750 MPa creep in a Ni-based single crystal superalloy with 4% Re addition. *Mater. Charact.* **2018**, *137*, 127–132. [CrossRef]
25. Guo, X.; He, H.; Chen, F.; Liu, J.; Li, W.; Zhao, H. Microstructural Degradation and Creep Property Damage of a Second-Generation Single Crystal Superalloy Caused by High Temperature Overheating. *Materials* **2023**, *16*, 1682. [CrossRef]
26. Shang, Z.; Niu, H.; Wei, X.; Song, D.; Zou, J.; Liu, G.; Liang, S.; Nie, L.; Gong, X. Microstructure and tensile behavior of nickel-based single crystal superalloys with different Re contents. *J. Mater. Res. Technol.* **2022**, *18*, 2458–2469. [CrossRef]
27. Körber, S.; Wolff-Goodrich, S.; Völkl, R.; Glatzel, U. Effect of Wall Thickness and Surface Conditions on Creep Behavior of a Single-Crystal Ni-Based Superalloy. *Metals* **2022**, *12*, 1081. [CrossRef]
28. Zhang, Y.; Zhang, J.; Li, P.; Jin, H.; Zhang, W.; Wang, Z.; Mao, S. Characterization of topologically close-packed phases and precipitation behavior of P phase in a Ni-based single crystal superalloy. *Intermetallics* **2020**, *125*, 106887. [CrossRef]
29. Lee, S.; Do, J.; Jang, K.; Jun, H.; Park, Y.; Choi, P. Promotion of topologically close-packed phases in a Ru-containing Ni-based superalloy. *Scr. Mater.* **2023**, *222*, 115041. [CrossRef]
30. Wang, Z.; Li, Y.; Zhao, H.; Chen, L.; Zhang, Z.; Shen, D.; Wang, M. Evolution of μ phase in a Ni-based alloy during long-term creep. *J. Alloys Compd.* **2019**, *782*, 1–5. [CrossRef]
31. Zhao, P.; Xie, G.; Chen, C.; Wang, X.; Zeng, P.; Wang, F.; Zhang, J.; Du, K. Interplay of chemistry and deformation-induced defects on facilitating topologically-close-packed phase precipitation in nickel-base superalloys. *Acta Mater.* **2022**, *236*, 118109. [CrossRef]
32. Sun, F.; Zhang, J. Topologically close-packed phase precipitation in Ni-based superalloys. *Adv. Mater. Res.* **2011**, *320*, 26–32. [CrossRef]
33. *ASTM E139-11*; Standard Test Methods for Conducting Creep, Creep-Rupture, and Stress-Rupture Tests of Metallic Materials. ASTM International: West Conshohocken, PA, USA, 2018.
34. Bürger, D.; Dlouhý, A.; Yoshimi, K.; Eggeler, G. How Nanoscale Dislocation Reactions Govern Low-Temperature and High-Stress Creep of Ni-Base Single Crystal Superalloys. *Crystals* **2020**, *10*, 134. [CrossRef]

Disclaimer/Publisher's Note: The statements, opinions and data contained in all publications are solely those of the individual author(s) and contributor(s) and not of MDPI and/or the editor(s). MDPI and/or the editor(s) disclaim responsibility for any injury to people or property resulting from any ideas, methods, instructions or products referred to in the content.

Article

Structure Evolution and Mechanical Properties of Sheet Al–2Cu–1.5Mn–1Mg–1Zn (wt.%) Alloy Designed for Al$_{20}$Cu$_2$Mn$_3$ Disperoids

Nikolay Belov [1], Torgom Akopyan [1,2,*], Kirill Tsydenov [1], Nikolay Letyagin [1,2] and Anastasya Fortuna [1]

[1] Department of Metal Forming, National University of Science and Technology MISiS, 4 Leninsky Pr., Moscow 119049, Russia; belov.na@misis.ru (N.B.); n.v.letyagin@gmail.com (N.L.)

[2] Sector of Scientific Activity, Moscow Polytechnic University, 38, Bolshaya Semyonovskaya Str., Moscow 107023, Russia

* Correspondence: akopyan.tk@misis.ru

Abstract: This work was focused on studying the possibility of increasing the strength of non-heat-treatable sheet alloy Al2Cu1.5Mn (wt.%) by the joint addition of 1% Mg and 1% Zn. The effect of these elements on the structure and mechanical properties of the new sheet Al2Cu1.5Mn alloy designed for Al$_{20}$Cu$_2$Mn$_3$ dispersoids has been studied by calculations and experimental methods. The obtained data on the phase composition, microstructure, and physical and mechanical properties of the new alloy for different processing routes (including hot rolling, cold rolling, and annealing) have been compared with those for the ternary Mg- and Zn-free alloy. It has been shown that the formation of nanosized Al$_{20}$Cu$_2$Mn$_3$ dispersoids (~7 vol.%) provides for the preservation of the non-recrystallized grain structure after annealing at up to 400 °C (3 h), while Mg and Zn have a positive effect on the strength due to the formation of alloyed aluminum solid solution. As a result, cold-rolled sheets of the Al2Cu1.5Mn1Mg1Zn model alloy showed a substantially higher strength performance after annealing at 400 °C in comparison with the ternary reference alloy. In particular, the UTS is ~360 vs. ~300 MPa, and the YS is 280 vs. 230 MPa. For the example of the Al2Cu1.5Mn1Mg1Zn model alloy, it has been shown that the system is promising for designing new heat-resistant alloys as a sustainable alternative to the 2xxx alloys. The new alloy has an advantage over the commercial alloys (particularly, 2219, 2024, 2014), not only in manufacturability but also in thermal stability. The sheet production cycle for the model alloy is much shorter because the stages of homogenization, solution treatment, and water quenching are excluded.

Keywords: Al–Cu–Mn–Mg–Zn system; structure evolution; phase composition; annealing; Al$_{20}$Cu$_2$Mn$_3$ dispersoids; mechanical properties; thermal stability

1. Introduction

Nowadays, aluminum ranks second in the world in terms of consumption among all metals, being second only to steel. In the coming decades, the demand for aluminum will continue to grow [1–3]. The latest developments in the automotive industry and the rapid growth of cities and other application domains of aluminum as a replacement for copper in electrical conductors will inevitably lead to an increase in aluminum consumption [4–7].

The use of aluminum alloys, in particular wrought ones, as a structural material has increased in recent years due to its manufacturability, high specific strength, significant ductility, excellent thermal conductivity, and attractive appearance [8–11]. Structural aluminum alloys are also recyclable, thus reducing carbon dioxide emissions [9,12–15]. Popular construction aluminum-based materials are 2xxx grade alloys, but homogenization annealing of ingot, heating the semi-finished wrought products up to 500–540 °C for solutionizing treatment followed by water quenching, and aging are required for the production of items from them [16–19]. Thus, the manufacturing routine for obtaining wrought semi-finished products is quite complex and requires special industrial equipment, which makes the final product expensive [20,21]

To reduce the cost of deformed semi-finished products, it is advisable to simplify and shorten the technological cycle of their production as much as possible [22–24], in particular, using alloys that do not require quenching (so-called non-heat treatable alloys). New alloys containing 1.5–2% Cu and 1.5–2% Mn [25–28] (hereinafter, wt%, unless otherwise indicated) were proposed as an alternative to branded heat-treatable alloys of the 2xxx series. Such alloys do not require homogenization and quenching, and hardening is achieved due to the formation of $Al_{20}Cu_2Mn_3$ dispersoids during annealing. These dispersoids provide higher thermal stability compared to grade alloys [29–31]. It was also shown [25] that the Al-2%Cu-2%Mn model alloy obtained in the form of cold-rolled sheets has higher thermal stability and better processability for slab rolling as compared to the AA2219 grade alloy. The proposed alloys have a tensile strength of about 300 MPa in the as-annealed state.

It was shown [32] that the presence of iron and silicon, which are the main impurities in aluminum alloys, in an amount of up to 0.5–0.6%, reduces the mechanical properties of the base model alloy, but only slightly. This provides the possibility of using secondary raw materials in preparation (in particular, canned scrap) contaminated with these elements. Since secondary raw materials, in addition to iron and silicon, also contain other impurities, in particular, magnesium and zinc (the main elements in the 5xxx and 7xxx series alloys), due account for the influence of these elements requires special attention. Therefore, the aim of this work was to study the effect of the joint introduction of 1% Mg and 1% Zn addition on the structure, phase composition, and mechanical properties of the non-heat treatable Al-2%Cu-1.5%Mn base wrought alloy. It is shown that Mg and Zn have a positive effect on the strength and heat resistance of the alloy. As a result, cold-rolled sheets of the new Al2Cu1.5Mn1Mg1Zn model alloy showed a substantially higher strength performance after annealing at 350–400 °C in comparison with the ternary reference Al2Cu1.5Mn alloy or industrial 2219 type alloys.

2. Experimental

The main test materials of this study were 2 model alloys containing 2%Cu and 1.5%Mn: the base reference alloy (hereinafter referred to as 0Mg0Zn) and the alloy containing 1%Mg and 1%Zn (hereinafter referred to as 1Mg1Zn). The composition of the 1Mg1Zn alloy was chosen according to the isothermal section of the Al–Cu–Mn–Mg–Zn system calculated at 2%Cu, 1.5%Mn, and 400 °C (Figure 1).

The alloys were melted in a resistance furnace (GRAFICARBO) using a graphite-chased crucible. To obtain the selected composition, we used pure metals (99.85% aluminum, 99.9% copper, 99.9% magnesium, 99.9% zinc, and Al–10%Mn master alloy. When the material batch had melted, we held it for about 10 min for homogenization and then poured it into a flat graphite mold 10 mm × 40 mm × 180 mm in size at 750 °C. The cooling rate during solidification was approximately 20 K/s. The chemical composition of the experimental alloys according to spectral analysis (Oxford Instruments, Oxfordshire, UK) is given in Table 1. It can be seen that the actual compositions were close to the target ones.

Table 1. Chemical compositions of experimental alloys.

Alloy Designation	Concentration, wt.%						
	Cu	Mn	Mg	Zn	Fe	Si	
0Mg0Zn	2.06	1.46	0.03	0.03	0.11	0.08	balance
1Mg1Zn	2.07	1.44	1.02	1.11	0.12	0.08	balance

Ingots of experimental alloys were subjected to hot rolling at 400 °C. Before rolling, the ingots were annealed at 400 °C and held for 1 h. Then, the ingots were rolled to a thickness of 2 mm (compression ratio of 80%). Then, the hot-rolled sheets were annealed at 350 °C for 3 h, and then cold-rolled to a thickness of 0.5 mm (compression ratio 75%).

The cold-rolled sheet products were prepared using a laboratory-scale rolling mill machine (Chinetti LM160). To evaluate the effect of annealing on the structure and hardness, the semi-finished products were further subjected to stepwise annealing in accordance with the processing route shown in Table 2.

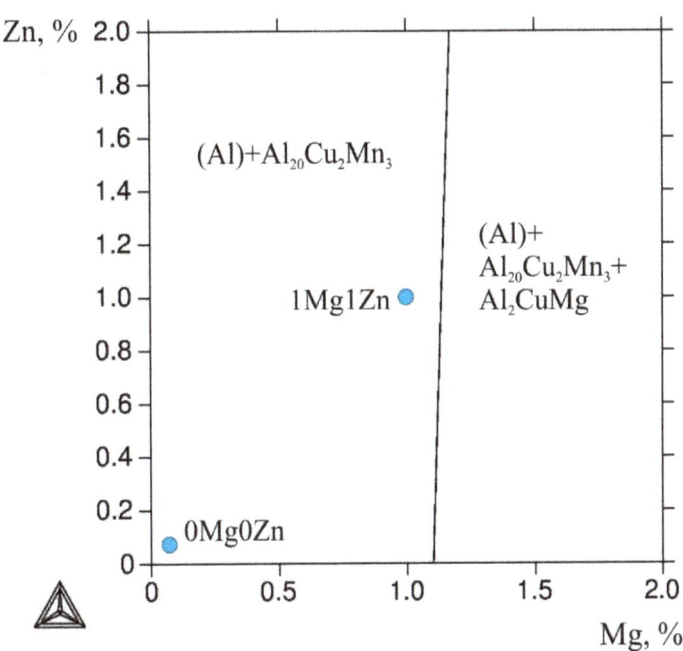

Figure 1. The isothermal section of Al–Cu–Mn–Mg–Zn phase diagram at 2%Cu, 1.5%Mn, and 400 °C with marked experimental alloys.

Table 2. Processing routes for experimental alloys.

Process	Obtained Product	Designation
Casting	Ingot 10 mm × 40 mm × 180 mm	F
Hot rolling (at 400 °C) of the foundry ingot (10 mm × 40 mm × 180 mm)	Sheet 2 mm in thickness	HR
HR + 300 °C, 3 h	Annealed hot rolled sheet	HR300
HR300 + 350 °C, 3 h		HR350
HR350 + 400 °C, 3 h		HR400
HR400 + 450 °C, 3 h		HR450
HR450 + 500 °C, 3 h		HR500
Cold rolling of the rolled sheet (from HR350, sheet 0.5 mm)	Sheet 0.5 mm in thickness	CR
CR + 350 °C, 3 h	Annealed cold rolled sheet	CR350
CR + 400 °C, 3 h		CR400

The microstructure was examined by optical microscopy (OM, Axio Observer MAT), transmission electron microscopy (TEM, JEM-2100), scanning electron microscopy (SEM, TESCAN VEGA 3), and electron microprobe analysis (EMPA, OXFORD AZtec). The samples were prepared using mechanical and electrolytic polishing. Electrolytic polishing was carried out at a voltage of 12 V in an electrolyte (6 C_2H_5OH, 1 $HClO_4$, and 1 glycerine). Initial microstructural observations were carried out using OM, and detailed studies were then performed using SEM and TEM. Thin foils for TEM were prepared by ion thinning with the PIPS technique (Precision Ion Polishing System, Gatan, Pleasanton, USA) and studied at 160 kV. X-ray diffraction (XRD) data were obtained using CuKα radiation and treated with a software package [33]. The objects of the XRD study were polished samples of the 0Mg0Zn and 1Mg1Zn alloys cut from part of the cold-rolled sheets in the state CR400 (Table 2).

The Vickers hardness (HV) was measured using a DUROLINE MH-6 setup (METKON Instruments, Bursa, TURKEY) with a load of 1 kg and a dwell time of 10 s. At least five measurements were performed for each sample. The specific electrical conductivity (EC) of the ingots and hot-rolled sheets (2 mm) was determined using the eddy current method with a VE-26NP eddy structure scope. Room-temperature tensile tests were conducted for the cold-rolled sheets (0.5 mm) using an Instron 5966 machine. The loading rate was 10 mm/min.

The phase composition of the Al–Cu–Mn–Mg–Zn (Fe,Si) system (isothermal sections, fractions of phases, and (Al) composition in the experimental alloys) was calculated using Thermo-Calc software (TTAL5 database [34].

3. Experimental Results

3.1. Phase Composition and Microstructure of the Ingots

The existence of a tiny quantity of Cu-containing phase crystals created as a result of nonequilibrium solidification is a common feature of the alloys' as-cast structures. These crystals are located along the boundaries of the dendritic cells in the primary crystals of the aluminum solid solution (hereinafter (Al)). In the reference alloy, they are represented by the Al_2Cu phase (Figure 2a), and in the 1Mg1Zn alloy, by the Al_2CuMg phase (Figure 2b). In addition, a small amount of Fe-containing phase crystals is present in the as-cast structure due to the presence of iron impurities in primary aluminum used for alloy preparation (Table 1). According to the EDS data, they are identified as $Al_{15}(Mn,Fe)_3Si_2$ in the base alloy (Figure 2a), and as $Al_6(Mn,Fe)$ in the 1Mg1Zn alloy (Figure 2b). Almost the entire amount of manganese in both alloys is in (Al). Zinc (in the 1Mg1Zn alloy) is completely dissolved in (Al), while magnesium and copper are partially dissolved (because they form eutectic particles of the Al_2Cu and Al_2CuMg phases). In general, the microstructure of the alloy with magnesium and zinc additions differs but only a little from the microstructure of the base alloy, the ingots of which, as was shown earlier [26–28,32], have a sufficiently high deformation processability.

3.2. Microstructure and Phase Composition of Hot-Rolled Sheets

Sheets of the alloys were obtained by hot rolling. Analysis of the microstructure of the hot-rolled sheets showed that the particles of Fe-containing phases formed during casting (Figure 2) and were preserved after rolling (Figure 3) due to the low solubility of iron in (Al). The number of Al_2Cu and Al_2CuMg particles somewhat decreased, since magnesium and copper partially dissolved in (Al) during heating at 400 °C (before and during rolling). In addition, heating at this temperature led to both partial decomposition of (Al) and the formation of $Al_{20}Cu_2Mn_3$ phase dispersoids, the size of which, according to [26–28,32], does not exceed 100 nm, and therefore they cannot be detected in Figure 3. Annealing of the hot-rolled sheets at 300 °C and 350 °C had almost no effect on their structure. After 3 h of annealing at 400 °C (state HR400, see Table 2), almost complete dissolution of copper, magnesium, and zinc (the last two elements in the 1Mg1Zn alloy) took place. XRD analysis confirms the presence of the $Al_{20}Cu_2Mn_3$ compound in both alloys in about the same quantity (Figure 4). With an increase in the annealing temperature to 500 °C, the phase composition of the alloys did not change qualitatively; however, the $Al_{20}Cu_2Mn_3$ dispersoids coarsened. Therefore, the latter is detected in the structure by the SEM method, which is shown in Figure 5.

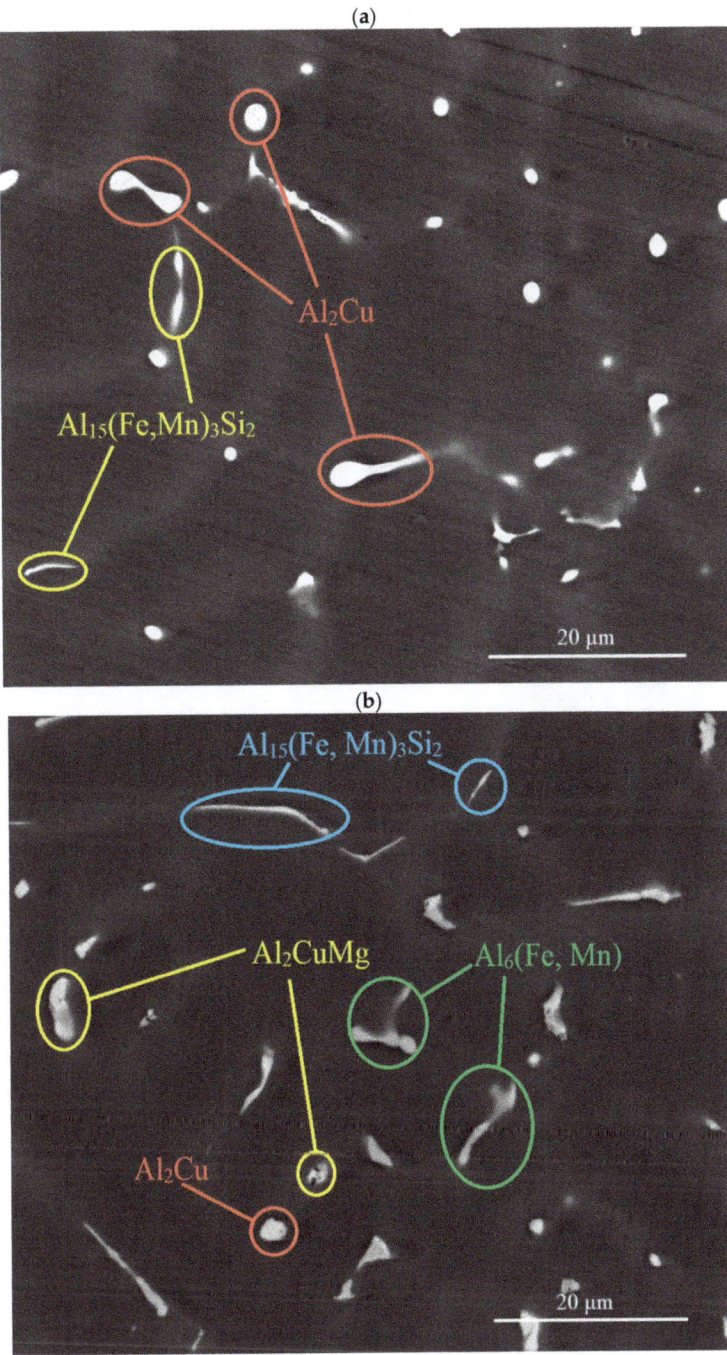

Figure 2. Microstructure of experimental alloys in ingots (F), SEM: (**a**) 0Mg0Zn, (**b**) 1Mg1Zn.

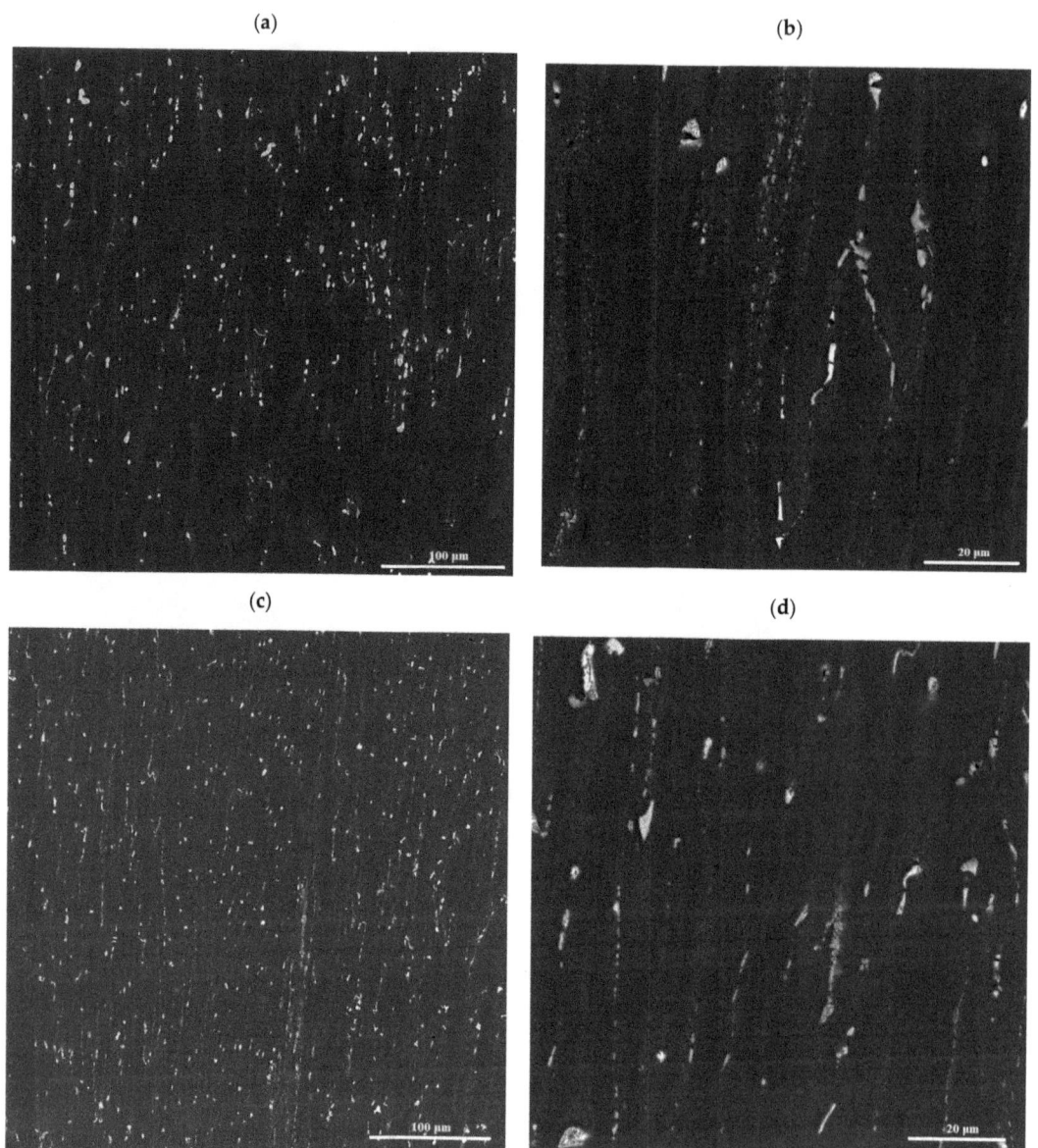

Figure 3. Microstructure of experimental alloys in hot rolled sheets (HR), SEM: (**a**,**b**) 0Mg0Zn (**c**,**d**) 1Mg1Zn, (**b**,**d**) are the magnified SEM images of (**a**,**b**).

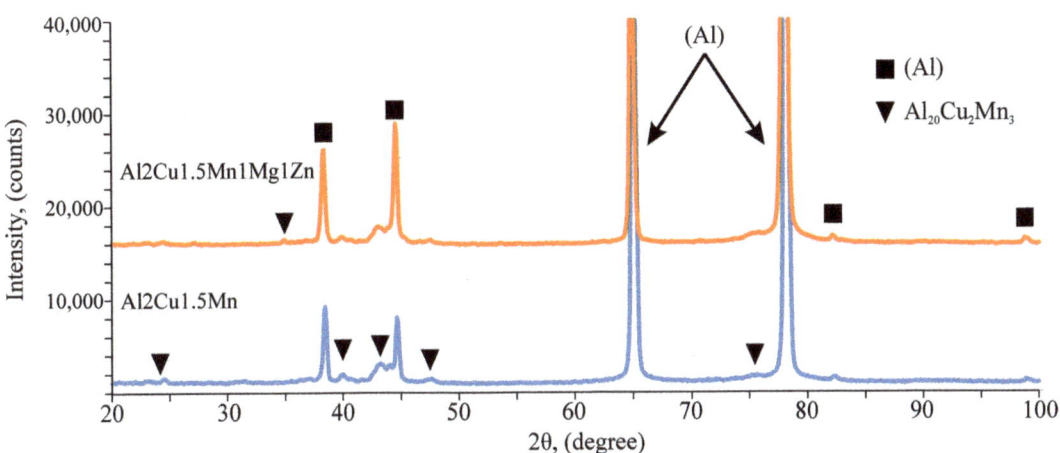

Figure 4. XRD data for experimental alloys in cold rolled sheets after annealing at 400 °C (CR400).

3.3. Hardness and Electrical Resistivity of Hot-Rolled Sheets

As can be seen from Figure 6a, magnesium and zinc additions have a hardening effect for all states of the processed alloy. The greatest difference in the hardness compared to the reference alloy is observed in the as-cast state (~20 HV). After hot rolling, it decreases to 13 HV, and after annealing at 300 °C to 7 HV. As the annealing temperature increases, the difference in the hardness values starts to increase, reaching 16 HV in the HR500 condition. The reason for the increase in hardness in the 1Mg1Zn alloy is explained in the Discussion section, where it is shown that the obtained result is due to the solid solution hardening.

Changes in the EC value agree well with the changes occurring in the microstructure, i.e., primarily dissolution and precipitation. Since manganese has the strongest influence on the EC, the difference between the alloys in the EC is smaller than the difference in the hardness. It can be seen from Figure 6b that both alloys have the lowest EC values in the as-cast state, since all the manganese is in (Al). After hot rolling and subsequent annealing at up to 450 °C inclusively, the concentration of Mn in (Al) decreases, and the EC values increase accordingly. The decrease in the EC in the HR500 state is caused by an increase in the concentration of Mn in (Al) at an elevated annealing temperature. The latter fact is consistent with previous studies of the Al–Cu–Mn alloys [26–28,32].

3.4. SEM and TEM Structure of Cold-Rolled Sheets

Taking into account that in order to increase the strength properties, it is necessary to obtain a minimum size of the $Al_{20}Cu_2Mn_3$ phase dispersoids (preferably not larger than 100 nm [26–28,32]), the initial hot-rolled sheets were preliminarily annealed at 350 °C for 3 h before cold rolling. This annealing proved to be sufficient to obtain high-quality cold-rolled sheets with a thickness of 0.5 mm (compression ratio 75%). Since inclusions of eutectic Cu-containing particles still remain in the structure of the hot-rolled sheets, they are detected in the structure after cold rolling along with the Fe-containing phases (Figure 7a,b). In this case, the morphology of these particles is improved due to their fragmentation during deformation. Figure 7c,d illustrate that both alloys, after annealing at 400 °C, have a non-recrystallized structure.

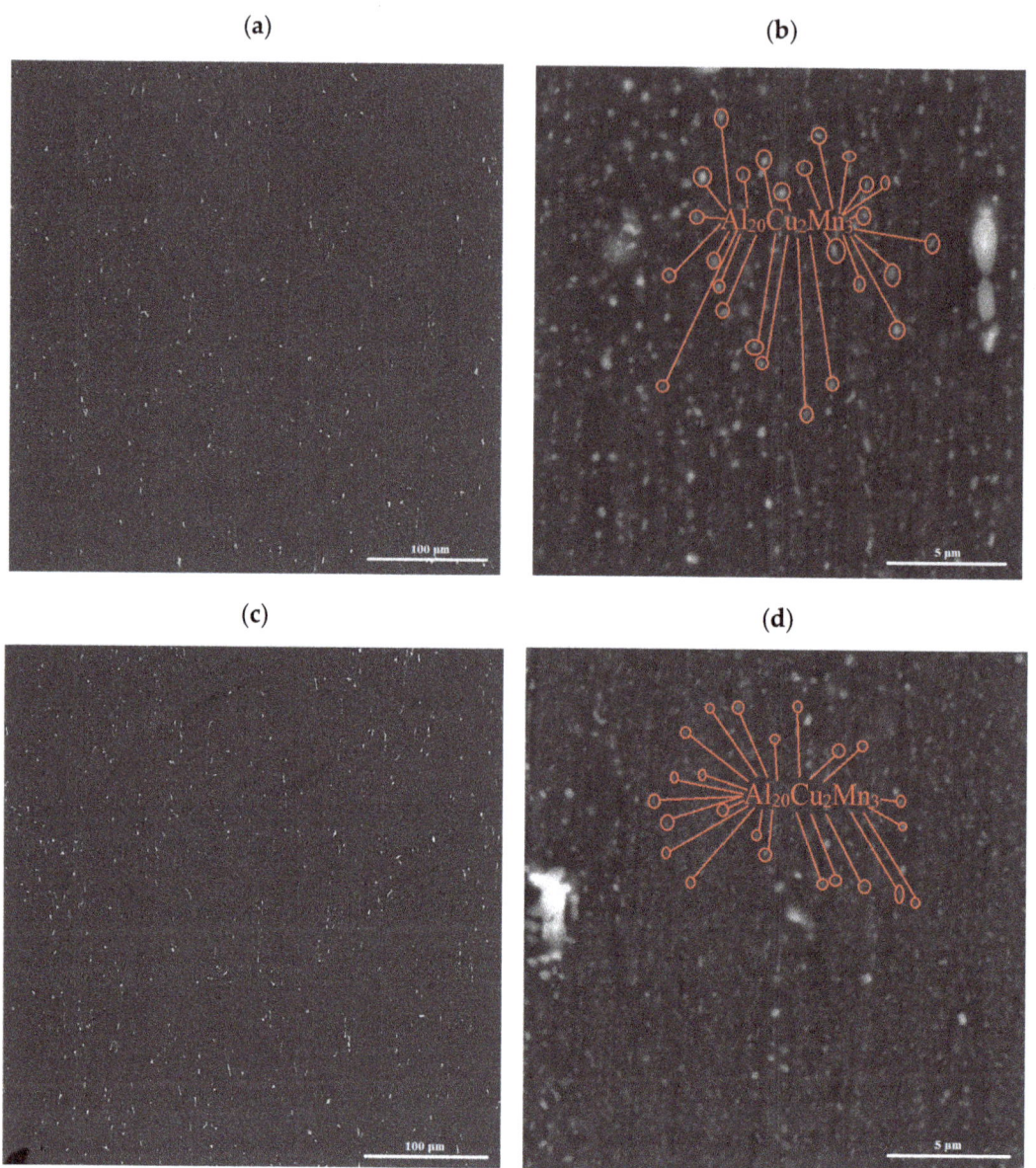

Figure 5. Microstructure of experimental alloys in hot rolled sheets after annealing at 500 °C (HR500), SEM: (**a**,**b**) 0Mg0Zn, (**c**,**d**) 1Mg1Zn, (**b**,**d**) are the magnified SEM images of (**a**,**b**).

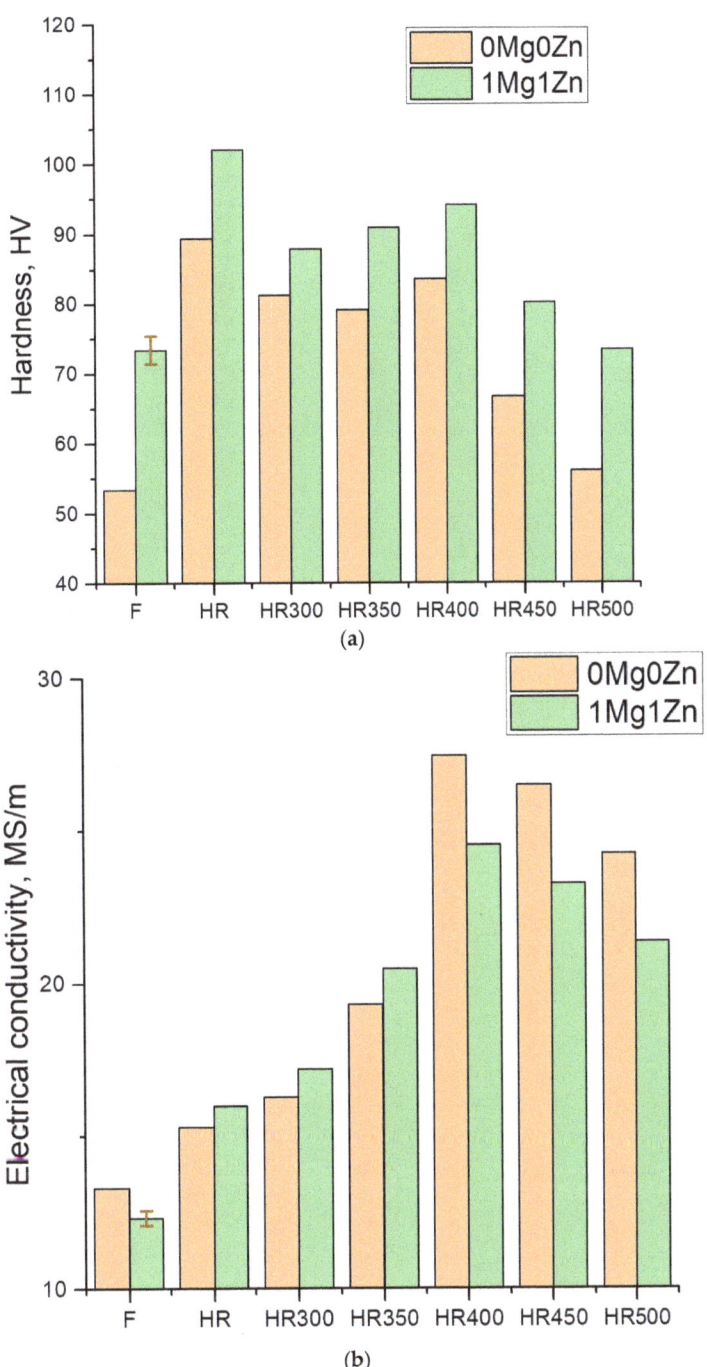

Figure 6. Change in (**a**) hardness (HV) and (**b**) electrical conductivity (EC) of hot rolled sheet alloys during stepwise annealing in accordance with the processing route shown in Table 2.

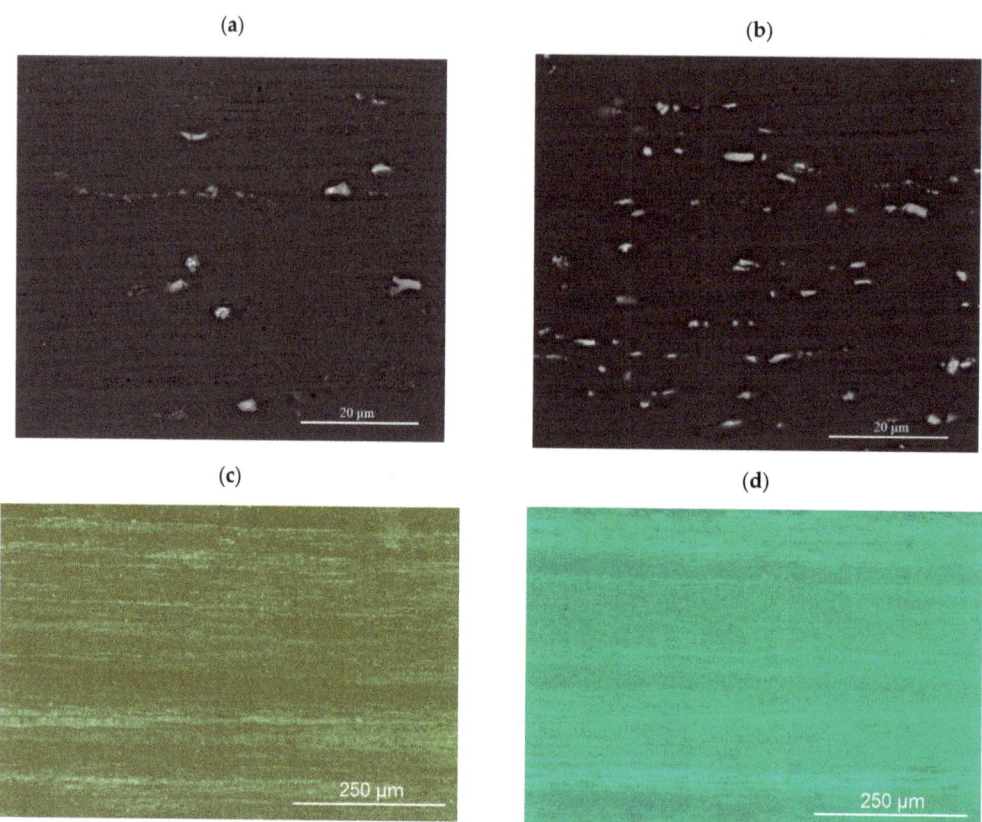

Figure 7. Microstructure of experimental alloys: (**a**,**c**) 0Mg0Zn CR, (**b**,**d**) 1Mg1Zn after cold rolling (**a**,**b**) and cold rolling and annealing at 400 °C (**c**,**d**). (**a**,**b**) SEM, (**c**,**d**) OM (polarized light).

Cold rolling leads to an increase in the dislocation density and the formation of a cell structure, which can be seen in the example of the 1Mg1Zn alloy in Figure 8a. Dispersoids of the $Al_{20}Cu_2Mn_3$ phase with a size not exceeding 100 nm (Figure 8b) are also detected. Annealing at 350 °C leads to the formation of subgrains about 1 μm in size with partial preservation of the cell dislocation structure (Figure 8c). The amount of $Al_{20}Cu_2Mn_3$ dispersoids increases, and they are more clearly detected (Figure 8d). After annealing at 400 °C, a completely non-recrystallized structure is preserved in both alloys (Figure 7c,d). The subgrain size in the 1Mg1Zn alloy remains at about 1 μm (Figure 8e), the dislocation density still being quite high in some regions (Figure 8f). The size of most dispersoids remains at less than 100 nm, with only a few of them growing up to 150–200 nm (Figure 8g). Obviously, $Al_{20}Cu_2Mn_3$ disperoids prevent recrystallization and subgrain growth (Figure 8h).

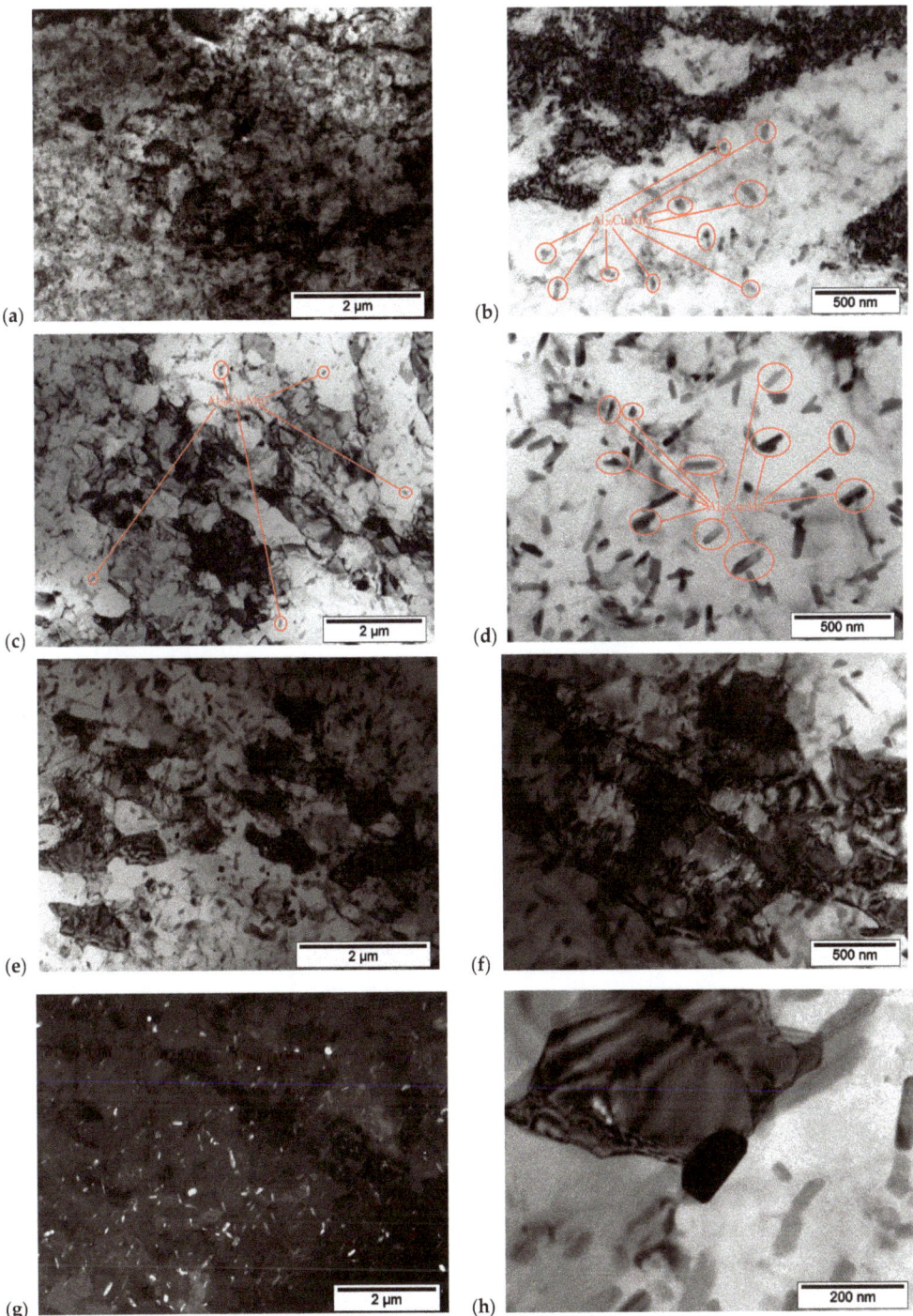

Figure 8. TEM structures of cold rolled sheet alloy 1Mg1Zn: (**a**,**b**) CR, (**c**,**d**) CR350, (**e**–**h**) CR400; (**a**–**e**,**g**,**h**) bright field, (**f**) dark field and diffraction patterns, (**b**,**d**,**h**,**f**) are the magnified TEM images of (**a**,**c**,**e**,**g**).

3.5. Mechanical Properties of Cold-Rolled Sheets

Table 3 presents the results of tensile testing of cold-formed sheets in the initial state (CR) and after single-stage 3-h annealing at 350 °C (CR350) and 400 °C (CR400). It can be seen from the results that magnesium and zinc addition in the CR state significantly increases the tensile strength and yield strength, while the elongation remains at the same low level as for the base alloy. However, the CR350 and CR400 states in which a relatively stable structure is formed are more indicative. Annealing reduces the strength and increases the ductility of both alloys; however, the degree of change varies greatly. The strength properties of the 1Mg1Zn alloy are approximately 15% higher than those in the base alloy with close plasticity (examples of engineering stress–strain curves are given in Figure 9). Since there is almost no difference in the mechanical properties between the CR350 and CR400 states, the latter seems to be preferable since it exhibits a greater stabilization of the structure.

Table 3. Mechanical properties of cold rolled sheets (0.5 mm).

Alloy [1]	Heat Treatment Process [2]	HV, MPa	UTS, MPa	YS, MPa	El, %
0Mg0Zn	CR	102.4	345	339	0.6
	CR350	84.4	297	238	4.8
	CR400	84.2	304	231	7.1
1Mg1Zn	CR	122.9	422	417	1.6
	CR350	96.1	346	279	5.5
	CR400	100.6	358	280	6.0
Average deviations		1.5	8	4	0.5

[1] See in Table 1, [2] see in Table 2.

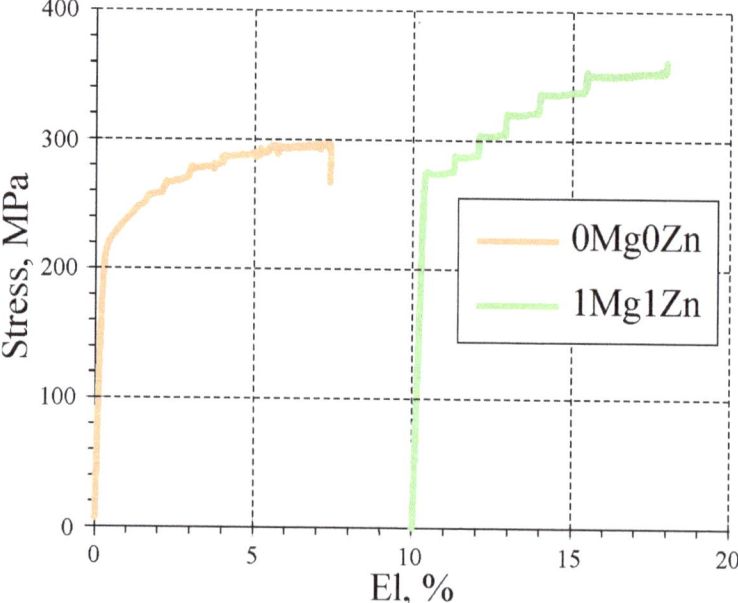

Figure 9. Typical tensile engineering stress–strain curves of experimental alloys obtained after cold rolling and annealing at 400 °C (CR400).

Fracture surface analysis of the alloys showed a dimple structure both in the initial and as-annealed states (Figure 10). However, the number of dimples in the as-annealed state (Figure 10c,d) is much larger, and their size is smaller, which is obviously associated with a greater deformation before fracture.

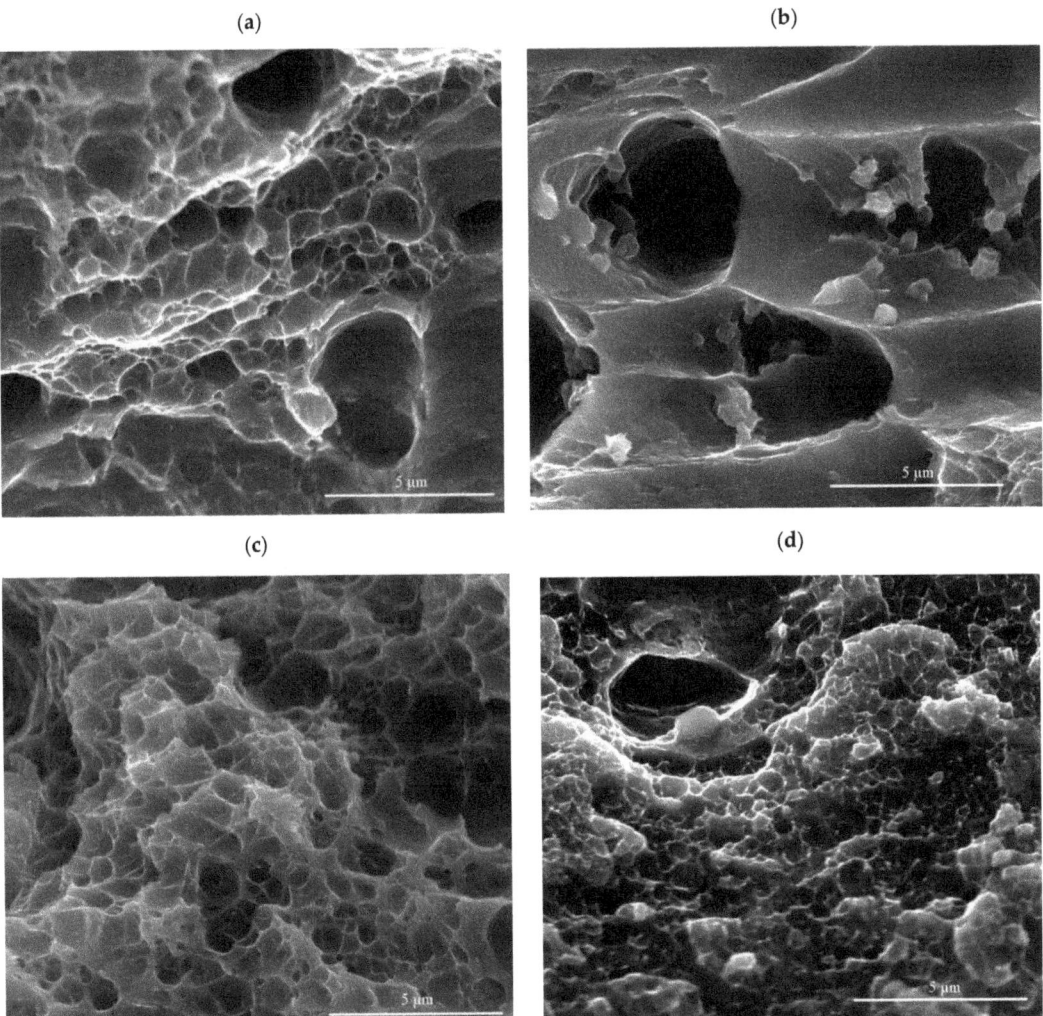

Figure 10. The fracture surface of cold rolled sheets after tensile test, SEM. (**a**,**b**) cold rolled, (**c**,**d**) annealed at 400 °C after cold rolling. (**a**,**c**) 0Mg0Zn, (**b**,**d**) 1Mg1Zn.

4. Discussion

The experimental results show that the joint addition of 1%Mg and 1%Zn to the base alloy significantly increases the UTS and YS of the cold-rolled sheets annealed at 400 °C (Figure 11). At the same time, the plasticity decreases, but only a little. To explain this effect, we analyzed the differences in the structure and phase composition between the 0Mg0Zn and 1Mg1Zn alloys.

Previous studies of the related Al–Cu–Mn alloys [25] suggest that annealing at 400 °C or higher makes it possible to obtain a close-to-equilibrium state. Calculations of the equilibrium phase composition were carried out for the test alloys (the iron impurity was not taken into account). As can be seen from Table 4, the amount of the $Al_{20}Cu_2Mn_3$ phase in all the alloys is approximately the same (about 6.5 wt.%). It follows from this fact that magnesium and zinc addition should not affect the number of nanosized dispersoids of this ternary compound, which are formed during the deformation heat treatment and determine the strength and heat resistance of the base alloy [25]. According to the calculations, magnesium and zinc in the 1Mg1Zn alloy are completely dissolved in (Al) at 400 °C or higher temperatures (Table 5). This is the main difference in the phase composition between the 0Mg0Zn and 1Mg1Zn alloys.

Table 4. Calculated fractions of phases for experimental alloys at various temperatures.

Alloy [1]	T, °C	Fractions of Precipitates, wt.%					
		Al_{20}	Al_{15}	Al_2Cu	Mg_2Si	S	(Al)
0Mg0Zn	350	6.63	0.99	0.45	–	–	balance
	400	6.55	0.95	–	–	–	balance
	450	6.25	0.92	–	–	–	balance
1Mg1Zn	350	6.66	0.94	–	0.04	1.30	balance
	400	6.51	0.97	–	0.02	0.15	balance
	450	6.18	0.99	–	–	–	balance

[1] See in Table 1.

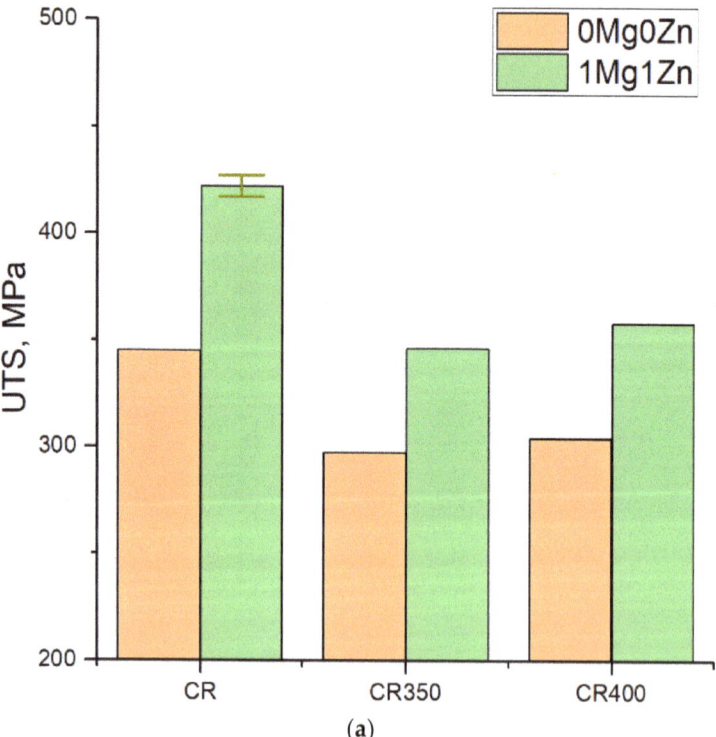

Figure 11. *Cont.*

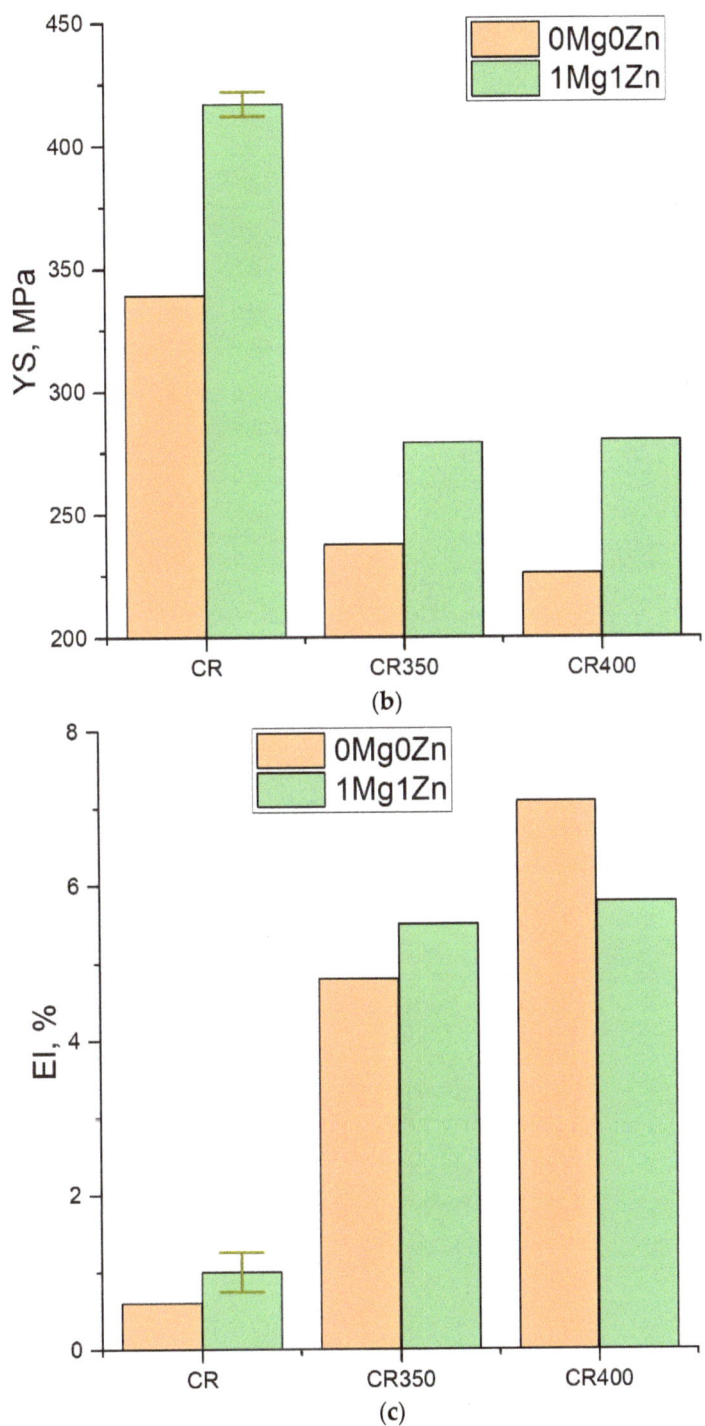

Figure 11. Comparison of mechanical properties: (a) UTS, (b) YS, (c) El.

Table 5. Calculated composition of aluminum solid solution for experimental alloys at various temperatures.

Alloy [1]	T, °C	Concentration in (Al), wt.%					
		Cu	Mn	Mg	Zn	Si	Fe
0Mg0Zn	350	0.87	0.02	0.03	0.03	<0.01	<0.01
	400	1.14	0.06	0.03	0.03	0.01	<0.01
	450	1.19	0.13	0.03	0.03	0.01	<0.01
1Mg1Zn	350	0.51	0.04	0.85	1.21	<0.01	<0.01
	400	1.09	0.05	1.07	1.20	<0.01	<0.01
	450	1.21	0.12	1.10	1.20	0.01	<0.01

[1] See in Table 1.

Assuming that different strengthening mechanisms, such as solid solution strengthening (σ_{ss}), grain boundary strengthening (σ_{gb}), dislocations strengthening (σ_p), and second phase strengthening (σ_{pp}) have independent contributions in the YS of the alloy, the latter can be calculated as follows [35]:

$$YS = \sigma_0 + \sigma_{ss} + \sigma_{gb} + \sigma_{pp}, \quad (1)$$

The base yield strength, σ_0 (~20 MPa).

The solid solution strengthening σ_{ss} caused by Zn and Mg can be determined as follows [36]:

$$\sigma_{ss} = k_i C_i^{2/3}, \quad (2)$$

where k_i is the coefficient describing the effect of atomic solute on solid solution strengthening; the corresponding strengthening coefficients of Zn, Mg, and Cu are 3.085, 20.081, and 12.431, respectively [37]; C_i is the weight percentage of atomic solute in the matrix. Taking into account that for the 1Mg1Zn alloy in question C_{Zn} = 1.20 wt.%, C_{Mg} = 1.07 wt.%, and C_{Cu} = 1.09 wt.% (Table 5), σ_{ssZn} is ~3.5 MPa, σ_{ssMg} is ~21 MPa, σ_{ssCu} is ~13.1 MPa, and the total σ_{ss} is ~37.6.

The contribution of grain boundary strengthening to the YS can be calculated using the Hall–Patch equation which can be expressed as [38].

$$\sigma_{gb} = \sigma_0 + kD^{-1/2}, \quad (3)$$

where σ_0 is the intrinsic resistance of the lattice to dislocation motion, which is approximately 20 MPa for most aluminum alloys; k = 0.14 MPa·m$^{1/2}$ is the Hall–Petch coefficient [38], and D is the average grain or sub-grain sizes. From the structural analysis data, the average size of sub-grain boundaries is 0.45 μm (Figure 8). From the latter fact, the value of σ_{gb} is ~228 MPa.

The strengthening effect caused by the Al$_{20}$ phase can be assessed using the following modified Orowan equations for rod-like particles of diameter D_r and length l_r (>>D_r) [39,40]:

$$\sigma_{pp} = 0.15 G \frac{b}{D_r} \left(f_v^{\frac{1}{2}} + 1.84 f_v + 1.84 f_v^{\frac{3}{2}} \right) \ln \frac{1.316 D_r}{r_0} \quad (4)$$

where G = 25.4 GPa is the shear modulus; b = 0.286 nm is the Burgers vector; f_v is the volume fraction of the dispersed phase; D_r is the rod diameter (according to metallographic analysis, D_r is 40 nm); and r_0 is the inner cut-off radius for the calculation of the dislocation line tension. According to the aforementioned calculation (Equation (4)), the increase in the yield strength is 10.5 MPa.

According to the calculations, the total value of the YS is 296 MPa, which is quite close to the experimental data of 280 MPa (taking into account the measurement error, Figure 12).

According to Table 3, the difference in the YS between the 0Mg0Zn and 1Mg1Zn alloys is about 42 MPa, which is very close to the calculated data of solid solution hardening: σ_{ss} is ~37.6 MPa. Thus, the results show that the addition of 1% magnesium and zinc to the base alloy can be considered a promising method of increasing the strength of rolled sheets while maintaining the main advantage, namely, the exclusion of homogenization and quenching operations from the process route.

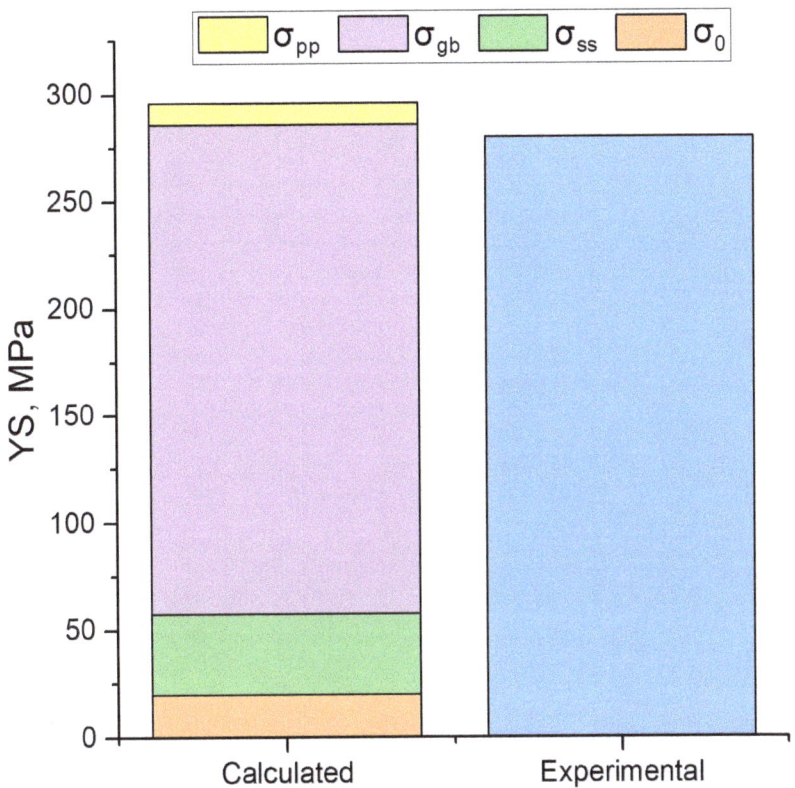

Figure 12. Comparison of calculated and measured yield strength for the cold-rolled 1Mg1Zn alloy and fraction of different strengthening mechanisms in the total YS.

The experimental results show that despite both the commercial (2xxx) and experimental alloys having the same alloying system (Al–Cu–Mn(–Mg)), their properties and microstructure constitutions are very different. This circumstance stems from the fact that the commercial alloys were designed for a maximum number of precipitates forming at aging [41,42], while the model alloy Al2Cu1.5Mn1Mg1Zn provides a high fraction (much higher than those for the commercial alloys) of $Al_{20}Cu_2Mn_3$ dispersoids.

Recently, several works were published where authors argued that the increase in the thermal stability of the 2xxx alloys is possible via microalloying with Ag [43], Sc, and Zr [44–46]. Although this way can lead to some increase in heat resistance through stabilization of the θ' phase, it is limited in terms of temperature. For instance, work [45] studied the influence of Sc and Zr micro additions on the high-temperature strength of the AA2219-type composition. However, no temperatures higher than 280 °C are considered in this study. Moreover, the high cost of Sc limits the wide industrial application of such alloys.

From the data obtained, one can conclude that the Al2Cu1.5Mn1Mg1Zn-based model alloy is very promising for the designing of new wrought heat-resistant aluminum alloys. The new alloy is notably superior to commercial ones (particularly, 2219, 2024, 2014) in thermal stability (toward softening at heating up to 400 °C) and provides a much shorter technological routine (the stages of homogenization, solution treatment, and water quenching are excluded) to obtain the wrought products. In addition, the new alloys can be prepared from secondary raw materials containing a typical set of elements (i.e., Cu, Mg, Zn, Mn, Fe, and Si) [32].

5. Summary

Sheet model alloys Al2Cu1.5Mn and Al2Cu1.5Mn1Mg1Zn (wt.%) designed for $Al_{20}Cu_2Mn_3$ dispersoids were compared in phase composition, microstructure, and mechanical properties after different processing routes (including hot rolling, cold rolling, and annealing). The conclusions are shown in the following points:

1. It was shown using calculation and experimental methods that the alloys contain approximately equal volume fractions (~7 vol.%) of $Al_{20}Cu_2Mn_3$ dispersoids after annealing at 350–450 °C. The main difference between them is in the composition of the aluminum solid solution (Al), because Mg and Zn are almost fully dissolved in (Al);

2. The formation of nanosized $Al_{20}Cu_2Mn_3$ dispersoids was found to provide for the preservation of the fiber-like (non-recrystallized) grain structure after annealing at up to 400 °C for (3 h), despite a high cold-rolling reduction ratio (80%);

3. The Al2Cu1.5Mn1Mg1Zn model alloy showed a substantially higher strength performance after annealing at 400 °C in comparison with the reference ternary alloy. In particular, the UTS is ~360 vs. ~300 MPa, and the YS is 280 vs. 230 MPa. This indicates a positive effect of Mg and Zn dissolved in (Al) on the strength;

4. Summing up the results, the Al2Cu1.5Mn1Mg1Zn model alloy shows potential to become the basis for designing new high-tech heat-resistant alloys as a sustainable alternative to the 2xxx alloys.

Author Contributions: Conceptualization, N.B.; investigation, K.T. and A.F.; methodology, N.L. and T.A.; supervision, N.B.; writing—original draft, N.B.; investigation, A.F., Writing—review and editing, T.A. All authors have read and agreed to the published version of the manuscript.

Funding: The authors gratefully acknowledge the financial support of the Russian Science Foundation (Project No. 20-19-00249-P) (Conceptualization, SEM, hardness and electrical conductivity tests, tensile tests, discussion) and financial support by the Moscow Polytechnic University within the framework of the grant named after Pyotr Kapitsa (TEM, Thermo-Calc calculations).

Data Availability Statement: The raw/processed data required to reproduce these findings cannotbe shared at this time due to technical or time limitations.

Acknowledgments: The results were obtained by using the equipment of the Center for Collective Use 'Materials Science and Metallurgy' with the financial support of the Ministry of Science and Higher Education of the Russian Federation (#075-15-2021-696).

Conflicts of Interest: The authors declare no conflict of interest.

References

1. Jaunky, V.C. Are Shocks to Aluminium Consumption Transitory or Permanent? *Rev. Appl. Econ.* **2013**, *9*, 21–37. [CrossRef]
2. Babcsán, N. *Aluminium Infinite Green Circular Economy–Theoretical Carbon Free Infinite Loop, Combination of Material and Energy Cycles*; Solutions for Sustainable Development; CRC Press: Boca Raton, FL, USA, 2019; pp. 205–210.
3. Brough, D.; Jouhara, H. The aluminium industry: A review on state-of-the-art technologies, environmental impacts and possibilities for waste heat recovery. *Int. J. Thermofluids.* **2020**, *1*, 100007. [CrossRef]
4. Ashkenazi, D. How aluminum changed the world: A metallurgical revolution through technological and cultural perspectives. *Technol. Forecast. Soc. Chang.* **2019**, *143*, 101–113. [CrossRef]
5. Pedneault, J.; Majeau-Bettez, G.; Pauliuk, S.; Margni, M. Sector-specific scenarios for future stocks and flows of aluminum: An analysis based on shared socioeconomic pathways. *J. Ind. Eco.* **2022**, *26*, 1728–1746. [CrossRef]
6. Ujah, C.O.; Popoola, A.P.I.; Popoola, O.M. Review on materials applied in electric transmission conductors. *J. Mater. Sci.* **2022**, *57*, 1581–1598. [CrossRef]
7. Zheng, K.; Politis, D.J.; Wang, L.; Lin, J. A review on forming techniques for manufacturing lightweight complex—Shaped aluminium panel components. *Int. J. Light. Mater. Manuf.* **2018**, *1*, 55–80. [CrossRef]

8. Kermanidis, A.T. Aircraft Aluminum Alloys: Applications and Future Trends. In *Revolutionizing Aircraft Materials and Processes*; Pantelakis, S., Tserpes, K., Eds.; Springer: Cham, Switzerland, 2020; pp. 21–55.
9. Yang, C.; Zhang, L.; Chen, Z.; Gao, Y.; Xu, Z. Dynamic material flow analysis of aluminum from automobiles in China during 2000-2050 for standardized recycling management. *J. Clean. Prod.* **2022**, *337*, 130544. [CrossRef]
10. Stemper, L.; Tunes, M.A.; Tosone, R.; Uggowitzer, P.J.; Pogatscher, S. On the potential of aluminum crossover alloys. *Prog. Mater. Sci.* **2022**, *124*, 100873. [CrossRef]
11. Asadikiya, M.; Yang, S.; Zhang, Y.; Lemay, C.; Apelian, D.; Zhong, Y. A review of the design of high-entropy aluminum alloys: A pathway for novel Al alloys. *J. Mater. Sci.* **2021**, *56*, 12093–12110. [CrossRef]
12. Raabe, D.; Ponge, D.; Uggowitzer, P.; Roscher, M.; Paolantonio, M.; Liu, C.; Antrekowitsch, H.; Kozeschnik, E.; Seidmann, D.; Gault, B.; et al. Making sustainable aluminium by recycling scrap: The science of "dirty" alloys. *Prog. Mater. Sci.* **2022**, *128*, 100947. [CrossRef]
13. Arowosola, A.; Gaustad, G. Estimating increasing diversity and dissipative loss of critical metals in the aluminum automotive sector. *Resour. Conserv. Recycl.* **2019**, *150*, 104382. [CrossRef]
14. Capuzzi, S.; Timelli, G. Preparation and melting of scrap in aluminum recycling: A review. *Metals* **2018**, *8*, 249. [CrossRef]
15. Niu, G.; Wang, J.; Ye, J.; Mao, J. Enhancing Fe content tolerance in A356 alloys for achieving low carbon footprint aluminum structure castings. *J. Mater. Sci. Technol.* **2023**, *161*, 180–191. [CrossRef]
16. Polmear, I.; StJohn, D.; Nie, J.F.; Qian, M. *Light Alloys: Metallurgy of the Light Metals*, 5th ed.; Elsevier, Butterworth-Heinemann: Oxford, UK, 2017; pp. 31–107. [CrossRef]
17. Mondol, S.; Alam, T.; Banerjee, R.; Kumar, S.; Chattopadhyay, K. Development of a high temperature high strength Al alloy by addition of small amounts of Sc and Mg to 2219 alloy. *Mater. Sci. Eng. A* **2017**, *687*, 221–231. [CrossRef]
18. He, H.; Yi, Y.; Huang, S.; Zhang, Y. Effects of cold predeformation on dissolution of second-phase Al2Cu particles during solution treatment of 2219 Al-Cu alloy forgings. *Mater. Charact.* **2018**, *135*, 18–24. [CrossRef]
19. Zuiko, I.; Kaibyshev, R. Aging behavior of an Al–Cu–Mg alloy. *J. Alloys Compd.* **2018**, *759*, 108–119. [CrossRef]
20. de Sousa Araujo, J.V.; Milagre, M.X.; Ferreira, R.O.; de Souza Carvalho Machado, C.; de Abreu, C.P.; Costa, I. Microstructural characteristics of the Al alloys: The dissimilarities among the 2XXX alloys series used in aircraft structures. *Metallogr. Microstruct. Anal.* **2020**, *9*, 744–758. [CrossRef]
21. Kumar, N.S.; Pramod, G.K.; Samrat, P.; Sadashiva, M. A critical review on heat treatment of aluminium alloys. *Mater. Today Proc.* **2022**, *58*, 71–79. [CrossRef]
22. Sirichaivetkul, R.; Limmaneevichitr, C.; Tongsri, R.; Kajornchaiyakul, J. Isothermal Investigation and Deformation Behavior during Homogenization of 6063 Aluminum Alloy. *J. Mater. Eng. Perform.* **2023**, *32*, 638–650. [CrossRef]
23. Rinderer, B. The metallurgy of homogenization. *Mater. Sci. Forum* **2011**, *693*, 264–275. [CrossRef]
24. Ber, L.B.; Kolobnev, N.I.; Tsukrov, S.L. *Heat Treatment of Aluminum Alloys*, 1st ed.; CRC Press: Boca Raton, FL, USA, 2020.
25. Belov, N.A.; Akopyan, T.K.; Shurkin, P.K.; Korotkova, N.O. Comparative analysis of structure evolution and thermal stability of experimental AA2219 and model Al-2wt.%Mn-2wt.%Cu cold rolled alloys. *J. Alloys Compd.* **2021**, *864*, 158823. [CrossRef]
26. Belov, N.A.; Alabin, A.N.; Matveeva, I.A. Optimization of phase composition of Al–Cu–Mn–Zr–Sc alloys for rolled products without requirement for solution treatment and quenching. *J. Alloys Compd.* **2014**, *583*, 206–213. [CrossRef]
27. Belov, N.A.; Korotkova, N.O.; Akopyan, T.K.; Tsydenov, K.A. Simultaneous Increase of Electrical Conductivity and Hardness of Al–1.5 wt.% Mn Alloy by Addition of 1.5 wt.% Cu and 0.5 wt.% Zr. *Metals* **2019**, *9*, 1246. [CrossRef]
28. Belov, N.A.; Akopyan, T.K.; Korotkova, N.O.; Timofeev, V.N.; Shurkin, P.K. Effect of cold rolling and annealing temperature on structure, hardness and electrical conductivity of rapidly solidified alloy of Al–Cu–Mn–Zr system. *Mater. Lett.* **2021**, *300*, 130199. [CrossRef]
29. Korotkova, N.O.; Shurkin, P.K.; Cherkasov, S.O.; Aksenov, A.A. Effect of Copper Concentration and Annealing Temperature on the Structure and Mechanical Properties of Ingots and Cold-Rolled Sheets of Al–2% Mn Alloy. *Russ. J. Non-Ferr. Met.* **2022**, *63*, 190–200. [CrossRef]
30. Dar, S.M.; Liao, H. Creep behavior of heat resistant Al–Cu–Mn alloys strengthened by fine (θ') and coarse ($Al_{20}Cu_2Mn_3$) second phase particles. *Mater. Sci. Eng. A* **2019**, *763*, 138062. [CrossRef]
31. Mikhaylovskaya, A.V.; Mukhamejanova, A.; Kotov, A.D.; Tabachkova, N.Y.; Prosviryakov, A.S.; Mochugovskiy, A.G. Precipitation Behavior of the Metastable Quasicrystalline I-Phase and θ'-Phase in Al-Cu-Mn Alloy. *Metals* **2023**, *13*, 469. [CrossRef]
32. Belov, N.A.; Cherkasov, S.O.; Korotkova, N.O.; Yakovleva, A.O.; Tsydenov, K.A. Effect of Iron and Silicon on the Phase Composition and Microstructure of the Al–2% Cu–2% Mn (wt%) Cold Rolled Alloy. *Phys. Met. Metallogr.* **2021**, *122*, 1095–1102. [CrossRef]
33. Shelekhov, E.V.; Sviridova, T.A. Programs for X-ray analysis of polycrystals. *Met. Sci. Heat Treat.* **2000**, *42*, 309–313. [CrossRef]
34. Thermo-Calc Software. Available online: http://www.thermocalc.com (accessed on 27 June 2023).
35. Tian, A.; Sun, L.; Deng, Y.; Yuan, M. Study of the Precipitation Kinetics, Microstructures, and Mechanical Properties of Al-Zn-Mg-xCu Alloys. *Metals* **2022**, *12*, 1610. [CrossRef]
36. Cinkilic, E.; Yan, X.; Luo, A.A. Modeling Precipitation Hardening and Yield Strength in Cast Al-Si-Mg-Mn Alloys. *Metals* **2020**, *10*, 1356. [CrossRef]
37. Dixit, M.; Mishra, R.S.; Sankaran, K.K. Structure–property correlations in Al 7050 and Al 7055 high-strength aluminum alloys. *Mater. Sci. Eng. A* **2008**, *478*, 163–172. [CrossRef]

38. Thangaraju, S.; Heilmaier, M.; Murty, B.S.; Vadlamani, S.S. On the Estimation of True Hall–Petch Constants and Their Role on the Superposition Law Exponent in Al Alloys. *Adv. Eng. Mater.* **2012**, *14*, 892–897. [CrossRef]
39. Zhu, A.W.; Starke, E.A. Strengthening effect of unshearable particles of finite size: A computer experimental study. *Acta Mater.* **1999**, *47*, 3263–3269. [CrossRef]
40. Starink, M.J.; Wang, S.C. A model for the yield strength of overaged Al–Zn–Mg–Cu alloys. *Acta Mater.* **2003**, *51*, 5131–5150. [CrossRef]
41. Çadırlı, E.; Kaya, H.; Büyük, U.; Üstün, E.; Gündüz, M. Effect of heat treatment on the microstructures and mechanical properties of Al–4Cu–1.5 Mg alloy. *Int. J. Metalcast.* **2022**, *16*, 1020–1033. [CrossRef]
42. Liang, S.S.; Wen, S.P.; Wu, X.L.; Huang, H.; Gao, K.Y.; Nie, Z.R. The synergetic effect of Si and Sc on the thermal stability of the precipitates in AlCuMg alloy. *Mater. Sci. Eng. A* **2020**, *783*, 139319. [CrossRef]
43. Zamani, M.; Toschi, S.; Morri, A.; Ceschini, L.; Seifeddine, S. Optimisation of heat treatment of Al–Cu–(Mg–Ag) cast alloys. *J. Therm. Anal. Calorim.* **2020**, *139*, 3427–3440. [CrossRef]
44. Kairy, S.K.; Rouxel, B.; Dumbre, J.; Lamb, J.; Langan, T.J.; Dorin, T.; Birbilis, N. Simultaneous improvement in corrosion resistance and hardness of a model 2xxx series Al-Cu alloy with the microstructural variation caused by Sc and Zr additions. *Corros. Sci.* **2019**, *158*, 108095. [CrossRef]
45. Mondol, S.; Kashyap, S.; Kumar, S.; Chattopadhyay, K. Improvement of high temperature strength of 2219 alloy by Sc and Zr addition through a novel three-stage heat treatment route. *Mater. Sci. Eng. A* **2018**, *732*, 157–166. [CrossRef]
46. Mondol, S.; Kumar, S.; Chattopadhyay, K. Effect of thermo-mechanical treatment on microstructure and tensile properties of 2219ScMg alloy. *Mater. Sci. Eng. A* **2019**, *759*, 583–593. [CrossRef]

Disclaimer/Publisher's Note: The statements, opinions and data contained in all publications are solely those of the individual author(s) and contributor(s) and not of MDPI and/or the editor(s). MDPI and/or the editor(s) disclaim responsibility for any injury to people or property resulting from any ideas, methods, instructions or products referred to in the content.

Review

Recent Advances in Additive Manufacturing of Soft Magnetic Materials: A Review

Bryan Ramiro Rodriguez-Vargas [1], Giulia Stornelli [1], Paolo Folgarait [2], Maria Rita Ridolfi [2], Argelia Fabiola Miranda Pérez [3] and Andrea Di Schino [1,*]

1. Dipartimento di Ingegneria, Università degli Studi di Perugia, Via G. Duranti 93, 06125 Perugia, Italy; bryanramiro.rodriguezvargas@studenti.unipg.it (B.R.R.-V.); giulia.stornelli@unipg.it (G.S.)
2. Seamthesis Srl, Via IV Novembre 156, 29122 Piacenza, Italy; paolo.folgarait@seamthesis.com (P.F.); mariarita.ridolfi@seamthesis.com (M.R.R.)
3. Department of Strategic Planning and Technology Management, Universidad Popular Autónoma del Estado de Puebla, 17 Sur, 901, Barrio de Santiago, Puebla 72410, Mexico
* Correspondence: andrea.dischino@unipg.it

Abstract: Additive manufacturing (AM) is an attractive set of processes that are being employed lately to process specific materials used in the fabrication of electrical machine components. This is because AM allows for the preservation or enhancement of their magnetic properties, which may be degraded or limited when manufactured using other traditional processes. Soft magnetic materials (SMMs), such as Fe–Si, Fe–Ni, Fe–Co, and soft magnetic composites (SMCs), are suitable materials for electrical machine additive manufacturing components due to their magnetic, thermal, mechanical, and electrical properties. In addition to these, it has been observed in the literature that other alloys, such as soft ferrites, are difficult to process due to their low magnetization and brittleness. However, thanks to additive manufacturing, it is possible to leverage their high electrical resistivity to make them alternative candidates for applications in electrical machine components. It is important to highlight the significant progress in the field of materials science, which has enabled the development of novel materials such as high-entropy alloys (HEAs). These alloys, due to their complex chemical composition, can exhibit soft magnetic properties. The aim of the present work is to provide a critical review of the state-of-the-art SMMs manufactured through different AM technologies. This review covers the influence of these technologies on microstructural changes, mechanical strengths, post-processing, and magnetic parameters such as saturation magnetization (M_S), coercivity (H_C), remanence (B_r), relative permeability (M_r), electrical resistivity (r), and thermal conductivity (k).

Keywords: additive manufacturing; soft magnetic materials; microstructure; magnetic properties

1. Introduction

In recent years, efficiency and sustainability in the energy sector (such as the oil and gas industry, aerospace, and automotive sector, specifically electric vehicle production) [1–7], along with the demand for a reduction in pollutant emissions, have driven the advancement of materials for such purposes. These advancements extend to manufacturing methods as well, most prominently additive manufacturing (AM), driven by the imperative to curtail pollutant emissions.

The AM process, known as "3D printing," is an innovative manufacturing process that allows the generation of components, layer by layer, directly from a Computer-Aided Design (CAD) model. Among the various qualities of this technology, there is the ability to produce complex geometries, which has generated considerable interest and development within various industries [8–13].

AM has facilitated the rapid creation of prototypes and a reduction in material waste. This can be justified from a comparative standpoint by considering the value of the metal in additive manufacturing versus traditional machining, considering factors such as product

performance, time savings, and production costs. As a result, AM is becoming increasingly competitive in terms of speed and cost. Consequently, it has emerged as an appealing option for various industries seeking efficient and cost-effective production methods, contributing to its rapid growth and implementation [14–18]. In addition to cost savings, AM has the potential to generate substantial energy savings in the final product compared to those manufactured using traditional machining methods. This is particularly relevant in industries that require lightweight electric machines, including transportation, renewable energies, passenger aircraft, and naval applications. Conventional manufacturing techniques often impose limitations on lightweight designs, but as mentioned earlier, AM has introduced unprecedented design freedom. New opportunities have arisen for the manufacturing and design of lightweight electric machines, which can enhance the efficiency and performance of these applications [19–27]. This requirement is particularly critical in several industrial sectors that consume a significant portion of the total electricity, with electric motors and machinery accounting for at least 65% of this consumption. For instance, an efficiency improvement of just 1% could result in a reduction in several million metric tons of hazardous emissions [28,29]. Therefore, by exploiting unconventional manufacturing techniques, such as AM, the researchers aim to overcome the limits of traditional processes and promote significant contributions to energy savings and environmental sustainability.

Another notable feature of this manufacturing process is the wide range of materials that can be used in the various components required by the industry, including polymers, biopolymers, metals, and composites, among others. This extensive material selection opens new possibilities for innovation and customization in various industrial applications. This is particularly relevant in the case of soft magnetic materials (SMMs) employed in electromagnetic devices used for energy conversion and generation [30–33].

SMMs with an iron base alloy with the addition of silicon (Si), nickel (Ni), and/or cobalt (Co) exhibit high saturation induction, low coercivity, high permeability, and low core loss. As a result, SMMs find widespread application in electric motors, generators, and transformers, ensuring optimal energy conversion and facilitating energy generation and transmission [28,34,35].

The importance of AM in the processing of soft magnetic materials can be understood by considering their conventional manufacturing methods. Traditionally, these materials have been fabricated in various forms, such as foils, plates, sheets, and bars, following a series of steps known as "ingot metallurgy practices," involving sequential processes of melting, iterative thermomechanical forming, and annealing treatments. Casting is the initial step, where the molten alloy is poured into molds to form ingots. These ingots are then subjected to thermomechanical forming processes, such as rolling, forging, or extrusion, to refine the microstructure and improve the mechanical properties of the alloys. After that, the alloys undergo annealing treatments to promote recrystallization and grain growth [28,36–38]. Such multi-step processing techniques, involved in ingot metallurgy practices, play a significant role in the magnetic behavior of these alloys. However, during conventional manufacturing processes, some solid-state disorder–order phases tend to form, reducing the workability of soft magnetic alloys and limiting the optimization of the chemical composition of the alloy [39–41]. For example, in high-silicon Fe–Si alloys, there is a tendency during cooling to form phases with oriented structures through lattice rearrangement into B2 and D03 phases [42,43]. This is known to increase material brittleness to the extent of limiting workability at low temperatures [39].

In this regard, by exploring new emerging manufacturing methods, such as additive manufacturing, it is possible to integrate or replace traditional processing techniques. This opens significant opportunities for the development of SMMs with improved performance and processability suitable for a wide range of electromagnetic applications, leading to improved efficiency.

AM processes, such as Laser Powder Bed Fusion (L-PBF), Selective Laser Melting (SLM), Directed Energy Deposition (DED), and Spark Plasma Sintering (SPS), among others, offer significant advantages in the production of various components, including the ferromagnetic core for the electrical motors. For Fe–Si alloys, additive technology provides an alternative to preprocessing variants with high silicon content. Indeed, the rapid cooling rates achieved during the laser melting process prevent the occurrence of the usual ordered–disordered phase transition in Fe–Si steels. Furthermore, in the conventional Fe–Si rolling and stacking process, the sheet sizes are limited by the available rolling mills as well as the chemical composition of the alloy itself, resulting in a limited ability to create complex geometries for the magnetic flux path. Additive technology offers greater design freedom, enabling the production of parts with complex geometries and intricate internal structures. This allows for the optimization of the magnetic flux path and, consequently, the electromagnetic performance of the device, leading to higher efficiency and improved energy conversion [44–47]. This technology has also led to the development of Fe–Co and Fe–Ni alloys, allowing precise control of the resulting magnetic properties, including high saturation magnetization and low iron losses [46,48,49]. Materials such as soft ferrites, which are unsuitable for use in rotors due to their brittleness and low magnetization, could be processed thanks to advancements in additive manufacturing. This allows for their utilization based on their high electrical resistance, employing them as insulating materials in electric cores [50]. Soft magnetic composites (SMCs) are novel materials characterized by low core loss, high saturation magnetization, and high electrical resistivity. They are composed of magnetic particles coated with insulation, allowing for the creation of three-dimensional flux paths. However, obtaining these geometries is limited by the capacity of the compaction process to achieve uniform pressure. As mentioned earlier, the capability of AM to fabricate complex geometries has facilitated the use and development of these materials. An example of this is the study carried out on the AM processing of FINEMET®, which is a trademark of a SMC with extremely high permeability and significantly high electrical resistivity. The utilization of techniques such as laser-engineered net shaping (LENS™) for its processing has led to the attainment of coercivity values exceeding 2000 A per meter [51–53]. By manipulating the chemical composition of materials and matching operating parameters (energy input, scanning speed, heat treatment, etc.), AM provides opportunities to tailor the magnetic characteristics of components, optimizing their performance in specific applications [54–58].

The state-of-the-art review of additive manufacturing in magnetic materials for this research covered a period of investigation up to the year 2023. Figure 1 illustrates the various alloys used as soft magnetic materials that were covered in this work, along with the semi-quantitative percentage of papers found in the literature review. It is noteworthy that Fe–Si and Fe–Ni alloys maintain their significance in the manufacturing of electromechanical devices for energy conversion and generation. However, SMCs, despite being relatively recent materials compared to others, exhibit a clear increasing trend in publications on the topic. It is important to mention that at the end of this review, special emphasis is placed on the immediate future of SMMs, which includes the exploration of high-entropy alloys (HEAs). These alloys are considered magnetic materials that, based on the control and manipulation of their chemical composition, can exhibit soft or hard magnetic properties depending on their application.

The importance of studying these soft magnetic materials lies in the impact of their favorable magnetic properties, which allow them to respond quickly to external magnetic fields and minimize energy losses due to magnetic hysteresis. Therefore, electromechanical devices that utilize soft magnetic materials can achieve high efficiency and performance. The objective of this document is to provide a summary of the recent advancements in utilizing soft magnetic materials in AM for the design of electrical machines. It will explore the advantages and limitations of processability for each alloy, as well as the AM technologies employed. This comprehensive overview aims to contribute to the advancement of this field by inspiring further research and fostering innovation.

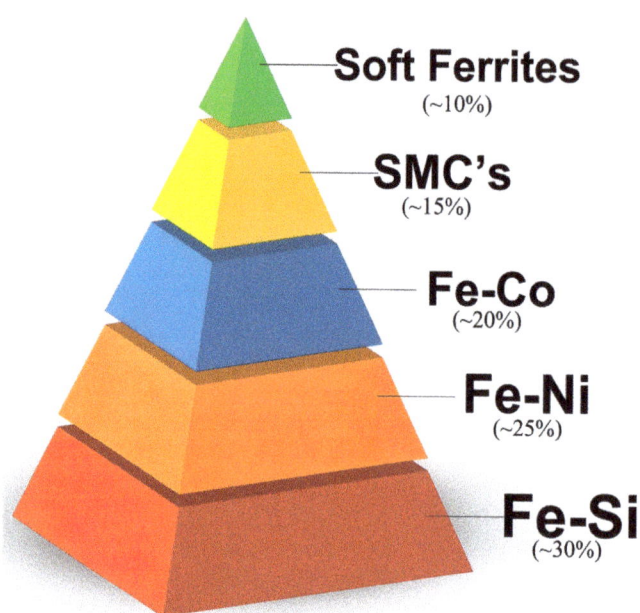

Figure 1. Summary of the soft magnetic materials utilized in the reviewed articles.

2. Overview of Soft Magnetic Materials

The term "magnetic materials" is commonly used to refer to ferromagnetic materials that exhibit strong attraction to magnets due to their spontaneous magnetization. These materials possess inherent magnetic properties that allow them to be easily magnetized and demagnetized [56–61]. This can be classified into three distinct groups: soft-magnetic materials, hard-magnetic materials, and superparamagnetic materials [61]. Figure 2 illustrates the qualitative distinctions among these groups of magnetic materials based on their magnetization curves. In the subsequent lines, a brief explanation of each of these materials will be provided to explain their differences. However, given the primary focus of this research, the emphasis of the following paragraphs and consecutive sections will be on soft magnetic materials.

Soft magnetic materials (SMMs; Figure 2A) are called "soft" because they can be magnetized with low excitation, reach high values of magnetic induction, possess high magnetic permeability, and are easily magnetized and demagnetized. They exhibit a magnetization curve that saturates at relatively low magnetic fields. SMMs are commonly used in applications where efficient magnetic flux circulation and low energy losses are essential, such as transformers, electric motors, and magnetic sensors. According to the international standard, SMMs are those materials with coercivity less than 1 kA/m [62–64].

Hard-magnetic materials (Figure 2B), also known as "permanent magnets," exhibit high coercivity and remanence, maintaining a significant portion of their magnetization even after the external magnetic field is removed. Hard-magnetic materials require high magnetic fields to reach saturation. These materials are used in devices requiring a stable and strong magnetic field, such as magnetic storage systems, magnetic resonance imaging (MRI) machines, and magnetic separators [65–67].

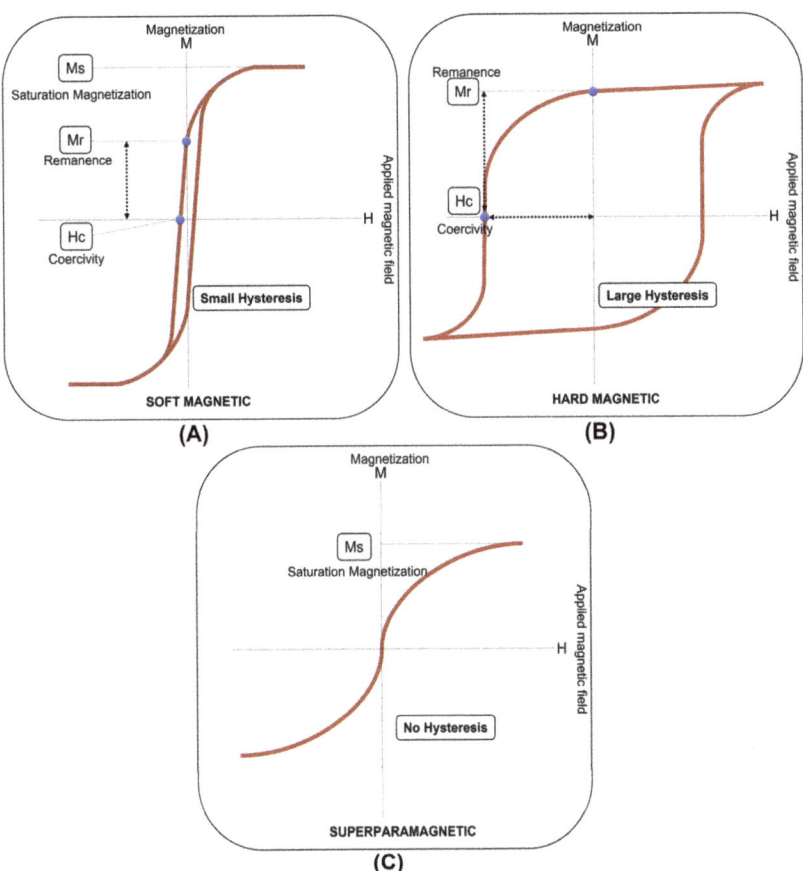

Figure 2. Magnetic behavior of different classes of ferromagnetic materials: (**A**) soft magnetic materials; (**B**) hard magnetic materials; and (**C**) superparamagnetic materials.

Superparamagnetic materials (Figure 2C) are characterized by their small particle size, typically on the nanoscale. Unlike hard-magnetic materials, superparamagnetic materials do not possess permanent magnetization. Instead, they exhibit a response to magnetic fields like SMMs. Superparamagnetic materials are widely used in biomedical applications, such as targeted drug delivery, MRI contrast agents, and magnetic hyperthermia for cancer therapy [68,69].

For the SMMs, there are two important parameters to consider: the remanent magnetization (M_r) and the coercive field (H_C). M_r represents the amount of residual magnetization in the material when the external magnetic field is removed. It indicates the material's ability to preserve magnetization. H_C indicates the material's resistance to demagnetization. It represents the magnetic field strength required to reverse the remanent magnetization and return the material to its demagnetized state. In addition, other significant parameters in magnetic materials include magnetic susceptibility (χ_m) and saturation magnetization (M_S). χ_m measures the extent to which a material can be magnetized in response to an applied magnetic field. It is proportional to the slope of the magnetization curve for magnetically isotropic materials. As the strength of the applied magnetic field increases, the material becomes more magnetized until it reaches saturation. Saturation occurs when all the magnetic domains of the material align with the applied field. Therefore, the induced magnetization reaches the M_S value, which represents the maximum magnetization that the material can achieve [50].

These parameters play a crucial role in characterizing and understanding the magnetic properties of materials. By measuring and analyzing these values, researchers can assess the material's magnetic behavior and determine its suitability for various applications [61,70–72]. For example, Fe–Si and Fe–Ni alloys exhibit high magnetic susceptibility and saturation magnetization while maintaining low remanence and coercivity. This characteristic makes them easily magnetized, highly responsive to magnetic fields, but susceptible to demagnetization. Additionally, silicon steel alloys provide a well-balanced solution in terms of cost and performance for low-frequency applications of 50–60 Hz. However, for higher operation frequencies in the kHz/MHz range, SMCs such as amorphous and nanocrystalline materials are preferred due to their lower losses, despite their higher cost. In industries where high power densities are required, such as aerospace, Fe–Co alloys are commonly utilized, which possess excellent magnetic properties [73–82]. The response of SMMs to external magnetic fields is complex and primarily dependent on the chemical composition of the material as well as on the microstructure and texture. It is important to carefully consider the specific requirements and operating conditions of the application to select the most suitable soft magnetic material. Factors such as cost, performance, frequency range, and power density warrant careful consideration in this selection process [83–86].

As mentioned so far, soft magnetic materials exhibit a wide range of options, with performance varying significantly. Table 1 presents a general overview of different SMMs, such as Fe–Si, Fe–Ni, Fe–Co, SMCs, and soft ferrites, along with their relevant magnetic properties for industrial use, such as saturation magnetization (M_S), relative permeability (μ_r), and core resistivity (r). These materials are manufactured in various forms using additive manufacturing techniques such as L-PBF, SLM, DED, and/or SPS. The information provided in this table is highly useful for understanding the differences in properties and characteristics among various soft magnetic materials, enabling proper selection for the design and manufacturing of electromagnetic devices.

Table 1. List of the general values of the main magnetic properties of SMMs utilized in the reviewed articles, together with the applied AM technique.

SMMs		Saturation Magnetization M_S (T)	Relative Permeability μ_r	Core Resistivity r ($\mu\Omega$cm)	AM	Ref.
Fe–Si	Fe–(2.9–3.7) Si	2.10	25	20–50	L-PBF, SLM	[87–94]
	Fe–(6.5–6.9) Si	1.9	11,000	85		[95–105]
	Fe–Ni	0.60–1.10	8000–100,000	-	L-PBF, SLM, DED	[106–117]
	Fe–Co	2.4–2.5	20,000–60,000	45	L-PBF, LENS	[118–125]
	SMCs	0.50–1.30	6000–100,000	120–130	SLM, DED, SPS	[52,126–131]
	Soft Ferrites	0.32–0.545	350–20,000	$10^7 \times 10^9$	L-PBF, DED	[132–136]

3. Fe–Si

Silicon steels (Fe–Si) have long been the standard industrial material for the manufacture of electric motor cores, generator cores, and transformer cores due to their remarkable electromagnetic properties and cost-effectiveness, making them highly competitive in the industry [39,50,137]. They are typically produced with a silicon content ranging from 2.0% to 7.0% wt. As the Si content increases, there is a substantial improvement in their soft magnetic characteristics. However, the material becomes more brittle, making it difficult to process using traditional methods [95,138]. This section is divided into two parts: the first one describes several recent studies on the processing of low-silicon Fe–Si alloys using different additive manufacturing processes. The second part focuses on high-silicon steels. In both parts, the aim is to illustrate the effect of process parameters such as energy source, scanning strategy, and post-processing techniques. These factors influence the microstructure, phase distribution, porosity, grain size, and texture of the materials, which in turn affect their magnetic and mechanical behavior.

In the study conducted by Andreiev et al. [87], they worked with a Fe–3.0%Si alloy processed by laser beam melting (LBM), aiming to evaluate a graded cross-section generated by incorporating grooves at various positions within the cross-section. Figure 3 shows the Computer-Aided Design (CAD) model of the samples, in which various slits were varied. In Figure 3a, the proposed five slits with sizes ranging from 50 μm to 250 μm are observed, while Figure 3b displays the specimen with a single slit of 50 μm, 100 μm, or 150 μm. Figure 3c illustrates the slit design on the ring used for assessing magnetic behavior for four different transverse sections. They found that the groove geometry can be controlled by adjusting the initial groove thickness defined in the CAD model, which should be greater than 150 μm. This control is further influenced by specific parameters within the lower layer (laser power—70 W and 100 W, along with a minimum requirement of eight processed layers or more). These conditions lead to the formation of grooves under conditions of porous structures or continuous gaps filled with unfused powder (approximately 390 μm). Additionally, a heat treatment of annealing at 550 °C for 2 h followed by air cooling was necessary to prevent bending of the walls between the outer grooves. The geometry and position of the grooves, whether on the internal or external surface or in a displaced manner, influence the power losses. This can be observed in the results, where grooves as continuous gaps on the external surfaces reduce the power losses from 19.7 W/kg to 11.2 W/kg at 50 Hz. However, the authors suggest that further reducing power losses below the values of conventionally laminated magnetic cores could entail several strategies for magnetic cores produced via this AM technique. These strategies include generating thinner grooves as nearly continuous gaps, using finer metallic powder (20 μm), increasing the silicon content in the alloy, and/or adapting the post-processing sequence.

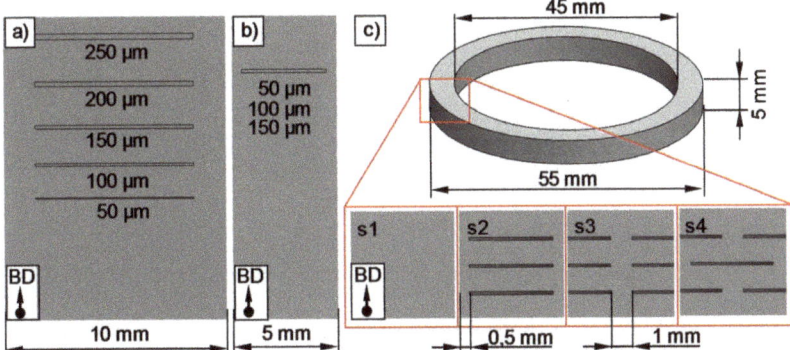

Figure 3. CAD-models of samples: (**a**) parallelepiped with five slits exhibiting constant thickness from 50 μm to 250 μm; (**b**) parallelepiped with one slit exhibiting constant thickness of 50 μm, 100 μm, or 150 μm; (**c**) ring specimen used for measuring the magnetic properties. BD: building direction. Reprinted from Andreiev et al. [87], with permission from Elsevier®.

Quercio et al. [88] fabricated Fe–2.9%Si samples using Laser Powder Bed Fusion (L-PBF) to evaluate the coercivity and permeability before and after annealing. For the coercivity test, rectangular and cylindrical bar-shaped components were fabricated. The results for the cylindrical shape showed a low coercivity value (69.1 A/m) even before treatment (87.7 A/m) compared to those obtained in the rectangular bar (212 A/m before treatment and 85.6 A/m after treatment). This behavior could be related to the geometric shape and/or printing conditions of a small diameter. The evaluation of relative permeability before heat treatment showed values of 748 and magnetic induction not exceeding 1.1 T. After heat treatment, the relative permeability increased to 3224, and the magnetic induction exhibited values higher than 1.2 T. These results indicate that the sample geometry and heat treatment directly influence the magnetic characteristics of the produced parts.

Selema et al. [89] investigated the manufacturing of Fe–3%Si samples using two different AM techniques: 3D micro-extrusion and L-PBF. The results suggest that 3D micro-extrusion, coupled with a subsequent sintering process, allows the production of magnetic materials with improved properties, approaching those of conventional electrical steel. The study highlights the potential of AM techniques to open new frontiers for the electromagnetic sector. Regarding the experimentation carried out for the same Fe–3%Si via L-PBF technology, different scanning strategies were used for producing thin-walled samples by Haines et al. [90]. The main purpose was to examine how the scanning strategy would impact the microstructure of the alloy after undergoing an annealing heat treatment (argon atmosphere at 1000 °C for 5, 60, and 240 min and 1200 °C for 5 and 60 min using a 10 °C/min ramp rate and furnace cooling). Figure 4 illustrates the specimen sizes (Figure 4A) and the three distinct scanning strategies proposed by the authors. These include longitudinal scanning along the length of the part (Figure 4B), transverse scanning across the width (Figure 4C), and a rotated approach with a 67° rotation in the scanning direction between each layer (Figure 4D). A prominent aspect of the execution of each scan, called "point skipping," is presented in Figure 5. This technique involves skipping two points every time the laser reverses during a scanning sequence. This technique helps reduce overheating in the area that has recently received an input of energy, resulting in improved process control and enhanced material properties. A comparison between the two scanning strategies revealed distinct differences in grain morphology and size. Specifically, the longitudinal scanning strategy leads to the formation of more equiaxed and/or misoriented grains with smaller dimensions than the transverse scanning strategy. On the other hand, the effect of annealing at different temperatures revealed distinct changes: at 1000 °C, the longitudinal samples experienced grain growth after 60 min, whereas at 1200 °C, grain growth occurred after 5 min. Additionally, the transverse scanning samples exhibited grain growth after 60 min. These findings highlight the temperature-dependent behavior of the material and the influence of the annealing and scanning strategies on its microstructure, indicating the importance of controlling the annealing parameters for achieving desired grain characteristics and optimizing material properties.

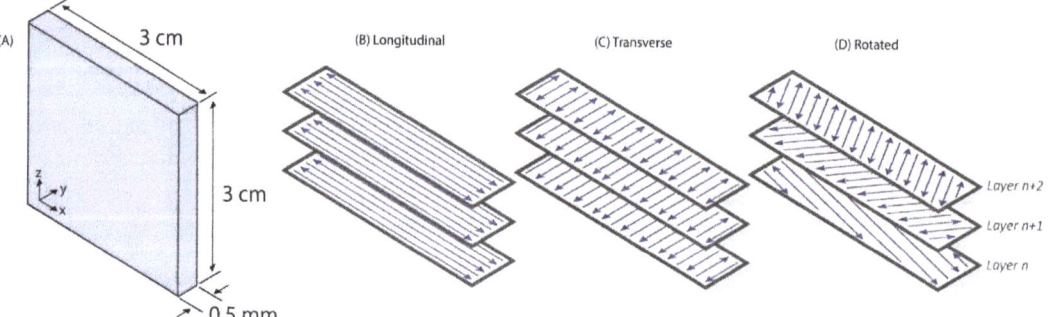

Figure 4. (**A**) Schematic illustration of the prismatic shape used for L-PBF processing; scan directions with reference to the top surface of the prism: (**B**) longitudinal; (**C**) transverse; and (**D**) rotated. Reprinted from Haines et al. [90], with permission from Elsevier®.

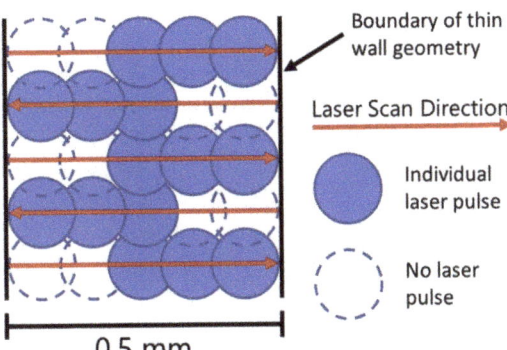

Figure 5. Schematic illustration of "point-skipping" during laser scanning. Reprinted from Haines et al. [90], with permission from Elsevier®.

Martin et al. [91] conducted a study focused on the fabrication of Fe–3%Si toroidal cores using an additive manufacturing technique like MIM-like (Metal Injection Molding), which offers a more cost-effective approach. The aim was to demonstrate the feasibility of producing dense metal parts in Fe–3%Si through this process. The results indicated a relatively uniform surface on the sintered sample, without any significant defects or open porosities that could typically occur at high temperatures. Therefore, using this technique, it is now possible to effectively solve specific applications that previously faced challenges in terms of material manufacturability or geometric complexity.

The works carried out by Tiismus et al. [92,137] present a comprehensive investigation of additive manufacturing using L-PBF to produce Fe–3.7%Si magnetic cores and subsequent applications. The study presented in [137] describes the adjustment of process parameters to achieve samples with appropriate density, surface roughness, and magnetic properties. Figure 6 provides an overview of the correlation between scanning parameters, with a focus on laser input energy, and the resulting relative density and surface roughness of the manufactured samples. In Figure 6a, it can be observed that the lowest relative density value (47.91%) is achieved at 250 W and 2 m/s, while at 300 W and 0.5 m/s (78 J/mm^3), a notably high relative density of 99.86% is attained. Figure 6b suggests that the average surface roughness of the samples (ranging from 6.8 to 18.2 μm for Ra and 36.1 to 85.2 μm for Rz) does not exhibit a monotonic increase with rising laser input energy density. Moreover, Figure 6c demonstrates that within the 20–50 J/mm^3 range, the energy supplied under specific parameter sets is insufficient to achieve uniform fusion (300 W, 1.5 m/s). As scanning speed is decreased, homogeneous fusion of samples is achieved (Figure 6d,e). Notably, it is important to highlight that with increased laser power (250–400 W, 0.25 m/s), samples experience deformation due to the heightened energy input (Figure 6f). XRD results indicate the formation of a single α-ferrite BCC phase, and the formation of the γ phase is suppressed with increasing silicon content during solidification. Finally, the authors conducted direct current measurements, obtaining hysteresis losses of 0.8 W/kg (W10,50), a maximum relative permeability of 8400, and magnetizations of 1.5 T at 1480 A/m and 1 T at 90 A/m. These results were correlated with values in the magnetic properties of conventional steels, showing equivalency to the latter. The application of these results can be visualized in further research [92], which focused on the development of an induction motor with L-PBF printed Fe–3.7%Si cores. The performance evaluation of the finished motor showed an output power of 68 W from the steel core. To address the challenge of minimizing eddy current losses, an innovative segregation strategy was implemented by creating a uniform and narrow air gap of 0.35 mm perpendicular to the core axes.

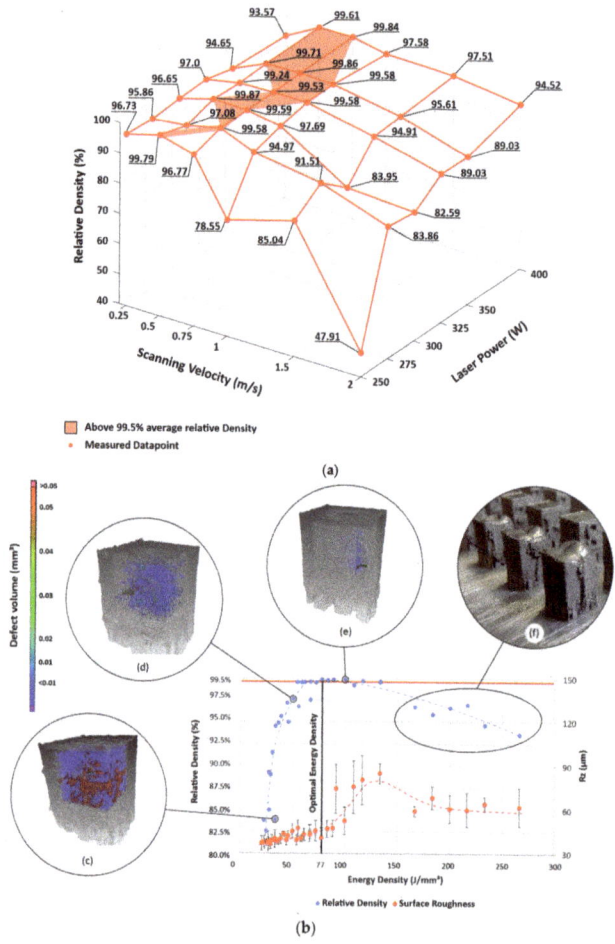

Figure 6. (**a**) Relationship between scanning parameters and relative density; (**b**) relative density and surface roughness as functions of laser input energy density; (**c**) sample at 300 W, 1.5 m/s; (**d**) sample at 300 W, 1 m/s; (**e**) sample at 300 W, 0.5 m/s; (**f**) samples at 400 W, 0.25 m/s. Reprinted from Tiismus et al. [137].

Additionally, lattice structures were added in these gaps to consolidate the printed part into a single core, preventing deformation and delamination during printing and promoting powder fusion between the layers. A thorough understanding of these effects is crucial for optimizing L-PBF process parameters and achieving the desired microstructural features and performance in 3D-printed Fe–Si alloys.

It is essential to mention the influence of chemical composition and microstructural changes on the magnetic properties of components manufactured through additive manufacturing. In a study conducted by Kumary et al. [93], the impact of Si and B content on the properties of Fe–Si–B alloys fabricated through binder jet printing was investigated. The Si content ranged from 3 to 5 wt.%, while the B content ranged from 0.0 to 0.25 wt.%. The Fe–5.0%Si alloy with 0.25% B, sintered at 1200 °C, demonstrated the highest magnetic relative permeability and the lowest intrinsic coercivity. The addition of B led to larger grain sizes with reduced porosity and increased density, attributed to the formation of the ferromagnetic lamellar phase Fe_2B at the grain boundaries.

Kang et al. [94] undertook a study on Fe–Ni–Si alloys with compositions of 54%, 42%, and 4% in weight proportions using L-PBF technology. The study explored the impact of the hot isostatic pressing (HIPing) process on a reduction in porosity, with a focus on eliminating small pores while larger pores remained even after the HIPing process. During HIPing, the acicular Ni/Si-rich structure, which forms in the L-PBF fabricated samples due to the rapid cooling rates, undergoes a transformation of equilibrium phases such as austenite, Ni_3Si, and $FeNi_3$. The study also assessed the mechanical properties of the samples, showing an increase in both the elastic modulus and strength, from about 11 GPa and 650 MPa up to 18 GPa and 900 MPa after HIPing. Furthermore, the fitted data indicated an increment in coercivity from approximately 1.8 kA/m up to 2.9 kA/m.

At the beginning of this section, a general overview is provided regarding the advantages and limitations of these alloys when increasing the silicon content. It is considered high-silicon steel when the percentage of this element is $\geq$4.5 wt.% [139,140]. This increase in Si provides the steel with remarkable magnetic properties such as high electrical resistivity, reduced magnetostriction, and magneto-crystalline anisotropy, making them ideal for the mentioned components [39,95,141,142]. The main limitation of high silicon steels lies in their brittleness when forming phases with ordered structure (B2 and D03) during the cooling stage of manufacturing, which reduces the ductility of the material and limits its machinability using traditional forming processes. These phases represent two types of ordered structures commonly formed in high-silicon-content steels. The B2 structure arises from the A2 (a disordered solid solution with a BCC lattice) by pairing different atoms from the nearest neighbors. With an increase in silicon content in the alloy, a phase transition towards the ordered D03 structure might occur through further arrangement among second-order neighboring atoms [42,143]. The following section will address various works on these materials using AM. We focus on the influence of process parameters, design or orientation of printed geometries, and heat treatment, among others, on changes in crystallographic texture, microstructure, hardness, and magnetic properties. These research studies demonstrate the significant impact of additive manufacturing in the processing of these materials.

The studies conducted by Garibaldi et al. [96,97] are a reference point for addressing this topic. In one of their research projects [96], they specifically investigated the effect of laser energy in the selective laser melting (SLM) process on the magnetic properties of high-silicon-content steel samples (6.9%). An interesting starting point in this study was the choice of the manufacturing orientation of the components. By constructing the samples with their bases parallel to the horizontal plane, it was possible to measure the magnetic properties of the material in its isotropy plane. The SLM process promotes the $\langle 001 \rangle$ crystallographic direction (easy magnetization axis), resulting in isotropic magnetic properties within the horizontal plane of an SLM sample. In contrast, the $\langle 111 \rangle$ direction (the hard direction of magnetization) tends to be outside the horizontal plane. Using the Williamson–Hall method, it was determined that annealing the samples at 700 °C for 5 h allowed for the complete elimination of residual stresses. However, Vickers microhardness (HV) analysis revealed a significant increase in HV after this treatment. This increase can be attributed to the redistribution of silicon-rich microsegregations and the relief of internal stresses. The results of the magnetic characterization showed that it was possible to achieve a maximum relative permeability of 5300, a coercivity of 49 A/m, and power losses of 4 W/kg using a laser power of 70 W and a scanning speed of 0.250 m/s. The authors provided a detailed explanation of these results based on the crystallographic texture in [95]. According to these studies, processing Fe–6.9%Si using SLM with low laser energies results in a fibrous $\langle 001 \rangle$ texture. As the laser power increases, the texture changes to a cubic texture due to epitaxial grain growth and a shift in the shape of the melt pool from a surface depression to a protrusion. As the texture becomes cubic, the easy magnetization axis aligns with the direction of the magnetic field only at approximately 45° to the laser scanning directions. This causes the horizontal plane to lose its isotropic nature, leading to an increase in hysteresis loss, a decrease in remanence (B_r), and a decrease in permeability as the alignment between the magnetic field and the ⟨001⟩ axes becomes less pronounced. Through the other investigation on Fe–6.9%Si samples processed using SLM [97], they examined the effect of annealing parameters on the microstructural and magnetic development of the components. They employed a laser power of 70 W, a laser scanning speed of 500 mm/s, and four different annealing temperatures (400 °C, 700 °C, 900 °C, and 1150 °C) in an argon atmosphere for 1 h, followed by cooling in the furnace. The microstructure of the samples prior to annealing indicated a grain size of 10–30 µm, which remained unchanged after annealing at 700 °C. However, after annealing at 900 °C, several significant aspects emerged in the investigation of these materials: (1) A partially recrystallized microstructure formed due to the release of residual stresses. (2) At this annealing temperature, the dissolution of Si led to the complete disappearance of fusion pool boundaries. Finally, the microstructure at 1150 °C exhibited a non-uniform grain size, with large equiaxed grains (up to 300 µm) and smaller grains (<30 µm). The authors suggested that the recrystallization process was incomplete. Figure 7 shows the XRD analysis conducted in this study, revealing a single BCC phase before and after annealing as well as a superlattice diffraction line related to the D03 phase ordering in the samples annealed at 900 °C and 1150 °C.

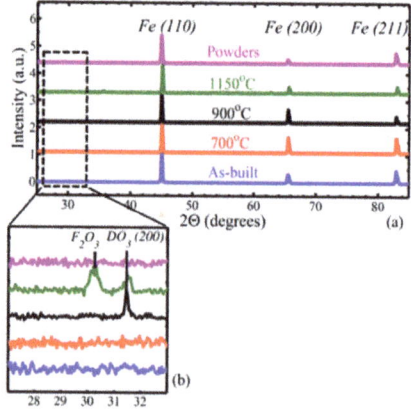

Figure 7. XRD before and after annealing at various temperatures. (**a**) complete XRD spectra; (**b**) amplified extracts of the spectra. Reprinted from Garibaldi et al. [97], with permission from Elsevier®.

The authors concluded that the high cooling rates experienced during SLM partially suppressed diffusion-driven ordering reactions based on the Fe–Si binary alloy phase diagram. The D03/B2 ordered phases in Fe–6.9wt% Si are thermodynamically stable below 830 °C. On the other hand, EBSD analysis indicated that annealing did not significantly alter the ⟨001⟩ texture generated by SLM along the build direction. The analysis of annealing effects and their relationship with magnetic properties through microstructural changes revealed that the maximum relative permeability increased (from 2000 to 24,000) with the increase in annealing temperature. This was associated with stress relief and grain growth effects. The coercivity values were observed at 16 A/m, achieved by increasing the annealing temperature, while the residual M_r decreased as the temperature increased.

This increase also leads to a variation in the maximum flux density (B_{MAX}). Furthermore, the microstructural analysis shows that the average grain size tends to increase slightly with the increase in $E_0{}^*$, while the coercivity (H_C) decreases as the average grain size increases. The evaluation of texture to assess the correlation between the average magnetocrystalline energy and the magnetic properties (H_C, B_{MAX}) demonstrates a weak effect of texture, which can be attributed to the high level of magnetization. This indicates that, in this study, the crystallographic texture has a secondary effect compared to density and grain size.

The works of Stornelli et al. [99–102] in these types of alloys, through microstructural and magnetic studies, have provided results of relevant importance. In [101], they conducted a study to identify the optimal process parameters for L-PBF of Fe–Si steels with Si contents of 3.0% and 6.5%, aiming to evaluate their effect on the density, microstructure, and electromagnetic properties of the printed samples. The authors found that to reduce the internal porosity fraction in the Fe–3.0%Si steel, the suitable process parameters were E = 250 J/m, v = 1 m/s, and P = 250 W (Figure 8b). The reported values for the Fe–6.5% steel were E = 200 J/m, v = 0.835 m/s, and P = 167 W (Figure 9b). In terms of relative density values, both steels exhibited values close to 1. Microstructural results showed that the printed samples maintained a columnar solidification microstructure aligned with the build direction due to the epitaxial growth of the previously consolidated material in the underlying layers. The influence of the process parameters can be observed based on the laser-specific energy (E), where low energy favors the formation of irregular-shaped pores (Figures 8a and 9a). This shape can be associated with voids between partially melted powder particles. On the other hand, high energy levels lead to spherical-shaped pores resulting from the capture of metal vapor and partially ionized gas in the liquid pool, which solidify into a spherical cavity upon rapid cooling. It is important to mention that crack formation was not observed in the Fe–3.0%Si steel. At the same time, an increase in E resulted in more noticeable crack formation in the Fe-6.5% samples (Figure 9c), which could be attributed to the high thermal gradients. As a future work, the authors suggest raising the platform temperature to reduce thermal stresses generated by the process's nature. Magnetic analysis was performed on different cross-sectional geometries of samples from both steels. This analysis revealed that the Fe-6.5% samples exhibited lower effects of eddy currents compared to the Fe–3.0%Si samples, resulting in an increase in magnetization capacity and a more than 50% reduction in power losses. The results obtained in this research demonstrate the ability to produce ferromagnetic cores with high Si content and attractive electromagnetic properties through AM, offering a competitive alternative for industrial applications.

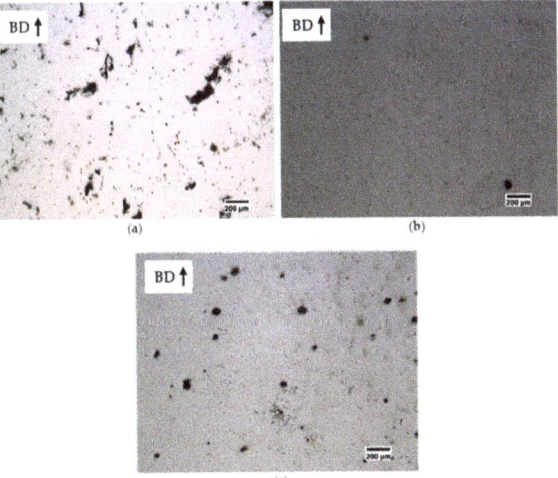

Figure 8. Effect of the specific laser energy (E) on the densification of Fe–3.0%Si steel samples. (**a**) E = 150 J/m, v = 0.5 m/s, P = 75 W, relative density 99.93%, and pores with irregular shape; (**b**) E = 250 J/m, v = 1 m/s, P = 250 W, relative density 99.99%; (**c**) E = 350 J/m, v = 0.5 m/s, P = 175 W, relative density 99.98%, and pores with spherical shape. Reprinted from Stornelli et al. [101].

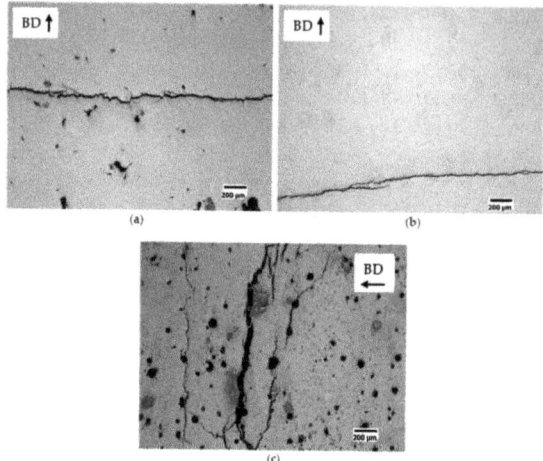

Figure 9. Effect of the specific laser energy (E) on the densification of Fe–6.5%Si samples. (**a**) E = 150 J/m, v = 0.5 m/s, P = 75 W, relative density = 99.93%, and pores with irregular shape; (**b**) E = 200 J/m, v = 0.835 m/s, P = 167 W, relative density = 99.99%; (**c**) E = 350 J/m, v = 0.5 m/s, P = 175 W, relative density = 99.98%, and pores with spherical shape. Reprinted from Stornelli et al. [101].

Seiler et al. [103] investigated the influence of high-frequency modulation through the combination of continuous laser power (high build speed) and pulsed laser power (reduction in input energy) in the L-PBF process for Fe–6.7%Si samples to evaluate its effect on the surface topography of the samples. The results of this research show incomplete fusion of the powder layers (with a thickness of 20 µm) when using laser power below 75 W. However, although increasing laser energy promotes the complete fusion of powder particles in the sample and the attainment of a nearly fully dense structure, temperature gradients and the increase in residual stresses generated by higher laser power contribute to the formation of cracks in the studied material.

An investigation made by Agapovichev et al. [104] through the use of the L-PBF process to produce the samples from Fe–6.5%Si powder shows that even at the best optimization of the process parameters, cracks may appear. In fact, carbon content was detected at the boundaries of the cracks. This suggests that one of the reasons for the crack formation is the presence of Fe_3C around the ordered αFe–Si (B2) + Fe_3Si(D03) phases.

Goll et al. [105] have conducted research on the processing of high-silicon steels using L-PBF for soft magnetic core materials such as Fe–6.7%Si. The objective was to reduce eddy current losses and achieve good magnetic performance. In this work, a laser power of 300 W and a laser scanning speed of 500 mm/s were used to reduce the number of pores, increase the grain size, and prevent crack formation. The research results indicate that a larger grain size can be obtained by using higher laser power due to the decrease in cooling rate. However, special attention must be given to avoiding/reducing crack formation, which becomes more pronounced at higher power levels due to increased residual stresses. Along with the mentioned parameters, maintaining the platform heating temperature at 400 °C allows for high electrical resistivity (0.8 µΩm), maximum permeability (31,000), minimum coercivity (16 A/m), and hysteresis losses (0.7 W/kg at 1 T and 50 Hz). Furthermore, a significant aspect of the study was focused on evaluating the design flexibility offered by AM, specifically in the context of component design. The researchers explored the utilization of topological structures, such as inner slits, to take advantage of AM's capability to create complex geometries. These topological structures were strategically incorporated into the components to interrupt the paths of electrical current, reducing eddy currents and minimizing the adverse effects of Foucault currents. This aspect of the study highlights the inherent advantage of AM, where the freedom in design allows for the incorporation of optimized structures that are challenging to achieve through traditional manufacturing methods.

The results from various investigations presented in this section provide insight into the scope of additive manufacturing in Fe–Si alloys. Different processes such as L-PBF, LBM, SLM, DED, and even innovative methods like MIM have been employed to fabricate components based on this alloy, encompassing various geometries. This showcases AM's capacity to produce components with intricate geometries. The impact of process parameters, such as scanning orientation, laser power, scanning speed, or their correlation through laser energy input, is observed in the resultant microstructure. This microstructure typically consists of columnar and/or equiaxed grains, with the possible presence of discontinuities like pores, cracks, or a lack of fusion. These aspects are linked to gas entrapment, excessive energy input, or insufficient energy for the complete fusing of the manufactured piece. Moreover, microstructural analyses were conducted using characterization techniques such as SEM, EBSD, or XRD to assess the distribution of chemical elements based on process parameters, crystalline orientation, and the presence of ordered B2/D03 phases that could affect the alloy's mechanical behavior. The evaluation of magnetic properties, including parameters like magnetic saturation and coercivity, reveals competitive outcomes when compared with values from Fe–Si alloys processed using traditional techniques. Notably, additive manufacturing demonstrates the capability to process alloys with high silicon content, allowing the creation of various geometries with remarkable magnetic and mechanical properties. Even though there are flaws in some of the results, there are still areas for improvement. For instance, post-processing conditions could be improved or fusion bed manipulation could be modified to achieve defect-free components or reduce their occurrence.

4. Fe–Ni

The iron–nickel alloys (Fe–Ni) can be classified based on their nickel content. Among them, "permalloys" stand out for their excellent soft magnetic properties, including low power losses, high permeability, and a favorable response to magnetic field annealing. Although permalloys offer significant advantages for high-frequency electrical machines, their implementation in electric motors is less frequent due to their higher cost compared to Fe–Si steel. Other important Fe–Ni alloys are the "supermalloys," characterized by their high nickel content of around 78%, 5.0% molybdenum, iron, and a small amount of manganese. They also possess exceptional magnetic properties, including superior permeability, low losses, and excellent response to magnetic field annealing. To optimize their permeability, precise heat treatment is required, involving specific temperatures and controlled cooling rates. Supermalloys may cost more than permalloys and silicon steels, but their exceptional properties make them a promising alternative for high-frequency electrical machines, where their superior performance justifies the investment [144–148].

Additive manufacturing techniques have been employed for the processing of Fe–Ni-based alloys [106–111]. The importance and notable characteristics of these alloys can be verified in various studies. Permalloy samples of Ni–Fe–Mo fabricated using SLM were produced with different process parameters to study the correlation between these parameters, the morphology of generated defects, and the magnetic properties, as can be observed in [112]. Laser power in the range of 160–240 W and scanning speeds of 600–1000 mm/s were utilized, with a hatch space of 0.1 mm and a layer thickness of 0.03 mm. This set of parameters was correlated through an equation to obtain input laser energy densities (E_V), resulting in values ranging from 59.26 to 111.1 J/mm^3. As E_V increases, there is an increase in magnetic saturation induction and magnetic permeability of the material, accompanied by a reduction in coercivity and magnetic losses. Figure 10 depicts the 3D reconstruction using X-ray computed tomography (XCT), an important aspect of this research.

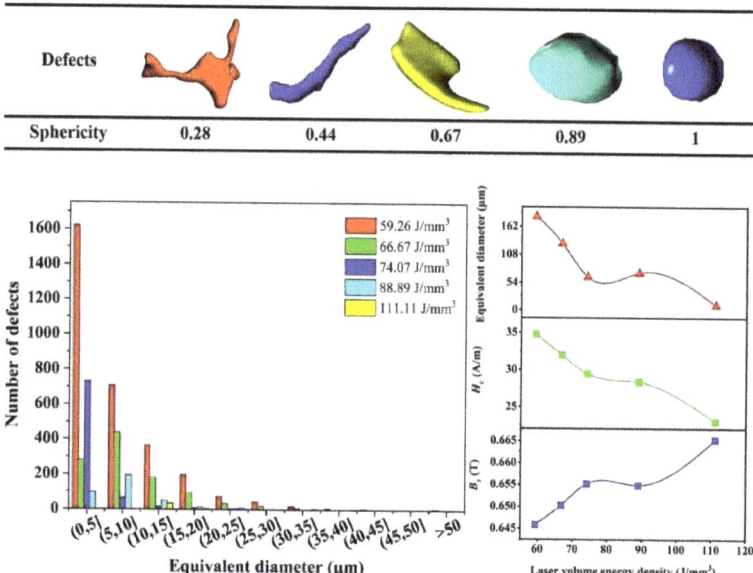

Figure 10. The sphericity of the typical defects, equivalent diameter distribution, and curves of H_c and B_s in the Ni–Fe–Mo samples fabricated at different laser volume energies. Reprinted from Yang et al. [112], with permission from Elsevier®.

The authors illustrate sphericity (S), which is calculated through a formula that relates the volume and surface area of the defect that the sample presents and whose result will oscillate between values of 0 and 1 (where 1 denotes higher sphericity). The outcomes indicate that with a low E_V value (59.26 J/mm³), microdefects displayed irregular shapes due to incomplete fusion. In contrast, with high E_V (111 J/mm³), these microdefects exhibited circular forms attributed to gas-induced porosities. The distribution of the equivalent diameter (D_{eq}) of defects in samples fabricated with varied E_V values is also shown, revealing a significant reduction in defect count as E_V increased. Finally, Figure 10 also presents the correlation between H_C and B_S with the D_{eq} of internal metallurgical defects at different E_V levels, demonstrating a clear decrease in H_C and an increase in B_S as the D_{eq} diminishes. The study by B. Li et al. [113] demonstrated the feasibility of producing modified honeycomb parts of Ni–15%Fe–5%Mo permalloy through L-PBF (Figure 11). The results highlighted the importance of optimizing process parameters and the unique microstructural features obtained, which could have significant implications for magnetic shielding applications. The as-printed permalloy bulk samples achieved a remarkable relative density exceeding 99.5%. The researchers successfully printed thin walls with varying thicknesses and different overhanging angles using L-PBF technology. Microscale or even submicron-scale cellular structures were observed within the grains, demonstrating growth through multiple molten pools. Notably, the modified honeycomb structures demonstrated preferential plastic deformation at the inclined corners. This characteristic allowed for better plastic deformation buffering effects and prevented instantaneous cracking.

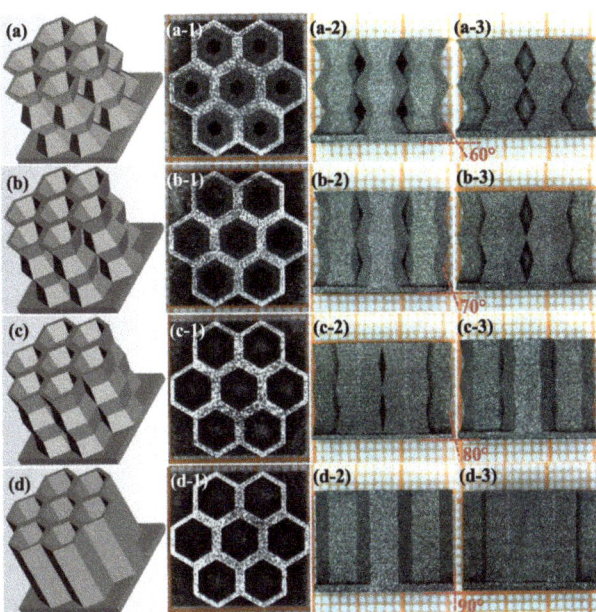

Figure 11. Macro-morphologies of modified honeycomb prototype parts with 0.5 mm walls and overhanging angles of 60° (**a**), 70° (**b**), 80° (**c**), and 90° (**d**) in 3D models (**a–d**), top views (**a-1–d-1**), front views (**a-2–d-2**), and side views (**a-3–d-3**). Reprinted from B. Li et al. [113], with permission from Elsevier®.

The studies conducted by Kim et al. [114,115] extensively demonstrate the effect of additive manufacturing parameters on the properties of these soft magnetic alloys. In [114], they conducted an evaluation of Fe–50%Ni samples manufactured using direct energy deposition based on lasers under various process conditions. These conditions include laser power ranging from 200 to 220 W, a powder feeding rate of 2.25–2.75 g/min, a laser scan speed of 100–1000 mm/s, and hatch spacing of 0.06–0.3 mm. The microstructural analysis revealed that the samples exhibited a single FCC phase with epitaxial columnar grains. Increasing the laser power to 220 W induced a change in the solidification mode, resulting in larger grain sizes. At this laser power level, the sample demonstrated favorable mechanical and magnetic properties, including a hardness value of 167 HV, a maximum tensile strength of 493 MPa, a yield strength of 315 MPa, a saturation magnetization of 151 emu/g, and a minimum coercivity of 3.16 Oe. These properties make the Fe–50%Ni alloy competitive with other soft magnetic alloys in terms of performance. In another study [115], three energy densities (4000, 4800, and 5867 J/g) based on the process parameters from the previous study were used. Subsequently, the samples were heated to 1200 °C (the temperature at which only the FCC phase exists) for 1 h and cooled in a furnace. The results revealed that after annealing, the sample manufactured with an energy density of 4800 J/g maintained the FCC phase structure. In comparison, a secondary ferrite phase (BCC microstructure) and lower microhardness values were observed in the samples manufactured with energy densities of 4000 and 5867 J/g, respectively. This reduction in microhardness is associated with grain growth, partial stress relaxation, recovery of the cellular structure, and especially the formation of a softer BCC phase. It is noteworthy that the sample processed at 4800 J/g and annealed exhibited outstanding values of coercivity, saturation magnetization, and Curie temperature of approximately 1.8 Oe, 170 emu/g, and 530 °C, respectively. This is attributed to the unique microstructure of the FCC phase, lower concentration of grain boundaries, and lower deformation level after annealing.

The crystallographic texture plays a crucial role in the magnetic behavior of materials, particularly in the case of soft magnetic materials processed through additive manufacturing. Post-processing treatments, such as heat treatments and orientation changes during component fabrication, offer a wide range of possibilities to control and optimize the crystallographic texture. This was demonstrated in the study conducted by Haftlang et al. [116] on the crystallographic texture in Fe–50%Ni permalloy samples using the DED process. To evaluate this topic, they employed two rotation strategies (67° and 90°) and processed the sample without rotation (NR). Additionally, after printing, the specimens underwent a heat treatment at 1200 °C for 60 min in a pure Ar atmosphere. The process parameters included laser power ranging from 200 to 220 W, powder feeding rate from 2.2 to 2.75 g/min, and laser scan speed from 100 to 1000 mm/s. Figure 12 presents SEM images and the results of Energy-Dispersive X-ray Spectroscopy (EDS) analysis conducted on the manufactured samples. The outcomes reveal a fluctuation in the Fe percentage contingent on the processing rotation (67° = 45 wt.% Fe, 90° = 55 wt.% Fe, and NR = 55 wt.% Fe), which can be attributed to Fe/Ni micro-segregation at the fusion pool boundaries. Moreover, Figure 12 demonstrates that subjecting the samples to subsequent heat treatment results in a uniform chemical composition, irrespective of the processing rotation. The effect of rotation was also observed in the microstructural evolution of the parts. Without rotation, a heterogeneous microstructure with fine, inclined columnar grains was obtained. In contrast, rotations of 67° and 90° resulted in grain sizes of 70 μm and 93 μm, respectively, with elongated and wavy structures due to the variation in heat flow direction. The results of the texture analysis conducted by the authors yielded important findings. The sample without rotation exhibited predominant Cube {001} ⟨100⟩ and Copper {112} ⟨111⟩ texture components. With a rotation of 67°, the sample shows a pronounced γ_γ-fiber ({111} ⟨uvw⟩) component at $\varphi_2 = 45°$ and a pronounced S_γ ({123} ⟨634⟩) component at $\varphi_2 = 65°$. In contrast, the sample with a 90° rotation exhibited predominant Cube {001} ⟨100⟩ and Brass {011} ⟨211⟩ texture components. It is important to mention that, based on their results, the authors indicated that the heat treatment did not significantly change the grain morphology of the printed samples. However, it did increase the intensities of the main texture components, while the intensities of non-ideal orientations decreased. Furthermore, its effect was manifested in its influence on magnetic properties, as the sample without rotation exhibited values of Hc = 1.7 Oe, Ms = 160 emu/g, and Tc = 540 °C.

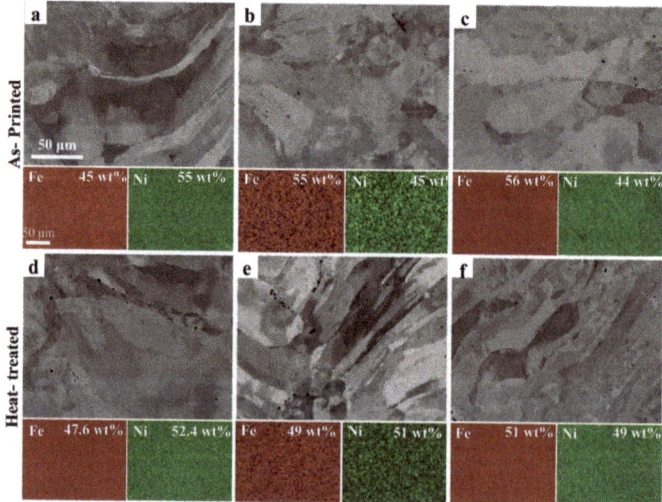

Figure 12. SEM images of DED samples with rotation: (**a**,**d**) 67°, (**b**,**e**) 90°, and (**c**,**f**) no rotation (black spots are metallic oxides trapped inside). Reprinted from Haftlang et al. [116], with permission from Elsevier®.

Mohamed et al. [117] evaluated the combined impact of subsequent heat treatment and control of crystallographic texture in a Ni–Fe–Mo alloy processed via L-PBF. They generated cubic samples that were tilted along the [103] and [104] directions with respect to the building direction, corresponding to 45° and 35°, respectively. This was achieved using a range of parameters that included laser power from 150 to 300 W, scan speed from 800 to 3500 mm/s, and scan spacing from 0.2 to 0.6 mm. Consequently, these parameters contributed to an energy density spectrum ranging from 0.95 to 6.25 J/mm². Four types of heat treatments were performed to evaluate their effect on different batches of parts. The first treatment (HT) involved heating to 1150 °C for 4 h at a rate of 2 °C/min in a hydrogen atmosphere. When it reached 550 °C, it was held for 1 h, then increased to 1150 °C and held for 4 h before being reduced to 200 °C at a rate of 0.8 °C/min. The second treatment, hot isostatic pressing (HIP), was carried out at 1230 °C and 120 MPa for 3 h (with simultaneous application of pressure and temperature) at a rate of 5 °C/min. The third and fourth treatments combined the first two, with only the sequence being varied: (3) HT + HIP and (4) HIP + HT. The results showed that the inclination of the build orientation promoted an improvement in the normal and transverse magnetic shielding characteristics in the main build directions. The samples exhibited recrystallization and grain growth, transitioning from columnar grains extending for 10 μm along the build direction to equiaxial grains after the application of heat treatments. Moreover, HT, HIP, and their combinations resulted in stress relief, a reduction in inclusions, and pore consolidation, directly influencing the magnetic properties of the samples in all build directions. This can be seen in the results of M_S and H_C for the samples subjected to HIP + HT. For example, specimens printed with a 35° orientation exhibited attractive values of M_S = 637 and H_C = 195, compared to values of M_S = 542 and H_C = 205 for 45° and M_S = 561 and H_C = 207 for 0° orientation samples.

As previously mentioned, Fe–Ni alloys stand out as SMM with suitable magnetic characteristics for applications requiring high frequency. The findings from the investigations explored in this section underscore the necessity of post-manufacturing heat treatment for Fe–Ni alloys involving conventional annealing, processes like hot isostatic pressing (HIP), or their combinations. The impact of printing orientation, laser power, and scanning speed emerges prominently in both magnetic behavior and microstructural development. Several research studies have indicated that as energy input increases, magnetic saturation and permeability increase while coercivity decreases. Furthermore, the analysis of crystallographic texture has revealed its correlation with printing orientation. This allows the identification of planes where the alloy's magnetic performance could achieve outstanding values. A vital element in analyzing these materials is defect reconstruction through XCT (X-ray computed tomography), as this enables the visualization and establishment of the relationship between the obtained discontinuities and the utilized parameters.

5. Fe–Co

Iron cobalt alloys (Fe–Co) possess excellent characteristics such as high saturation magnetization, high permeability, low magnetic losses, and a high Curie temperature, making them valuable in applications that prioritize thermal stability. It is important to highlight the highest magnetic saturation value for this alloy (approximately 2.4 T), surpassing other soft magnetic iron alloys. However, the conventional production of Fe–Co iron cores faces challenges such as the high cost of cobalt, the limited workability of the iron–cobalt alloy, and the necessity for heat treatment during manufacturing [55,111]. The mechanical strength, iron losses, and magnetic permeability of Fe–Co alloys can be controlled through various means. One approach is to modify the ratio of cobalt and iron as well as add other alloying elements. Another method involves modifying the temperature cycle during the annealing heat treatment. These operations allow for the obtaining of Fe–Co alloys in different forms, such as "Permendur," "Supermendur," and "Hiperco," with compositions very similar to each other [118–121]. By exploring innovative approaches and techniques, such as AM, it is possible to overcome these issues, open new possibilities for optimizing the manufacturing process, and exploit the full potential of Fe–Co alloys in various applications.

The work presented by Lindroos et al. [122] aims to investigate the effect of process parameters in Laser Powder Bed Fusion (L-PBF) and post-manufacturing heat treatment on Fe–Co binary and ternary alloys (Fe–35%Co, Fe–50%Co, Fe–49%Co–2%V, and Fe–49%Co–2%V–0.1%Nb). The study also focuses on the development of mesostructures to mitigate eddy currents. This paper mentions the use of a laser power of 200 W, a scanning speed of 775 mm/s, and a layer thickness of 80 µm, which resulted in a reduction in porosity to below 0.1% in all cases. The effect of heat treatment varies for each material. For the Fe–Co–V ternary alloys, an optimized two-stage heat treatment (pre-annealing at 700 °C for 2 h followed by annealing at 820 °C for 10 h) generates a structure with uniformly large grain sizes. In contrast, conventional heat treatment (annealing at 820 °C for 4 h) leads to bimodal grain growth. It is noteworthy that the size of the printed samples plays a crucial role in the final microstructural and magnetic characteristics after annealing, as it affects the differences in the thermal history generated by the L-PBF process. For example, in large samples (20 × 20 × 65 mm), the microstructure consists of approximately 90% large grains, compared to 65% large grains in smaller samples (10 × 10 × 35 mm). Additionally, the specimen with a higher percentage of large grains achieves high magnetic induction, while the sample with fewer large grains exhibits low permeability. However, varying the annealing holding time (24 h) shows that it is possible to obtain almost identical microstructures in both sample sizes, leading to good magnetic performance. The addition of Nb to the Fe–Co–V alloy helps reduce grain growth. The addition of 0.1% Nb to the L-PBF printed samples results in a 50% smaller grain size compared to samples processed without this element. On the other hand, in the Fe–35%Co and Fe–50%Co binary alloys, the effect of different heat treatments is less significant, and the grain size remains almost unchanged when using either conventional or optimized heat treatment. The manuscript discusses the study of two mesoscale structure designs to limit eddy currents: (1) Slotted structures and (2) Topological optimization (TO) in defining the optimal cross-sectional shape for magnetic flux. The effect of these designs is exemplified by the coercivity values of Fe–Co–V alloys, which are 118 A/m for slotted structures and 74 A/m for TO, while the permeability values are 5148 and 3489, respectively. The recent study by Riipen et al. [123] investigated the influence of heat treatment and Nb addition on the chemical composition of a Fe–Co–V alloy processed by L-PBF. The authors used a laser power of 200 W, a scan speed of 775 mm/s, a hatch spacing of 80 µm, and a build platform temperature of 200 °C. The annealing heat treatment consisted of heating at 800 °C and 850 °C for 1 h, 10 h, and 24 h, followed by furnace cooling (cooling rate of 100 °C/h). The results of the microstructural evaluation show an irregular, small grain structure in the untreated samples. In contrast, the annealed samples show a bimodal grain structure even after 24 h, but it is noticeable that the proportion of small grains decreases with increasing annealing temperature and time. Furthermore, the influence of Nb can be observed in grain growth. In the sample with Nb addition, annealed at 800 °C and 850 °C after 24 h, the average grain size was smaller (45 µm) than in the sample without Nb (210 µm) at the same annealing temperature, but whose growth stopped after 10 h. Based on these results and investigations, the authors indicate the ability of Nb to inhibit grain growth. This effect is achieved by binding the particles at the grain boundaries through the precipitation of nano-sized particles in this zone. This precipitation is induced by a slight difference in chemical composition. The evaluation of the magnetic properties shows that heat treatment at 800 °C for 24 h gives optimum values of relative permeability (15,000) and coercivity (20 A/m) for the alloy with Nb addition. In contrast, for the alloy without Nb, these values are obtained at a higher annealing temperature (850 °C for 24 h).

The fabrication of Fe–49%Co–2%V alloys was performed by Nartu et al. [124] using a laser-engineered net shaping (LENS) system. Figure 13 presents the distinct microhardness values achieved in the samples of this study, along with the Inverse Pole Figure (IPF) maps and their relationship with grain size. It is shown that an increase in laser power at both scanning speeds leads to a reduction in microhardness. Through the EBSD results derived from the IPF maps, the authors observe that the increase in grain size is responsible for the microhardness behavior. As the laser power increases from 200 W (189 J/mm^2) to 400 W (94 J/mm^2) at both scanning speeds, the grain size varies from 52 to 74 µm. The samples underwent two types of heat treatment: (1) A single-step annealing at 950 °C for 30 min followed by water quenching; and (2) a two-step treatment consisting of a first step at 950 °C for 30 min followed by rapid water quenching, followed by a second step at 500 °C for 50 h followed by rapid water quenching.

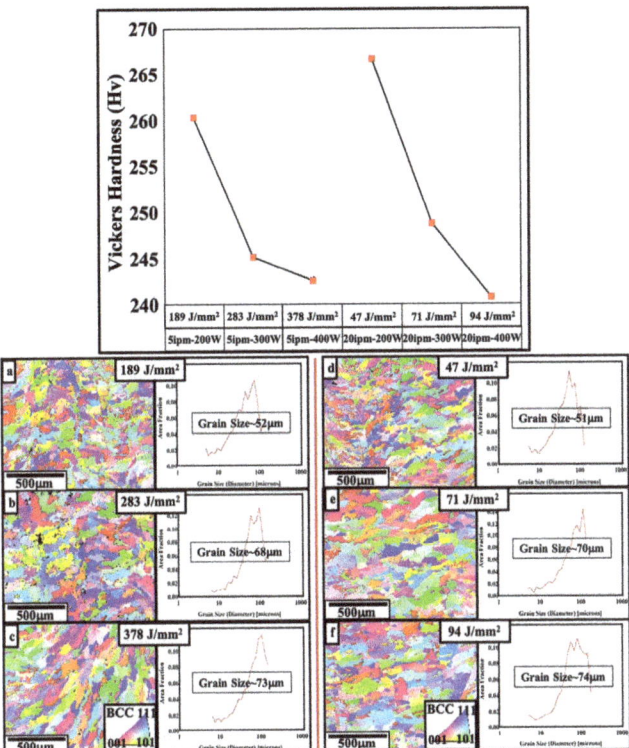

Figure 13. Vickers microhardness measurements for all six as-processed samples in Fe–Co–2V alloy, Inverse Pole Figure (IPF), and grain size distribution maps for (**a**) 189 J/mm^2, (**b**) 283 J/mm^2, (**c**) 378 J/mm^2, (**d**) 47 J/mm^2, (**e**) 71 J/mm^2, and (**f**) 94 J/mm^2. Reprinted from Nartu et al. [124], with permission from Elsevier®.

It was found that the single-step heat treatment promotes a refined grain size due to grain recovery and recrystallization of the BCC grains, resulting in increased H_C values. The reduction in H_C observed in the two-step treatment led the researchers to perform a TEM study. In this study, they mention that in the two-step treatment condition, substantially larger ordered B2 domains exist, separated by antiphase domain (APD) boundaries. Conversely, in the single-step treatment condition, the samples exhibit weakly ordered nanoscale B2 domains homogeneously distributed within a disordered BCC matrix.

The influence of AM process parameters on the microstructure and atomic ordering of Fe–Co alloys is extensively addressed in the works of Kustas et al. [118,125]. The authors utilized the LENS technique to process Fe–Co–1.5%V alloy, known for its poor workability. One of the investigations [118] provides a detailed characterization of the correlation between the manufacturing process, microstructure, and magnetic properties of the alloy. In the initial study, they explored a range of laser powers from 150 to 450 W, scanning speeds from 2.5 to 10 mm/s, and inter-layer time intervals (t_l) from 0.4 to 11 s. The microstructural characterization of the as-built samples revealed a fine, equiaxed grain morphology contrary to the expected columnar grain structure. The authors suggest several theories for this microstructural behavior: (1) The repetitive layer-by-layer heating process reduces thermal gradients in subsequent layers, promoting equiaxed solidification instead of columnar. (2) The refinement could be attributed to the inherent phase transformations of Fe–Co alloys during repetitive heating and cooling. The rapid cooling along the γ/α phase boundary prompts the nucleation of a large number of α-BCC grains from the parent γ-FCC solidification structure. This process produces finely equiaxed grains. (3) The presence of oxide particles in the microstructures, which would increase the number of heterogeneous nucleation sites during solidification. However, it is mentioned that this microstructure, based on its grain size and texture, could promote higher creep resistance (via Hall–Petch strengthening), improving the machinability of these alloys and enhancing their magnetic performance. The microstructure of the specimens after annealing at 840 °C for 2 h under vacuum consists of a bimodal, equiaxed, and coarse-grained morphology. The results of the atomic ordering analysis indicate that when a t_l of 0.4 s is employed, the relative ordering value is 65 ± 3.2%, while with a t_l of 11 s, the ordering value is 43 ± 3.8%. The latter t_l value corresponds to faster cooling rates and higher thermal gradients. Therefore, at shorter layer intervals (0.4 s), the high value of the atomic ordering parameter is due to increased heat retention and slower cooling rates during the process. A more detailed analysis of this behavior can be found in [125], where they investigated the formation of B2-ordered phases in Fe–Co–1.5%V alloys, continuing with the use of the LENS process. The aim of this work was to suppress atomic ordering to tailor the microstructure for electromagnetic applications. They employed process parameters that allowed for cooling rates in the range of 10^3–10^4 °C/s. This is because cooling rates are typically suitable for suppressing the disorder-to-order phase transformation in conventionally processed Fe–Co alloys. Through EBSD, it is clear that the processed samples have a homogeneous and fine-grained structure with a predominantly equiaxed morphology. The analysis of ordering as a function of cooling rate indicates that the ordering range decreases from 0.72 ± 0.028 to 0.49 ± 0.029 as the cooling rate increases from 1875 ± 282 °C/s to 5764 ± 628 °C/s. However, more detailed analysis results show that a fraction of the ordered B2 phase is still present under conditions where a completely disordered BCC phase is expected. The authors suggest that the reheating of the material in previously solidified layers at high temperatures and the high atomic mobility promote B2 ordering.

The results of these studies reveal that, through the analysis and optimization of AM process parameters, it is feasible to discern their impact on the microstructure of Fe–Co alloys and their influence on magnetic properties. Moreover, alterations in chemical composition and the inclusion of elements like niobium provide insights into controlling atomic structure. This holds potential value not only for these specific systems of soft magnetic alloys but also for other analogous alloy systems.

6. Soft Magnetic Composites (SMCs)

The significant advancements in materials science over the past decades, particularly in the control and manipulation of chemical composition, the feasibility of operating at nanoscales, and progress in processing techniques, have led to the emergence of soft magnetic composites (SMCs) [34,35]. These alloys consist of micrometer- or even nanometer-sized particles of iron, Fe–Si, and Fe–Co, among others, coated with an insulating material and are typically consolidated through high-pressure processes. These materials are considered essential in the fabrication of electrical cores, sensors, and other energy conversion devices. This significance is due to their useful properties, such as low core loss, high saturation magnetization, and high electrical resistivity. These properties are directly related to their unique construction. The insulating coating on the powder particles allows for reduced losses from induced currents by increasing electrical resistance. Moreover, the compressed particles maintain tiny air gaps, resulting in increased resistivity [149–151].

Additional benefits of SMCs arise from their processing using techniques such as additive manufacturing. The flexibility to shape complex geometries greatly expands their potential applications in the mentioned components [50,111]. A brief overview of various works on AM in these materials is described below.

Borkar et al. [126] processed a Fe–Si–B–Nb–Cu alloy using laser additive manufacturing, varying the proportion of Si and B. Microstructural analysis revealed that when employing low contents of these elements, the microstructure consisted of dendritic α-Fe_3Si grains. The presence of B and Nb in the interdendritic regions promoted the formation of Fe_3B grains. As the percentages of B and Si increased, Fe_3B grains were not observed. However, a eutectic phase enriched in Fe and Nb was formed alongside α-Fe_3Si grains.

Gheiratmand et al. [52] used the Spark Plasma Sintering (SPS) process to develop the FINEMET® alloy samples ($Fe_{73.5}Si_{13.5}B_9Nb_3Cu_1$). These samples are characterized by the random distribution of FeSi nanograins in an amorphous matrix, which enables their attractive soft magnetic properties such as high magnetization, low coercivity, and high Curie temperature. Based on microstructural analysis, the authors demonstrated the presence of the FeSi phase in the amorphous matrix with a grain size of 9 nm and an 84 wt.%. The evaluation of magnetic properties indicated magnetization values of 122.29 emu/g and coercivity of 123 A/m.

Studies conducted by Conteri et al. [127] involving the processing of a $Fe_{73.5}Si_{13.5}B_9Nb_3Cu_1$ alloy using the LENS process revealed its semi-hard magnetic properties with good saturation magnetization and coercivity. The microstructure consisted of dendritic α-Fe_3Si grains, with B and Nb in the interdendritic regions forming Fe_3B grains, along with the presence of $Fe_{23}B_6$ grains at the α-Fe_3Si/Fe_3B interface. The results indicated that the scan speed substantially influenced grain size control.

The continued advancement of these materials and additive manufacturing will continue to contribute to increased efficiency in the components where they are applied. Additionally, the use of these materials based on the utilization of nanoparticles would further enhance their magnetic performance. By reducing the particle size, the losses due to eddy currents can be minimized to the point of being negligible. This highlights the potential for achieving optimal magnetic properties and improved overall performance in various applications [128–131].

7. Soft Ferrites

Soft magnetic ferrites, developed in the 1940s by J.L. Noek [152], are predominantly ceramic materials composed of Fe_3O_4 and other elements such as Ni-Zn, Mn-Zn, or Co-Zn. These materials exhibit useful magnetic properties suitable for various high-frequency applications (e.g., transformer cores, inductive toroidal cores in antennae, and different microwave components) due to their high resistivity and low magnetic saturation. Despite the benefits, their processability can be challenging due to their fragility, difficulty in achieving complex shapes, dimensional tolerances, and limitations in size and density [132,133]. However, advancements in additive manufacturing have opened possibilities for their easy processing and subsequent use in electric machines. This can be observed in the studies conducted by Liu et al. [134,135], who developed a UV-curable Ni–Zn paste and fabricated components using a direct-extrusion 3D printer, exhibiting, in some situations, magnetic properties (relative permeability ranging from 63 to 103 and resonance frequency exceeding 30 MHz) similar to those of commercial Ni–Zn ferrite cores. Andrews et al. [136] developed the toroid with Ni–Zn–Cu/Fe-oxide ceramic powder using L-PBF and achieved high permeability values in the fabricated samples. Additionally, the study determined that the proper selection of parameters (process and powder fabrication), as well as heat treatment and weight percentage of ferrite loading, are crucial for reducing porosity and improving magnetic properties.

8. Novel Materials: High-Entropy Alloys (HEAs)

As seen so far, soft magnetic materials are essentially composed of two chemical components (e.g., Fe–Si, Fe–Co, or Fe–Ni), which, based on their microstructure, allow them to obtain the previously described magnetic properties for use in electrical devices. However, the constant advancement in materials science and engineering has led to the development of materials based on a greater number of chemical elements, such as high-entropy alloys (HEAs) [153,154]. These alloys exhibit unique characteristics that distinguish them from traditional alloys. They possess a unique crystal structure, which can be body-centered cubic (BCC), face-centered cubic (FCC), hexagonal close-packed (HCP), or combinations thereof. By varying their composition, grain size, and precipitate density, among others, they can exhibit remarkable properties such as high corrosion and wear resistance, high strength, and ductility even at elevated temperatures [155–158]. Furthermore, these alloys stand out for their ability to vary their chemical composition and adapt to processing conditions, positioning them as magnetic materials with wide application potential. These conditions allow HEAs to exhibit characteristics of both hard and soft magnetic materials, which gives them a promising future in the field of manufacturing motor cores, transformers, power converters, sensors, information system components, etc. [29,159–161]. In this regard, the addition of elements with a high magnetic moment, such as Fe, Si, Ni, and/or Co, and the resulting changes they can induce in the distortion of their crystal lattice, improve electrical resistivity and reduce losses due to eddy currents. This renders HEAs candidates for novel soft magnetic materials [161–163]. Conversely, the addition of elements such as Al, Ga, Nd, Ti, etc. to these alloys promotes a hard magnetic behavior [159,164,165]. In this section, given the research theme, we will focus solely on the processing of HEAs as materials with soft magnetic properties.

Radhakrishnan et al. [166] investigated the influence of varying Cr content on the soft magnetic properties of $CoCr_XFeNi$ alloy, where x indicates the atomic percentage of Cr (0–24%), fabricated using laser-directed energy deposition (L-DED). The results of the evaluation of magnetic properties indicate that saturation magnetization and Curie temperature decrease as the percentage of Cr increases. At the same time, coercivity showed values in the range of 1.3–2 Oe.

Zhou et al. [167] processed a $FeCoCrNiC_{0.05}$ HEA using the SLM process with a laser power of 200–400 W and different laser scanning speeds (800, 1200, 1600, and 2000 mm/s). The evaluation of density indicates that at 800 mm/s and 200 W, the relative density has a value of 99%, while at high scanning speeds and low power, the values of relative density oscillate between 87 and 92%, which increases as the power increases. The effect of these parameters is also observed in the grain size. When the laser power increases from 250 W to 400 W, the grain size increases from 42 μm to 50 μm. On the other hand, the addition of carbon to the alloy indicates that the samples containing C and higher density had a yield strength of 650 MPa and a total elongation of 13.5%. Similar studies were conducted by Park et al. [168], who investigated the mechanical behavior of a CoCrFeMnNi alloy with the addition of 1% carbon using the SLM process. The results demonstrate that SLM enables the attainment of superior tensile properties in this alloy. This is attributed to multiple reinforcing mechanisms such as solid solution strengthening, grain refinement, increased dislocation density, and nanoprecipitate formation. These research efforts highlight the dependence of the microstructure and mechanical properties of HEAs on SLM process conditions, providing valuable insights for optimizing the additive manufacturing of HEAs.

Nartu et al. [169] conducted a comparative study on the mechanical and magnetic properties of CoFeNi alloys processed via laser-engineered net shaping (LENS) and conventionally processed (90% cold-rolled and 90% cold-rolled plus annealed) cast alloys. The LENS-processed sample exhibited a significantly larger grain size (99 μm) compared to the cold-rolled and annealed cast sample (16 μm). Furthermore, the LENS-manufactured sample demonstrated optimal saturation magnetization and coercivity, although its permeability was lower than that of its cast counterpart. On the other hand, severe plastic deformation from the 90% cold rolling of the cast alloy did not significantly affect the Ms and Hc values.

In another study, Yang et al. [170] utilized laser additive manufacturing (LAM) to process an alloy $(Fe_{60}Co_{35}Ni_5)_{78}Si_6B_{12}Cu_1Mo_3$ using different laser powers (2000, 2500, and 3000 W). Microstructural analysis revealed that the samples consisted of columnar and equiaxed dendrites of the α-Fe–(Co, Ni, Si) phase, along with the presence of secondary phases (Fe_3B) due to increased diffusion during multiple reheating cycles, resulting in enhanced magnetocrystalline anisotropy and eventual magnetic hardening. The higher saturation magnetization (179–199 emu/g) was observed with increasing laser power, combined with the addition of Cr and increased Fe–Si phase content.

9. Summary and Outlook

In this paper, recent advances in the processing of soft magnetic materials through additive manufacturing processes are reviewed. The diversity of materials available for each industry's demand and their constant development through variations in chemical composition, manufacturing, and post-processing, among others, are noteworthy. The impossibility of having a single material that encompasses all the requirements of an electrical machine compels scientists and engineers to design materials that are adaptable to industrial needs and/or select from those available, justifying their cost based on performance.

Despite existing since the 1940s, soft ferrites are continuously being adapted to current industry demands through new processing approaches, as they are an excellent choice for applications where high power densities are not required. Soft magnetic composites (SMCs) and high-entropy alloys (HEAs) with soft magnetic properties are shown to be the most promising materials in this field for use in electrical machines. This is due to their remarkable properties, such as good deformability, high-temperature mechanical strength, fatigue, and corrosion resistance, which are a result of the combination of chemical elements, grain size modification, precipitate formation, and grain orientation, among others. Additionally, their unique nanostructure and the ability to work with extremely thin laminations and geometric variations have allowed for low energy losses due to eddy currents while maintaining suitable M_s values. However, Fe–Si, Fe–Ni, and Fe–Co alloys remain the most widely used materials in these applications, as they combine notable magnetic properties such as high resistivity, permeability, magnetic saturation, and low energy loss. Their microstructural characteristics, mechanical properties, and technological impact have been extensively studied, leading to continuous innovation and development. Among these materials, Fe–Si alloys are the most commercially affordable and widely available, making them highly suitable options for electrical machines.

The constant evolution of these materials is a reality that, based on improved control over the microstructure and chemical composition of existing and emerging materials, has allowed for increased magnetic properties. This review shows that optimizing the structure within magnetic materials will require greater use of advanced characterization tools, such as electron microscopy, X ray characterization, and magnetic force microscopy, among others. In addition to material optimization, ongoing activities in research groups worldwide indicate that additive manufacturing is a viable manufacturing process for developing components based on these materials while maintaining or enhancing their magnetic properties and reducing processing costs. Many soft magnetic materials, such as Fe–Ni, Fe–Co, Fe–Si, and SMCs, have been successfully fabricated using additive manufacturing, particularly with processes like L-PBF, SLM, DED, and LENS. The effect of process parameters, such as laser speed, laser power, printing direction, and subsequent heat treatment, has been discussed as determining factors for the microstructure, mechanical properties, and magnetic properties of the final product.

Furthermore, additive manufacturing has enabled the implementation of novel and radical design concepts for magnetic components, allowing for the spread of SMMs in applications where their use was previously limited. The comprehensive review in this work provides interesting data that could contribute to the future of AM technology. This review illustrated the need for the development of other AM components in the future, such as thermal sensors capable of providing on-site diagnostics during printing, temperature-controlled construction platforms to manage thermal gradients during printing, and high-temperature ovens for automated post-printing heat treatments.

Finally, studying and presenting the various advancements and discoveries in soft magnetic materials and their application in additive manufacturing to the scientific community allowed for contrasting research results and identifying different practices, techniques, and promising materials for use in magnetic components, electric motors, transformers, sensors, etc. Moreover, this work can serve as an analytical point to identify knowledge gaps and areas for future research in this field, guiding researchers toward areas that require more effort and where significant advancements can be achieved. It is now the responsibility of the community of scientists, engineers, and other professionals to enhance these materials and their processing techniques through scientific knowledge to meet the requirements of the next generation of electric machines.

Author Contributions: Conceptualization, B.R.R.-V., G.S., and M.R.R.; methodology, B.R.R.-V. and G.S.; formal analysis, B.R.R.-V.; writing—original draft preparation, B.R.R.-V.; writing—review and editing, G.S., A.D.S., M.R.R., and A.F.M.P.; supervision, P.F. and A.D.S. All authors have read and agreed to the published version of the manuscript.

Funding: The research was funded by Regione Umbria in the framework of "Piano Sviluppo e Coesione FSC ex DGR n. 251/2021—Avviso ricerca 2020—Determinazione Dirigenziale n. 13637 del 27/12/2022—codice CUP I39J20002770008."

Conflicts of Interest: The authors declare no conflict of interest.

Abbreviations

AM	Additive Manufacturing
APD	Antiphase Domain
B	Flux Density
BCC	Body-centered Cubic
BJP	Binder Jetting Technology
B_r	Remanence
CAD	Computed-Aided Design
DED	Directed Energy Deposition
E	Specific Energy/Specific Laser Energy
EDS	Energy-dispersive X-ray Spectroscopy
EBSD	Electron Backscatter Diffraction
E_V	Input laser energy densities
FCC	Face Centered Cubic
H_C	Coercivity
HCP	Hexagonal Close-packed
HEAs	High-entropy Alloys
HIPing	Hot Isostatic Pressing
HT	Heat Treatment
HV	Hardness Vickers
IPF	Inverse Pole Figures
k	Thermal Conductivity
LBM	Laser Beam Melting
L-DED	Laser-directed Energy Deposition
LENS	Laser-engineered Net Shaping
L-PBF	Laser Powder Bed Fusion
MIM	Metal Injection Molding
M_r	Relative Permeability

MRI	Magnetic Resonance Imaging
Ms or Bs	Saturation Magnetization
r	Electrical Resistivity
SEM	Scanning Electron Microscopy
SLM	Selective Laser Melting
SMCs	Soft Magnetic Composites
SMMs	Soft Magnetic Materials
SPS	Spark Plasma Sintering
TEM	Transmission electron microscopy
TO	Topological Optimization
XCT	X-ray computed tomography
X_m	Magnetic Susceptibility
XRD	X-ray Diffraction

References

1. Stornelli, G.; Gaggiotti, M.; Mancini, S.; Napoli, G.; Rocchi, C.; Tirasso, C.; Di Schino, A. Recrystallization and Grain Growth of AISI 904L Super-Austenitic Stainless Steel: A Multivariate Regression Approach. *Metals* **2022**, *12*, 200. [CrossRef]
2. Stornelli, G.; Di Schino, A.; Mancini, S.; Montanari, R.; Testani, C.; Varone, A. Grain Refinement and Improved Mechanical Properties of EUROFER97 by Thermo-Mechanical Treatments. *Appl. Sci.* **2021**, *11*, 10598. [CrossRef]
3. Di Schino, A.; Testani, C. Corrosion Behavior and Mechanical Properties of AISI 316 Stainless Steel Clad Q235 Plate. *Metals* **2020**, *10*, 552. [CrossRef]
4. Di Schino, A.; Gaggiotti, M.; Testani, C. Heat Treatment Effect on Microstructure Evolution in a 7% Cr Steel for Forging. *Metals* **2020**, *10*, 808. [CrossRef]
5. Stornelli, G.; Gaggia, D.; Rallini, M.; Di Schino, A. Heat Treatment Effect on Maraging Steel Manufactured by Laser Powder Bed Fusion Technology: Microstructure and Mechanical Properties. *Acta Metall. Slovaca* **2021**, *27*, 122–126. [CrossRef]
6. Miranda-Pérez, A.F.; Rodríguez-Vargas, B.R.; Calliari, I.; Pezzato, L. Corrosion Resistance of GMAW Duplex Stainless Steels Welds. *Materials* **2023**, *16*, 1847. [CrossRef]
7. Rodriguez, B.R.; Miranda, A.; Gonzalez, D.; Praga, R.; Hurtado, E. Maintenance of the Austenite/Ferrite Ratio Balance in GTAW DSS Joints Through Process Parameters Optimization. *Materials* **2020**, *13*, 780. [CrossRef]
8. Wang, J.; Zhang, Y.; Aghda, N.H.; Pillai, A.R.; Thakkar, R.; Nokhodchi, A.; Maniruzzaman, M. Emerging 3D Printing Technologies for Drug Delivery Devices: Current Status and Future Perspective. *Adv. Drug Deliv. Rev.* **2021**, *174*, 294–316. [CrossRef]
9. Yin, H.; Qu, M.; Zhang, H.; Lim, Y. 3D Printing and Buildings: A Technology Review and Future Outlook. *Technol. Archit. Des.* **2018**, *2*, 94–111. [CrossRef]
10. Lee, J.-Y.; An, J.; Chua, C.K. Fundamentals and Applications of 3D Printing for Novel Materials. *Appl. Mater. Today* **2017**, *7*, 120–133. [CrossRef]
11. Gao, W.; Zhang, Y.; Ramanujan, D.; Ramani, K.; Chen, Y.; Williams, C.B.; Wang, C.C.L.; Shin, Y.C.; Zhang, S.; Zavattieri, P.D. The Status, Challenges, and Future of Additive Manufacturing in Engineering. *Comput. Aided Des.* **2015**, *69*, 65–89. [CrossRef]
12. González-Henríquez, C.M.; Sarabia-Vallejos, M.A.; Sanz-Horta, R.; Rodriguez-Hernandez, J. Additive Manufacturing of Polymers: 3D and 4D Printing, Methodologies, Type of Polymeric Materials, and Applications. In *Macromolecular Engineering*; John Wiley & Sons, Ltd: Hoboken, NJ, USA, 2022; pp. 1–65. ISBN 9783527815562.
13. Mahmood, A.; Akram, T.; Chen, H.; Chen, S. On the Evolution of Additive Manufacturing (3D/4D Printing) Technologies: Materials, Applications, and Challenges. *Polymers* **2022**, *14*, 4698. [CrossRef]
14. Jung, S.; Kara, L.B.; Nie, Z.; Simpson, T.W.; Whitefoot, K.S. Is Additive Manufacturing an Environmentally and Economically Alternative for Mass Production? *Environ. Sci. Technol.* **2023**, *57*, 6373–6386. [CrossRef]
15. Attaran, M. The Rise of 3-D Printing: The Advantages of Additive Manufacturing over Traditional Manufacturing. *Bus. Horiz.* **2017**, *60*, 677–688. [CrossRef]
16. Javaid, M.; Haleem, A.; Singh, R.P.; Suman, R.; Rab, S. Role of Additive Manufacturing Applications towards Environmental Sustainability. *Adv. Ind. Eng. Polym. Res.* **2021**, *4*, 312–322. [CrossRef]
17. Ford, S.; Despeisse, M. Additive Manufacturing and Sustainability: An Exploratory Study of the Advantages and Challenges. *J. Clean. Prod.* **2016**, *137*, 1573–1587. [CrossRef]
18. Prashar, G.; Vasudev, H.; Bhuddhi, D. Additive Manufacturing: Expanding 3D Printing Horizon in Industry 4.0. *Int. J. Interact. Des. Manuf.* **2022**, 1–15. [CrossRef]
19. Vafadar, A.; Guzzomi, F.; Rassau, A.; Hayward, K. Advances in Metal Additive Manufacturing: A Review of Common Processes, Industrial Applications, and Current Challenges. *Appl. Sci.* **2021**, *11*, 1213. [CrossRef]
20. Mehrpouya, M.; Dehghanghadikolaei, A.; Fotovvati, B.; Vosooghnia, A.; Emamian, S.S.; Gisario, A. The Potential of Additive Manufacturing in the Smart Factory Industrial 4.0: A Review. *Appl. Sci.* **2019**, *9*, 3865. [CrossRef]
21. May, G.; Psarommatis, F. Maximizing Energy Efficiency in Additive Manufacturing: A Review and Framework for Future Research. *Energies* **2023**, *16*, 4179. [CrossRef]

22. Altıparmak, S.C.; Xiao, B. A Market Assessment of Additive Manufacturing Potential for the Aerospace Industry. *J. Manuf. Process* **2021**, *68*, 728–738. [CrossRef]
23. Naseer, M.U.; Kallaste, A.; Asad, B.; Vaimann, T.; Rassõlkin, A. A Review on Additive Manufacturing Possibilities for Electrical Machines. *Energies* **2021**, *14*, 1940. [CrossRef]
24. Rejeski, D.; Zhao, F.; Huang, Y. Research Needs and Recommendations on Environmental Implications of Additive Manufacturing. *Addit. Manuf.* **2018**, *19*, 21–28. [CrossRef]
25. du Plessis, A.; Razavi, S.M.J.; Benedetti, M.; Murchio, S.; Leary, M.; Watson, M.; Bhate, D.; Berto, F. Properties and Applications of Additively Manufactured Metallic Cellular Materials: A Review. *Prog. Mater. Sci.* **2022**, *125*, 100918. [CrossRef]
26. Buchanan, C.; Gardner, L. Metal 3D Printing in Construction: A Review of Methods, Research, Applications, Opportunities and Challenges. *Eng. Struct.* **2019**, *180*, 332–348. [CrossRef]
27. Najmon, J.C.; Raeisi, S.; Tovar, A. 2—Review of Additive Manufacturing Technologies and Applications in the Aerospace Industry. In *Additive Manufacturing for the Aerospace Industry*; Froes, F., Boyer, R., Eds.; Elsevier: Amsterdam, The Netherlands, 2019; pp. 7–31. ISBN 978-0-12-814062-8.
28. Kustas, A.B.; Susan, D.F.; Monson, T. Emerging Opportunities in Manufacturing Bulk Soft-Magnetic Alloys for Energy Applications: A Review. *JOM* **2022**, *74*, 1306–1328. [CrossRef]
29. Gutfleisch, O.; Willard, M.A.; Brück, E.; Chen, C.H.; Sankar, S.G.; Liu, J.P. Magnetic Materials and Devices for the 21st Century: Stronger, Lighter, and More Energy Efficient. *Adv. Mater.* **2011**, *23*, 821–842. [CrossRef]
30. Żukowska, M.; Rad, M.A.; Górski, F. Additive Manufacturing of 3D Anatomical Models—Review of Processes, Materials and Applications. *Materials* **2023**, *16*, 880. [CrossRef]
31. Lehmhus, D.; Aumund-Kopp, C.; Petzoldt, F.; Godlinski, D.; Haberkorn, A.; Zöllmer, V.; Busse, M. Customized Smartness: A Survey on Links between Additive Manufacturing and Sensor Integration. *Procedia Technol.* **2016**, *26*, 284–301. [CrossRef]
32. Del Rosario, M.; Heil, H.S.; Mendes, A.; Saggiomo, V.; Henriques, R. The Field Guide to 3D Printing in Optical Microscopy for Life Sciences. *Adv. Biol.* **2022**, *6*, 2100994. [CrossRef] [PubMed]
33. Liu, C.; Tian, W.; Kan, C. When AI Meets Additive Manufacturing: Challenges and Emerging Opportunities for Human-Centered Products Development. *J. Manuf. Syst.* **2022**, *64*, 648–656. [CrossRef]
34. Silveyra, J.M.; Ferrara, E.; Huber, D.L.; Monson, T.C. Soft Magnetic Materials for a Sustainable and Electrified World. *Science* **2018**, *362*, eaao0195. [CrossRef]
35. Shokrollahi, H.; Janghorban, K. Soft Magnetic Composite Materials (SMCs). *J. Mater. Process. Technol.* **2007**, *189*, 1–12. [CrossRef]
36. Babu, S.S. Thermo-Mechanical Processing. In *Encyclopedia of Materials: Metals and Alloys*; Caballero, F.G., Ed.; Elsevier: Oxford, UK, 2022; pp. 27–38. ISBN 978-0-12-819733-2.
37. 6—Production and Casting of Aerospace Metals. In *Introduction to Aerospace Materials*; Mouritz, A.P. (Ed.) Woodhead Publishing: Sawston, UK, 2012; pp. 128–153. ISBN 978-1-85573-946-8.
38. Litovchenko, I.; Akkuzin, S.; Polekhina, N.; Almaeva, K.; Moskvichev, E. Structural Transformations and Mechanical Properties of Metastable Austenitic Steel under High Temperature Thermomechanical Treatment. *Metals* **2021**, *11*, 645. [CrossRef]
39. Lemke, J.N.; Simonelli, M.; Garibaldi, M.; Ashcroft, I.; Hague, R.; Vedani, M.; Wildman, R.; Tuck, C. Calorimetric Study and Microstructure Analysis of the Order-Disorder Phase Transformation in Silicon Steel Built by SLM. *J. Alloys Compd.* **2017**, *722*, 293–301. [CrossRef]
40. Ming, K.; Jiang, S.; Niu, X.; Li, B.; Bi, X.; Zheng, S. High-Temperature Strength-Coercivity Balance in a FeCo-Based Soft Magnetic Alloy via Magnetic Nanoprecipitates. *J. Mater. Sci. Technol.* **2021**, *81*, 36–42. [CrossRef]
41. Kuznetsov, D.D.; Kuznetsova, E.I.; Mashirov, A.V.; Loshachenko, A.S.; Danilov, D.V.; Mitsiuk, V.I.; Kuznetsov, A.S.; Shavrov, V.G.; Koledov, V.V.; Ari-Gur, P. Magnetocaloric Effect, Structure, Spinodal Decomposition and Phase Transformations Heusler Alloy Ni-Mn-In. *Nanomaterials* **2023**, *13*, 1385. [CrossRef]
42. Cava, R.D.; Botta, W.J.; Kiminami, C.S.; Olzon-Dionysio, M.; Souza, S.D.; Jorge, A.M.; Bolfarini, C. Ordered Phases and Texture in Spray-Formed Fe–5wt%Si. *J. Alloys Compd.* **2011**, *509*, S260–S264. [CrossRef]
43. Jang, P.; Lee, B.; Choi, G. Effects of Annealing on the Magnetic Properties of Fe–6.5%Si Alloy Powder Cores. *J. Appl. Phys.* **2008**, *103*, 07E743. [CrossRef]
44. Urban, N.; Meyer, A.; Leckel, M.; Leder, M.; Franke, J. Additive Manufacturing of an Electric Drive a Feasability Study. In Proceedings of the 2018 International Symposium on Power Electronics, Electrical Drives, Automation and Motion (SPEEDAM), Amalfi, Italy, 20–22 June 2018; pp. 1327–1331.
45. Yang, S.; Tang, Y.; Zhao, Y.F. A New Part Consolidation Method to Embrace the Design Freedom of Additive Manufacturing. *J. Manuf. Process.* **2015**, *20*, 444–449. [CrossRef]
46. Atzeni, E.; Salmi, A. Economics of Additive Manufacturing for End-Usable Metal Parts. *Int. J. Adv. Manuf. Technol.* **2012**, *62*, 1147–1155. [CrossRef]
47. Morales, C.; Merlin, M.; Fortini, A.; Fortunato, A. Direct Energy Depositions of a 17-4 PH Stainless Steel: Geometrical and Microstructural Characterizations. *Coatings* **2023**, *13*, 636. [CrossRef]
48. Krings, A.; Cossale, M.; Tenconi, A.; Soulard, J.; Cavagnino, A.; Boglietti, A. Magnetic Materials Used in Electrical Machines: A Comparison and Selection Guide for Early Machine Design. *IEEE Ind. Appl. Mag.* **2017**, *23*, 21–28. [CrossRef]
49. Mazeeva, A.K.; Staritsyn, M.V.; Bobyr, V.V.; Manninen, S.A.; Kuznetsov, P.A.; Klimov, V.N. Magnetic Properties of Fe–Ni Permalloy Produced by Selective Laser Melting. *J. Alloys Compd.* **2020**, *814*, 152315. [CrossRef]

50. Lamichhane, T.N.; Sethuraman, L.; Dalagan, A.; Wang, H.; Keller, J.; Paranthaman, M.P. Additive Manufacturing of Soft Magnets for Electrical Machines—A Review. *Mater. Today Phys.* **2020**, *15*, 100255. [CrossRef]
51. Yoshizawa, Y.; Oguma, S.; Yamauchi, K. New Fe-based Soft Magnetic Alloys Composed of Ultrafine Grain Structure. *J. Appl. Phys.* **1988**, *64*, 6044–6046. [CrossRef]
52. Gheiratmand, T.; Madaah Hosseini, H.R.; Davami, P.; Sarafidis, C. Fabrication of FINEMET Bulk Alloy from Amorphous Powders by Spark Plasma Sintering. *Powder Technol.* **2016**, *289*, 163–168. [CrossRef]
53. Geng, J.; Nlebedim, I.C.; Besser, M.F.; Simsek, E.; Ott, R.T. Bulk Combinatorial Synthesis and High Throughput Characterization for Rapid Assessment of Magnetic Materials: Application of Laser Engineered Net Shaping (LENS^{TM}). *JOM* **2016**, *68*, 1972–1977. [CrossRef]
54. Moghimian, P.; Poirié, T.; Habibnejad-Korayem, M.; Zavala, J.A.; Kroeger, J.; Marion, F.; Larouche, F. Metal Powders in Additive Manufacturing: A Review on Reusability and Recyclability of Common Titanium, Nickel and Aluminum Alloys. *Addit. Manuf.* **2021**, *43*, 102017. [CrossRef]
55. Pham, T.; Kwon, P.; Foster, S. Additive Manufacturing and Topology Optimization of Magnetic Materials for Electrical Machines—A Review. *Energies* **2021**, *14*, 283. [CrossRef]
56. Selema, A.; Ibrahim, M.N.; Sergeant, P. Metal Additive Manufacturing for Electrical Machines: Technology Review and Latest Advancements. *Energies* **2022**, *15*, 1076. [CrossRef]
57. Mohd Yusuf, S.; Cutler, S.; Gao, N. Review: The Impact of Metal Additive Manufacturing on the Aerospace Industry. *Metals* **2019**, *9*, 1286. [CrossRef]
58. Ladani, L.; Sadeghilaridjani, M. Review of Powder Bed Fusion Additive Manufacturing for Metals. *Metals* **2021**, *11*, 1391. [CrossRef]
59. Nadzharyan, T.A.; Shamonin, M.; Kramarenko, E.Y. Theoretical Modeling of Magnetoactive Elastomers on Different Scales: A State-of-the-Art Review. *Polymers* **2022**, *14*, 4096. [CrossRef] [PubMed]
60. Bastola, A.K.; Hossain, M. A Review on Magneto-Mechanical Characterizations of Magnetorheological Elastomers. *Compos. B Eng.* **2020**, *200*, 108348. [CrossRef]
61. Kim, Y.; Zhao, X. Magnetic Soft Materials and Robots. *Chem. Rev.* **2022**, *122*, 5317–5364. [CrossRef]
62. Zhukov, A.; Ipatov, M.; Corte-León, P.; Legarreta, L.G.; Churyukanova, M.; Blanco, J.M.; Gonzalez, J.; Taskaev, S.; Hernando, B.; Zhukova, V. Giant Magnetoimpedance in Rapidly Quenched Materials. *J. Alloys Compd.* **2020**, *814*, 152225. [CrossRef]
63. Klenam, D.E.P.; Egowan, G.; Bodunrin, M.O.; van der Merwe, J.W.; Rahbar, N.; Soboyejo, W. Complex Concentrated Alloys: A Cornucopia of Possible Structural and Functional Applications. In *Comprehensive Structural Integrity*, 2nd ed.; Aliabadi, M.H.F., Soboyejo, W.O., Eds.; Elsevier: Oxford, UK, 2023; pp. 50–90. ISBN 978-0-323-91945-6.
64. Jiles, D.C. Recent Advances and Future Directions in Magnetic Materials. *Acta Mater.* **2003**, *51*, 5907–5939. [CrossRef]
65. Nakamura, H. The Current and Future Status of Rare Earth Permanent Magnets. *Scr. Mater.* **2018**, *154*, 273–276. [CrossRef]
66. Coey, J.M.D. Hard Magnetic Materials: A Perspective. *IEEE Trans. Magn.* **2011**, *47*, 4671–4681. [CrossRef]
67. Goll, D.; Kronmüller, H. High-Performance Permanent Magnets. *Naturwissenschaften* **2000**, *87*, 423–438. [CrossRef]
68. Jeong, U.; Teng, X.; Wang, Y.; Yang, H.; Xia, Y. Superparamagnetic Colloids: Controlled Synthesis and Niche Applications. *Adv. Mater.* **2007**, *19*, 33–60. [CrossRef]
69. Sezer, N.; Arı, İ.; Biçer, Y.; Koç, M. Superparamagnetic Nanoarchitectures: Multimodal Functionalities and Applications. *J. Magn. Magn. Mater.* **2021**, *538*, 168300. [CrossRef]
70. Wei, X.; Jin, M.-L.; Yang, H.; Wang, X.-X.; Long, Y.-Z.; Chen, Z. Advances in 3D Printing of Magnetic Materials: Fabrication, Properties, and Their Applications. *J. Adv. Ceram.* **2022**, *11*, 665–701. [CrossRef]
71. Jordán, D.; González-Chávez, D.; Laura, D.; León Hilario, L.M.; Monteblanco, E.; Gutarra, A.; Avilés-Félix, L. Detection of Magnetic Moment in Thin Films with a Home-Made Vibrating Sample Magnetometer. *J. Magn. Magn. Mater.* **2018**, *456*, 56–61. [CrossRef]
72. Nisticò, R.; Cesano, F.; Garello, F. Magnetic Materials and Systems: Domain Structure Visualization and Other Characterization Techniques for the Application in the Materials Science and Biomedicine. *Inorganics* **2020**, *8*, 6. [CrossRef]
73. Berthod, P. High-Temperature Extreme Alloys. In *Encyclopedia of Materials: Metals and Alloys*; Caballero, F.G., Ed.; Elsevier: Oxford, UK, 2022; pp. 311–322. ISBN 978-0-12-819733-2.
74. Shahrubudin, N.; Lee, T.C.; Ramlan, R. An Overview on 3D Printing Technology: Technological, Materials, and Applications. *Procedia Manuf.* **2019**, *35*, 1286–1296. [CrossRef]
75. Krings, A.; Boglietti, A.; Cavagnino, A.; Sprague, S. Soft Magnetic Material Status and Trends in Electric Machines. *IEEE Trans. Ind. Electron.* **2017**, *64*, 2405–2414. [CrossRef]
76. Ruiz, D.; Venegas-Rebollar, V.; Anaya-Ruiz, G.; Moreno-Goytia, E.; Rodriguez-Rodriguez, J. Design and Prototyping Medium-Frequency Transformers Featuring a Nanocrystalline Core for DC–DC Converters. *Energies* **2018**, *11*, 2081. [CrossRef]
77. van Niekerk, D.; Schoombie, B.; Bokoro, P. Design of an Experimental Approach for Characterization and Performance Analysis of High-Frequency Transformer Core Materials. *Energies* **2023**, *16*, 3950. [CrossRef]
78. Chakraborty, S.; Mandal, K.; Sakar, D.; Cremaschi, V.J.; Silveyra, J.M. Dynamic Coercivity of Mo-Doped FINEMETs. *Phys. B Condens. Matter* **2011**, *406*, 1915–1918. [CrossRef]
79. Zhao, T.; Chen, C.; Wu, X.; Zhang, C.; Volinsky, A.A.; Hao, J. FeSiBCrC Amorphous Magnetic Powder Fabricated by Gas-Water Combined Atomization. *J. Alloys Compd.* **2021**, *857*, 157991. [CrossRef]

80. Zhang, D.; Su, Y.; Yang, X.; Sun, H.; Guo, Z.; Wang, B.; Ma, C.; Dong, Z.; Zhu, L. Texture and Magnetic Property of Fe-6.5 Wt% Si Steel Strip with Cu-Rich Particles Modification. *ACS Omega* **2023**, *8*, 8461–8472. [CrossRef]
81. Leuning, N.; Jaeger, M.; Schauerte, B.; Stöcker, A.; Kawalla, R.; Wei, X.; Hirt, G.; Heller, M.; Korte-Kerzel, S.; Böhm, L.; et al. Material Design for Low-Loss Non-Oriented Electrical Steel for Energy Efficient Drives. *Materials* **2021**, *14*, 6588. [CrossRef] [PubMed]
82. Ouyang, G.; Chen, X.; Liang, Y.; Macziewski, C.; Cui, J. Review of Fe-6.5 wt%Si High Silicon Steel—A Promising Soft Magnetic Material for Sub-KHz Application. *J. Magn. Magn. Mater.* **2019**, *481*, 234–250. [CrossRef]
83. Haines, M.P.; Rielli, V.V.; Primig, S.; Haghdadi, N. Powder Bed Fusion Additive Manufacturing of Ni-Based Superalloys: A Review of the Main Microstructural Constituents and Characterization Techniques. *J. Mater. Sci.* **2022**, *57*, 14135–14187. [CrossRef]
84. Parivendhan, G.; Cardiff, P.; Flint, T.; Tukovic, Z.; Obeidi, M.; Brabazon, D.; Ivankovic, A. A Numerical Study of Processing Parameters and Their Effect on the Melt-Track Profile in Laser Powder Bed Fusion Processes. *Addit. Manuf.* **2023**, *67*, 103482. [CrossRef]
85. Xue, M.; Chen, X.; Ji, X.; Xie, X.; Chao, Q.; Fan, G. Effect of Particle Size Distribution on the Printing Quality and Tensile Properties of Ti-6Al-4V Alloy Produced by LPBF Process. *Metals* **2023**, *13*, 604. [CrossRef]
86. Averardi, A.; Cola, C.; Zeltmann, S.E.; Gupta, N. Effect of Particle Size Distribution on the Packing of Powder Beds: A Critical Discussion Relevant to Additive Manufacturing. *Mater. Today Commun.* **2020**, *24*, 100964. [CrossRef]
87. Andreiev, A.; Hoyer, K.-P.; Dula, D.; Hengsbach, F.; Haase, M.; Gierse, J.; Zimmer, D.; Tröster, T.; Schaper, M. Soft-Magnetic Behavior of Laser Beam Melted FeSi3 Alloy with Graded Cross-Section. *J. Mater. Process. Technol.* **2021**, *296*, 117183. [CrossRef]
88. Quercio, M.; Galbusera, F.; Pošković, E.; Franchini, F.; Ferraris, L.; Canova, A.; Gruosso, G.; Demir, A.G.; Previtali, B. Functional Characterization of L-PBF Produced FeSi2.9 Soft Magnetic Material. In Proceedings of the 2022 International Conference on Electrical Machines (ICEM), Valencia, Spain, 5–8 September 2022; pp. 531–537.
89. Selema, A.; Beretta, M.; Van Coppenolle, M.; Tiismus, H.; Kallaste, A.; Ibrahim, M.N.; Rombouts, M.; Vleugels, J.; Kestens, L.A.I.; Sergeant, P. Evaluation of 3D-Printed Magnetic Materials For Additively-Manufactured Electrical Machines. *J. Magn. Magn. Mater.* **2023**, *569*, 170426. [CrossRef]
90. Haines, M.P.; List, F.; Carver, K.; Leonard, D.N.; Plotkowski, A.; Fancher, C.M.; Dehoff, R.R.; Babu, S.S. Role of Scan Strategies and Heat Treatment on Grain Structure Evolution in Fe-Si Soft Magnetic Alloys Made by Laser-Powder Bed Fusion. *Addit. Manuf.* **2022**, *50*, 102578. [CrossRef]
91. Martin, V.; Gillon, F.; Najjar, D.; Benabou, A.; Witz, J.-F.; Hecquet, M.; Quaegebeur, P.; Meersdam, M.; Auzene, D. MIM-like Additive Manufacturing of Fe3%Si Magnetic Materials. *J. Magn. Magn. Mater.* **2022**, *564*, 170104. [CrossRef]
92. Tiismus, H.; Kallaste, A.; Naseer, M.U.; Vaimann, T.; Rassõlkin, A. Design and Performance of Laser Additively Manufactured Core Induction Motor. *IEEE Access* **2022**, *10*, 50137–50152. [CrossRef]
93. Kumari, G.; Pham, T.Q.; Suen, H.; Rahman, T.; Kwon, P.; Foster, S.N.; Boehlert, C.J. Improving the Soft Magnetic Properties of Binder Jet Printed Iron-Silicon Alloy through Boron Addition. *Mater. Chem. Phys.* **2023**, *296*, 127181. [CrossRef]
94. Kang, N.; Li, Q.; El Mansori, M.; Yao, B.; Ma, F.; Lin, X.; Liao, H. Laser Powder Bed Fusion Processing of Soft Magnetic Fe–Ni–Si Alloys: Effect of Hot Isostatic Pressing Treatment. *Chin. J. Mech. Eng. Addit. Manuf. Front.* **2022**, *1*, 100054. [CrossRef]
95. Garibaldi, M.; Ashcroft, I.; Simonelli, M.; Hague, R. Metallurgy of High-Silicon Steel Parts Produced Using Selective Laser Melting. *Acta Mater.* **2016**, *110*, 207–216. [CrossRef]
96. Garibaldi, M.; Ashcroft, I.; Hillier, N.; Harmon, S.A.C.; Hague, R. Relationship between Laser Energy Input, Microstructures and Magnetic Properties of Selective Laser Melted Fe-6.9%wt Si Soft Magnets. *Mater. Charact.* **2018**, *143*, 144–151. [CrossRef]
97. Garibaldi, M.; Ashcroft, I.; Lemke, J.N.; Simonelli, M.; Hague, R. Effect of Annealing on the Microstructure and Magnetic Properties of Soft Magnetic Fe-Si Produced via Laser Additive Manufacturing. *Scr. Mater.* **2018**, *142*, 121–125. [CrossRef]
98. Zaied, M.; Ospina-Vargas, A.; Buiron, N.; Favergeon, J.; Fenineche, N.-E. Additive Manufacturing of Soft Ferromagnetic Fe 6.5%Si Annular Cores: Process Parameters, Microstructure, and Magnetic Properties. *IEEE Trans. Magn.* **2022**, *58*, 2001420. [CrossRef]
99. Di Schino, A.; Stornelli, G. Additive Manufacturing: A New Concept for End Users. The Case of Magnetic Materials. *Acta Metall. Slovaca* **2022**, *28*, 208–211. [CrossRef]
100. Faba, A.; Riganti Fulginei, F.; Quondam Antonio, S.; Stornelli, G.; Di Schino, A.; Cardelli, E. Hysteresis Modelling in Additively Manufactured FeSi Magnetic Components for Electrical Machines and Drives. *IEEE Trans. Ind. Electron.* **2023**, 1–9. [CrossRef]
101. Stornelli, G.; Faba, A.; Di Schino, A.; Folgarait, P.; Ridolfi, M.R.; Cardelli, E.; Montanari, R. Properties of Additively Manufactured Electric Steel Powder Cores with Increased Si Content. *Materials* **2021**, *14*, 1489. [CrossRef]
102. Stornelli, G.; Ridolfi, M.R.; Folgarait, P.; De Nisi, J.; Corapi, D.; Repitsch, C.; Di Schino, A. Feasibility Assessment of Magnetic Cores through Additive Manufacturing Techniques (Studio Di Fattibilità Della Fabbricazione Di Nuclei Ferromagnetici Attraverso Tecniche Di Manifattura Additiva). *Metall. Ital.* **2021**, *113*, 50–63.
103. Seiler, N.; Kolb, D.; Schanz, J.; Goll, D.; Riegel, H. Influence of Modulated Laser Power on the Surface Topography and Microstructure in the Laser Powder Bed Fusion Process. *Procedia CIRP* **2022**, *111*, 693–696. [CrossRef]
104. Agapovichev, A.V.; Khaimovich, A.I.; Erisov, Y.A.; Ryazanov, M. V Investigation of Soft Magnetic Material Fe-6.5Si Fracture Obtained by Additive Manufacturing. *Materials* **2022**, *15*, 8615. [CrossRef]
105. Goll, D.; Schuller, D.; Martinek, G.; Kunert, T.; Schurr, J.; Sinz, C.; Schubert, T.; Bernthaler, T.; Riegel, H.; Schneider, G. Additive Manufacturing of Soft Magnetic Materials and Components. *Addit. Manuf.* **2019**, *27*, 428–439. [CrossRef]

106. Chaudhary, V.; Sai Kiran Kumar Yadav, N.M.; Mantri, S.A.; Dasari, S.; Jagetia, A.; Ramanujan, R.V.; Banerjee, R. Additive Manufacturing of Functionally Graded Co–Fe and Ni–Fe Magnetic Materials. *J. Alloys Compd.* **2020**, *823*, 153817. [CrossRef]
107. Chaudhary, V.; Borkar, T.; Mikler, C.V.; Gwalani, B.; Choudhuri, D.; Soni, V.; Alam, T.; Ramanujan, R.V.; Banerjee, R. Additively Manufactured Functionally Graded FeNi Based High Entropy Magnetic Alloys. In Proceedings of the 2018 IEEE International Magnetics Conference (INTERMAG), Singapore, 23–27 April 2018; p. 1.
108. Yi, M.; Xu, B.-X.; Gutfleisch, O. Computational Study on Microstructure Evolution and Magnetic Property of Laser Additively Manufactured Magnetic Materials. *Comput. Mech.* **2019**, *64*, 917–935. [CrossRef]
109. Mikler, C.V.; Chaudhary, V.; Borkar, T.; Soni, V.; Choudhuri, D.; Ramanujan, R.V.; Banerjee, R. Laser Additive Processing of Ni-Fe-V and Ni-Fe-Mo Permalloys: Microstructure and Magnetic Properties. *Mater. Lett.* **2017**, *192*, 9–11. [CrossRef]
110. Mikler, C.V.; Chaudhary, V.; Borkar, T.; Soni, V.; Jaeger, D.; Chen, X.; Contieri, R.; Ramanujan, R.V.; Banerjee, R. Laser Additive Manufacturing of Magnetic Materials. *JOM* **2017**, *69*, 532–543. [CrossRef]
111. Chaudhary, V.; Mantri, S.A.; Ramanujan, R.V.; Banerjee, R. Additive Manufacturing of Magnetic Materials. *Prog. Mater. Sci.* **2020**, *114*, 100688. [CrossRef]
112. Yang, J.; Zhu, Q.; Wang, Z.; Xiong, F.; Li, Q.; Yang, F.; Li, R.; Ge, X.; Wang, M. Effects of Metallurgical Defects on Magnetic Properties of SLM NiFeMo Permalloy. *Mater. Charact.* **2023**, *197*, 112672. [CrossRef]
113. Li, B.; Zhang, W.; Fu, W.; Xuan, F. Laser Powder Bed Fusion (L-PBF) 3D Printing Thin Overhang Walls of Permalloy for a Modified Honeycomb Magnetic-Shield Structure. *Thin-Walled Struct.* **2023**, *182*, 110185. [CrossRef]
114. Kim, E.S.; Haftlang, F.; Ahn, S.Y.; Kwon, H.; Gu, G.H.; Kim, H.S. Mechanical and Magnetic Properties of Soft Magnetic Fe–Ni Permalloy Produced by Directed Energy Deposition Processes. *J. Mater. Sci.* **2022**, *57*, 17967–17983. [CrossRef]
115. Kim, E.S.; Haftlang, F.; Ahn, S.Y.; Gu, G.H.; Kim, H.S. Effects of Processing Parameters and Heat Treatment on the Microstructure and Magnetic Properties of the In-Situ Synthesized Fe-Ni Permalloy Produced Using Direct Energy Deposition. *J. Alloys Compd.* **2022**, *907*, 164415. [CrossRef]
116. Haftlang, F.; Kim, E.S.; Kim, H.S. Crystallographic-Orientation-Dependent Magnetic Properties of Fe–Ni Permalloy in-Situ Alloyed Using Additive Manufacturing. *J. Mater. Process. Technol.* **2022**, *309*, 117733. [CrossRef]
117. Mohamed, A.E.-M.A.; Zou, J.; Sheridan, R.S.; Bongs, K.; Attallah, M.M. Magnetic Shielding Promotion via the Control of Magnetic Anisotropy and Thermal Post Processing in Laser Powder Bed Fusion Processed NiFeMo-Based Soft Magnet. *Addit. Manuf.* **2020**, *32*, 101079. [CrossRef]
118. Kustas, A.B.; Susan, D.F.; Johnson, K.L.; Whetten, S.R.; Rodriguez, M.A.; Dagel, D.J.; Michael, J.R.; Keicher, D.M.; Argibay, N. Characterization of the Fe-Co-1.5V Soft Ferromagnetic Alloy Processed by Laser Engineered Net Shaping (LENS). *Addit. Manuf.* **2018**, *21*, 41–52. [CrossRef]
119. Sundar, R.S.; Deevi, S.C. Soft Magnetic FeCo Alloys: Alloy Development, Processing, and Properties. *Int. Mater. Rev.* **2005**, *50*, 157–192. [CrossRef]
120. Sreenivasulu, G.; Laletin, U.; Petrov, V.M.; Petrov, V.V.; Srinivasan, G. A Permendur-Piezoelectric Multiferroic Composite for Low-Noise Ultrasensitive Magnetic Field Sensors. *Appl. Phys. Lett.* **2012**, *100*, 173506. [CrossRef]
121. Gould, H.L.B.; Wenny, D.H. Supermendur: A New Rectangular-Loop Magnetic Material. *Electr. Eng.* **1957**, *76*, 208–211. [CrossRef]
122. Lindroos, T.; Riipinen, T.; Metsä-Kortelainen, S.; Pippuri-Mäkeläinen, J.; Manninen, A. Lessons Learnt—Additive Manufacturing of Iron Cobalt Based Soft Magnetic Materials. *J. Magn. Magn. Mater.* **2022**, *563*, 169977. [CrossRef]
123. Riipinen, T.; Pippuri-Mäkeläinen, J.; Que, Z.; Metsä-Kortelainen, S.; Antikainen, A.; Lindroos, T. The Effect of Heat Treatment on Structure and Magnetic Properties of Additively Manufactured Fe-Co-V Alloys. *Mater. Today Commun.* **2023**, *36*, 106437. [CrossRef]
124. Nartu, M.S.K.K.Y.; Dasari, S.; Sharma, A.; Chaudhary, V.; Varahabhatla, S.M.; Mantri, S.A.; Ivanov, E.; Ramanujan, R.V.; Dahotre, N.B.; Banerjee, R. Reducing Coercivity by Chemical Ordering in Additively Manufactured Soft Magnetic Fe–Co (Hiperco) Alloys. *J. Alloys Compd.* **2021**, *861*, 157998. [CrossRef]
125. Kustas, A.B.; Fancher, C.M.; Whetten, S.R.; Dagel, D.J.; Michael, J.R.; Susan, D.F. Controlling the Extent of Atomic Ordering in Intermetallic Alloys through Additive Manufacturing. *Addit. Manuf.* **2019**, *28*, 772–780. [CrossRef]
126. Borkar, T.; Conteri, R.; Chen, X.; Ramanujan, R.V.; Banerjee, R. Laser Additive Processing of Functionally-Graded Fe–Si–B–Cu–Nb Soft Magnetic Materials. *Mater. Manuf. Process.* **2017**, *32*, 1581–1587. [CrossRef]
127. Conteri, R.; Borkar, T.; Nag, S.; Jaeger, D.; Chen, X.; Ramanujan, R.V.; Banerjee, R. Laser Additive Processing of Fe-Si-B-Cu-Nb Magnetic Alloys. *J. Manuf. Process.* **2017**, *29*, 175–181. [CrossRef]
128. Dobrzanski, L.A.; Nowosielski, R.; Konieczny, J.; Przybył, A.; Wysłocki, J. Structure and Properties of Nanocrystalline Soft Magnetic Composite Materials with Silicon Polymer Matrix. *J. Magn. Magn. Mater.* **2005**, *290–291*, 1510–1512. [CrossRef]
129. Nowosielski, R.; Wysłocki, J.J.; Wnuk, I.; Gramatyka, P. Nanocrystalline Soft Magnetic Composite Cores. *J. Mater. Process. Technol.* **2006**, *175*, 324–329. [CrossRef]
130. Shokrollahi, H.; Janghorban, K. The Effect of Compaction Parameters and Particle Size on Magnetic Properties of Iron-Based Alloys Used in Soft Magnetic Composites. *Mater. Sci. Eng. B* **2006**, *134*, 41–43. [CrossRef]
131. Chicinas, I.; Geoffroy, O.; Isnard, O.; Pop, V. Soft Magnetic Composite Based on Mechanically Alloyed Nanocrystalline Ni3Fe Phase. *J. Magn. Magn. Mater.* **2005**, *290–291*, 1531–1534. [CrossRef]
132. Rimal, H.P.; Stornelli, G.; Faba, A.; Cardelli, E. Macromagnetic Approach to the Modeling in Time Domain of Magnetic Losses of Ring Cores of Soft Ferrites in Power Electronics. *IEEE Trans. Power Electron.* **2023**, *38*, 3559–3568. [CrossRef]

133. Sharma, K.; Aggarwal, N.; Kumar, N.; Sharma, A. A Review Paper: Synthesis Techniques and Advance Application of Mn-Zn Nano-Ferrites. *Mater. Today Proc.* 2023, in press. [CrossRef]
134. Liu, L.; Ge, T.; Ngo, K.D.T.; Mei, Y.; Lu, G.-Q. Ferrite Paste Cured with Ultraviolet Light for Additive Manufacturing of Magnetic Components for Power Electronics. *IEEE Magn. Lett.* 2018, 9, 5102705. [CrossRef]
135. Liu, L.; Ge, T.; Yan, Y.; Ngo, K.D.T.; Lu, G.-Q. UV-Assisted 3D-Printing of Soft Ferrite Magnetic Components for Power Electronics Integration. In Proceedings of the 2017 International Conference on Electronics Packaging (ICEP), Yamagata, Japan, 19–22 April 2017; pp. 447–450.
136. Andrews, C.E.; Chatham, M.P.; Dorman, S.F.; McCue, I.D.; Sopcisak, J.J.; Taheri, M.L. Additive Manufacturing of NiZnCu-Ferrite Soft Magnetic Composites. *J. Mater. Res.* 2021, 36, 3579–3590. [CrossRef]
137. Tiismus, H.; Kallaste, A.; Vaimann, T.; Lind, L.; Virro, I.; Rassõlkin, A.; Dedova, T. Laser Additively Manufactured Magnetic Core Design and Process for Electrical Machine Applications. *Energies* 2022, 15, 3665. [CrossRef]
138. Komatsubara, M.; Sadahiro, K.; Kondo, O.; Takamiya, T.; Honda, A. Newly Developed Electrical Steel for High-Frequency Use. *J. Magn. Magn. Mater.* 2002, 242–245, 212–215. [CrossRef]
139. Haiji, H.; Okada, K.; Hiratani, T.; Abe, M.; Ninomiya, M. Magnetic Properties and Workability of 6.5% Si Steel Sheet. *J. Magn. Magn. Mater.* 1996, 160, 109–114. [CrossRef]
140. Takada, Y.; Abe, M.; Masuda, S.; Inagaki, J. Commercial Scale Production of Fe-6.5 Wt. % Si Sheet and Its Magnetic Properties. *J. Appl. Phys.* 1988, 64, 5367–5369. [CrossRef]
141. Cullity, B.D.; Graham, C.D. *Introduction to Magnetic Materials*; John Wiley & Sons, Inc.: Hoboken, NJ, USA, 2008; ISBN 9780470386323.
142. Bozorth, R.M. IronSilicon Alloys. In *Ferromagnetism*; Wiley-IEEE Press: Hoboken, NJ, USA, 1978; pp. 67–101.
143. Ustinovshikov, Y.; Sapegina, I. Morphology of Ordering Fe-Si Alloys. *J. Mater. Sci.* 2004, 39, 1007–1016. [CrossRef]
144. Ghassemi, M. High Power Density Technologies for Large Generators and Motors for Marine Applications with Focus on Electrical Insulation Challenges. *High Volt.* 2020, 5, 7–14. [CrossRef]
145. Maklakov, S.S.; Naboko, A.S.; Maklakov, S.A.; Bobrovskii, S.Y.; Polozov, V.I.; Zezyulina, P.A.; Osipov, A.V.; Ryzhikov, I.A.; Rozanov, K.N.; Filimonov, D.F.; et al. Amorphization of Thin Supermalloy Films Ni79Fe17Mo4 with Oxygen during Magnetron Sputtering. *J. Alloys Compd.* 2021, 854, 157097. [CrossRef]
146. Li, B.; Fu, W.; Xu, H.; Qian, B.; Xuan, F. Additively Manufactured Ni-15Fe-5Mo Permalloy via Selective Laser Melting and Subsequent Annealing for Magnetic-Shielding Structures: Process, Micro-Structural and Soft-Magnetic Characteristics. *J. Magn. Magn. Mater.* 2020, 494, 165754. [CrossRef]
147. Schönrath, H.; Spasova, M.; Kilian, S.O.; Meckenstock, R.; Witt, G.; Sehrt, J.T.; Farle, M. Additive Manufacturing of Soft Magnetic Permalloy from Fe and Ni Powders: Control of Magnetic Anisotropy. *J. Magn. Magn. Mater.* 2019, 478, 274–278. [CrossRef]
148. Waeckerlé, T.; Demier, A.; Godard, F.; Fraisse, H. Evolution and Recent Developments of 80%Ni Permalloys. *J. Magn. Magn. Mater.* 2020, 505, 166635. [CrossRef]
149. Hamler, A.; Goričan, V.; Šuštaršič, B.; Sirc, A. The Use of Soft Magnetic Composite Materials in Synchronous Electric Motor. *J. Magn. Magn. Mater.* 2006, 304, e816–e819. [CrossRef]
150. Yoshida, S.; Mizushima, T.; Hatanai, T.; Inoue, A. Preparation of New Amorphous Powder Cores Using Fe-Based Glassy Alloy. *IEEE Trans. Magn.* 2000, 36, 3424–3429. [CrossRef]
151. Hasegawa, R. Design and Fabrication of New Soft Magnetic Materials. *J. Non Cryst. Solids* 2003, 329, 1–7. [CrossRef]
152. Louis, S.J.; Jan, B.; Willem, L.M. Manganese Zinc Ferrite Core. US Patent 2,551,711, 8 May 1951.
153. Yeh, J.-W.; Chen, S.-K.; Lin, S.-J.; Gan, J.-Y.; Chin, T.-S.; Shun, T.-T.; Tsau, C.-H.; Chang, S.-Y. Nanostructured High-Entropy Alloys with Multiple Principal Elements: Novel Alloy Design Concepts and Outcomes. *Adv. Eng. Mater.* 2004, 6, 299–303. [CrossRef]
154. Cantor, B.; Chang, I.T.H.; Knight, P.; Vincent, A.J.B. Microstructural Development in Equiatomic Multicomponent Alloys. *Mater. Sci. Eng. A* 2004, 375–377, 213–218. [CrossRef]
155. Gan, G.-Y.; Ma, L.; Luo, D.-M.; Jiang, S.; Tang, B.-Y. Influence of Al Substitution for Sc on Thermodynamic Properties of HCP High Entropy Alloy Hf0.25Ti0.25Zr0.25Sc0.25-XAlx from First-Principles Investigation. *Phys. B Condens. Matter* 2020, 593, 412272. [CrossRef]
156. Alijani, F.; Reihanian, M.; Gheisari, K. Study on Phase Formation in Magnetic FeCoNiMnV High Entropy Alloy Produced by Mechanical Alloying. *J. Alloys Compd.* 2019, 773, 623–630. [CrossRef]
157. Guo, S.; Liu, C.T. Phase Stability in High Entropy Alloys: Formation of Solid-Solution Phase or Amorphous Phase. *Prog. Nat. Sci. Mater. Int.* 2011, 21, 433–446. [CrossRef]
158. Wu, Z.; Wang, C.; Zhang, Y.; Feng, X.; Gu, Y.; Li, Z.; Jiao, H.; Tan, X.; Xu, H. The AC Soft Magnetic Properties of FeCoNixCuAl ($1.0 \leq x \leq 1.75$) High-Entropy Alloys. *Materials* 2019, 12, 4222. [CrossRef]
159. Na, S.-M.; Lambert, P.K.; Jones, N.J. Hard Magnetic Properties of FeCoNiAlCu$_x$Ti$_x$ Based High Entropy Alloys. *AIP Adv.* 2021, 11, 015210. [CrossRef]
160. Miracle, D.B.; Senkov, O.N. A Critical Review of High Entropy Alloys and Related Concepts. *Acta Mater.* 2017, 122, 448–511. [CrossRef]
161. Chen, C.; Zhang, H.; Fan, Y.; Zhang, W.; Wei, R.; Wang, T.; Zhang, T.; Li, F. A Novel Ultrafine-Grained High Entropy Alloy with Excellent Combination of Mechanical and Soft Magnetic Properties. *J. Magn. Magn. Mater.* 2020, 502, 166513. [CrossRef]

162. Zhao, C.C.; Inoue, A.; Kong, F.L.; Zhang, J.Y.; Chen, C.J.; Shen, B.L.; Al-Marzouki, F.; Greer, A.L. Novel Phase Decomposition, Good Soft-Magnetic and Mechanical Properties for High-Entropy (Fe0.25Co0.25Ni0.25Cr0.125Mn0.125)100−B (x = 9–13) Amorphous Alloys. *J. Alloys Compd.* **2020**, *843*, 155917. [CrossRef]
163. Zhang, Q.; Xu, H.; Tan, X.H.; Hou, X.L.; Wu, S.W.; Tan, G.S.; Yu, L.Y. The Effects of Phase Constitution on Magnetic and Mechanical Properties of FeCoNi(CuAl) (x = 0–1.2) High-Entropy Alloys. *J. Alloys Compd.* **2017**, *693*, 1061–1067. [CrossRef]
164. Feng, X.; Zheng, R.; Wu, Z.; Zhang, Y.; Li, Z.; Tan, X.; Xu, H. Study on a New High-Entropy Alloy Nd20Pr20La20Fe20Co10Al10 with Hard Magnetic Properties. *J. Alloys Compd.* **2021**, *882*, 160640. [CrossRef]
165. Chen, H.; Gou, J.; Jia, W.; Song, X.; Ma, T. Origin of Hard Magnetism in Fe-Co-Ni-Al-Ti-Cu High-Entropy Alloy: Chemical Shape Anisotropy. *Acta Mater.* **2023**, *246*, 118702. [CrossRef]
166. Radhakrishnan, M.; McKinstry, M.; Chaudhary, V.; Nartu, M.S.K.K.Y.; Krishna, K.V.M.; Ramanujan, R.V.; Banerjee, R.; Dahotre, N.B. Effect of Chromium Variation on Evolution of Magnetic Properties in Laser Direct Energy Additively Processed CoCrxFeNi Alloys. *Scr. Mater.* **2023**, *226*, 115269. [CrossRef]
167. Zhou, R.; Liu, Y.; Zhou, C.; Li, S.; Wu, W.; Song, M.; Liu, B.; Liang, X.; Liaw, P.K. Microstructures and Mechanical Properties of C-Containing FeCoCrNi High-Entropy Alloy Fabricated by Selective Laser Melting. *Intermetallics* **2018**, *94*, 165–171. [CrossRef]
168. Park, J.M.; Choe, J.; Kim, J.G.; Bae, J.W.; Moon, J.; Yang, S.; Kim, K.T.; Yu, J.-H.; Kim, H.S. Superior Tensile Properties of 1%C-CoCrFeMnNi High-Entropy Alloy Additively Manufactured by Selective Laser Melting. *Mater. Res. Lett.* **2020**, *8*, 1–7. [CrossRef]
169. Nartu, M.S.K.K.Y.; Jagetia, A.; Chaudhary, V.; Mantri, S.A.; Ivanov, E.; Dahotre, N.B.; Ramanujan, R.V.; Banerjee, R. Magnetic and Mechanical Properties of an Additively Manufactured Equiatomic CoFeNi Complex Concentrated Alloy. *Scr. Mater.* **2020**, *187*, 30–36. [CrossRef]
170. Yang, X.; Cui, X.; Jin, G.; Liu, J.; Chen, Y.; Liu, Z. Soft Magnetic Property of (Fe60Co35Ni5)78 Si6B12Cu1Mo3 Alloys by Laser Additive Manufacturing. *J. Magn. Magn. Mater.* **2018**, *466*, 75–80. [CrossRef]

Disclaimer/Publisher's Note: The statements, opinions and data contained in all publications are solely those of the individual author(s) and contributor(s) and not of MDPI and/or the editor(s). MDPI and/or the editor(s) disclaim responsibility for any injury to people or property resulting from any ideas, methods, instructions or products referred to in the content.

Article

Effect of Tempering Time on Carbide Evolution and Mechanical Properties of a Nb-V-Ti Micro-Alloyed Steel

Qian Zhao [1], Zhixia Qiao [1,*] and Ji Dong [2,*]

1 School of Mechanical Engineering, Tianjin University of Commerce, Tianjin 300134, China
2 School of Mechanical Engineering, Tianjin Sino-German University of Applied Sciences, Tianjin 300350, China
* Correspondence: qzhxia@tjcu.edu.cn (Z.Q.); dongji@tsguas.edu.cn (J.D.)

Abstract: The evolution of the microstructure, the precipitation behavior, and the mechanical performances of Nb-V-Ti micro-alloyed steel prepared under different tempering time were studied using transmission electron microscopy (TEM), X-ray diffraction (XRD), and mechanical tests. It was found that the width of the martensite laths increases with the increasing tempering time. Several kinds of carbides, including M_3C, M_2C, $M_{23}C_6$, M_7C_3, and MC particles, were identified after tempering. The MC carbides remain stable during tempering, but the transformation behavior of other carbides was identified. The transformation sequence can be summarized as: $M_3C \rightarrow M_2C \rightarrow M_7C_3 \rightarrow M_{23}C_6$. The strength decreases and the Charpy impact toughness increases gradually with the increase in the tempering time. The ultimate strength (UTS) decreases from 1231 to 896 MPa, and the yield strength (YS) decreases from 1138 to 835 MPa. The $-40\ °C$ Charpy impact toughness increases from 20 to 61 J as the tempering time increases from 10 min to 100 h. The evolution of carbides plays an important role in their mechanical performances.

Keywords: carbide; tempering time; Nb-V-Ti micro-alloyed steel; mechanical performance

1. Introduction

With the rapid development of modern industry and technology, the exploitation and utilization of crude oil and natural gas have been gradually increasing [1–3]. In the past several decades, oil and gas exploration has been extended to abyssal regions [4]. To enhance the transportation efficiency, a larger diameter and a higher operation pressure have been adopted. This requires an excellent combination of a high strength and a good toughness for pipeline steels [5–8]. Therefore, micro-alloyed components are added to pipeline steels to improve their mechanical performance through grain refinement and precipitation strengthening [9–12]. Nowadays, micro-alloyed ultra-high-strength pipeline steel is indispensable to the development of offshore oil and gas exploitation.

The conventional heat treatment of micro-alloyed steels consists of quenching and tempering. Quenching and tempering at different temperatures for different times can provide a beneficial combination of a microstructure and precipitates [13,14]. Multi-phase microstructures and different types of carbides can further meet the requirements for excellent mechanical properties [15–17].

The precipitation of fine particles during tempering plays an important role in improving the strength and toughness of micro-alloyed ultra-high-strength steels. Much research on the evolution of carbide precipitates under different tempering times has been reported. Li et al. [1] investigated the evolution of precipitates in a G18CrMo2-6 steel during tempering at 680 °C for up to 100 h. It was found that M_3C carbides were transformed from M-A particles first, and then MC carbides precipitated. M_3C particles refined and spheroidized gradually, and $M_{23}C_6$ coarsened with increasing the tempering time. Moon et al. [18] clearly showed the transformation sequence of carbide precipitates in Cr-Mo API steels. They also found that the precipitation sequence of carbides during tempering at 650 °C was: MC +

$M_3C \rightarrow MC \rightarrow MC + M_7C_3 + M_{23}C_6$. Tao et al. [19] investigated the evolution of different carbides in X12CrMoWVNbN10-1-1 steel after tempering and found that $M_{23}C_6$ possesses a higher thermal stability than M_7C_3. The precipitation sequence of carbides can be summarized as: $M_3C \rightarrow M_7C_3 \rightarrow M_{23}C_6$. On the contrary, Asadabad et al. [20] reported that part of $M_{23}C_6$ transformed into M_7C_3 in 4.5Cr-2W-0.25V-0.1C steel. Janovec [21] reported that the Cr content determines the transformation sequence of precipitates in Cr-alloyed steels. With the addition of 1wt.% Cr, the precipitation sequence of Cr-Mo-V steels was proven to be $M_{23}C_6 \rightarrow M_7C_3$ [18]. In middle- or high-Cr steels, Jia et al. found that four kinds of precipitates, M_2X, M_3C, $M_{23}C_6$, and M_7C_3, existed in 9CrMoCoB (CB2) steel [22], and the thermal stability of M_7C_3 was lower than $M_{23}C_6$ [23–25]. However, the transformation sequence of the carbide precipitates of Nb-V-Ti micro-alloyed ultra-high-strength steel during high-temperature tempering is still not clear, and it should have more attention paid to it.

This paper focused on the carbide evolution of Nb-V-Ti micro-alloyed steel during different tempering times. The microstructure and precipitates under different tempering times were analyzed. Further, the mechanical properties of the explored steel under different tempering times were evaluated, and the correlation between the evolution of the microstructure, the precipitation behavior, and the mechanical performances was also discussed in detail.

2. Experimental

The chemical composition of the investigated steel was 0.25C-0.29Si-0.4Mn-0.97Cr-0.88Mo-0.015Nb-0.089V-0.026Ti (wt.%). In this study, the investigated steels were austenitized at 1000 °C for 30 min, quenched with water, and then tempered at 600 °C for 10 min, 30 min, 1 h, 2 h, 3 h, 5 h, 10 h, 25 h, 50 h, and 100 h, respectively. All the specimens were cooled in water to room temperature after tempering.

The thin-foil specimens were used for TEM characterization, and they were electropolished in a double-jet electropolishing device with a solution of 7% perchloric acid and 93% ethanol at −20 °C. The morphologies, sizes, distributions, and chemical compositions of the precipitates were identified using TEM (JEM-2100F, JEOL, Tokyo, Japan). The experimental steels were immersed in hydrochloric acid for 15 days, and then the precipitated particles were extracted by centrifuge. Hydrochloric acid was removed by continuously adding alcohol to the solution during extraction. The precipitated particles were then dried in a drying cabinet at 50 °C for more than 5 h. The phase composition and the various carbides formed during tempering were determined by XRD (D8 Advanced, Bruker, Karlsruhe, Germany).

Tensile tests were performed on the tempered specimens at room temperature. The gauge length of the specimens was 15 mm. An Instron 5565 (Instron, Boston, MA, USA) tensile machine was used for tensile test with a strain rate of 1.0×10^{-3} s^{-1}. The standard CVN samples were machined to 10 mm × 10 mm × 55 mm with a notch depth of 2 mm. The tensile and the Charpy impact toughness experiments were repeated three times to ensure the accuracy of the results. The Charpy impact toughness tests were carried out at −40 °C using an Instron MPX (Instron, Boston, MA, USA) impact tester. The samples were cooled in a cold box with air as the cooling medium. Micro-hardness testing was conducted using a MH-6 type (Shanghai Everone Precision Instruments, Shanghai, China) Vickers hardness tester with a maximum load of 100 g for 5 s.

3. Results and Discussion

3.1. Microstructure Evolution under Different Tempering Times

Figure 1 shows the TEM images of samples tempered at 600 °C at different times. The martensite lath with dislocations indicated by arrows can be seen in the microstructure. With a prolonged tempering time, the width of martensite lath increased, and the mean widths of martensite lath were determined to be 149, 199, 247, 282, 305, and 388 nm for the tempering times of 10 min, 30 min, 5 h, 10 h, 50 h, and 100 h, respectively. The lath width of the martensite was obtained by measuring about 200 martensites. The density of the dislocation was reduced with the extension of the tempering time, which was due to the recovery of the martensite lath.

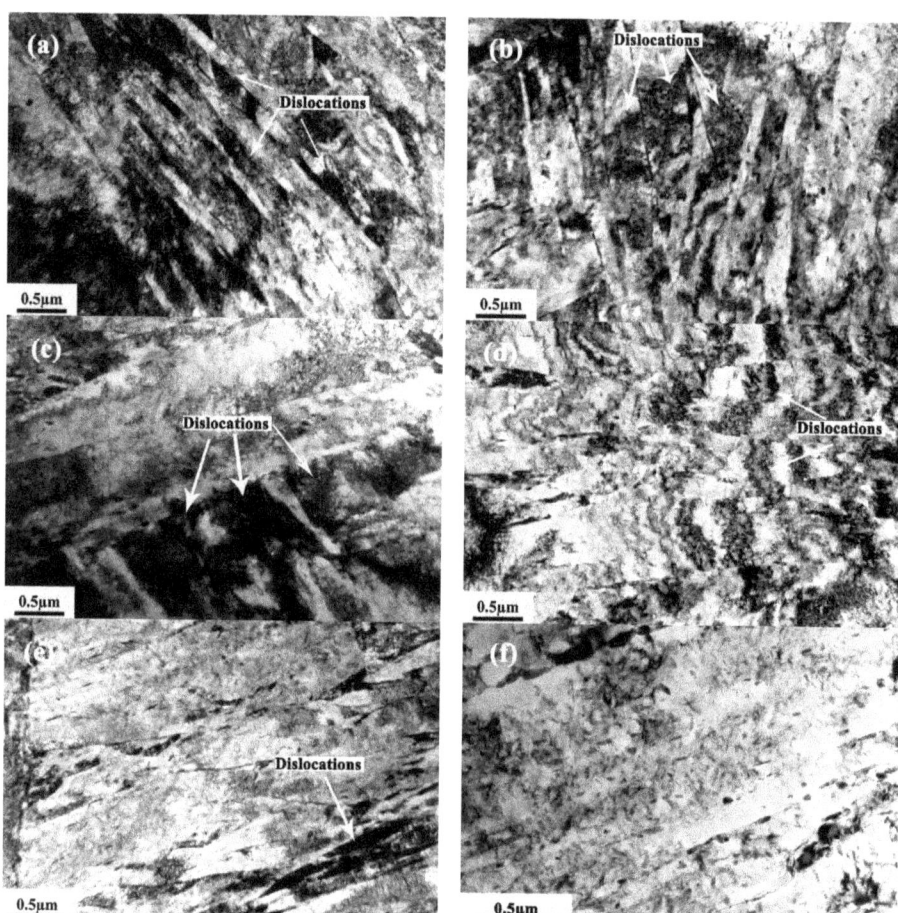

Figure 1. TEM micrographs of steel specimens after being tempered at 600 °C for different periods: (**a**) 10 min, (**b**) 30 min, (**c**) 5 h, (**d**) 10 h, (**e**) 50 h, and (**f**) 100 h.

The microstructure of the retained austenite in the samples tempered at 600 °C is shown in Figure 2. The retained austenites are marked by white arrows. The retained austenite films were continuously distributed between the martensite lathes when the tempering time was 10 min (Figure 2a), and then the films became discontinuous with a prolonged tempering time (Figure 2b,c). After tempering for 100 h, few films could be seen between the martensite laths (Figure 2d). The mean thickness of the retained austenite film decreased with the increase in tempering time, and the average thickness of the retained austenite films tempered for 10 min, 30 min, and 5 h was about 57, 35, and 27 nm, respectively.

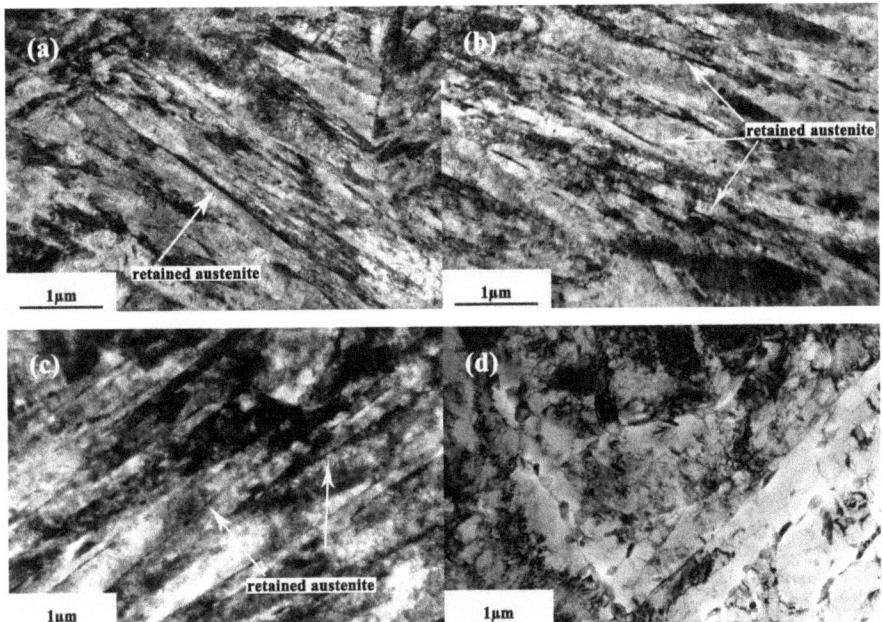

Figure 2. TEM micrographs of retained austenite in the steel specimens after tempering at 600 °C for different tempering periods: (**a**) 10 min, (**b**) 30 min, (**c**) 5 h, and (**d**) 100 h.

3.2. Precipitation Behavior of Carbide under Different Tempering Times

The TEM micrographs of samples under different tempering times were examined. After tempering for 10 min, needle-like particles could be observed in the interior of the martensite lath. As shown by the black oval in Figure 3a, the mean length of the needle-like precipitates was about 94.2 nm. The sizes of the carbides were obtained by measuring about 400 carbides. According to the corresponding electron diffraction pattern results shown in Figure 3b, the needle-like precipitates possessed an orthorhombic crystal structure. Based on the EDS results (Figure 3c), the carbides contained a high content of Fe and a small amount of Mn and Cr. Referring to the PDF No. 75-0910, the carbide was M_3C with $a = 4.518$ nm, $b = 5.069$ nm, $c = 6.736$ nm, and $\alpha = \beta = \gamma = 90°$.

Spherical nano-sized precipitates with a mean diameter of about 4.89 nm and square particles with a mean dimension of 42.45 mm were also observed, as shown in Figure 4a,d. The EDS analysis results combined with the SAD identifications are shown in Figure 4b–e. These two kinds of particles both had an f.c.c. structure. The nano-sized spherical particles contained Nb and a small amount of Ti and V, and the corresponding PDF card was 47-1418. The square precipitates contained Ti with a small amount of Nb and V, and the corresponding PDF retrieval number was 65-7931. The Fe, Mn, and Cr elements were derived from the matrix. Therefore, these two types of carbides were identified as Nb-rich MC (I) carbide and Ti-rich MC (II) carbide.

When the tempering time was prolonged to 30 min, the needle-like M_3C carbides dissolved into the matrix during tempering and acted as nucleation sites for rod-like particles, as denoted by the arrow in Figure 5a. The EDS results (Figure 5d) indicated that the rod-like particles contained a high iron content and Mo with a tiny amount of Cr and Mn. The SAD pattern shown in Figure 5b matched well with the electron diffraction pattern of M_2C, and the PDF retrieval number was 35-0787. The mean length of the rod-like M_2C particles was about 59.04 nm. Irregular-shaped particles precipitated along the lath or grain boundaries, as marked by the white oval in Figure 5a. The EDS analysis results and the SAD identifications are shown in Figure 5c–e. These precipitates had a complex cubic structure and contained Fe and a certain amount of Cr and Mn. The corresponding PDF retrieval number was 78-1499, and this type of precipitate was recognized as $M_{23}C_6$. The mean size of $M_{23}C_6$ was around 46.77 nm.

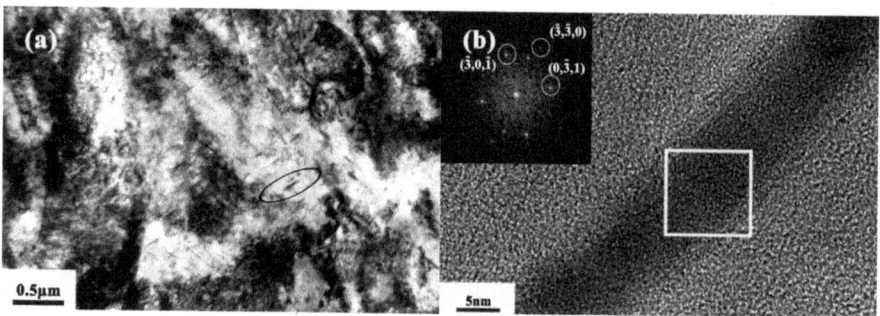

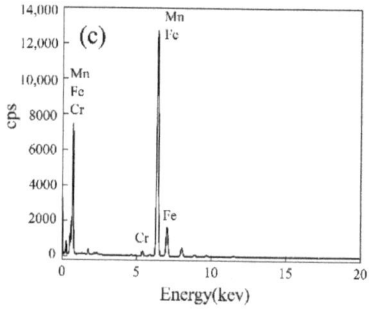

Figure 3. (a) TEM micrographs of M_3C after tempering at 600 °C for 10 min; (b,c) corresponding SAD pattern and EDS analysis of M_3C precipitates, respectively.

The TEM images of the specimens after being tempered for 5 h are shown in Figure 6. It was obvious that more M_2C particles precipitated, and the angle between the particles was about 60°. In addition, some of the M_2C particles connected and formed long-line precipitates, as demonstrated by the white oval in Figure 6a and the black arrow in Figure 6b. The EDS analysis results and the diffraction pattern of these precipitates are indicated in Figure 6c,d. The needle-like precipitates had a hexagonal structure and mainly contained Fe and certain amounts of Cr and Mo. The corresponding PDF retrieval number was 05-0720; thus, the needle-like precipitates were identified to be M_7C_3 carbide.

The TEM images of the samples tempered at 600 °C for 100 h are demonstrated in Figure 7. After tempering for 100 h, the density of the carbides obviously increased, compared with the specimen tempered for 5 h. Most of the needle-like M_7C_3 particles disappeared, and some of the needle-like M_7C_3 particles transformed into rod-like carbides, as shown in Figure 7a. The corresponding electron diffraction pattern is shown in Figure 7b. This type of carbide had a complex cubic structure, which was consistent with the previously identified $M_{23}C_6$. This result showed that the M_7C_3-type precipitates transformed into $M_{23}C_6$ with a prolonged tempering time. Most of the coarse $M_{23}C_6$ carbides precipitated at the grain boundaries (Figure 7c) with an average size of 86.83 nm. Two kinds of MC carbide particles remained stable during the 100 h tempering, as shown in Figure 7c,d. The diameters of the MC particles presented almost no significant change when compared with the specimen tempered at 600 °C for 10 min. The average size of the MC(I) and MC(II) was about 5.02 and 54.34 nm after tempering for 100 h, respectively.

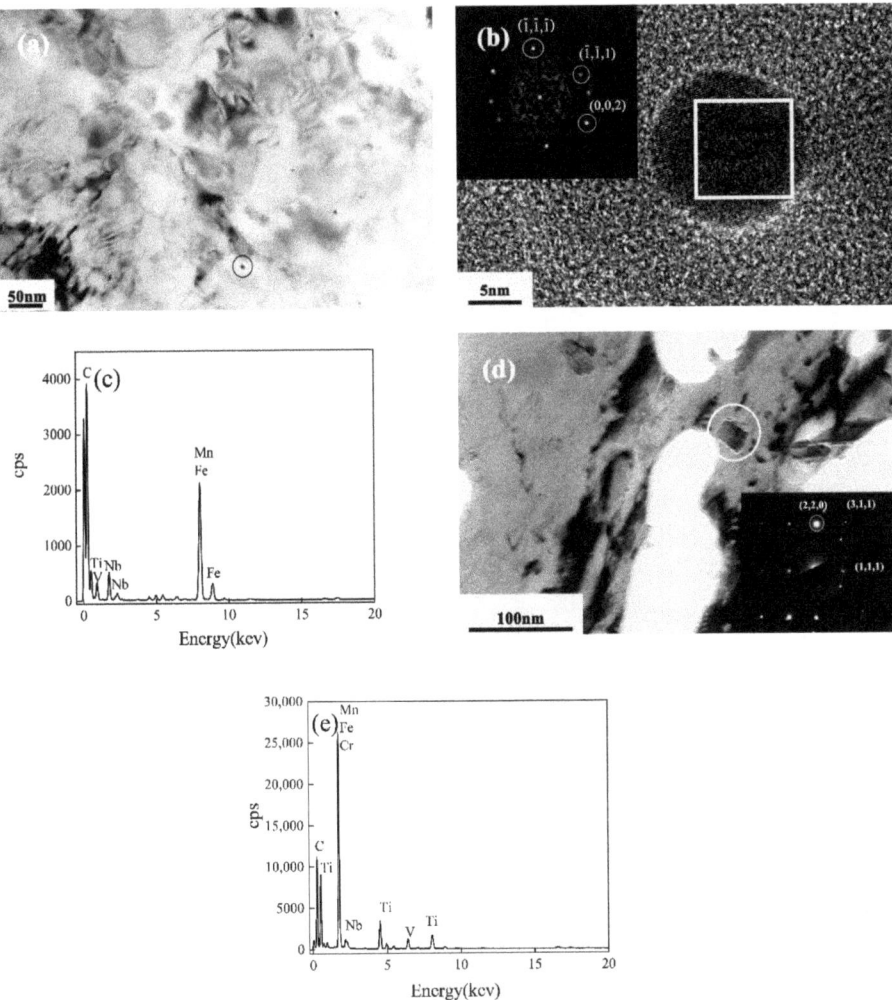

Figure 4. (**a**–**c**) TEM micrographs, corresponding SAD pattern, and EDS analysis of Nb-rich MC (I) carbide; (**d**,**e**) TEM micrographs, corresponding SAD pattern, and EDS analysis of Ti-rich MC (II) particles after tempering at 600 °C for 10 min.

Meta-stable carbides precipitated during tempering can be gradually replaced by more stable precipitates. The precipitation sequence of different types of carbides relates to the chemical composition of steel, the relative diffusion coefficient of the different alloying atoms, and the nucleation sites of the matrix [19]. Therefore, it is essential to comprehend the transformation sequence of carbides. Several kinds of carbides after tempering at 600 °C for different times were identified above. The transformation sequence of the carbides at different stages is summarized as follows.

Stage 1 (tempered for 10 min): three kinds of carbides (M_3C, MC(I), and MC(II)) were found after tempering at 600 °C for 10 min. The quick precipitation of the M_3C particles could be due to a carbon-diffusion-controlled reaction [26]. Even the meta-stable M_3C precipitated before the other carbides due to the relatively rapid diffusion of interstitial carbon, while the substitutionary atoms remained indiffusible [27]. Some researchers [28] have suggested that smaller spherical MC(I)-type carbide particles precipitate in the ferrite region following a K-S relationship with ferrite. MC(II)-type carbides precipitate in austenite regions, exhibiting a cube–cube K-S relationship with the matrix. MC-type second-phase particles result in the strengthening of steel by forming fine and densely distributed carbides [29].

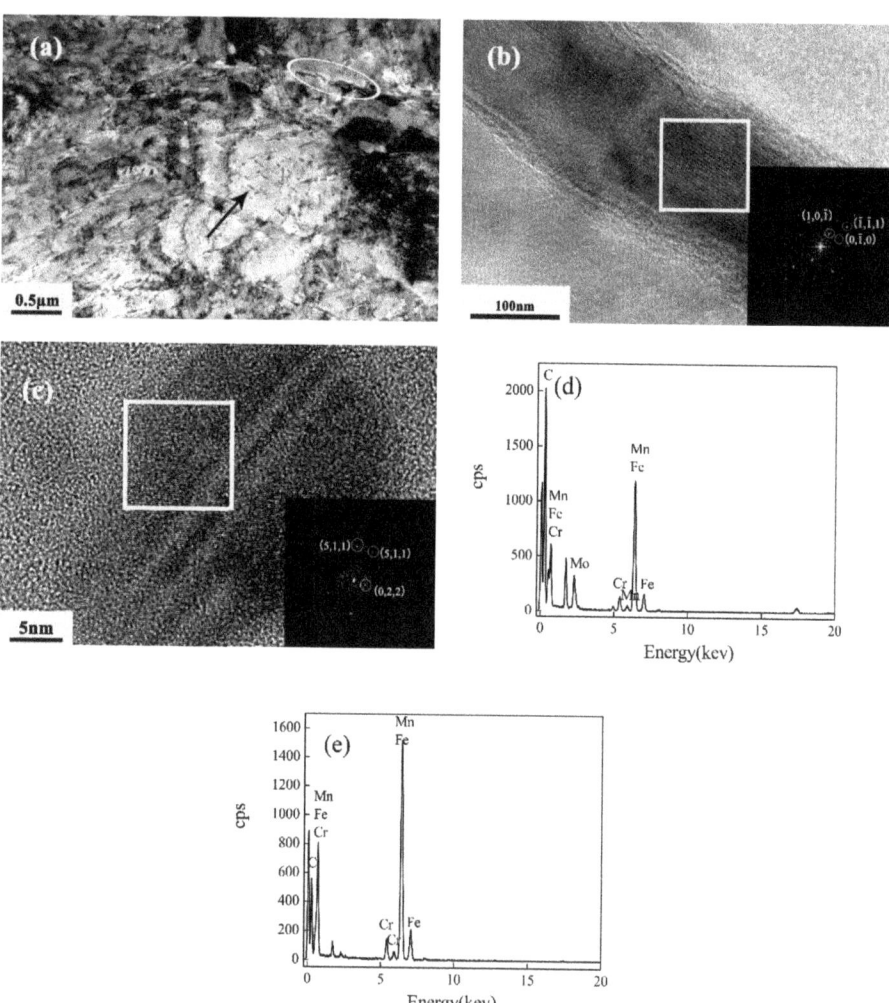

Figure 5. (**a**,**b**,**d**) TEM micrographs, corresponding SAD pattern, and EDS analysis of M_2C particles; (**c**,**e**) TEM micrographs, corresponding SAD pattern, and EDS analysis of $M_{23}C_6$ precipitates after tempering at 600 °C for 30 min.

Stage 2 (tempered for 30 min): the M_3C carbides disappeared, and the M_2C and $M_{23}C_6$ carbides precipitated. The formation of M_2C and $M_{23}C_6$ was controlled by substitutionary diffusion, and they hardly precipitated during the early tempering process since the substitutionary alloying components could not sufficiently diffuse during the relatively short tempering process [22]. As the tempering time was prolonged, the substitutionary alloying components could diffuse, which caused the formation of M_2C and $M_{23}C_6$ carbides. The precipitation temperature of the M_2C particles was approximately between 500 and 600 °C, which was mainly determined by the diffusion of Mo [22]. The formation of the M_2C carbides led to the dissolution of the M_3C carbides, which resulted in a secondary hardening effect in the tempered martensitic steel [30]. Fine and dispersive M_2C can improve the toughness of steel, since the precipitation of M_2C results in the dissolution of brittle M_3C carbides [21].

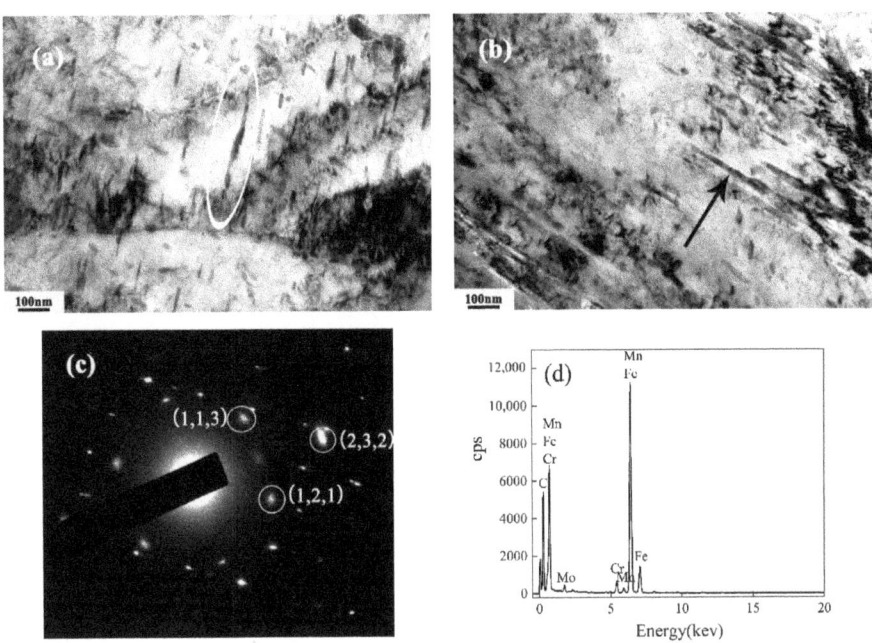

Figure 6. (**a**,**b**) TEM micrographs of steel specimen after tempering at 600 °C for 5 h; (**c**,**d**) corresponding SAD pattern and EDS analysis results of M_7C_3 particles.

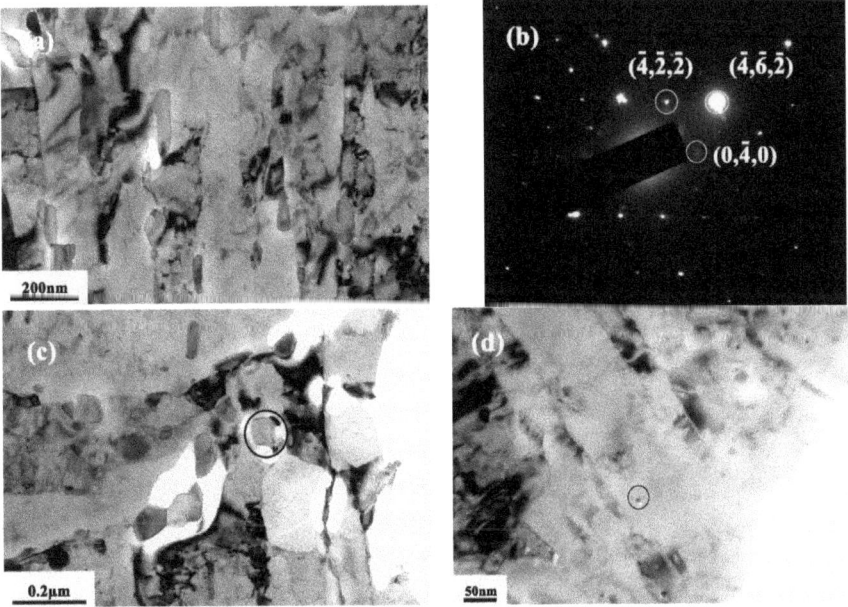

Figure 7. (**a**,**c**,**d**) TEM micrographs of the specimen after tempering at 600 °C for 100 h; (**b**) corresponding SAD pattern of $M_{23}C_6$ particles.

Stage 3 (tempered for 5 h): part of the M_2C dissolved into the matrix and then transformed into M_7C_3 particles. The M_2C in the experimental steel was a meta-stable transition phase. Previous studies have found that M_7C_3 nucleates on M_2C and grows inward into the M_2C through an in situ nucleation mechanism during tempering at 600 °C [31].

Stage 4 (tempered for 100 h): Hou et al. [32] found that the preferential nucleation of M_7C_3 occurred at dislocations and sub-grain boundaries, while $M_{23}C_6$ nucleated on grain and twin boundaries. As the tempering time was prolonged, the dislocation density decreased gradually. The recombination of the sub-grain boundaries caused a decrease in the sub-grain boundaries [33]. Thus, the number of M_7C_3 nuclei decreased and the $M_{23}C_6$ coarsened progressively with the elimination of the M_7C_3 particles. This indicates that the M_7C_3 carbides transformed into $M_{23}C_6$ with the increase in the tempering time [34,35]. The nucleation-free energy of $M_{23}C_6$ and M_7C_3 is -1.82 and -10.43 kJ/mol, respectively [36]. The higher the nucleation-free energy of precipitates, the more stable they are. The nucleation-free energy of $M_{23}C_6$ and M_7C_3 demonstrates that the M_7C_3 precipitated first and then transformed into $M_{23}C_6$. Two kinds of MC particles were also observed in the samples tempered for 100 h. This indicates that the MC(I) and MC(II) precipitates were stable when tempering at 600 °C for different times. During the prolonged tempering process, the precipitates in the specimen consisted of M_3C, M_2C, $M_{23}C_6$, M_7C_3, MC(I), and MC(II) precipitates. The transformation sequence of the other four types of carbides tempered at 600 °C for a long time can be summarized as follows: $M_3C \rightarrow M_2C \rightarrow M_7C_3 \rightarrow M_{23}C_6$.

To further investigate the evolution of the different types of carbides, their extraction from hydrochloric acid was studied by XRD. The XRD patterns of the carbide particles of the specimens tempered at 600 °C under different times are illustrated in Figure 8. M_3C and two types of MC were identified in the sample tempered for 10 min. With increasing the tempering time to 30 min, M_2C, $M_{23}C_6$, and two types of MC were positively identified. After tempering for 100 h at 600 °C, M_2C, M_7C_3, $M_{23}C_6$, and two types of MC carbide were identified. The XRD analysis showed that the peak intensity of the extracted particles was much greater than that of the specimens with shorter tempering times. This means that the number of precipitates obviously increased with increasing the tempering time, and the main carbide particles were $M_{23}C_6$.

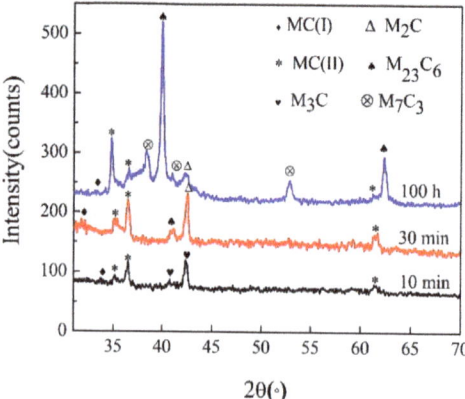

Figure 8. X-ray diffraction patterns of the carbides extracted from investigated steel after being tempered at 600 °C for different lengths of time.

The TEM images of the M_2C carbides after tempering at 600 °C are shown in Figure 9. It was confirmed that fine M_2C carbides were dispersedly distributed in the sample tempered for 30 min. The angle between the particles was about 60°, and the mean length of the M_2C carbides was about 59.04 nm. With increasing the tempering time to 1 h, the M_2C particles precipitated adequately, and the average length increased to 64.78 nm. The increase in the average length of the M$_2$C carbides was attributed to the adequate diffusion of Mo atoms and their combination with carbon to form larger-scale M$_2$C. As the tempering time was prolonged to 10 h, part of the M$_2$C changed into M$_7$C$_3$, and the average length reached a peak value of around 102.48 nm. For the experimental steel tempered for a long time (100 h) at 600 °C, the mean length decreased to 88.32 nm.

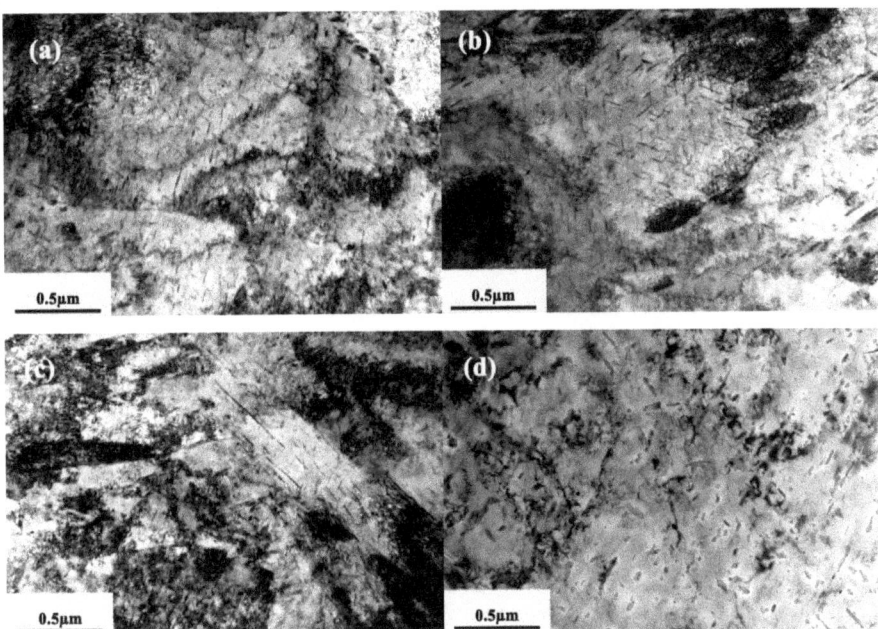

Figure 9. TEM micrographs of M$_2$C precipitates after tempering at 600 °C for different lengths of time: (**a**) 30 min; (**b**) 1 h; (**c**) 10 h; and (**d**) 100 h.

The mean length of the M$_7$C$_3$ particles and the average diameter of the M$_{23}$C$_6$ particles were also measured statistically. The variation in the mean size of the precipitates with the tempering time is given in Figure 10. It can be seen that the mean diameter of the M$_{23}$C$_6$ particles increased gradually with the tempering time. The average length of the M$_2$C and M$_7$C$_3$ particles first increased and then decreased. The variation in the size of carbides can be explained based on coarsening and evolution of the precipitates. The isothermal coarsening equation considering a constant volume fraction of carbides can be expressed as [37]:

$$d^3 - d_0^3 = kt \tag{1}$$

where d and d_0 are the mean diameter of the carbides at different tempering times, k is a constant, and t is the tempering time. Thus, the size of the M$_{23}$C$_6$ carbides increased with the tempering time. In addition, the decreased length of the M$_2$C and M$_7$C$_3$ particles at longer tempering times was correlated with the evolution of the carbide. Since relatively longer M$_2$C and M$_7$C$_3$ particles are apt to transform into M$_7$C$_3$ and M$_{23}$C$_6$, the sizes of the untransformed M$_2$C and M$_7$C$_3$ carbides were relatively small. With a prolonged tempering time, the amount of transformed carbides increased gradually. Therefore, the mean length of the M$_2$C and M$_7$C$_3$ particles decreased when the tempering time was extended to 10 h and 50 h, respectively.

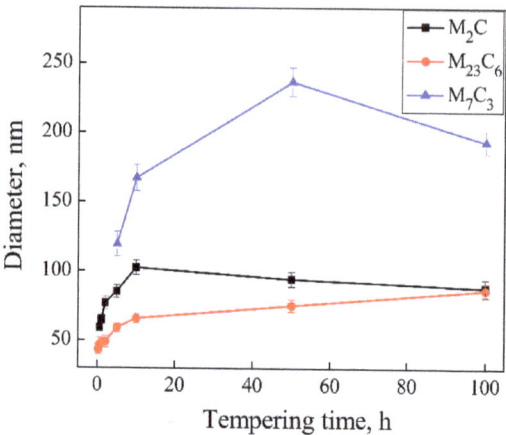

Figure 10. Mean size of precipitates as a function of time during tempering at 600 °C.

3.3. Mechanical Properties under Different Tempering Times

The stress–strain curve of the explored steel under different tempering times is demonstrated in Figure 11. The tensile property data can be seen in Table 1, including the YS, UTS, and total elongation (TE). With the increase in the tempering time, the YS and UTS decreased gradually. The UTS decreased from 1231 to 896 MPa, and the YS decreased from 1138 to 835 MPa.

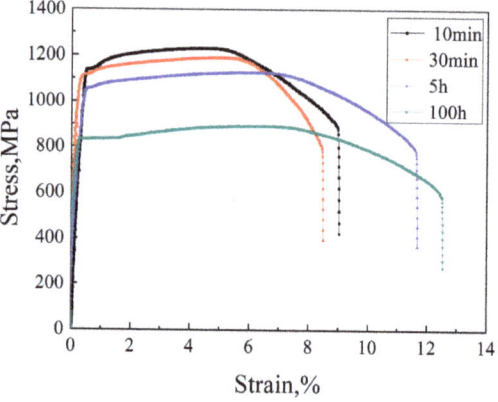

Figure 11. The stress–strain curves of the investigated steel during tempering at 600 °C.

Table 1. Tensile properties of the investigated steel during tempering at 600 °C.

Tempering Time	YS, MPa	UTS, MPa	TE, %
10 min	1138 ± 16	1231 ± 20	8.8 ± 0.3
30 min	1112 ± 12	1193 ± 18	8.4 ± 0.2
5 h	1055 ± 14	1132 ± 21	11.5 ± 0.5
100 h	835 ± 11	896 ± 15	12.3 ± 0.6

Zhang et al. [38] showed that the minimum structural unit that could effectively control the strength of 25CrMo48V micro-alloyed steel was the width of the martensitic lath. The relationship between the width of the martensitic lath and the strength conforms to the Langford–Cohen formula:

$$\sigma_y = \sigma_0 + k_y d^{-1} \qquad (2)$$

where k_y is the structural constant and d is the average lath width. In the Langford–Cohen model, the strength of the material is directly proportional to the reciprocal of the lath width.

With the prolongation of the tempering time, the width of the martensite lath increased and the dislocation decreased gradually. The recovery phenomenon was obviously enhanced, which resulted in matrix softening. Matrix softening was the main reason for the strength reduction. It is generally accepted that σ_0 is mainly composed of Peierls stress, precipitation strengthening, and solution strengthening [39,40]. That is to say, the strength of the sample was determined not only by the martensitic lath width and dislocation density but also by precipitation strengthening and solution strengthening. After tempering for 100 h, the tensile strength was about 900 MPa. The high strength was attributed to the precipitation strengthening from the nano-sized Nb-rich MC (I) particles. According to the Abshy–Oroman mechanism, nano-size precipitates with a size of less than 10 nm can provide about 200 MPa [41].

The −40 °C Charpy impact property of the specimens tempered under different times is shown in Figure 12. It was concluded that the Charpy impact toughness increased gradually with the increase in the tempering time. When the tempering time was extended from 10 min to 100 h, the Charpy impact toughness increased from 20 to 61 J.

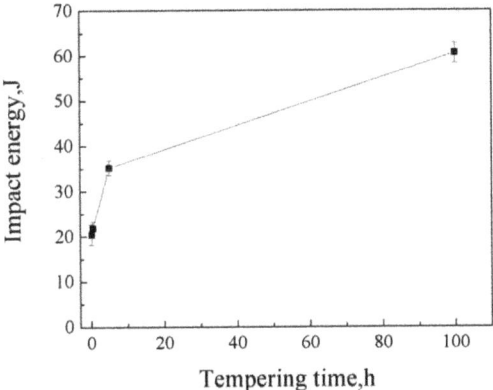

Figure 12. Change in the Charpy impact properties in investigated steel during tempering at 600 °C.

This phenomenon mainly depended on the following factors. One was matrix softening. With the increase in the tempering time, the recovery of the matrix was considerable, and the dislocation density decreased gradually. Furthermore, the increase in the number of precipitates meant that the interstitial atoms dissolved in the matrix precipitated gradually, and the supersaturation decreased, resulting in the softening of the metal matrix [42]. The other was the fact that some small-angle lath boundaries evolved gradually into large-angle boundaries. Several studies have indicated [43] that when cleavage cracks attempt to propagate through a large-angle boundary, the crystal orientation and crack direction may change, and then the growth of cracks may eventually be stopped. Consequently, an increasing amount of large-angle boundaries would lead to a larger deflection of cleavage cracks, and a higher toughness could be obtained.

The correlation between the micro-hardness and tempering time is depicted in Figure 13. The Vickers hardness first increased rapidly to the peak value when the tempering time was 1 h, and it then reduced remarkably after tempering for 5 h. Finally, the hardness decreased monotonically with increasing the tempering time to 100 h. This trend was due to two competitive mechanisms that occurred during the tempering process. The strengthening was due to the precipitation of the carbides, and the softening was due to the recovery of the martensitic matrix. As described above, when the tempering time was between 10 min and 1 h, the hardness increased steadily due to the transformation from M_3C to M_2C. The precipitation of M_2C had a strong secondary hardening effect on the matrix, which results in the increasing hardness during tempering for 1 h. Then, the hardness decreased significantly when the tempering time was 5 h, since most of the M_2C transformed into long-line M_7C_3. The secondary hardening effect was weakened, and precipitate particles were gradually coarsened, which had a detrimental impact on the Vickers hardness. With the increase in the tempering time, the coarsening of the carbides resulted in a reduction in the dislocation pile-up effect [44]. Martinez-de-Guerenu et al. [45] concluded that the recovery percentage of steels during tempering could be estimated by the following equation:

$$1 - R_y = b - a\ln t \tag{3}$$

where R_y is the recovery percentage, which is related to the dislocation density variation, t is the tempering time, and a and b are constants, which are related to the tempering temperature. During the continuous tempering, the recovery of dislocations in the matrix changed with the prolonging of the tempering time, which led to a gradual decrease in the Vickers hardness. It can be concluded that the variation in the Vickers hardness should be derived from the evolution and coarsening of the precipitate and the recovery of the lath martensite.

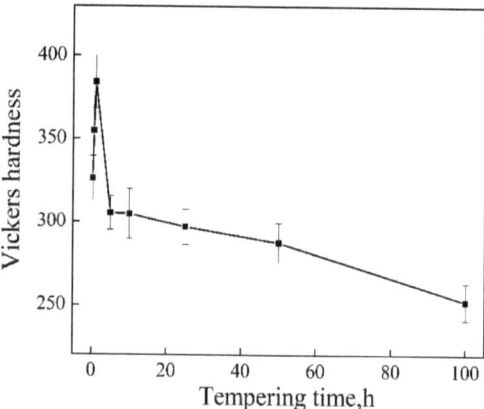

Figure 13. Changes in the Vickers hardness during tempering at 600 °C.

4. Conclusions

The evolution behaviors of the microstructure and precipitates and their effect on the mechanical properties of Nb-V-Ti micro-alloyed steel under different tempering times were explored, and the main conclusions obtained are shown as follows:

1. The width of martensite laths increased and the retained austenite and dislocation decreased gradually with the prolongation of the tempering time.
2. The evolution sequence of the carbides during tempering at 600 °C for the different times was identified as: $M_3C \rightarrow M_2C \rightarrow M_7C_3 \rightarrow M_{23}C_6$. The other two kinds of MC carbides remained stable during the tempering process.
3. The strength decreased and the Charpy impact toughness increased gradually with the prolongation of the tempering time. The Vickers hardness increased remarkably as the tempering time was extended to 1 h and then decreased sharply with a further increase in the tempering time up to 5 h. When the tempering time was between 5 and 100 h, the Vickers hardness values decreased gradually.

Author Contributions: Q.Z. designed the research, performed the research, analyzed the data, and wrote the paper. Z.Q. conceived of the study and designed the study. J.D. developed the idea for the study and collected the data. All authors have read and agreed to the published version of the manuscript.

Funding: This research was funded by National Natural Science Foundation of China grant number 52004181 and the Tianjin Jinnan District Science and Technology Plan Project grant number 20220110.

Data Availability Statement: Data presented in this article are available at request from the corresponding author.

Acknowledgments: The authors are grateful to the National Natural Science Foundation of China (grant nos. 52004181) and the Tianjin Jinnan District Science and Technology Plan Project (grant no. 20220110) for grant and financial support.

Conflicts of Interest: The authors declare no conflict of interest.

References

1. Li, Z.J.; Xiao, N.M.; Li, D.X.; Zhang, J.Y.; Luo, Y.J.; Zhang, R.X. Effect of microstructure evolution on strength and impact toughness of G18CrMo2-6 heat-resistant steel during tempering. *Mater. Sci. Eng. A* **2014**, *604*, 103–110. [CrossRef]
2. Sun, J.; Lian, F.L.; Sun, Y.; Wang, Y.J.; Guo, S.W.; Liu, Y.N. The Fine Grain Effect on a New Carbide Fe_4C_3 Formed in Pipeline Steel X80. *JOM* **2023**, *75*, 417–427. [CrossRef]
3. Li, X.; Shi, L.; Liu, Y.; Gan, K.; Liu, C. Achieving a desirable combination of mechanical properties in HSLA steel through step quenching. *Mater. Sci. Eng. A* **2020**, *772*, 138683. [CrossRef]
4. Mandal, A.; Barik, R.K.; Bhattacharya, A.; Chakrabarti, D.; Davis, C. The Correlation Between Bending, Tensile and Charpy Impact Properties of Ultra-high-Strength Strip Steels. *Metall. Mater. Trans. A* **2023**, 1–24. [CrossRef]
5. Shao, Y.; Liu, C.; Yan, Z.; Li, H.; Liu, Y. Formation mechanism and control methods of acicular ferrite in HSLA steels: A review. *J. Mater. Sci. Technol.* **2018**, *34*, 737–744. [CrossRef]
6. Kan, L.; Ye, Q.; Wang, Z.; Zhao, T. Improvement of strength and toughness of 1 GPa Cu-bearing HSLA steel by direct quenching. *Mater. Sci. Eng. A* **2022**, *855*, 143875. [CrossRef]
7. Liu, G.; Li, Y.; Liao, T.; Wang, S.; Lv, B.; Guo, H.; Huang, Y.; Yong, Q.; Mao, X. Revealing the precipitation kinetics and strengthening mechanisms of a 450 MPa grade Nb-bearing HSLA steel. *Mater. Sci. Eng. A* **2023**, *884*, 145506. [CrossRef]
8. Kar, S.; Srivastava, V.C.; Mandal, G.K. Low-density nano-precipitation hardened Ni-based medium entropy alloy with excellent strength-ductility synergy. *J. Alloys Compd.* **2023**, *963*, 171213. [CrossRef]
9. Huang, W.; Lei, L.; Fang, G. Optimizing Heat Treatment to Improve the Microstructures and Mechanical Properties of 5CrNiMoV Steel. *Metals* **2023**, *13*, 1263. [CrossRef]
10. Shi, L.; Yan, Z.; Liu, Y.; Zhang, C.; Qiao, Z.; Ning, B.; Li, H. Improved toughness and ductility in ferrite/acicular ferrite dual-phase steel through intercritical heat treatment. *Mater. Sci. Eng. A* **2014**, *590*, 7–15. [CrossRef]
11. Dudko, V.; Belyakov, A.; Molodov, D.; Kaibyshev, R. Microstructure evolution and pinning of boundaries by precipitates in a 9 pct Cr heat resistant steel during creep. *Metall. Mater. Trans. A* **2013**, *44*, 162–172. [CrossRef]
12. Chen, H.; Zhang, C.; Zhu, J.; Yang, Z.; Ding, R.; Zhang, C.; Yang, Z. Austenite/ferrite interface migration and alloying elements partitioning: An overview. *Acta Metall. Sin.* **2017**, *54*, 217–227.
13. Dong, J.; Zhou, X.; Liu, Y.; Li, C.; Liu, C.; Li, H. Effects of quenching-partitioning-tempering treatment on microstructure and mechanical performance of Nb-V-Ti microalloyed ultra-high strength steel. *Mater. Sci. Eng. A* **2017**, *690*, 283–293. [CrossRef]
14. Roy, S.; Romualdi, N.; Yamada, K.; Poole, W.; Militzer, M.; Collins, L. The Relationship Between Microstructure and Hardness in the Heat-Affected Zone of Line Pipe Steels. *JOM* **2022**, *74*, 2395–2401. [CrossRef]
15. Wu, H.; Ju, B.; Tang, D.; Hu, R.; Guo, A.; Kang, Q.; Wang, D. Effect of Nb addition on the microstructure and mechanical properties of an 1800 MPa ultrahigh strength steel. *Mater. Sci. Eng. A* **2015**, *622*, 61–66. [CrossRef]
16. Mandal, A.; Bandyopadhyay, T.K. Effect of tempering on microstructure and tensile properties of niobium modified martensitic 9Cr heat resistant steel. *Mater. Sci. Eng. A* **2015**, *620*, 463–470. [CrossRef]
17. Shen, Y.; Liu, H.; Shang, Z.; Xu, Z. Precipitate phases in normalized and tempered ferritic/martensitic steel P92. *J. Nucl. Mater.* **2015**, *465*, 373–382. [CrossRef]
18. Moon, J.; Choi, J.; Han, S.K.; Huh, S.; Kim, S.J.; Lee, C.H.; Lee, T.H. Influence of precipitation behavior on mechanical properties and hydrogen induced cracking during tempering of hot-rolled API steel for tubing. *Mater. Sci. Eng. A* **2016**, *652*, 120–126. [CrossRef]
19. Tao, X.G.; Han, L.Z.; Gu, J.F. Effect of tempering on microstructure evolution and mechanical properties of X12CrMoWVNbN10-1-1 steel. *Mater. Sci. Eng. A* **2014**, *618*, 189–204. [CrossRef]
20. Stornelli, G.; Tselikova, A.; Mirabile Gattia, D.; Mortello, M.; Schmidt, R.; Sgambetterra, M.; Testani, C.; Zucca, G.; Di Schino, A. Influence of vanadium micro-alloying on the microstructure of structural high strength steels welded joints. *Materials* **2023**, *16*, 2897. [CrossRef]

21. Janovec, J.; Svoboda, M.; Kroupa, A.; Výrostková, A. Thermal-induced evolution of secondary phases in Cr–Mo–V low alloy steels. *J. Mater. Sci.* **2006**, *41*, 3425–3433. [CrossRef]
22. Janovec, J.; Svoboda, M.; Výrostková, A.; Kroupa, A. Time–temperature–precipitation diagrams of carbide evolution in low alloy steels. *Mater. Sci. Eng. A* **2005**, *402*, 288–293. [CrossRef]
23. Jia, C.; Liu, Y.; Liu, C.; Li, C.; Li, H. Precipitates evolution during tempering of 9CrMoCoB (CB2) ferritic heat-resistant steel. *Mater. Charact.* **2019**, *152*, 12–20. [CrossRef]
24. Janovec, J.; Vyrostkova, A.; Svoboda, M. Influence of tempering temperature on stability of carbide phases in 2.6 Cr–0.7 Mo–0.3 V steel with various carbon content. *Metall. Mater. Trans. A* **1994**, *25*, 267–275. [CrossRef]
25. Li, C.; Duan, R.; Fu, W.; Gao, H.; Wang, D.; Di, X. Improvement of mechanical properties for low carbon ultra-high strength steel strengthened by Cu-rich multistructured precipitation via modification to bainite. *Mater. Sci. Eng. A* **2021**, *817*, 141337. [CrossRef]
26. Thomson, R.C.; Bhadeshia, H.K.D.H. Atom probe and STEM studies of carbide precipitation in 2sol14Cr1Mo steel. *Appl. Surf.* **1993**, *67*, 334–341. [CrossRef]
27. Jain, D.; Isheim, D.; Hunter, A.H.; Seidman, D.N. Multicomponent high-strength low-alloy steel precipitation-strengthened by sub-nanometric Cu precipitates and M2C carbides. *Metall. Mater. Trans. A* **2016**, *47*, 3860–3872. [CrossRef]
28. Lee, W.B.; Hong, S.G.; Park, C.G.; Park, S.H. Carbide precipitation and high-temperature strength of hot-rolled high-strength, low-alloy steels containing Nb and Mo. *Metall. Mater. Trans. A* **2002**, *33*, 1689–1698. [CrossRef]
29. Fedoseeva, A.; Dudova, N.; Glatzel, U.; Kaibyshev, R. Effect of W on tempering behaviour of a 3% Co modified P92 steel. *J. Mater. Sci.* **2016**, *51*, 9424–9439. [CrossRef]
30. Chakraborty, G.; Das, C.R.; Albert, S.K.; Bhaduri, A.K.; Paul, V.T.; Panneerselvam, G.; Dasgupta, A. Study on tempering behaviour of AISI 410 stainless steel. *Mater. Charact.* **2015**, *100*, 81–87. [CrossRef]
31. Dong, J.; Zhou, X.; Liu, Y.; Li, C.; Liu, C.; Guo, Q. Carbide precipitation in Nb-V-Ti microalloyed ultra-high strength steel during tempering. *Mater. Sci. Eng. A* **2017**, *683*, 215–226. [CrossRef]
32. Hou, Z.; Hedström, P.; Chen, Q.; Xu, Y.; Wu, D.; Odqvist, J. Quantitative modeling and experimental verification of carbide precipitation in a martensitic Fe–0.16 wt% C–4.0 wt% Cr alloy. *Calphad* **2016**, *53*, 39–48. [CrossRef]
33. Kasana, S.S.; Pandey, O.P. Effect of electroslag remelting and homogenization on hydrogen flaking in AMS-4340 ultra-high-strength steels. *Int. J. Min. Met. Mater.* **2019**, *26*, 611–621. [CrossRef]
34. Li, J.; Zhang, C.; Liu, Y. Influence of carbides on the high-temperature tempered martensite embrittlement of martensitic heat-resistant steels. *Mater. Sci. Eng. A* **2016**, *670*, 256–263. [CrossRef]
35. Kolluri, M.; Martin, O.; Naziris, F.; D'Agata, E.; Gillemot, F.; Brumovsky, M.; Ulbricht, A.; Autio, J.-M.; Shugailo, A.; Horvath, A. Structural MATerias research on parameters influencing the material properties of RPV steels for safe long-term operation of PWR NPPs. *Nucl. Eng. Des.* **2023**, *406*, 112236. [CrossRef]
36. Tao, X.; Gu, J.; Han, L. Characterization of precipitates in X12CrMoWVNbN10-1-1 steel during heat treatment. *J. Nucl. Mater.* **2014**, *452*, 557–564. [CrossRef]
37. Maddi, L.; Deshmukh, G.S.; Ballal, A.R.; Peshwe, D.R.; Paretkar, R.K.; Laha, K.; Mathew, M.D. Effect of Laves phase on the creep rupture properties of P92 steel. *Mater. Sci. Eng. A* **2016**, *668*, 215–223. [CrossRef]
38. Zhang, C.; Wang, Q.; Ren, J.; Li, R.; Wang, M.; Zhang, F.; Yan, Z. Effect of microstructure on the strength of 25CrMo48V martensitic steel tempered at different temperature and time. *Mater. Des.* **2012**, *36*, 220–226. [CrossRef]
39. Caballero, F.G.; Santofimia, M.J.; García-Mateo, C.; Chao, J.; De Andres, C.G. Theoretical design and advanced microstructure in super high strength steels. *Mater. Des.* **2009**, *30*, 2077–2083. [CrossRef]
40. Dai, L.; Liu, Z.; Yu, L.; Liu, Y.; Shi, Z. Microstructural characterization of Mg–Al–O rich nanophase strengthened Fe–Cr alloys. *Mater. Sci. Eng. A* **2020**, *771*, 138664. [CrossRef]
41. Kamikawa, N.; Sato, K.; Miyamoto, G.; Murayama, M.; Sekido, N.; Tsuzaki, K.; Furuhara, T. Stress–strain behavior of ferrite and bainite with nano-precipitation in low carbon steels. *Acta Mater.* **2015**, *83*, 383–396. [CrossRef]
42. Yan, P.; Liu, Z.; Bao, H.; Weng, Y.; Liu, W. Effect of tempering temperature on the toughness of 9Cr–3W–3Co martensitic heat resistant steel. *Mater. Des.* **2014**, *54*, 874–879. [CrossRef]
43. Li, X.; Cai, Z.; Hu, M.; Li, K.; Hou, M.; Pan, J. Effect of NbC precipitation on toughness of X12CrMoWNbVN10-1-1 martensitic heat resistant steel for steam turbine blade. *J. Mater. Res. Technol.* **2021**, *11*, 2092–2105. [CrossRef]
44. Zhang, X.; Li, H.; Zhan, M.; Zheng, Z.; Gao, J.; Shao, G. Electron force-induced dislocations annihilation and regeneration of a superalloy through electrical in-situ transmission electron microscopy observations. *J. Mater. Sci. Technol.* **2020**, *36*, 79–83. [CrossRef]
45. Martınez-de-Guerenu, A.; Arizti, F.; Gutiérrez, I. Recovery during annealing in a cold rolled low carbon steel. Part II: Modelling the kinetics. *Acta Mater.* **2004**, *52*, 3665–3670. [CrossRef]

Disclaimer/Publisher's Note: The statements, opinions and data contained in all publications are solely those of the individual author(s) and contributor(s) and not of MDPI and/or the editor(s). MDPI and/or the editor(s) disclaim responsibility for any injury to people or property resulting from any ideas, methods, instructions or products referred to in the content.

Communication

Hierarchical Multiple Precursors Induced Heterogeneous Structures in Super Austenitic Stainless Steels by Cryogenic Rolling and Annealing

Duo Tan [1,*], Bin Fu [1,*], Wei Guan [1], Yu Li [1], Yanhui Guo [1], Liqun Wei [1] and Yi Ding [2]

1. School of Materials Science and Engineering, Shanghai Institute of Technology, Shanghai 201418, China
2. Baowu Special Metallurgy Co., Ltd., Shanghai 200940, China
* Correspondence: tanduo2023@163.com (D.T.); bin_fu@163.com (B.F.)

Abstract: Multiple deformed substructures including dislocation cells, nanotwins (NTs) and martensite were introduced in super austenitic stainless steels (SASSs) by cryogenic rolling (Cryo-R, 77 K/22.1 mJ·m^{-2}). With the reduction increasing, a low stacking fault energy (SFE) and increased flow stress led to the activation of secondary slip and the occurrence of NTs and martensite nanolaths, while only dislocation tangles were observed under a heavy reduction by cold-rolling (Cold-R, 293 K/49.2 mJ·m^{-2}). The multiple precursors not only possess variable deformation stored energy, but also experience competition between recrystallization and reverse transformation during subsequent annealing, thus contributing to the formation of a heterogeneous structure (HS). The HS, which consists of bimodal-grained austenite and retained martensite simultaneously, showed a higher yield strength (~1032 MPa) and a larger tensile elongation (~9.1%) than the annealed coarse-grained Cold-R sample. The superior strength–ductility and strain hardening originate from the synergistic effects of grain refinement, dislocation and hetero-deformation-induced hardening.

Keywords: super austenitic stainless steel; cryogenic rolling; heterogeneous structure; microstructure; mechanical properties

Citation: Tan, D.; Fu, B.; Guan, W.; Li, Y.; Guo, Y.; Wei, L.; Ding, Y. Hierarchical Multiple Precursors Induced Heterogeneous Structures in Super Austenitic Stainless Steels by Cryogenic Rolling and Annealing. *Materials* 2023, 16, 6298. https://doi.org/10.3390/ma16186298

Academic Editors: Andrea Di Schino and Claudio Testani

Received: 18 August 2023
Revised: 10 September 2023
Accepted: 18 September 2023
Published: 20 September 2023

Copyright: © 2023 by the authors. Licensee MDPI, Basel, Switzerland. This article is an open access article distributed under the terms and conditions of the Creative Commons Attribution (CC BY) license (https://creativecommons.org/licenses/by/4.0/).

1. Introduction

Super austenitic stainless steels (SASSs) have been widely used in extremely harsh service environments such as flue gas desulfurization, seawater desalination, and the petrochemical industry [1,2]. However, the yield strength (YS) of SASS is relatively low due to its large grain size and low dislocation density, which strongly limit its further applications. Microstructure refinement is an effective method to improve its YS, and some severe plastic deformation (SPD) methods such as dynamic plastic deformation (DPD) [3,4] and surface mechanical grinding treatment (SMGT) [5,6] are often conducted to obtain the nanostructure; however, they are always accompanied by a significant deterioration in ductility. On this basis, the strategy of heterogeneous structures (HSs), such as bimodal, harmonic and hierarchical lamellar structures, was proposed to achieve a good combination of strength and ductility [7–9]. Generally, the commonly available approach to obtain bulk HS in conventional austenitic stainless steels (ASSs) is through annealing after severe cold deformation [10]. The heterogeneity was obtained due to a competitive process between the reverse transformation of strain-induced martensite and the recrystallization of a dislocation-cell-type structure [3], thus resulting in a hierarchical distribution of grain sizes, from tens of manometers to several micro-meters. However, the deformation mode of the studied SASS at an ambient temperature is mainly dislocation-dominated shear band, owing to a heavy alloying of Mo and N and increased stacking fault energy (SFE) [11,12] during cold deformation. The homogeneous rolled substructure hardly contributes to the HS formation due to a similar deformation stored energy for the nucleation and growth of recrystallization [13]. Additionally, dislocation annihilation during recovery or recrystallization impedes structural refinement by walls or cells [14], resulting in low YS.

Recently, an innovative cryogenic rolling (Cryo-R) procedure was performed in conventional ASSs to obtain nanostructures with high YS [15]. An effective inhibition of dislocation recovery during Cryo-R contributed to a larger stored energy and higher driving force for the nucleation of recrystallization compared to the traditional Cold-R [16]. However, most investigations have ignored the advantage of Cryo-R in activating multiple deformed substructures, including dislocation tangles, mechanical twins and deformation-induced martensite (DIM) [12,17], due to a temperature-controlled variable SFE, since the diversity of precursor is a crucial factor in obtaining a heterogeneous grain-size distribution. Firstly, the recrystallization dynamics are closely related to the nucleation site, such as the dislocation cell and twin boundaries [18–20]. Some studies have shown that the nanocrystalline nucleation rate in the grain boundary, nanotwin and shear band regions of precursors was faster than that in the dislocation region of coarse crystals during annealing [21,22]. Moreover, the reversed austenite grains may inherit the lath morphology of the DIM after the reverse transformation [23] compared to the equiaxed recrystallized grains. However, little attention has been paid to the effects of deformed precursors with complex substructures in obtaining HS. The motivation of the present study is to investigate the relationship among the HSs, deformation modes and the evolution of multiple precursors during Cryo-R, which provides a guiding design of high-strength SASS for extreme applications.

2. Materials and Methods

In this work, the chemical composition of SASS was listed as follows: 0.024 C, 0.38 Si, 0.98 Mn, 0.027 P, 0.004 S, 16.9 Cr, 13.23 Ni, 5.3 Mo, 0.13 N and balanced Fe, wt. %. The cast ingots were cut into 5 × 30 × 100 mm samples and then heated in a muffle furnace at 1200 °C for 2 h for the solution treatment, followed by water cooling. A multi-pass Cryo-R procedure was conducted in a rolling mill (ø180 mm) with different total reductions of 30%, 50%, and 70%. Before Cryo-R, the iron oxide scale on the surface was removed with coarse sandpaper, and a soak for 15 min was made at a cryogenic temperature. For comparison, Cold-R samples with a 70% reduction according to the same rolling parameters were prepared. Moreover, the 70% rolled samples were subsequently annealed at 700 °C for 10 min. The X-ray (XRD, D/max 2200 PC; Shimadzu, Kyoto, Japan) analysis was conducted by Cu target Kα rays with a scan rate of 2°/min from 40° to 100°. The Vickers hardness values were performed on a 402SXV hardness tester for 5 s with 300 g of load retention. For transmission electron microscope (TEM) observation, the samples were thinned by double-spray electrolysis, and the double-spray liquid was a perchloric acid alcohol solution (voltage and temperature were 30 V and −30 °C, respectively). Then, the samples were observed by FEI TECNAI TEM (FEI, Lexington, KY, USA). The electron backscatter diffraction (EBSD) characterizations of the RD-ND surface were investigated by MIR3 high-resolution scanning electron microscopy at a 20 kV acceleration voltage.

3. Results and Discussion

Figure 1 shows the evolution of the XRD patterns and Vicker hardness values of SASS before and after Cryo/Cold-R with different reductions. The diffraction peaks of all samples showed a single austenitic phase structure, which possesses a strong mechanical stability even at cryogenics. However, the DIM was observed in the Cryo-R samples by TEM analysis later; this inaccuracy was mainly due to its low volume fraction and nano-lath size. Moreover, the hardness value showed a monotonous increasing tendency with the cryogenic reduction increasing. However, it increased slowly during the early rolling process but showed a high increasing rate after the 30% reduction, which demonstrated the change in deformation modes. In addition, the hardness of the Cryo-R sample was obviously higher than that of the Cold-R sample at an equivalent strain of 70%, indicating that the former possessed a higher deformation stored energy due to the inhibition of dislocation recovery.

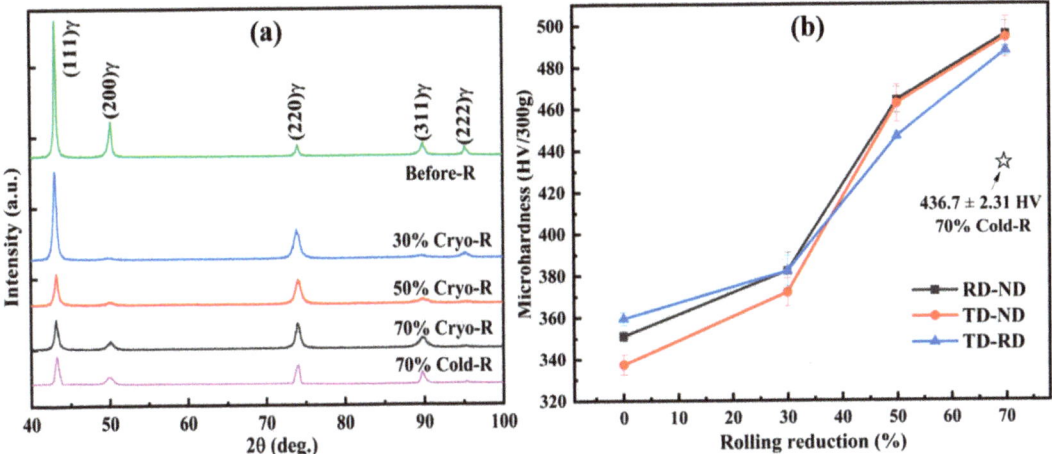

Figure 1. (a) The XRD patterns and (b) Vickers hardness values of SASS before and after Cryo/Cold-R with different reductions.

Figure 2 depicts the typical TEM images of the Cryo-R SASS sample after 30% and 50% reductions. At low strains, a large number of slip bands were accompanied with dislocation cells or walls, and the motion of dislocations continuously crossed the matrix [24]. The corresponding selected area electron diffraction spots (SAED) shown in Figure 2a illustrate a single austenitic structure in this area, indicating that the martensitic transformation hardly occurs in the early cryogenic deformation stage. Furthermore, several nanotwins (NTs) were also observed, which was confirmed by SAED in Figure 2b. At low temperatures, the motion of dislocations was suppressed and instead the formation of twins was strongly promoted due to a decreased SFE [25]. According to the statistics, the average thickness of T/M lamellae was ~24.64 nm. With strain progressing to 50%, the further evolution of dislocation walls and cells not only inhibited dislocation slip, but also promoted an effective grain refinement [26]. More importantly, the enhanced nucleation driving force led to a significantly increased number density of NTs (Figure 2d), and the average size of T/M lamellar thickness was reduced to ~16.43 nm.

When the rolling reduction increased to 70%, the austenitic grains were strongly refined; the ring-shaped SAED (Figure 3a) also indicated that the number of high-angle grain boundaries [27] obviously increased with a high degree of lattice distortion [28]. In addition, some special parallelogram regions of shear band interactions could be observed, and their interactions acted as the nucleation sites of the deformation induced α' martensite, which possessed a block morphology with a grain size of several micro meters.

Moreover, some high-density secondary NTs could be observed between slip bands, and they showed a very thin T/M lamellar thickness of ~4.84 nm. The generation of secondary NTs was mainly caused by the strain gradient between adjacent grain boundaries [29]. Simultaneously, some nano-lath ε martensite also formed in the deformed ultra-fine grains (Figure 3c), which might be attributed to a limited dislocation storage and motion ability. In contrast, the formation of NTs or martensite was hardly observed in the Cold-R samples and only a large number of dislocation structures was evident in the grain interior.

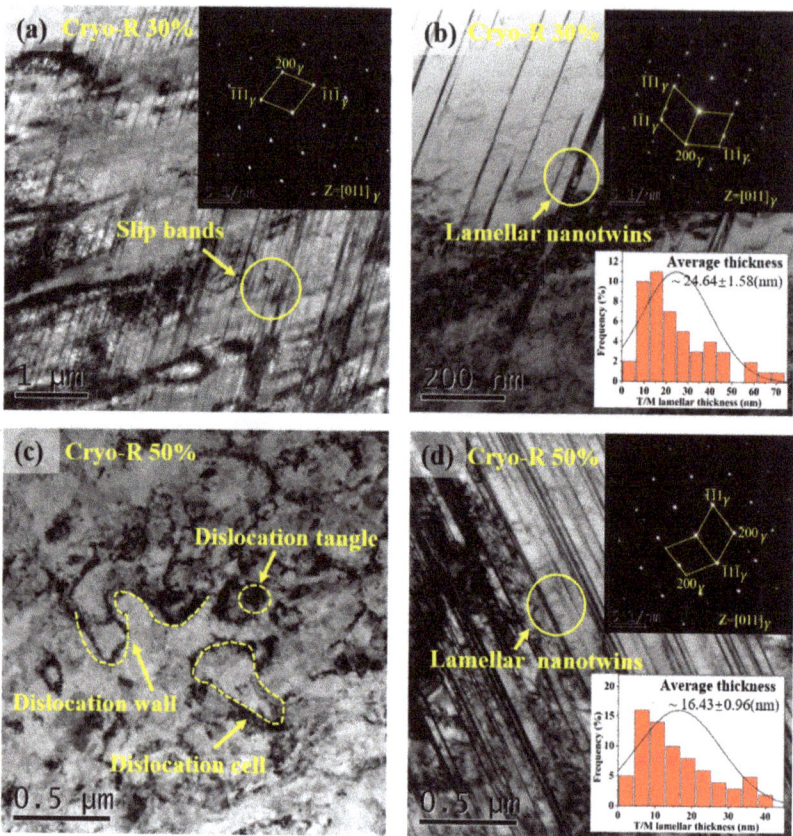

Figure 2. TEM micrographs of the SASSs after Cryo-R: (**a**) bright field image (BF) showing slip bands with 30% reduction; (**b**) BF showing lamellar NTs with 30% reduction; (**c**) BF showing dislocation structural with 50% reduction; (**d**) BF showing the lamellar NTs with 50% reduction.

As illustrated above, a hierarchical multiple deformed substructure, combined with dislocation cells, NTs, ε martensite and α′ martensite, could be found in the SASS during the cryogenic rolling process. The dislocation cells and walls dominate at low strains and the enhanced planar slip leads to the formation of slip bands. With the deformation progressing to a medium stage, a limited dislocation motion ability and increased flow stress result in the occurrence of high-density deformation twins and secondary slip bands, contributing to an obvious grain-refinement effect. Furthermore, the strain-induced α′ martensite nucleates in the interactions of shear bands and some stress-assisted ε martensite forms within the dislocation cell boundaries at large strains. The generation of such a multi-stage deformation mechanism is closely related to the temperature-controlled SFE and the conditions for the twin and martensite formation. Firstly, using the previously reported thermodynamic model by Saeed-Akbari et al. [30], and considering the change in SFE Γ by temperature besides alloy composition and grain size, Equation (1) is expressed as follows:

$$\Gamma = 2\rho\Delta G^{\gamma \to \varepsilon} + 2\sigma^{\gamma/\varepsilon} + 2\rho\Delta G_{ex} \qquad (1)$$

where ρ is the molar surface density along {111} planes and $\Delta G^{\gamma \to \varepsilon}$ and $\sigma^{\gamma/\varepsilon}$ are the free energy change and interfacial energy; ΔG_{ex} is the excess free energy due to the grain-size effect. Thus, the SFE values of the studied SASS are calculated as ~49.2 mJ·m^{-2} and ~22.1 mJ·m^{-2} at 293 K and 77 K, respectively. Moreover, according to the classic dislocation theory, the critical shear stress (CRSS) for the formation of the deformation twin can be expressed as follows [31]:

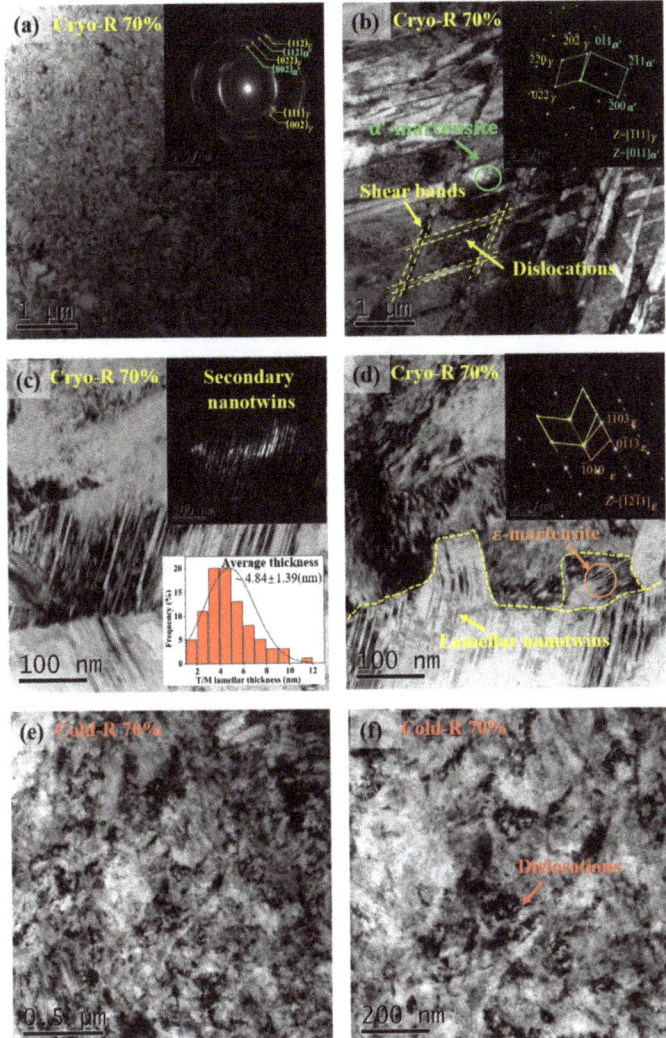

Figure 3. TEM micrographs of Cryo/Cold-R sample with 70% reduction: (**a**) BF and corresponding SAED; (**b**) BF showing the shear bands, α′ martensite; (**c**) BF showing the lamellar NTs, ε-martensite; (**d**) TEM micrograph of Cold-R sample with 70% reduction; (**e**) BF showing the matrix; (**f**) BF showing the dislocations.

$$\tau_{twin} = \frac{2\Gamma}{b_p} \quad (2)$$

where Γ is the SFE value, and b_p is the Burgers vector of partial dislocations. Also, the driving force for the martensitic formation can be expressed as follows:

$$-\Delta G^{\gamma \to \varepsilon} + \tau_{MT} s V_m = \int_\gamma^\varepsilon T dS \quad (3)$$

where τ_{MT} is the CRSS for the onset of $\gamma \to \varepsilon$ transformation, s is the homogeneous transformation shear strain, V_m is the molar volume of austenite, $\int_\gamma^\varepsilon TdS$ is the change in entropy during $\gamma \to \varepsilon$ transformation. The parameter details are referred to in the previous work [32]. Thus, the CRSS values of twinning and martensite can be calculated as ~671 MPa and ~630 MPa at 293 K, and ~301 MPa and ~350 MPa at 77 K, respectively; that is, a lower CRSS results in the sequence of nanotwins and martensite when deformed at cryogenics, while they hardly form at RT due to the high CRSS values.

Figure 4 shows the microstructural characteristic of annealed SASS after Cryo-R and Cold-R, including the inverse pole figures (IPFs), phase distribution maps and grain-size distributions from EBSD. A partial recrystallization process exists in both samples due to an insufficient annealing time. The annealed Cryo-R samples show a complex phase constitution of retained martensite and recrystallized and reversed austenite, which possesses a bimodal grain-size distribution, that is, ultrafine grains (~0.57 μm) and coarse grains (~2 μm). The reverse transformation of strain-induced martensite cannot be completed due to a low migration distance of γ/α' phase boundary during the short-term annealing at 700 °C. However, the annealed Cold-R samples show a fully austenitic microstructure, consisting of coarse recovered grains and fine recrystallized grains, and their average grain size is higher than that of annealed Cryo-R samples. Compared with the Cold-R samples, the annealed Cryo-R samples show a larger heterogeneity and wider grain-size distributions; this phenomenon can be attributed to the hierarchical multiple precursors which experience competition among reverse transformation, recrystallization and recovery. For instance, the reversed austenite is prone to inherit the nano-lath morphology of martensite [33], while the growth of recrystallized grains in the deformation twin is strongly suppressed by the twin plane [34], and they exhibit a significantly smaller grain size than those nucleating in the dislocation cell boundaries. Moreover, the Cryo-R process leads to higher deformation stored energy than the Cold-R at an equivalent strain [35], which facilitates the improvement in the nucleation rate and contributes to a smaller average grain size [29].

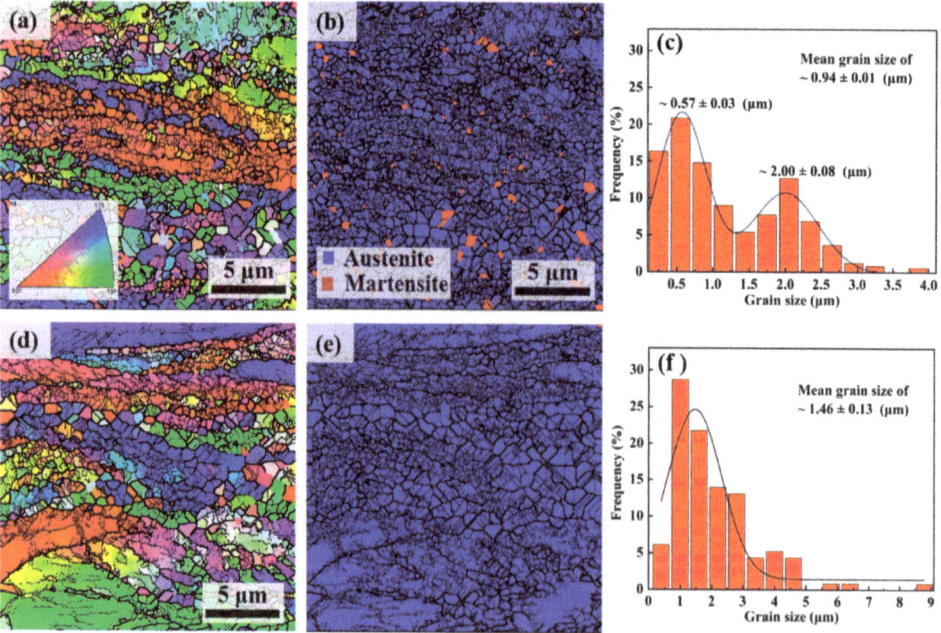

Figure 4. IPF maps, phase distribution maps and grain-size distribution of annealed SASS after (**a**–**c**) Cryo-R and (**d**–**f**) Cold-R.

Figure 5 shows the engineering stress–strain and corresponding strain-hardening curves of the rolled and annealed SASS specimens. For the rolled specimens (a and c), the Cryo-R samples show higher YS and ultimate tensile strength (UTS) values but similar total elongation (TEL) compared with those of the Cold-R samples, which originate from a higher dislocation density (see the Supplementary Material) due to the inhibition of dislocation recovery at cryogenics. Moreover, the annealed Cold-R samples show an improved TEL value under the sacrifice of tensile strength compared with the rolled state, that is, the strength–ductility trade-off. However, the annealed Cryo-R sample shows simultaneously higher UTS and TEL values (~1556.36 MPa, ~9.1%) than those of the rolled state (~1448 MPa, ~4.93%), and the YS only decreases a little. Moreover, although the strain-hardening (SH) rate showed a decreasing tendency with deformation progressing, the SH value of the annealed Cryo-R sample was significantly higher than that of the annealed Cold-R sample during the whole tensile stage. As shown in Figure 5c, the superior strength–ductility synergy of the annealed Cryo-R sample is mainly attributed to multiple strengthening mechanisms [36,37], including the grain boundary strengthening, dislocation strengthening, and solid solution strengthening. The detailed calculations can be seen in the Supplementary Materials. Moreover, a significant hetero-deformation-induced hardening exists among the heterogeneous domains, contributing to the additional strain-hardening ability for the improvement of strain hardening and tensile ductility. In contrast, the annealed Cold-R samples show small microstructural heterogeneity and a large average grain size, thus exhibiting limited strengthening modes and unattractive strength–ductility combinations.

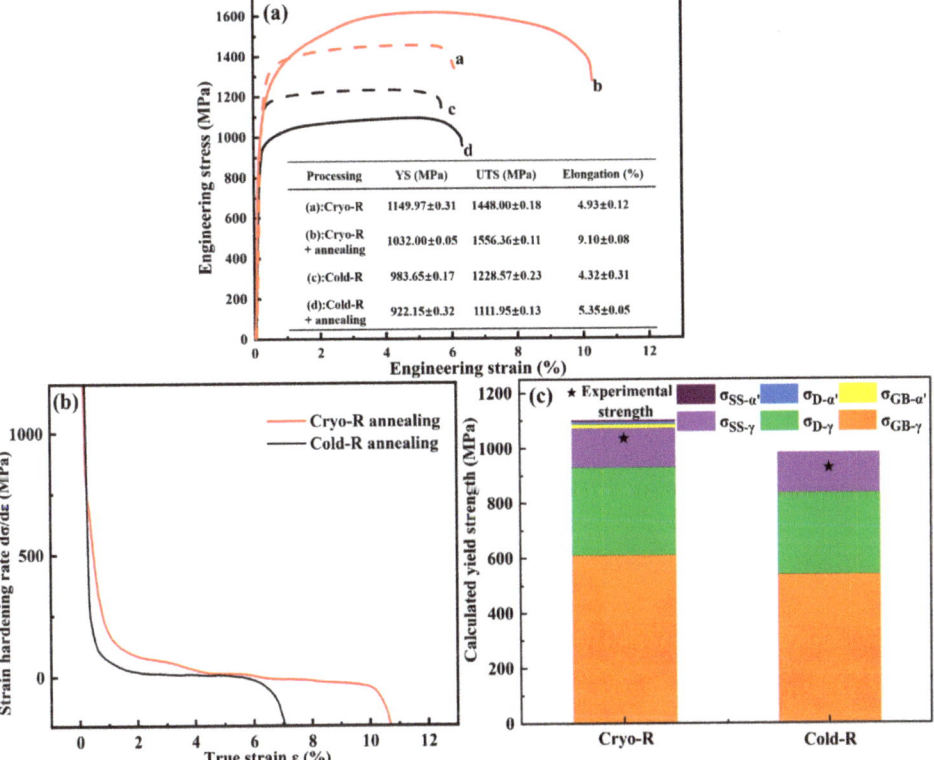

Figure 5. (**a**) Engineering stress–strain curves of the rolled and annealed SASS with Cryo-R and Cold-R processes (their tensile properties are shown in the inset table); (**b**) strain-hardening rate curve of Cryo-R/Cold-R annealing specimens; (**c**) calculated results of different strengthening mechanisms.

4. Conclusions

In summary, a good combination of strength and ductility was achieved in SASS by Cryo-R and subsequent annealing. The Cryo-R process resulted in the formation of a hierarchical multiple deformed substructure, including the dislocation cells, nanotwins and deformation-induced martensite, due to a low SFE and inhibited dislocation recovery. After annealing, the multiple precursors play an important role in obtaining the final heterogeneous structure, overcoming the strength–ductility trade-off. The high YS was mainly attributed to the ultra-fine microstructure, dislocation strengthening and retained martensite. Simultaneously, the hetero-deformation-induced hardening led to the improvement in strain hardening and tensile ductility.

Supplementary Materials: The following supporting information can be downloaded at: https://www.mdpi.com/article/10.3390/ma16186298/s1.

Author Contributions: Conceptualization, B.F., Y.L., W.G., Y.D. and D.T.; methodology, B.F., Y.G. and D.T.; validation, D.T.; formal analysis, B.F., Y.L., L.W. and D.T.; investigation, B.F.; data curation, D.T.; writing—original draft preparation, D.T.; writing—review and editing, B.F. and Y.L.; project administration, B.F. All authors have read and agreed to the published version of the manuscript.

Funding: This research received no external funding.

Institutional Review Board Statement: Not applicable.

Informed Consent Statement: Not applicable.

Data Availability Statement: The data are included in the text.

Conflicts of Interest: The authors declare that they have no known competing financial interest or personal relationships that could have appeared to influence the work reported in this paper.

References

1. Sunny, K.T.; Naik, K.N. A systematic review about welding of super austenitic stainless steel. *Mater. Today Proc.* **2021**, *47*, 41–44. [CrossRef]
2. Xin, H.; Zhuyu, W.; Lin, W.; Chen, C.; Fucheng, Z.; Wei, Z. Effect of pre-deformation on hot workability of super austenitic stainless steel. *J. Mater. Res. Technol.* **2022**, *16*, 66–75.
3. Zhang, Z.B.; Mishin, O.V.; Tao, N.R.; Pantleon, W. Microstructure and annealing behavior of a modified 9Cr–1Mo steel after dynamic plastic deformation to different strains. *J. Nucl. Mater.* **2015**, *458*, 458–463. [CrossRef]
4. Yan, F.K.; Liu, G.Z.; Tao, N.R.; Lu, K. Strength and ductility of 316L austenitic stainless steel strengthened by nano-scale twin bundles. *Acta Mater.* **2012**, *60*, 1059–1071. [CrossRef]
5. Tianyi, S.; Zhongxia, S.; Jaehun, C.; Jie, D.; Tongjun, N.; Yifan, Z.; Bo, Y.; Dongyue, X.; Jian, W.; Haiyan, W.; et al. Ultra-fine-grained and gradient FeCrAl alloys with outstanding work hardening capability. *Acta Mater.* **2021**, *215*, 215–223.
6. Li, W.L.; Tao, N.R.; Lu, K. Fabrication of a gradient nano-micro-structured surface layer on bulk copper by means of a surface mechanical grinding treatment. *Scr. Mater.* **2008**, *59*, 546–549. [CrossRef]
7. Li, J.; Mao, Q.; Nie, J.; Huang, Z.; Wang, S.; Li, Y. Impact property of high-strength 316L stainless steel with heterostructures. *Mater. Sci. Eng. A* **2019**, *754*, 457–460. [CrossRef]
8. Huang, C.X.; Hu, W.P.; Wang, Q.Y.; Wang, C.; Yang, G.; Zhu, Y.T. An Ideal Ultrafine-Grained Structure for High Strength and High Ductility. *Mater. Res. Lett.* **2014**, *3*, 88–94. [CrossRef]
9. Roy, B.; Kumar, R.; Das, J. Effect of cryorolling on the microstructure and tensile properties of bulk nano-austenitic stainless steel. *Mater. Sci. Eng. A* **2015**, *631*, 241–247. [CrossRef]
10. Wu, X.; Yang, M.; Yuan, F.; Wu, G.; Wei, Y.; Huang, X.; Zhu, Y. Heterogeneous lamella structure unites ultrafine-grain strength with coarse-grain ductility. *Proc. Natl. Acad. Sci. USA* **2015**, *112*, 14501–14505. [CrossRef]
11. Kalsar, R.; Ray, R.K.; Suwas, S. Effects of alloying addition on deformation mechanisms, microstructure, texture and mechanical properties in Fe-12Mn-0.5C austenitic steel. *Mater. Sci. Eng. A* **2018**, *729*, 385–397. [CrossRef]
12. Achmad, T.L.; Fu, W.; Chen, H.; Zhang, C.; Yang, Z.-G. Effects of alloying elements concentrations and temperatures on the stacking fault energies of Co-based alloys by computational thermodynamic approach and first-principles calculations. *J. Alloys Compd.* **2017**, *694*, 1265–1279. [CrossRef]
13. Fu, B.; Ge, Y.; Guan, W.; Guo, Y.; Wang, Z.; Ding, Y. Maintaining microstructure ultra-refinement in the austenitic stainless steel by inducing χ phase precipitating through severe cold rolling and annealing. *Mater. Res. Express* **2022**, *9*, 046515. [CrossRef]
14. Shu, J.; Wu, Y.; Ji, Y.; Chen, M.; Wu, H.; Gao, Y.; Wei, L.; Zhao, L.; Huo, T.; Liu, R. A new electrochemical method for simultaneous removal of Mn(2+) and NH(4)(+)-N in wastewater with Cu plate as cathode. *Ecotoxicol. Environ. Saf.* **2020**, *206*, 111341. [CrossRef]

15. Ye, K.L.; Luo, H.Y.; Lv, J.L. Producing Nanostructured 304 Stainless Steel by Rolling at Cryogenic Temperature. *Mater. Manuf. Process.* **2014**, *29*, 754–758. [CrossRef]
16. Xiong, Y.; Yue, Y.; Lu, Y.; He, T.; Fan, M.; Ren, F.; Cao, W. Cryorolling impacts on microstructure and mechanical properties of AISI 316 LN austenitic stainless steel. *Mater. Sci. Eng. A* **2018**, *709*, 270–276. [CrossRef]
17. Seol, J.-B.; Jung, J.E.; Jang, Y.W.; Park, C.G. Influence of carbon content on the microstructure, martensitic transformation and mechanical properties in austenite/ε-martensite dual-phase Fe–Mn–C steels. *Acta Mater.* **2013**, *61*, 558–578. [CrossRef]
18. Lee, T.-H.; Kim, S.-J.; Jung, Y.-C. Crystallographic Details of Precipitates in Fe-22Cr-21Ni-6Mo(N) Superaustenitic Stainless Steels Aged at 900 °C. *Metall. Mater. Trans. A* **2000**, *31*, 13. [CrossRef]
19. Li, Z.; Fu, L.; Peng, J.; Zheng, H.; Shan, A. Effect of annealing on microstructure and mechanical properties of an ultrafine-structured Al-containing FeCoCrNiMn high-entropy alloy produced by severe cold rolling. *Mater. Sci. Eng. A* **2020**, *786*, 139446. [CrossRef]
20. Wang, H.; Shuro, I.; Umemoto, M.; Ho-Hung, K.; Todaka, Y. Annealing behavior of nano-crystalline austenitic SUS316L produced by HPT. *Mater. Sci. Eng. A* **2012**, *556*, 906–910. [CrossRef]
21. Zheng, C.; Liu, C.; Ren, M.; Jiang, H.; Li, L. Microstructure and mechanical behavior of an AISI 304 austenitic stainless steel prepared by cold- or cryogenic-rolling and annealing. *Mater. Sci. Eng. A* **2018**, *724*, 260–268. [CrossRef]
22. Li, J.; Gao, B.; Huang, Z.; Zhou, H.; Mao, Q.; Li, Y. Design for strength-ductility synergy of 316L stainless steel with heterogeneous lamella structure through medium cold rolling and annealing. *Vacuum* **2018**, *157*, 128–135. [CrossRef]
23. Mallick, P.; Tewary, N.K.; Ghosh, S.K.; Chattopadhyay, P.P. Microstructure-tensile property correlation in 304 stainless steel after cold deformation and austenite reversion. *Mater. Sci. Eng. A* **2017**, *707*, 488–500. [CrossRef]
24. Wang, W.; Yuan, F.; Jiang, P.; Wu, X. Size effects of lamellar twins on the strength and deformation mechanisms of nanocrystalline hcp cobalt. *Sci. Rep.* **2017**, *7*, 9550–9560. [CrossRef]
25. Wang, H.T.; Tao, N.R.; Lu, K. Strengthening an austenitic Fe–Mn steel using nanotwinned austenitic grains. *Acta Mater.* **2012**, *60*, 9–16. [CrossRef]
26. Zhao, L.; Qi, X.; Xu, L.; Han, Y.; Jing, H.; Song, K. Tensile mechanical properties, deformation mechanisms, fatigue behaviour and fatigue life of 316H austenitic stainless steel: Effects of grain size. *Fatigue Fract. Eng. Mater. Struct.* **2020**, *44*, 533–550. [CrossRef]
27. Singh, R.; Sachan, D.; Verma, R.; Goel, S.; Jayaganthan, R.; Kumar, A. Mechanical behavior of 304 Austenitic stainless steel processed by cryogenic rolling. *Mater. Today Proc.* **2018**, *5*, 16880–16886. [CrossRef]
28. Korznikova, G.; Mironov, S.; Konkova, T.; Aletdinov, A.; Zaripova, R.; Myshlyaev, M.; Semiatin, S. EBSD Characterization of Cryogenically Rolled Type 321 Austenitic Stainless Steel. *Metall. Mater. Trans.* **2018**, *49*, 16–23. [CrossRef]
29. Yan, F.K.; Tao, N.R.; Archie, F.; Gutierrez-Urrutia, I.; Raabe, D.; Lu, K. Deformation mechanisms in an austenitic single-phase duplex microstructured steel with nanotwinned grains. *Acta Mater.* **2014**, *81*, 487–500. [CrossRef]
30. Saeed-Akbari, A.; Imlau, J.; Prahl, U.; Bleck, W. Derivation and Variation in Composition-Dependent Stacking Fault Energy Maps Based on Subregular Solution Model in High-Manganese Steels. *Metall. Mater. Trans. A* **2009**, *40*, 3076–3090. [CrossRef]
31. Huabing, L.; Zhouhua, J.; Hao, F.; Shucai, Z.; Peide, H.; Wei, Z.; Guoping, L.; Guangwei, F. Effect of Temperature on the Corrosion Behaviour of Super Austenitic Stainless Steel S32654 in Polluted Phosphoric Acid. *Int. J. Electrochem. Sci.* **2015**, *10*, 4832–4848.
32. Andersson, M.; Stalmans, R.; Gren, J.A. Unified Thermodynamic analysis of the stressassisted $\gamma \rightarrow \varepsilon$ martensitic transformation in Fe±Mn±Si alloys. *Acta Mater.* **1998**, *46*, 8. [CrossRef]
33. Wu, X.; Zhu, Y. Heterogeneous materials: A new class of materials with unprecedented mechanical properties. *Mater. Res. Lett.* **2017**, *5*, 527–532. [CrossRef]
34. Shi, Y.; Zhou, J. Analysis of mechanical behaviour of single-phase austenitic stainless steels based on microstructure deformation mechanism. *Micro Nano Lett.* **2019**, *14*, 1024–1028. [CrossRef]
35. Li, Y.S.; Tao, N.R.; Lu, K. Microstructural evolution and nanostructure formation in copper during dynamic plastic deformation at cryogenic temperatures. *Acta Mater.* **2008**, *56*, 230–241. [CrossRef]
36. Ma, K.; Wen, H.; Hu, T.; Topping, T.D.; Isheim, D.; Seidman, D.N.; Lavernia, E.J.; Schoenung, J.M. Mechanical behavior and strengthening mechanisms in ultrafine grain precipitation-strengthened aluminum alloy. *Acta Mater.* **2014**, *62*, 141–155. [CrossRef]
37. Wen, H.; Topping, T.D.; Isheim, D.; Seidman, D.N.; Lavernia, E.J. Strengthening mechanisms in a high-strength bulk nanostructured Cu–Zn–Al alloy processed via cryomilling and spark plasma sintering. *Acta Mater.* **2013**, *61*, 2769–2782. [CrossRef]

Disclaimer/Publisher's Note: The statements, opinions and data contained in all publications are solely those of the individual author(s) and contributor(s) and not of MDPI and/or the editor(s). MDPI and/or the editor(s) disclaim responsibility for any injury to people or property resulting from any ideas, methods, instructions or products referred to in the content.

Article

The Properties and Microstructure of Na₂CO₃ and Al-10Sr Alloy Hybrid Modified LM6 Using Ladle Metallurgy Method

Mhd Noor Ervina Efzan * and Hao Jie Kong [†]

Faculty of Engineering and Technology (FET), Multimedia University (MMU), Ayer Keroh 5450, Malaysia
* Correspondence: ervina.noor@mmu.edu.my
[†] The Current Affiliation: School of Materials Science and Engineering, Dongguan University of Technology, Dongguan 523808, China.

Abstract: In this work, Al-10Sr alloy and Na₂CO₃ were added to LM6 (reference alloy) as hybrid modifiers through ladle metallurgy. The microstructure enhancement was analyzed using an optical microscope (OM). The results were further confirmed with Scanning Electron Microscope (SEM) and Energy Dispersive X-ray (EDX) spectroscopy. The results showed that Na₂CO₃ and Al-10Sr alloy successfully hybrid modified the sharp needle-like eutectic Si into fibrous eutectic Si. Soft primary Al dendrites were also discovered after the hybrid modification. The formation of β-Fe flakes was suppressed, and α-Fe sludge was transformed into Chinese script morphology. A 2.13% density reduction was recorded. A hardness test was also performed to investigate the mechanical improvement of the hybrid-modified LM6. 2.3% of hardness reduction was recorded in the hybrid-modified LM6 through ladle metallurgy. Brittle cracks were not observed, while ductile pile-ups were the main features that appeared on the indentations of hybrid-modified LM6, indicating a brittle to ductile transformation after hybrid modification of LM6 by Na₂CO₃ and Al-10Sr alloy through ladle metallurgy.

Keywords: ladle; aluminium; LM6; microstructure; alloy

Citation: Ervina Efzan, M.N.; Kong, H.J. The Properties and Microstructure of Na₂CO₃ and Al-10Sr Alloy Hybrid Modified LM6 Using Ladle Metallurgy Method. *Materials* **2023**, *16*, 6780. https://doi.org/10.3390/ma16206780

Academic Editors: Andrea Di Schino and Claudio Testani

Received: 7 July 2023
Revised: 7 September 2023
Accepted: 7 September 2023
Published: 20 October 2023

Copyright: © 2023 by the authors. Licensee MDPI, Basel, Switzerland. This article is an open access article distributed under the terms and conditions of the Creative Commons Attribution (CC BY) license (https://creativecommons.org/licenses/by/4.0/).

1. Introduction

Aluminium-Silicon (Al-Si) alloy has been widely used as a lightweight material in aerospace and automotive industries [1–6]. However, Al-Si alloy is brittle due to the sharp needle-like Si morphology in the microstructure [7,8]. Efforts have been made recently to reduce the brittleness of Al-Si alloy through different techniques such as the alloying method, controlling the overheat temperature and cooling rate, performing iso-thermal mechanical stirring during the semi-solidified state, and the incorporation of ultrasonic vibration during the solidification process [9–11]. Researchers found out that the addition of modifiers such as Sodium (Na) and Strontium (Sr) can transform the needle-like Silicon (Si) into fibrous-like Si morphology [12,13]. Owing to the demanding requirements of the industries, Lu et al. [12] proposed that the modification solely based on a single type of modifier is unsatisfactory. For instance, Na is a fast-acting modifier, able to produce an immediate modification on the Si morphology.

Nevertheless, due to its high reactivity, its potency as an effective modifier diminishes over a long melting period. On the other hand, Sr can remedy its deficiency by serving as a slow-acting modifier with more pronounced modification power over a longer melt-holding period. Therefore, the combined addition of modifiers (hybrid modification) is considered important.

Na and Sr are very reactive. They oxidize rapidly when exposed to the atmosphere, especially at an elevated temperature. Thus, Na and Sr are seldom added as pure elements. Sr is normally added as an Al-Sr system alloy [12,13]. Meanwhile, Na is usually added as an Al-Si-Na system alloy [13] or Sodium Floride (NaF) [12]. Sodium Carbonate (Na₂CO₃) has been widely used as a flux in treating insoluble silicates. However, its potential to serve as a modifying flux in Al-Si alloy is not reported. Conventionally, modifiers are added inside the melting furnace [14]. Therefore, a long melt-holding period is required for hybrid modifiers' complete dissolution and homogenization. The prolonged melt holding period and high melting temperature in the melting furnace significantly reduce the availability of potent modifiers due to rapid oxidization. Thus, hybrid modifiers will be added in a ladle (ladle metallurgy) in this study. Dube et al. [15] developed a methodology to produce aluminium alloy via a ladle metallurgical process. According to the authors, ladle metallurgy can reduce the risk of contamination, especially if two or more alloys are produced. Besides, a well-insulated ladle can allow alloying operation to be carried out without any external heat input, which is particularly important during the alloying of Magnesium (Mg), Copper (Cu), and Silicon (Si) that involves endothermic dissolution.

Despite the early introduction of ladle metallurgy in the smelting of aluminium alloys, scarce study has been reported on the addition of modifiers and grain refiners through ladle metallurgy. The most recent study of modification through ladle metallurgy in aluminium alloys can be dated back to 2006, reported by Pena and Lozano [7]. In that study, additional modifiers Ti and Sr improve mechanical performance after the modification. Scarce literature on ladle metallurgy might be due to the preference for stream inoculation in the aluminium smelting industry. In-stream inoculation, modifiers and grain refiners are added before molten metal enters the casting moulds [16,17]. The addition is made in the launders between treatment vessels and casting stations. Through this method, the fading of modifiers and grain refiners can be reduced. Even though stream inoculation improves the modification and grain refinement of aluminium alloys, the modifiers and grain refiners should possess a high dissolution rate so that complete dissolution can be achieved. Besides, the pouring temperature of the molten metal should be high enough for the melt to flow smoothly through the launder. Higher pouring temperature will lead to the oxidation of modifiers and grain refiners. To solve these issues, the addition of hybrid modifiers through ladle metallurgy will be studied in this work.

This work aims to determine the influences of Al-Sr alloy and Na_2CO_3 as hybrid modifiers in density, microstructure, and hardness of LM6 alloy via ladle metallurgy.

2. Experimental Works

LM6 alloy, Al-10Sr alloy, and anhydrous Na_2CO_3 (99.5% pure) powder were used to prepare the hybrid modified Al-12Si-0.02Sr-0.02Na system alloy. The LM6 alloy was in the form of an ingot, conforming to the British Standards 1490 [18]. The LM6 alloy and Al-10Sr alloy were purchased from FES Foundry Equipment Supply (M) Sdn. Bhd. while the Na_2CO_3 powder was acquired from Comak Chemical Limited.

Before melting, small pieces of LM6 were cut with a dimension of 2.5 cm^3. Acetone was used to remove the dirt and impurities thoroughly. A digital weighing balance, Dragon 303, Mettler Toledo, was used to measure the mass of the cut LM6 and hybrid modifiers. Ladle metallurgy, or mixing alloying elements in a ladle [7], was incorporated into the melting process, as shown in Figure 1. LM6 was melted in an electrical resistance furnace, K 4/13, Nabertherm, according to the temperature profile in Figure 2. The melt was brought to a melting temperature of 900 °C in 2 h and soaked for another 30 min for homogenization. Ladle metallurgy was performed in the following sequence: charging hybrid modifiers in the ladle, followed by pouring molten LM 6 from the furnace into the ladle after slag skimming and stirring for 10 s. The melt containing molten LM 6 and hybrid modifiers in the ladle was immediately stirred for 5 s. Afterwards, re-melt was performed for 15 min at 900 °C to enhance the compositional homogeneity of the alloy. The melting ended with the casting process, where the re-melted alloy was poured into a preheated (100 °C) metallic (ASSAB 8407) mold and left for solidification.

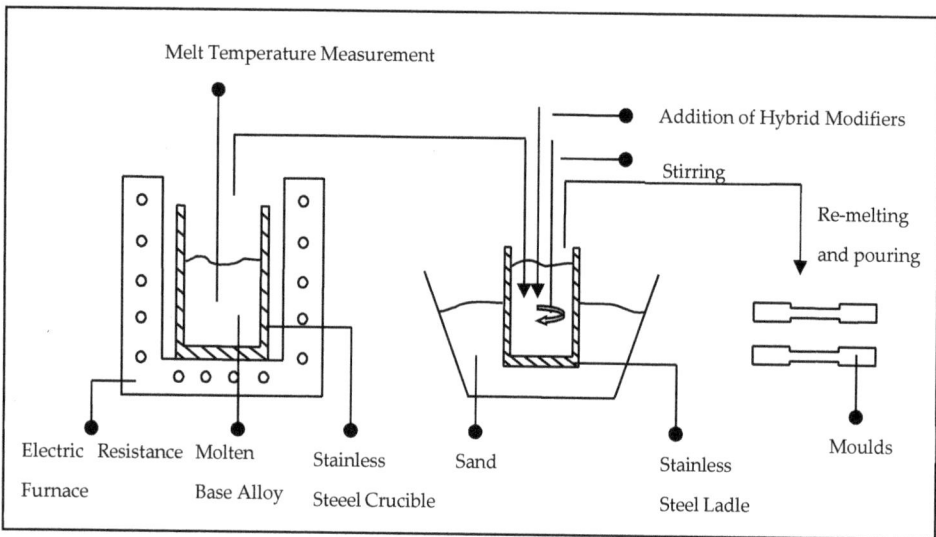

Figure 1. A schematic showing the ladle metallurgy melting sequence.

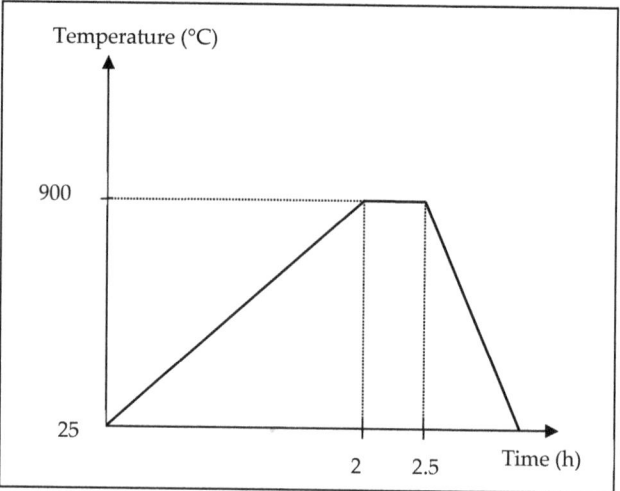

Figure 2. Temperature profile of the melting process.

Samples were ground using silicon carbide papers and polished to a mirror-like surface to remove scratches and artefacts. The density of the samples was measured using an electronic densimeter, MD-300S, Alta Mirage. For metallurgical analysis, etching was performed using Keller Reagent for 5 seconds to reveal the microstructure. Microstructure observation was performed under a reflected high-power optical microscope, Meiji Techno Japan, equipped with bright field imaging. The results were further analyzed under a scanning electron microscope (SEM), EVO 50, Carl Zeiss AG, in secondary electron (SE) mode. Energy Dispersive X-ray (EDX) was also performed at 1 k magnification to confirm the elements in the microstructure.

Vickers hardness was performed by applying a load of 2 kg with 15 s of dwell time to evaluate the hardness of the samples. An average hardness was obtained by averaging the hardness gathered from five indentations.

3. Results and Discussions

Density. Density is one of the important parameters during the selection of materials. Light (low-density) yet high-strength materials are often desirable in the aerospace and automotive industries. The unmodified LM6 recorded a density of 2.717 g/cm³. The hybrid-modified LM6 achieved a density of 2.659 g/cm³, equivalent to 2.13% reduction. The reduced density of LM6 after the combined addition of Na_2CO_3 and Al-Sr alloy can be owing to the formation of porosity. The addition of Sr has often been associated with the formation of porosity in Al-Si alloys [19]. The pores expand due to the lowered pressure and result in a much more porous sample than under atmospheric conditions of solidification.

Microstructure. Figures 3 and 4 show the microstructure of unmodified LM6 and the Al-12Si-0.02Sr-0.02Na system alloy (hybrid modified LM6) at 10 and 50 magnifications.

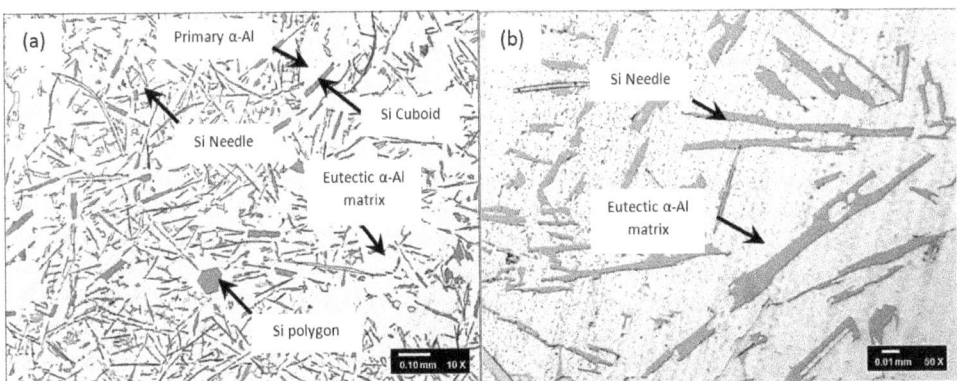

Figure 3. Microstructures of unmodified LM6 through ladle metallurgy at (**a**) 10× and (**b**) 50×.

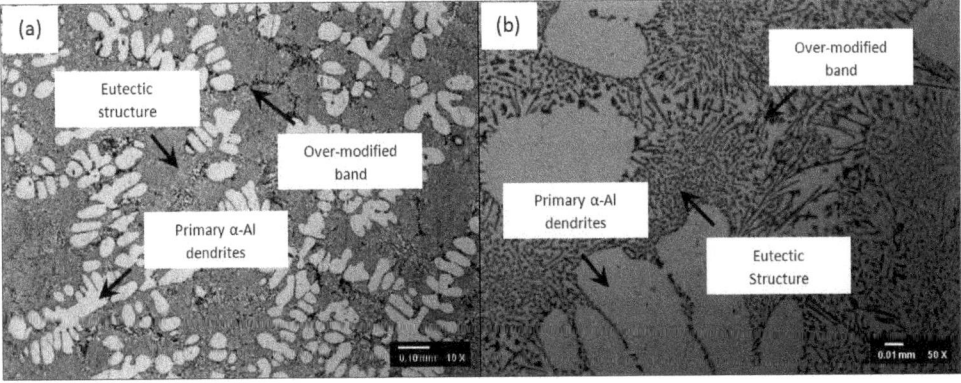

Figure 4. Microstructures of hybrid modified LM6 with Al-Sr alloy and Na_2CO_3 and through ladle metallurgy at (**a**) 10× and (**b**) 50×.

In the absence of a hybrid modifier, the unmodified LM6 consists of Al-rich (bright zone) and Si-rich (dark zone) phases, as demonstrated in Figure 3. The dark Si-rich phase resembles itself in needle, cuboid, and irregular polygon morphologies. The Si-rich needles combine with the Al-rich phase, forming a eutectic structure consisting of alternating layers of Si-rich needle-like lamellar with an Al-rich matrix. The Si needles have significantly been blamed as stress concentrators that result in the brittleness of the Al-Si alloys [8,20]. On the other hand, the Si-rich cuboids are suggested to be the primary Si phase. This is supported by Pena and Lozano [7]. In their work, they reported that these primary Si-rich cuboids nucleate the primary α-Al phase on some occasions.

After the addition of Na$_2$CO$_3$ and Al-Sr alloy to the LM6, bright dendrites are observed in Figure 4. However, these dendrites do not appear in the unmodified LM6. They have a similar morphology as the primary Al phase reported in a study on the modification of Al-10Si alloy with 0.0025% Sr carried out by Sun et al. [21]. It is, therefore, a posteriori to deduce that the bright dendrites are indeed the primary Al phase. Another discernible feature that results from the hybrid modification will be the fine black dot dots embedded in the bright Al-rich phase, as shown in Figure 3b. These fine dot-dots were also discovered by Lu et al. [12] in a Na-Sr modified Al-10Si alloy. The authors regarded these dot-dots as modified eutectic Si phase. In an attempt to have a better view of the morphology of the modified eutectic Si phase, Milenkovic et al. [22] performed a deep etching on a modified Al-Si eutectic. They found that the modified eutectic Si phase consists of a network of interconnected fine coral-like fibrous structures. Accordingly, in the current work, bands of coarse Si structures were also detected along the eutectic grain boundaries, as illustrated in Figure 3. The occurrence of these coarse Si bands is accounted for by the over-modification of the eutectic Si phase [12].

SEM-EDX. The SEM-EDX images of the unmodified LM6 and hybrid modified LM6 at 500 and 1 k magnifications are presented in Figures 5 and 6, respectively.

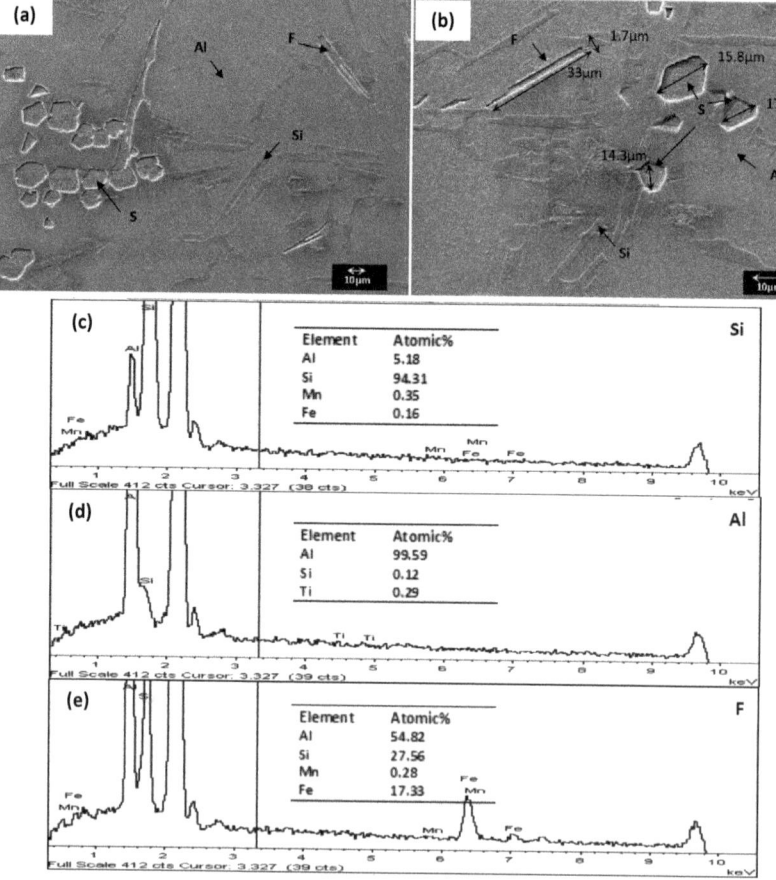

Figure 5. The microstructure of unmodified LM6 under SEM at (**a**) 500 (**b**) 1 k magnification 15 kV and (**c–e**) the corresponding EDX spectrums [F: Flake-like, S: Sludge-like, Al: Aluminium-rich phase, Si: Silicon-rich phase].

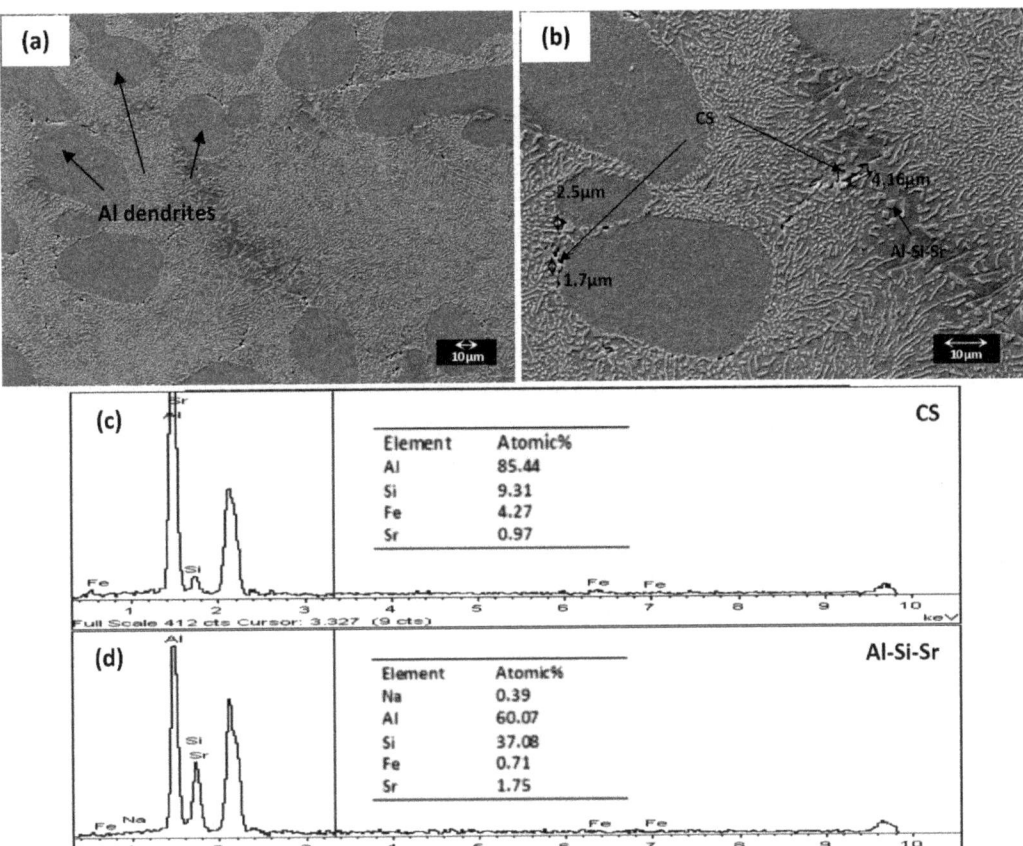

Figure 6. The microstructure of Na-Sr hybrid modified LM6 under SEM at (**a**) 500 (**b**) 1 k magnification 15 kV and (**c**,**d**) the corresponding EDX spectrums [Al: Aluminium, CS: Chinese Script].

From Figure 6, bright needles can be observed distributed evenly across the dark matrix in the unmodified LM6. Literature has described these bright needles as coarse Si-rich plates embedded in the eutectic Al-rich matrix [7,8,23]. Based on the EDX analysis, the bright needles contain approximately 94 wt.% of Si (Figure 5c), while the dark matrix consists of 99 wt.% of Al (Figure 5d). This confirms that the bright needles and dark matrix are Si rich needles and Al-rich eutectic matrix, respectively. Apart from the Si-rich needles and Al-rich eutectic matrix, two other phases with different morphologies, the flake (F) and the polyhedral sludge (S), can also be identified in Figure 5. The flake-like phase has a length and width of approximately 33 and 1.7 μm, respectively, whereas the sludge-like phase has an average size of approximately 15.8 μm (measured according to the maximum line segment). The EDX spectrum of the flake-like phase is shown in Figure 5e, revealing a significant amount of Al (55 at.%), Si (28 at.%), and Fe (17 at.%), along with a small amount of Mn (0.28 at.%). This elemental composition is close to that of the commonly reported β-Fe needles with a stoichiometry ratio of $Al_5(Fe,Mn)Si$ [24,25]. Biswas et al. [26] reported that the sludge-like phase has a stoichiometry composition of $Al_6(Mn,Fe)$. According to Ashtari et al. [27], in the presence of Mn and Fe, the solidification path of Al-Si alloy will follow the following sequence: (1) the solidification of Al-dendrites, (2) the formation of Chinese script or sludge-like α-Fe phase, and (3) the formation of needle-like β-Fe and α-Fe phases. The presence of flake-like and sludge-like Fe-bearing phases in this study supports the aforementioned viewpoint.

On the other hand, dark primary Al dendrites and bright fine fibrous eutectic Si can be observed in Figure 6 after the combined addition of Na_2CO_3 and Al-Sr alloy. The formation of primary Al dendrites is perhaps related to the constitutional undercooling owing to Sr that suppresses the eutectic temperature and shifts the eutectic composition to the right side of the Al-Si binary phase diagram [23]. The bright, long needle-like eutectic Si is no longer observed owing to the successful eutectic hybrid modification of Na-Sr. Besides, instead of the flake-like and sludge-like phases, fine (approximately 2–4 µm) Chinese script (CS) phase can be seen segregating at the eutectic boundary and near the Al dendrites. The segregation of Chinese script at the eutectic boundary and Al dendrites was also discovered by Timpel et al. [28]. The authors suggested that in the presence of Sr, Fe solutes will be rejected in front of the solid/liquid interface during the eutectic solidification, leading to the formation of the Fe-rich phase at the boundaries.

Additionally, Al dendrites can serve as the obstacles or preferred site of perturbation during eutectic grain growth. This results in the segregation of the Fe-rich phase at the Al dendrites. An EDX spectrum was taken at the fine Chinese script phase. Due to the fineness of the Chinese script, the EDX analysis becomes more difficult. The EDX spectrum of the Chinese script is presented in Figure 6c, revealing the peak of Al (85 at.%), Si (9 at.%), Fe (4 at.%) and Sr (1 at.%), close to the reported α-Fe phase [29]. This indicates that Na-Sr transforms the eutectic Si to fibrous morphology and suppresses the formation of β-Fe while transforming the sludge-like α-Fe to Chinese script α-Fe. The study has been reporting that the morphology of α-Fe depends on the nucleation and growth velocity [30]. Low growth velocity will produce sludge-like α-Fe, while Chinese script α-Fe favours high growth velocity. It is therefore proposed that the Na-Sr hybrid modification increases the undercooling of the Fe phase, further speeds up its growth velocity, and results in the formation of the Chinese script α-Fe phase.

Moreover, the suppression of β-Fe after adding Sr has also been reported [27,31]. It is believed that Sr can adsorb at the α-Fe/liquid interface, preventing the diffusion of Si into the α-Fe that is necessary for the formation of β-Fe. There is also another explanation given by Samuel et al. [31], proposing that the suppression of the β-Fe phase is attributed to the rejection of Si from the β-Fe needles after the addition of Sr. Al-Si-Sr phase will form at the surrounding Sr and Si-rich area. In this study, a small polyhedral phase can be identified at the eutectic grain boundary. Through EDX analysis (Figure 6d), the small polyhedral phase contains approximately 60 at.% of Al, 37 at.% of Si, 2 at.% of Sr, and 1 at.% of Fe, suggesting the presence of the Al-Si-Sr-Fe phase.

Hardness. Hardness provides a relatively cheap, convenient, and fast non-destructive assessment of the ability of a material to withstand plastic deformation. As highlighted in Figure 7, the hardness of LM6 decreases after the hybrid is modified by Na_2CO_3 and Al-Sr alloy. Before hybrid modification, 72.76 Hv was recorded. Its hardness reduces slightly to 71.10 Hv, equivalent to a 2.3% reduction after hybrid modification. Hardness reduction was also reported by [32,33] after the addition of Sr to Al-Si alloys.

The hardness reduction is believed to be very much related to the solid solution softening. The presence of Si atoms in the Al matrix has been known to increase the hardness of the Al-rich matrix through solid solution strengthening [34]. Therefore, the softening of the Al-rich matrix is most probably attributed to the incorporation of Na and Sr atoms. Besides, the hybrid-modified LM6 has not undergone any annealing or heat treatment process that relieves the strain, resulting in the softening of the Al-rich matrix. Atomic size misfit [35,36] and atomic modulus misfit [36] can be used to explain the solid solution softening results from the addition of Na and Sr atoms in an Al-rich matrix. The introduction of solute atoms having different atomic radii than that of the solvent atoms will lead to atomic size misfit. Substitution of solvent atoms by smaller solute atoms will induce compressive lattice strain, while the substitution of solvent atoms with larger solute atoms will cause tensile lattice strain. These lattice strains serve as barriers to dislocation motion and harden the solid solution. The hardness enhancement and reduction of solid solution are also governed by the shear modulus of the solute and solvent atoms. The substitution of solvent atoms by the solute atoms with shear modulus differs from that of solvent atoms. It will form local 'hard' and 'soft' spots due to the difference in elastic strain energy, leading to atomic modulus misfit [37].

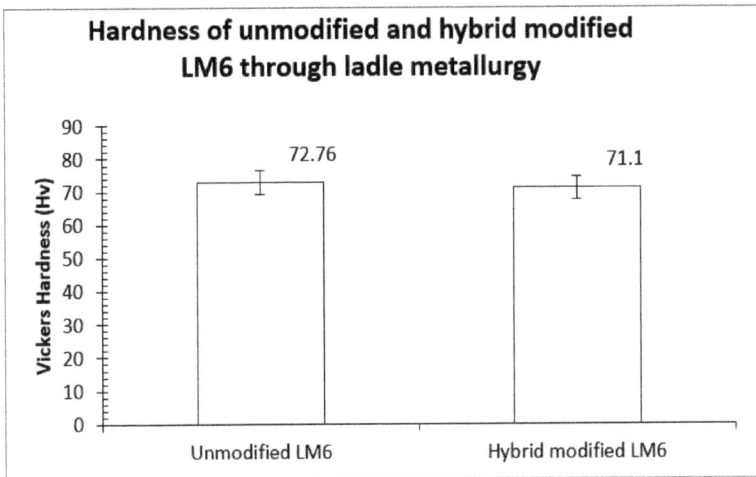

Figure 7. Vickers hardness of unmodified and hybrid modified LM6 with Na_2CO_3 and Al-10Sr alloy.

To provide a quick analysis of the effects of atomic size and modulus misfits on the hardness of the Al-rich matrix, the rule of mixture [38] is applied to determine the 'effective' radius and shear modulus of the Al-rich matrix after the addition of Na and Sr atoms. The corresponding 'effective' radius and shear modulus are tabulated in Table 1. The hardness, radius and shear modulus of pure Al are also included in the table as a reference. The hardness of LM6 is greater than the hardness of pure Al due to the compressive strain induced by the smaller Si atoms, as reflected in the reduced 'effective' radius. Besides, the presence of Si atoms in the Al increases the 'effective' shear modulus. The reduction in the 'effective' radius and increment in 'effective' shear modulus resulted in the hardening of LM6. On the other hand, after the addition of Na and Sr atoms to the LM6, the 'effective' radius increases by 0.029% while the 'effective' shear modulus reduces by 0.017%. The relief of compressive strain and reduction of 'effective' shear modulus attributed to the presence of Na and Sr atoms leading to the reduced hardness of LM6.

Table 1. Vickers hardness, Effective Atomic Radius, and Effective Shear Modulus of Pure Al, Unmodified LM 6, and Hybrid modified LM 6.

	Vickers Hardness (Hv)	Effective Atomic Radius (nm)	Effective Shear Modulus (GPa)
Pure Al	* 17.02	* 0.14300	* 26.00
Unmodified LM 6	72.76	0.13959	33.36
Hybrid modified LM 6	71.10	0.13963	33.35

* According to data published in Materials Science and Engineering [39]. Calculations of effective atomic radius and effective shear modulus are based on data published in Materials Science and Engineering [39].

However, it can be argued that atomic size misfit and modulus misfit only have slight influences in reducing the hardness of hybrid-modified LM6. The attempt to understand the reduction of hardness solely based on the atomic size and modulus misfit seems insufficient. The Na and Sr atoms may interact with vacancies, resulting in a net reduction of available defects that serve as dislocation barriers [37,40]. It was reported that a reduction in vacancy concentration can lower the hardness of an alloy. Shen and Koch [38] also discovered that the hardness of an alloy can be attributable to the combined effects resulting from the softening or hardening of solid solution and grain boundary. In this work, the reduction of hardness can also be related to the presence of the relatively soft primary dendritic Al phase in the microstructure of hybrid-modified LM6. Na and Sr atoms modify the microstructure of LM6, promoting the formation of primary Al dendrites that are not observable in unmodified LM6. This leads to the reduced volume fraction of the eutectic structure that contains hard Si particles. Similar instances were also reported in a study by Xu et al. [41]. In that study, the hardness was dependent on the percentage of the eutectic structure. The hardness reduces as the fraction of the eutectic structure decreases. Another possible reason leading to the reduction of the hardness of the hybrid-modified LM6 can be attributed to the formation of pores that serve as sites of stress concentration that subsequently reduce the hardness of LM6. The hardness reduction agrees with the decrease in density after the addition of Na_2CO_3 and Al-Sr alloy in this work.

To gain more insights into the mechanical behavior of hybrid-modified and unmodified LM6, the fracture features of their indentations after the hardness tests were studied. Figure 8a reveals the formation of intergranular cracks at the edge of the eutectic Si needles, indicating the brittle nature of unmodified LM6. Intra-granular cracks and trans-phase cracks were also observed. Contrarily, after hybrid modification (Figure 8b), cracks were not detected. Pile-ups of primary Al dendrites at the edge of indentation can be observed. These 'pile-up's indicate the hybrid-modified LM6 experienced great plastic deformation after the hardness test. The transformation of brittle to ductile fracture mode can be related to the suppression of sharp brittle β-Fe flakes and the formation of a more favourable Chinese script phase after the hybrid modification of LM6 by Na_2CO_3 and Al-10Sr alloy.

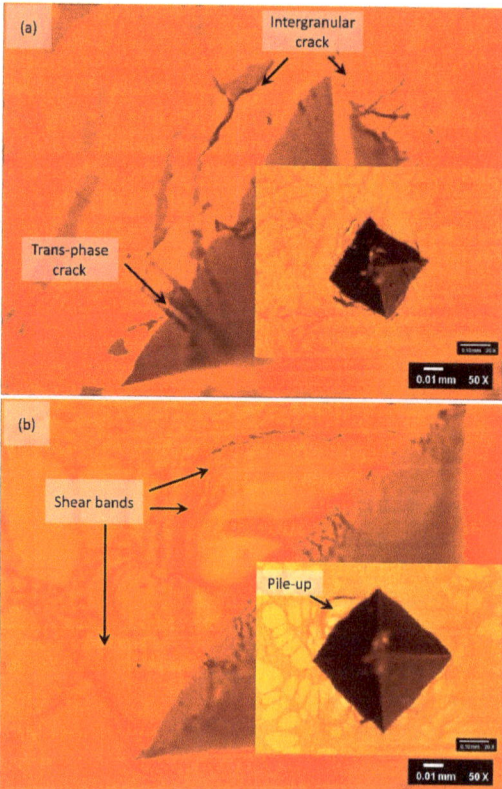

Figure 8. Vickers hardness indentations of (**a**) unmodified LM6 and (**b**) hybrid modified LM6 with Na_2CO_3 and Al-Sr alloy at 2 kg load.

4. Conclusions

Hybrid modification of LM6 by Na_2CO_3 and Al-10Sr alloy through ladle metallurgy provides a means of transforming the needle-like eutectic Si to fibrous eutectic Si. However, bands of over-modification were observed along the eutectic grain boundaries. Primary Al dendrites were also formed after the hybrid modification. The hybrid modification of LM6 by Na_2CO_3 and Al-10Sr alloy suppressed the formation of a flake-like β-Fe phase while transforming the sludge-like α-Fe to Chinese script morphology. The density of hybrid-modified LM6 was reduced by 2.13%. A slight hardness reduction (2.3%) was recorded after the hybrid modification. Elastic atomic size and modulus misfit could not describe the hardness reduction persuasively. The increased porosity after the hybrid modification by Na_2CO_3 and Al-10Sr alloy might be another explanation for the hardness reduction. From the observations on the fracture surfaces of the indentations, hybrid modification by Na_2CO_3 and Al-10Sr alloy through ladle metallurgy can improve the ductility of LM6 through brittle to ductile transformation.

Author Contributions: Conceptualization, M.N.E.E.; Methodology, M.N.E.E.; Writing—original draft, M.N.E.E. and H.J.K.; Writing—review & editing, M.N.E.E. and H.J.K.; Supervision, M.N.E.E. All authors have read and agreed to the published version of the manuscript.

Funding: This research received no external funding.

Conflicts of Interest: The authors declare no conflict of interest.

References

1. Efzan, M.N.E.; Kong, H.; Kok, C. Review: Effect of Alloying Element on Al-Si Alloys. *Adv. Mater. Res.* **2014**, *845*, 355–359. [CrossRef]
2. Zuo, M.; Zhao, D.; Teng, X.; Geng, H.; Zhang, Z. Effect of P and Sr complex modification on Si phase in hypereutectic Al–30Si alloys. *Mater. Des.* **2013**, *47*, 857–864. [CrossRef]
3. Efzan, M.N.E.; Syazwani, N.S.; Abdullah, M.M.A.B. Fabrication Method of Aluminum Matrix Composite (AMCs): A Review. *Key Eng. Mater.* **2016**, *700*, 102–110. [CrossRef]
4. Padmanabhan, S.; Boopathi, B.; Kumar, K.M.; Deepak, J.R.; Kumar, V.; Baskar, S. Investigation of alloy materials on a sports bike wheel rim designs using finite element analysis. *Multidiscip. Sci. J.* **2023**, *6*, 2024013. [CrossRef]
5. Kumaraswamy, J.; Kumar, V.; Purushotham, G. A review on mechanical and wear properties of ASTM a 494 M grade nickel-based alloy metal matrix composites. *Mater. Today Proc.* **2021**, *37*, 2027–2032. [CrossRef]
6. Kumaraswamy, J.; Kumar, V.; Purushotham, G.G.; Suresh, R. Thermal analysis of nickel alloy/Al2O3/TiO2 hybrid metal matrix composite in automotive engine exhaust valve using FEA method. *J. Therm. Eng.* **2021**, *7*, 415–428. [CrossRef]
7. Suárez-Peña, B.; Asensio-Lozano, J. Microstructure and mechanical property developments in Al–12Si gravity die castings after Ti and/or Sr additions. *Mater. Charact.* **2006**, *57*, 218–226. [CrossRef]
8. El Sebaie, O.; Samuel, A.; Samuel, F.; Doty, H. The effects of mischmetal, cooling rate and heat treatment on the eutectic Si particle characteristics of A319.1, A356.2 and A413.1 Al–Si casting alloys. *Mater. Sci. Eng. A* **2008**, *480*, 342–355. [CrossRef]
9. Xu, C.; Jiang, Q. Morphologies of primary silicon in hypereutectic Al–Si alloys with melt overheating temperature and cooling rate. *Mater. Sci. Eng. A* **2006**, *437*, 451–455. [CrossRef]
10. Sukumaran, K.; Pai, B.; Chakraborty, M. The effect of isothermal mechanical stirring on an Al–Si alloy in the semisolid condition. *Mater. Sci. Eng. A* **2004**, *369*, 275–283. [CrossRef]
11. Lin, C.; Wu, S.; Lu, S.; An, P.; Wan, L. Effects of ultrasonic vibration and manganese on microstructure and mechanical porperties of hypereutectic Al–Si alloys with 2%Fe. *Intermetallics* **2013**, *32*, 176–183. [CrossRef]
12. Lu, L.; Nogita, K.; Dahle, A.K. Combining Sr and Na additions in hypoeutectic Al–Si foundry alloys. *Mater. Sci. Eng. A* **2005**, *399*, 244–253. [CrossRef]
13. Wang, W. *Na, Sr and Sb interactions in Al-Si Alloy Melts and Their Effects on Modification*; Canadian Theses Service: Oxford, UK, 1991.
14. Dunville, B.T.; Koch, F.P.; Malliris, R.J.; Setzer, W.C.; Young, D.K. Aluminium Master Alloys Containing Strontium, Boron, and Silicon for Grain Refining and Modifying Aluminium Alloys. U.S. Patent 5230754 A, 27 July 1993.
15. Dube, G.; Gariepy, B.; Pare, J. Method of Alloying. Aluminium. Patent EP0260930 A1, 23 March 1988.
16. Beddoes, J.; Bibby, M. *Principles of Metal Manufacturing Processes*; Elsevier Butterworth-Heinemann: Oxford, UK, 2003. [CrossRef]
17. Efzan, M.E.; Syazwani, N.S.; Abdullah, M.M.A.B. Microstructure and mechanical properties of fly ash particulate reinforced in LM6 for energy enhancement in automotive applications. *IOP Conf. Ser. Mater. Sci. Eng.* **2016**, *133*, 012046. [CrossRef]
18. BS 1490:1988; Specification for Aluminium and Aluminium Alloy Ingots and Castings for General Engineering Purposes. British Standard: London, UK, 1988; pp. 1–99.

19. Liu, L.; Samuel, A.M.; Samuel, F.H.; Doty, H.W.; Valtierra, S. Influence of oxides on porosity formation in Sr-treated Al-Si casting alloys. *J. Mater. Sci.* **2003**, *38*, 1255–1267. [CrossRef]
20. Kobayashi, T.; Toda, H. Strength and Fracture of Aluminium Alloys. *Mater. Sci. Eng. A* **2000**, *280*, 8–16. [CrossRef]
21. Sun, Y.; Pang, S.-P.; Liu, X.-R.; Yang, Z.-R.; Sun, G.-X. Nucleation and growth of eutectic cell in hypoeutectic Al-Si alloy. *Trans. Nonferrous Met. Soc. China* **2011**, *21*, 2186–2191. [CrossRef]
22. Milenkovic, S.; Dalbert, V.; Marinkovic, R.; Hassel, A.W. Selective matrix dissolution in an Al–Si eutectic. *Corros. Sci.* **2009**, *51*, 1490–1495. [CrossRef]
23. Makhlouf, M.M.; Guthy, H.V. The aluminium-silicon eutectic reaction: Mechanisms and crystallography. *J. Light Met.* **2001**, *1*, 199–218. [CrossRef]
24. Irizalp, S.G.; Saklakoglu, N. Effect of Fe-rich intermetallics on the microstructure and mechanical properties of thixoformed A38 aluminium alloy. *Eng. Sci. Technol. Int. J.* **2014**, *17*, 58–62.
25. Li, L.; Zhou, R.; Jiang, Y.; Wang, X.; Zhou, R. Effect of manganese on the formation of Fe-rich phases in electromagnetic stirred hypereutectic Al-22Si alloy with 2% Fe. *Trans. Indian Inst. Met.* **2014**, *67*, 861–867. [CrossRef]
26. Biswas, P.; Patra, S.; Mondal, M.K. Effects of Mn addition on microstructure and hardness of Al-12.6Si alloy. *IOP Conf. Ser. Mater. Sci. Eng.* **2018**, *338*, 012043. [CrossRef]
27. Ashtari, P.; Tezuka, H.; Sato, T. Influence of Sr and Mn additions on intermetallic compound morphologies in Al-Si-Cu-Fe cast alloys. *Mater. Trans.* **2003**, *44*, 2611–2616. [CrossRef]
28. Timpel, M.; Wanderka, N.; Grothausmann, R.; Banhart, J. Distribution of Fe-rich phases in eutectic grains of Sr-modified Al-10 wt.% Si–0.1 wt.% Fe casting alloy. *J. Alloys Compd.* **2013**, *558*, 18–25. [CrossRef]
29. Suárez-Peña, B.; Asensio-Lozano, J. Influence of Sr modification and Ti grain refinement on the morphology of Fe-rich precipitates in eutectic Al–Si die cast alloys. *Scr. Mater.* **2006**, *54*, 1543–1548. [CrossRef]
30. Verma, A.; Kumar, S.; Grant, P.; O'reilly, K. Influence of cooling rate on the Fe intermetallic formation in an AA6063 Al alloy. *J. Alloys Compd.* **2013**, *555*, 274–282. [CrossRef]
31. Samuel, F.H.; Samuel, A.M.; Doty, H.W.; Valtierra, S. Decomposition of Fe-intermetallics in Sr-modified cast 6XXX type aluminium alloys for automotive skin. *Metall. Mater. Trans. A* **2011**, *32*, 2061–2075. [CrossRef]
32. Liu, G.; Li, G.; Cai, A.; Chen, Z. The influence of Strontium addition on wear properties of Al-20 wt% Si alloys under dry reciprocating slicing condition. *Mater. Des.* **2011**, *32*, 121–126. [CrossRef]
33. Ibrahim, M.F.; Samuel, E.; Samuel, A.M.; Ahmari, A.M.A.; Samuel, F.H. Metallurgical parameters controlling the mi-crostructure and hardness of Al-Si-Cu-Mg base alloys. *Mater. Des.* **2011**, *32*, 2130–2142. [CrossRef]
34. Gao, B.; Hao, Y.; Zhuang, W.; Tu, G.; Shi, W.; Li, S.; Hao, S.; Dong, C.; Li, M. Study on continuous solid solution of Al and Si elements of a high current pulsed electron beam treated hypereutectic Al17.5Si alloy. *Phys. Procedia* **2011**, *18*, 187–192. [CrossRef]
35. Pike, L.; Chang, Y.; Liu, C. *Point Defect Concentrations and Solid Solution Hardening in NiAl with Fe Additions*; United States Government: Washington, DC, USA, 1997. [CrossRef]
36. Gao, L.; Chen, R.S.; Han, E.H. Solid solution strengthening behaviours in binary Mg-Y single phase alloys. *J. Alloys Compd.* **2009**, *472*, 234–240. [CrossRef]
37. Jordan, J.; Deevi, S. Vacancy formation and effects in FeAl. *Intermetallics* **2003**, *11*, 507–528. [CrossRef]
38. Shen, T.; Koch, C. Formation, solid solution hardening and softening of nanocrystalline solid solutions prepared by mechanical attrition. *Acta Mater.* **1996**, *44*, 753–761. [CrossRef]
39. Callister, W.D.; Rethwisch, D.G. *Materials Science and Engineering*, 8th ed.; John Wiley & Sons, Inc.: Singapore, 2011.
40. Pike, L.; Chang, Y.; Liu, C. Solid-solution hardening and softening by Fe additions to NiAl. *Intermetallics* **1997**, *5*, 601–608. [CrossRef]
41. Xu, G.; Kutsuna, M.; Liu, Z.; Sun, L. Characteristic behaviours of clad layer by a multi-layer laser cladding with powder mixture of Stellite-6 and tungsten carbide. *Surf. Coat. Technol.* **2006**, *201*, 3385–3392. [CrossRef]

Disclaimer/Publisher's Note: The statements, opinions and data contained in all publications are solely those of the individual author(s) and contributor(s) and not of MDPI and/or the editor(s). MDPI and/or the editor(s) disclaim responsibility for any injury to people or property resulting from any ideas, methods, instructions or products referred to in the content.

Article

A Phase Field Study of the Influence of External Loading on the Dynamics of Martensitic Phase Transformation

Genggen Liu, Jiao Man *, Bin Yang, Qingtian Wang and Juncheng Wang

School of Mechanical Engineering, Xinjiang University, Urumqi 830054, China; liugg1345@163.com (G.L.)
* Correspondence: manjiao@xju.edu.cn

Abstract: An elastoplastic phase field model was employed for simulations to investigate the influence of external loading on the martensitic phase transformation kinetics in steel. The phase field model incorporates external loading and plastic deformation. During the simulation process, the authenticity of the phase field model is ensured by introducing the relevant physical parameters and comparing them with experimental data. During the calculations, loads of various magnitudes and loading conditions were considered. An analysis and discussion were conducted concerning the volume fraction and phase transition temperature during the phase transformation process. The simulation results prominently illustrate the preferential orientation of variants under different loading conditions. This model can be applied to the qualitative phase transition evolution of Fe-Ni alloys, and the crystallographic parameters adhere to the volume expansion effect. It is concluded that uniaxial loading promotes martensitic phase transformation, while triaxial compressive loading inhibits it. From a dynamic perspective, it is demonstrated that external uniaxial loading accelerates the kinetics of martensitic phase transformation, with uniaxial compression being more effective in accelerating the phase transformation process than uniaxial tension. When compared to experimental data, the simulation results provide evidence that under the influence of external loading, the martensitic phase transformation is significantly influenced by the applied load, with the impact of external loading being more significant than that of plastic effects.

Keywords: phase field simulation; martensitic phase transformation; phase transition kinetics; martensitic variant; phase field model

1. Introduction

The martensitic phase exhibits ideal mechanical performance; its microstructure and composition determine the mechanical properties of steel, making it one of the crucial constituent phases in high-strength steel. The high strength and hardness of the martensitic phase can be attributed to the solid solution strengthening by carbon atoms and the complex martensitic microstructure formed through rapid, diffusionless phase transformation. Consequently, many scholars have conducted in-depth experimental and theoretical research on the microstructure of martensite and the mechanical properties of martensitic steels [1–4]. In order to gain a profound understanding of the relationship between microstructure and properties in steel, it is imperative to delve into the processes of martensitic phase transformation under various complex conditions and the evolution of martensitic microstructures.

At present, phase field methods, as a powerful tool for predicting microstructural evolution, are widely applied in materials' solidification [5] and solid-state phase transformations [6,7]. Especially in martensitic phase transformations, refined theories and various phase field models can accurately predict the evolution of microstructures during the martensitic phase transformation process. In this study, numerical simulation methods [8] have been employed to investigate martensitic phase transformations. In engineering applications, there are often situations where experiments are impractical. Numerical simulation, with its specific computational techniques, can replicate complex processes.

Compared to experimental research, numerical simulation methods offer advantages such as cost-effectiveness, the ability to simulate conditions that are not achievable through experimental means, and comprehensive data collection. However, in the current state of numerical simulation, simplifications are often made to boundary conditions and material properties during the simulation and analysis process. The analysis results can significantly impact structural discretization, leading to varying outcomes and precision levels. The model employed in this study applies to the qualitative phase transition evolution of Fe-Ni alloys, as the crystallographic parameters conform to the volume expansion effect. Using Fe-Ni alloys as the prototype, we contrast experimental data with simulations to elucidate the mechanisms of external loading on martensitic phase transformation. However, due to variations in alloy parameters, the effects may differ, primarily concerning the crystallographic parameters of the martensitic phase transformation. The phase transition evolution is associated with its chemical free energy parameters, such as the Landau free energy coefficient (a_0) and the phase transition latent heat (Q). The influence of tensile and compressive effects on phase transformation is related to crystallographic parameters. The numerical simulation of martensitic phase transformations originated from the pioneering work of scholars such as Khachaturyan and Wang [9]. Building upon the foundation of micromechanics theory and inclusion physics, Khachaturyan and his colleagues proposed a phase field model for martensitic phase transformations, known as the phase field micromechanical model for martensitic phase transformations. In this model, a time-dependent Ginzburg–Landau (TDGL) equation is employed to simulate the evolution of martensite in single crystals. Artemev and his colleagues introduced a three-dimensional phase field model based on phase field microelasticity theory to investigate the influence of applied stress on the cubic-to-tetragonal martensitic phase transformation [10]. Yamanaka and other researchers [11] used an elastoplastic phase field model to study the effects of plastic deformation on martensitic phase transformations. Subsequently, Yeddu and colleagues built upon Khachaturyan's elastoplastic phase field microelasticity theory, developing a phase field model that incorporates plastic deformation and anisotropic properties. This model was used to investigate autocatalysis in martensitic phase transformations and classical features of martensitic microstructures [12]. Currently, the phase field method based on microelasticity theory has been demonstrated to effectively simulate martensitic phase transformations under various complex conditions in single crystals.

Stress-assisted martensitic phase transformations are commonly observed in high-strength steel materials. They find extensive applications because they can enhance the mechanical properties of materials by forming different martensitic structures under external loads. While numerous scholars have conducted extensive research on stress-assisted martensitic phase transformations [13,14], there have been relatively few reports on using phase field methods to simulate martensitic phase transformations under different external loading conditions, specifically to investigate the associated dynamic characteristics of phase transformations. Therefore, further in-depth research is needed to explore the microstructural evolution, phase transformation processes, and phase transformation temperatures under different external loading states.

This study employed a phase field method to conduct an in-depth investigation into the martensitic phase transformation of Fe-Ni single crystals under the influence of external loading conditions. This study employed a coupled elastic–plastic phase field model incorporating external loads to simulate martensitic phase transformation under various conditions, including no external load, uniaxial loading, and triaxial compression loading. The study conducted an analysis and discussion on the evolution of microstructures of variants during the phase transformation process, as well as changes in volume fractions. At the same time, this study compared the effects of external loads and plasticity on martensitic phase transformation. Statistical analysis was performed on the data related to elastic strain energy and equivalent plastic strain during the phase transformation process. Researchers such as Patel and Hagiwara [15,16] have experimentally investigated the martensitic start temperature (Ms) changes and martensite volume fraction in Fe-Ni alloys under different external loading conditions. In this study, our phase field model successfully predicted the trend of M_s temperature changes. This not only provides a reference for understanding the mechanism of external loading effects on martensitic phase transformations but also plays a significant role in controlling martensitic phase transformations and gaining deeper insights into the process of martensitic phase transformations.

2. The Elastoplastic Phase Field Model

The microstructure formed during martensitic phase transformations can be described using a set of long-range order (LRO) parameters, which represent changes in crystal symmetry during the phase transformation process. The phase field model employed in this paper is constructed based on the research of Wang, Chen, Yeddu, and others [9,12,17]. In solid-state phase transformations, the Gibbs free energy (G) is defined as the sum of three energies, namely, the chemical free energy (G_{chem}), gradient energy (G_{grad}), and elastic strain energy (G_{el}).

$$G = G_{chem} + G_{grad} + G_{el} \quad (1)$$

The system's Gibbs free energy governs the evolution of martensitic microstructures. For diffusionless phase transformations, the time-dependent Ginzburg–Landau (TDGL) dynamic equation, as proposed by Allen-Cahn, can be employed to describe this process:

$$\frac{\partial \eta_p(r,t)}{\partial t} = -\sum_{q=1}^{n} L_{pq} \frac{\delta G}{\delta \eta_p(r,t)} \quad (2)$$

Here, $r(x,y,z)$ is a vector representing spatial coordinates in Cartesian coordinates, η is the long-range order parameter or phase field variable, with the austenite phase being 0 and martensite phase being 1 in this model. L_{pq} represents the dynamic coefficient indicating interface mobility and is assumed to be isotropic. In this context, n equals 3, signifying that typical martensitic microstructures in Fe-Ni alloys can exhibit three different orientations or variants.

In the context of martensitic phase transformations, for the local free energy density, the Landau free energy is typically defined by a fourth-order Landau polynomial [18]:

$$f(\eta_p) = \frac{A}{2}\sum_{p=1}^{n} \eta_p^2 - \frac{B}{3}\sum_{p=1}^{n} \eta_p^3 + \frac{C}{4}\sum_{p=1}^{n} \eta_p^4 + \frac{D}{2}\sum_{p=1}^{n} \eta_p^2 \left(\sum_{q=1}^{n,q \neq p} \eta_q^2\right) \quad (3)$$

The chemical driving force and the energy barriers between different phases or variants determine the coefficients A, B, C, and D. Since this type of function does not explicitly represent thermodynamic or physical variables, this study refers to the work of Tae Wook Heo and Long-Qing Chen [19], which explicitly combines temperature and latent heat for the transformation. In other words, the function of chemical free energy is as follows:

$$G_{chem} = \int f(\eta_p, T) dV$$
$$= \int \left(\left[a_0 + \frac{3Q \cdot (T - T_0)}{T_0} \right] \cdot \sum_p \eta_p^2 - \left[2a_0 + \frac{2Q \cdot (T - T_0)}{T_0} \right] \cdot \sum_p \eta_p^3 + a_0 \cdot \sum_p \eta_p^4 \right) dV \quad (4)$$

Here, T represents the undercooling temperature, T_0 is the stress-free equilibrium temperature, and a_0 is an empirical parameter. The dimensionless local free energy as a function of the order parameter is schematically illustrated in Figure 1.

Figure 1, in the dimensionless local free energy schematic, clearly indicates that the phase transition proceeds in the direction of lower energy. Temperature values in Figure 1 are specified as T = 260 K, 280 K, 300 K, 320 K, and 340 K, with T_0 = 405 K.

The gradient energy is defined as the sum of energy contributions arising from the non-uniformity of the order parameter, as shown below [20]:

$$G_{grad} = \int \left[\frac{1}{2} \sum_{p=1}^{n} \beta_{ij}(p) \nabla_i \eta_p \nabla_i \eta_p \right] dV \quad (5)$$

where β_{ij} is the gradient coefficient matrix determined by the interfacial energy and interface width. In this study, it is assumed that the interface properties are isotropic.

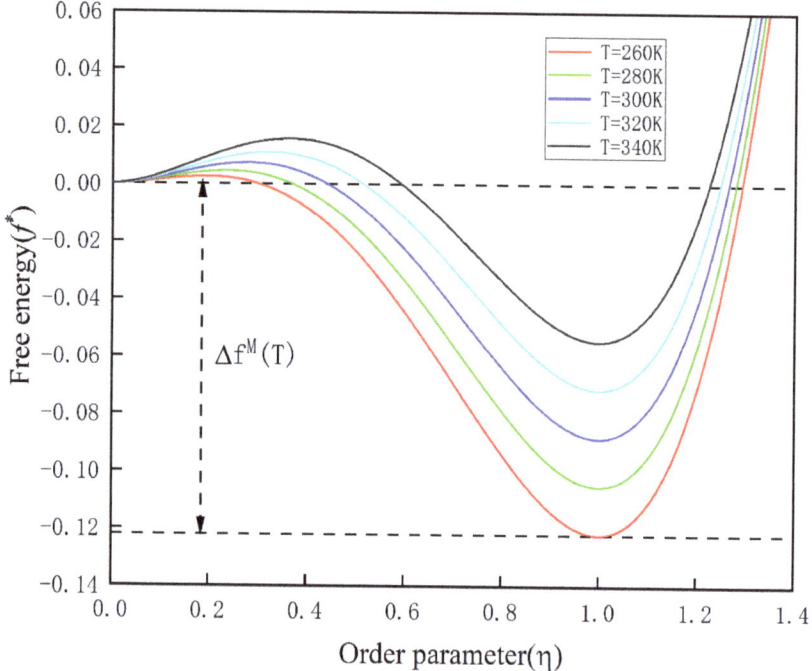

Figure 1. Schematic diagram of the local free energy function for displacement-type phase transformations at different temperatures.

The elastic strain energy for a mixture system with arbitrary parent and martensite phases is given by [21]

$$G_{el} = \frac{1}{2}\int C_{ijkl}\varepsilon_{ij}^{el}(r)\varepsilon_{kl}^{el}(r)dV \tag{6}$$

where C_{ijkl} is the fourth-order tensor of elastic constants, and $\varepsilon_{ij}^{el}(r)$ represents the elastic strain tensor. The elastic strain tensor $\varepsilon_{ij}^{el}(r)$ is defined as the difference between the total strain tensor $\varepsilon_{ij}^{tot}(r)$ and the eigenstrain tensor $\varepsilon_{ij}^{0}(r)$:

$$\varepsilon_{ij}^{el}(r) = \varepsilon_{ij}^{tot}(r) - \varepsilon_{ij}^{0}(r) \tag{7}$$

The total strain tensor $\varepsilon_{ij}^{tot}(r)$ is defined as the sum of the homogeneous strain tensor $\bar{\varepsilon}_{ij}^{tot}(r)$ and the inhomogeneous strain tensor $\delta\varepsilon_{ij}^{tot}(r)$:

$$\varepsilon_{ij}^{tot}(r) = \bar{\varepsilon}_{ij}^{tot}(r) + \delta\varepsilon_{ij}^{tot}(r) \tag{8}$$

When the macroscopic shape of the system remains fixed during the phase transformation process, the homogeneous strain tensor $\bar{\varepsilon}_{ij}^{tot}(r)$ is defined as follows:

$$\bar{\varepsilon}_{ij}^{tot}(r) = 0 \tag{9}$$

The inhomogeneous strain is defined as the deviation from the homogeneous strain and does not affect the macroscopic deformation. The inhomogeneous strain is represented by the elastic displacement field $u_i(r)$:

$$\delta\varepsilon_{ij}^{tot}(r) = \frac{1}{2}\left(\frac{\partial u_i}{\partial r_j} + \frac{\partial u_j}{\partial r_i}\right) \tag{10}$$

Assuming mechanical equilibrium is reached, the elastic solution is obtained by solving the following mechanical equilibrium equations:

$$\nabla_j \sigma_{ij} = \nabla_j\left[C_{ijkl} \cdot \left(\bar{\varepsilon}_{kl}^{tot}(r) + \delta\varepsilon_{kl}^{tot}(r) - \varepsilon_{kl}^0(r)\right)\right] = 0 \tag{11}$$

The local stress σ_{ij} in the equation is calculated using Hooke's law. To describe the elastoplastic deformation leading to martensitic phase transformations, plastic strain is introduced in the model. The eigenstrain is defined as the sum of phase transformation strain $\varepsilon_{ij}^t(r)$ and plastic strain $\varepsilon_{ij}^p(r)$:

$$\varepsilon_{ij}^0(r) = \varepsilon_{ij}^t(r) + \varepsilon_{ij}^p(r) \tag{12}$$

According to the microelasticity theory, the phase transformation strain $\varepsilon_{ij}^t(r)$ is a linear combination of the stress-free phase transformation strain $\varepsilon_{ij}^{00}(p)$ and the corresponding phase field variable η_p:

$$\varepsilon_{ij}^t(r) = \sum_{p=1}^{n} \eta_p(r,t)\varepsilon_{ij}^{00}(p) \tag{13}$$

Here, $\varepsilon_{ij}^{00}(p)$ is the stress-free phase transformation strain of a variant, calculated based on lattice parameters and the orientation relationship between the parent phase and martensite phase [22]:

$$\varepsilon_{ij}^{00}(p) = \left(F_p^T F_p - I\right)/2 \tag{14}$$

where F_p is the deformation gradient tensor for the martensitic phase transformation. The deformation gradient tensor, F_p, is computed based on relevant crystallographic representation theory, utilizing the Bain strain for the transformation from a cubic phase to a tetragonal phase, and the stress-free phase transformation strain $\varepsilon_{ij}^{00}(p)$ can be given by the Bain strain as follows:

$$\varepsilon_{ij}^{00}(1) = \begin{bmatrix} \varepsilon_3 & 0 & 0 \\ 0 & \varepsilon_1 & 0 \\ 0 & 0 & \varepsilon_1 \end{bmatrix} \varepsilon_{ij}^{00}(2) = \begin{bmatrix} \varepsilon_1 & 0 & 0 \\ 0 & \varepsilon_3 & 0 \\ 0 & 0 & \varepsilon_1 \end{bmatrix} \varepsilon_{ij}^{00}(3) = \begin{bmatrix} \varepsilon_1 & 0 & 0 \\ 0 & \varepsilon_1 & 0 \\ 0 & 0 & \varepsilon_3 \end{bmatrix} \tag{15}$$

where $\varepsilon_1 = \left(a_{bcc} - \frac{\sqrt{2}}{2}a_{fcc}\right)/\frac{\sqrt{2}}{2}a_{fcc}$ and $\varepsilon_3 = \left(c_{bcc} - a_{fcc}\right)/a_{fcc}$, where ε_1 and ε_3 are related to the lattice parameters of the parent phase $\left(a_{fcc}\right)$ and martensitic phase (a_{bcc}, c_{bcc}). Based on the above derivation, the elastic strain energy can be rewritten using Equation (6):

$$G_{el} = \frac{1}{2}\int C_{ijkl}\left(\frac{1}{2}\varepsilon_{ij}^{tot}(r)\varepsilon_{kl}^{tot}(r) - \varepsilon_{ij}^{tot}(r)\varepsilon_{kl}^0(r) + \frac{1}{2}\varepsilon_{ij}^0(r)\varepsilon_{kl}^0(r)\right)dV \tag{16}$$

When the local von Mises stress reaches the yield stress σ_y, the material begins to undergo plastic deformation. The yield criterion can be determined using the von Mises yield criterion, which can be expressed by the following equation to determine if the equivalent stress σ_{eq} has reached the yield limit:

$$\sigma_{eq}^2 - \sigma_y^2 = \frac{1}{2}(\sigma_{xx} - \sigma_{yy})^2 + \frac{1}{2}(\sigma_{yy} - \sigma_{zz})^2 + \frac{1}{2}(\sigma_{xx} - \sigma_{zz})^2 + 3\left(\sigma_{xy}^2 + \sigma_{yz}^2 + \sigma_{xz}^2\right) - \sigma_y^2 \geq 0 \tag{17}$$

where σ_{eq} represents the von Mises equivalent stress, and σ_y represents the yield stress. In order to model martensitic phase transformations under external loading conditions, the additional Gibbs energy induced by externally applied stresses needs to be included in the system's Gibbs free energy, as shown in [23].

$$G = G_{chem} + G_{grad} + G_{el} + G_{appl} \tag{18}$$

where G_{chem}, G_{grad}, and G_{el} are the chemical, gradient, and elastic parts of the Gibbs free energy. G_{appl} is the additional Gibbs free energy induced by externally applied stresses. Therefore, G_{appl} can be expressed as follows:

$$G_{appl} = -\sigma_{ij}^{appl}\varepsilon_{ij}^{t}(r) \tag{19}$$

where σ_{ij}^{appl} is the externally applied stress tensor, represented by the Cauchy stress tensor as follows:

$$\sigma_{ij}^{appl} = \begin{bmatrix} \sigma_{xx} & \sigma_{xy} & \sigma_{xz} \\ \sigma_{yx} & \sigma_{yy} & \sigma_{yz} \\ \sigma_{zx} & \sigma_{zy} & \sigma_{zz} \end{bmatrix} \tag{20}$$

From Equation (19), it can be observed that the applied stress affects the phase transformation strain $\varepsilon_{ij}^{t}(r)$, which in turn influences the eigenstrain $\varepsilon_{ij}^{0}(r)$ in the elastic strain energy G_{el} in Equation (18).

3. Simulation Parameters and Conditions

In this study, physical parameters for Fe-Ni alloys were selected based on [19,24]. The relevant physical parameters are shown in Table 1.

Table 1. Physical parameters of Fe-Ni alloy.

Physical Parameter	Notation	Numerical Value
Latent heat for the transformation	$Q/(\text{J}\cdot\text{m}^{-3})$	3.5×10^8
Stress-free equilibrium temperature	$T_0/(\text{K})$	405
Shear modulus	$G/(\text{Pa})$	2.8×10^{10}
Martensite start temperature	$M_s/(\text{K})$	223
Characteristic energy	$E_0/(\text{J}\cdot\text{m}^{-3})$	1.026×10^7
Poisson's ratio	v	0.375
Bain strain tensors 1	ε_1	0.1322
Bain strain tensors 2	ε_3	-0.1994

In the phase field calculations, a semi-implicit Fourier spectral method proposed by Chen [17] was used to solve the dynamic equations. The simulation system had a grid size of 65 × 65 × 65. To study the growth process of martensitic phase transformations, a small cubic martensite nucleus with a side length of 1.6 μm was assumed to pre-exist at the center of the simulation domain. A single crystal grain undergoing martensitic phase transformation was considered the simulation region and had a physical size of approximately 16 μm. The iso-surface of the phase field variable (g = 0.7) is shown in all figures. For simplicity, Equation (7) in this paper is solved in dimensionless form, with dimensionless parameters as listed in Table 2.

Table 2. Dimensionless parameters in the simulation.

Dimensionless Parameter	Numerical Value
a_0^*	60.02
Q^*	34.11
$\Delta x^* = \Delta y^* = \Delta z^*$	1.0
Δt^*	0.01

Here, Q^* and a_0^* are dimensionless parameters for the phase transition latent heat and the Landau free energy coefficient, obtained by dividing these quantities by a characteristic energy, where the characteristic energy is represented as $E_0 = 1.026 \times 10^7$. The dimensionless time $\Delta t^* = 0.01$ corresponds to a real time of approximately 0.975 nanoseconds. In the study, the parameter t^* represents non-dimensional time, and the colors of the phase field variable, such as red, blue, and green, correspond to martensitic variants 1, 2, and 3, respectively. The model employs stress boundary conditions to solve mechanical problems. External stresses are selected so that they remain smaller than the yield stresses of both the austenite and martensite phases. Different simulations were conducted under various loading conditions in the phase field simulations presented in this paper. To compare with relevant experimental results and validate them, uniaxial tensile and uniaxial compressive loads of 100 MPa, 130 MPa, and 150 MPa were applied along the [100] direction. The triaxial compressive loading conditions correspond to isotropic hydrostatic pressure, where equal pressure is applied from all sides. To match the experimental conditions in the reference literature, triaxial compressive loads of 100 MPa, 130 MPa, and 150 MPa were used. The phase field model was implemented as code in Matlab R2021b.

4. Simulation Results and Analysis

4.1. Martensitic Phase Transformation without External Loading

The evolution of martensite volume fraction during martensitic phase transformation without external loading is shown in Figure 2. The three-dimensional microstructure at $t^* = 0$, $t^* = 35$, $t^* = 55$, and $t^* = 65$ is depicted in the figure.

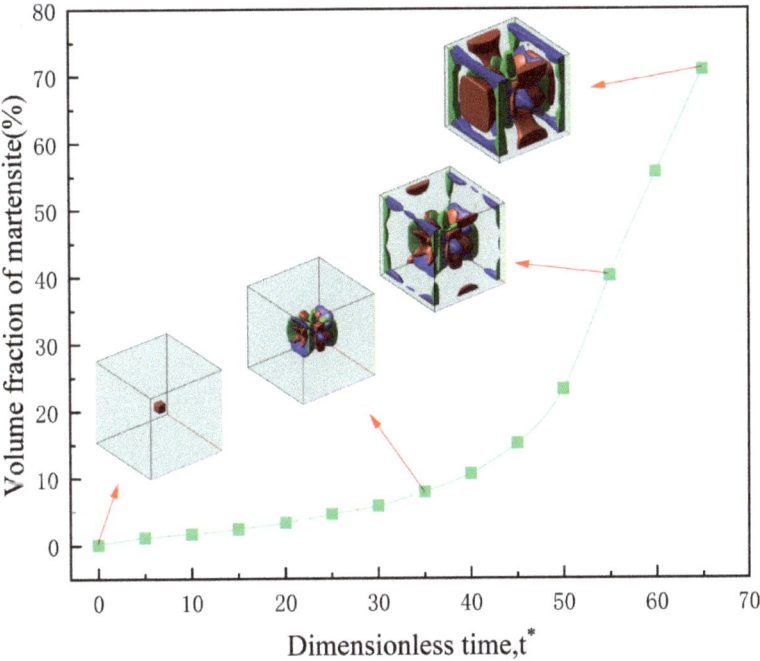

Figure 2. Variation in martensite volume fraction during phase transformation without loading.

From Figure 2, it can be observed that the martensite volume fraction increases with time, and it increases rapidly after $t^* = 50$. From the three-dimensional microstructure images, it is evident that as the phase transformation progresses, the small cubic martensite nucleus pre-existing at the center of the simulation domain gradually transforms into different martensite variants. Growing martensite can induce other unstable martensite nucleus embryos to transform into stable nuclei and start growing, a phenomenon known as self-catalytic nucleation [25]. Martensite nuclei create stress fields around them during growth, and to reach a stable state, nucleation of new variants is required to reduce strain energy. This evolutionary process reduces the stress generated around growing martensite due to self-catalytic nucleation.

4.2. Microstructure Evolution of Martensitic Phase Transformation under External Loading

The microstructure obtained under uniaxial tensile loading with σ = 100 MPa is shown in Figure 3.

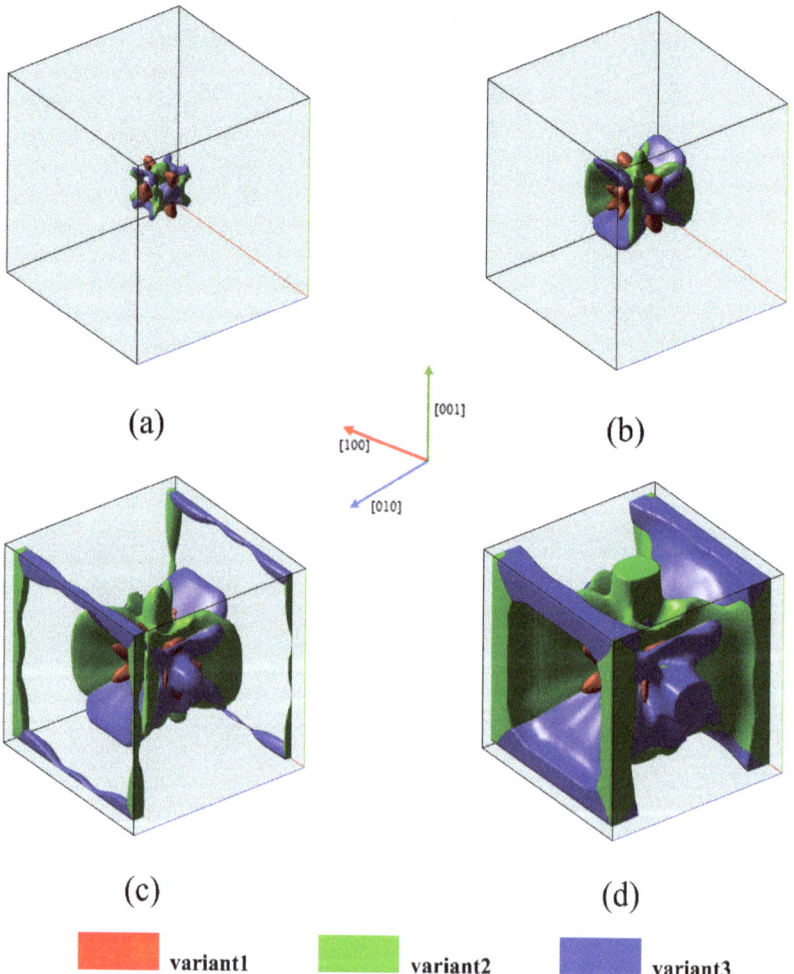

Figure 3. Three-dimensional microstructure evolution of martensitic variants under uniaxial tensile loading: (**a**) t* = 5; (**b**) t* = 25; (**c**) t* = 45; (**d**) t* = 60.

As depicted in the figure, the final microstructure is primarily composed of two variants, variant 2 and variant 3. The evolution is driven by the energy imposed externally, aiming to minimize the Gibbs free energy, which means the evolution proceeds in the direction of lower energy. According to the phase field model, variant 1 is controlled by the Bain strain tensor, which compresses along the [100] direction but stretches along the [010] and [001] direction. Under uniaxial tensile loading along the [100] direction, G_{appl} increases, leading to the suppression of variant 1. However, variants 2 and 3 controlled by the Bain strain tensor experience a decrease in G_{appl} under uniaxial tensile loading along the [100] direction. Therefore, the formation of variants 2 and 3 is promoted.

The microstructure obtained under uniaxial compressive loading with $\sigma = -100$ MPa is shown in Figure 4. As seen in Figure 4, under uniaxial compressive loading, the formation of variant 1 is favorable. In contrast to uniaxial tensile loading, compressive loads promote the formation of martensite variants that experience the maximum compression along the loading direction, which minimizes G_{appl}. Therefore, variant 1 is promoted to grow under uniaxial compressive loading conditions.

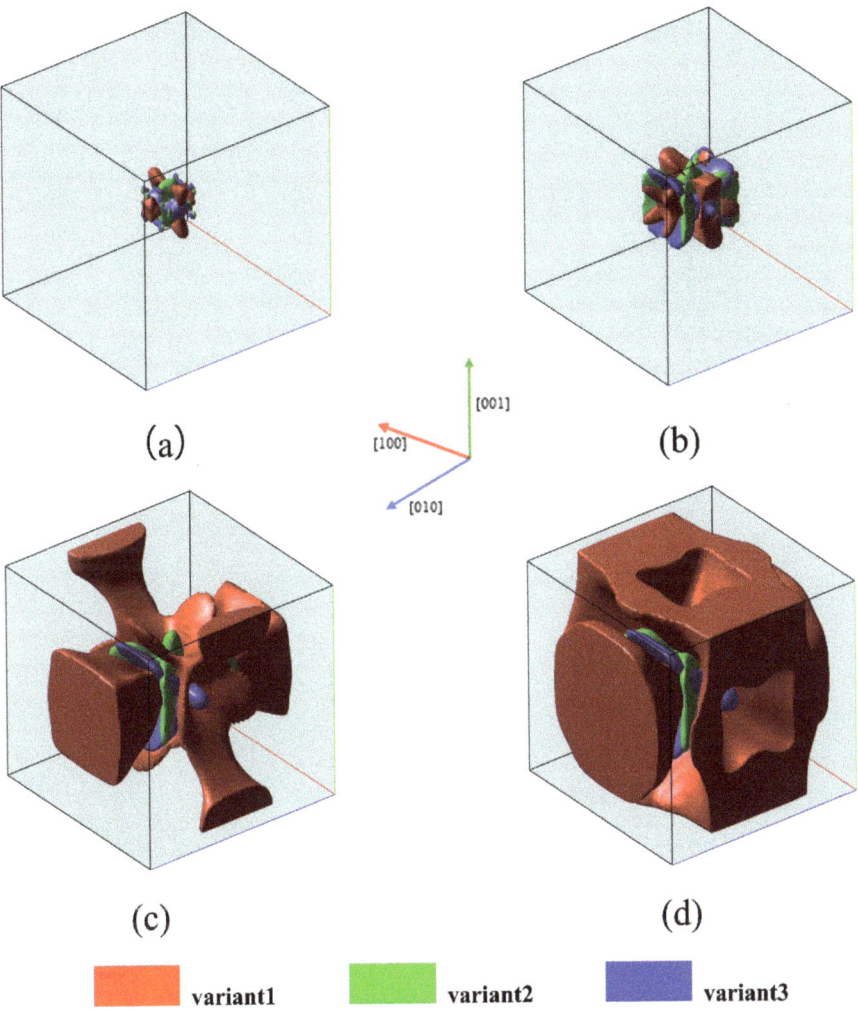

Figure 4. Three-dimensional microstructure evolution of martensitic variants under uniaxial compression loading: (a) t* = 5; (b) t* = 25; (c) t* = 45; (d) t* = 60.

The microstructure obtained under triaxial compressive loading with $\sigma = -100$ MPa is shown in Figure 5. Figure 5 shows that unlike uniaxial tension and uniaxial compression, triaxial compressive loading does not result in different martensite microstructures. Instead, it resembles the martensite microstructure obtained without external loading. Comparing Figure 5 with Figure 2, it can be observed that the evolution process under triaxial compressive loading lags behind that under no external loading, and all variants are inhibited to varying degrees.

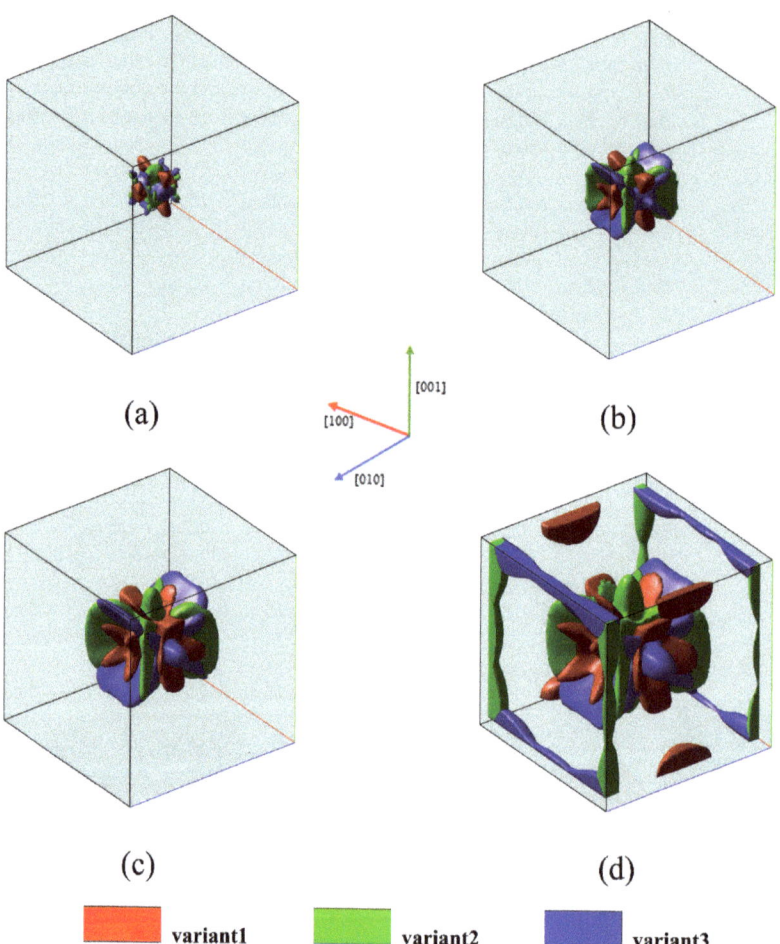

Figure 5. Three-dimensional microstructural evolution of martensite variants under triaxial compressive loading: (**a**) t* = 5; (**b**) t* = 25; (**c**) t* = 45; (**d**) t* = 60.

Depending on the martensite variants favored under different stress conditions, different stress states can lead to complex martensite microstructures. Martensitic phase transformations under loading conditions exhibit variant selection behavior, and this selection behavior is related to the magnitude and direction of external loading; the loading direction determines the types of variants that are favored, and the magnitude of the load influences the extent of variant selection. Due to the different contributions of external energy terms to the total system free energy, this also results in diversity in variant selection orientations during different loading processes. It can be said that applying external loads favors the formation of martensite variants that reduce the Gibbs free energy in the direction of the applied load, which is the common Magee effect [26,27].

4.3. Dynamics Analysis of Martensitic Phase Transformation under External Loading and Plastic Deformation

In order to visually observe the variation in martensite content under uniaxial loading, this study explores the effects of martensite variants and the total martensite volume fraction over time under uniaxial tensile and compressive loading. Figures 6 and 7 illustrate the relationship between external loading, evolution time, and volume fraction.

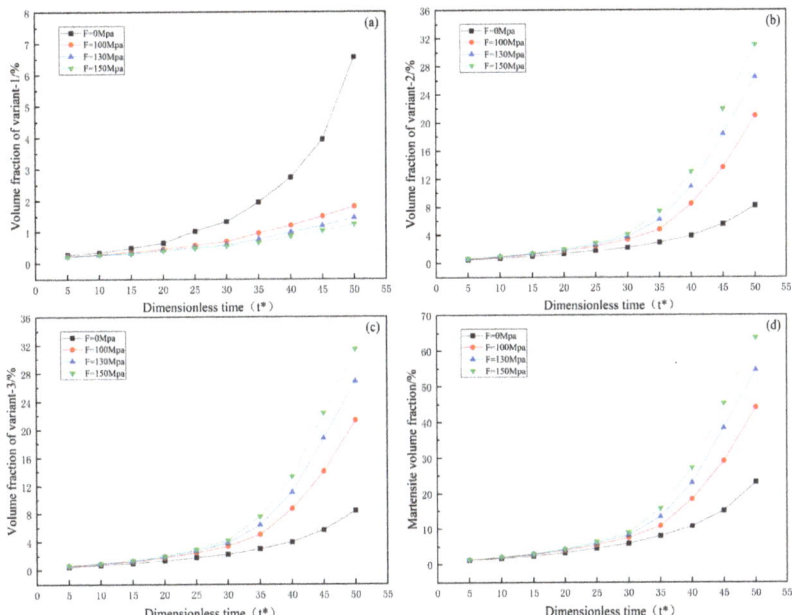

Figure 6. Evolution of martensite volume fraction with respect to time under uniaxial tension: (**a**–**c**) volume fractions of variants 1, 2, and 3; (**d**) total martensite volume fraction.

Figures 6 and 7 show that both martensite variants and the total volume fraction of martensite increase with increasing evolution time under both uniaxial tensile and compressive loading. Furthermore, the influence on volume fraction becomes more significant with increasing external loading. Under tensile loading, variant 1 decreases with increasing external loading, while variants 2 and 3 increase with increasing external loading. The opposite trend is observed under compressive loading. Therefore, as explained in the study by Yeddu et al. [28], externally applied loads may contribute to forming certain variants. It is worth noting that from Figure 7a, under compressive loading, variant 1 shows rapid growth after $t^* = 30$. Under tensile loading, there is also a tendency for the rapid growth of variants 2 and 3, but it remains lower than the growth rate of variant 1 under compressive loading.

Figure 8 shows that, compared to no external loading, the martensite volume fractions under uniaxial tensile and compressive loading are ultimately greater. Therefore, uniaxial tensile and compressive loading both accelerate martensitic phase transformation. At the beginning of martensitic phase transformation, the volume fraction under uniaxial compressive loading is lower than that under uniaxial tensile loading. However, as the evolution progresses, it starts to increase rapidly after $t^* = 35$ and surpasses the uniaxial tensile loading, ultimately resulting in a higher martensite volume fraction under uniaxial compressive loading. Traditional thermodynamics cannot reveal dynamic aspects of phase transformation, such as the evolution process. In this study, the phase field method reveals that uniaxial compressive loading accelerates martensitic phase transformation more effectively than uniaxial tensile loading from a dynamic perspective. This phenomenon is in excellent agreement with the experimental results of Hagiwara et al. [16].

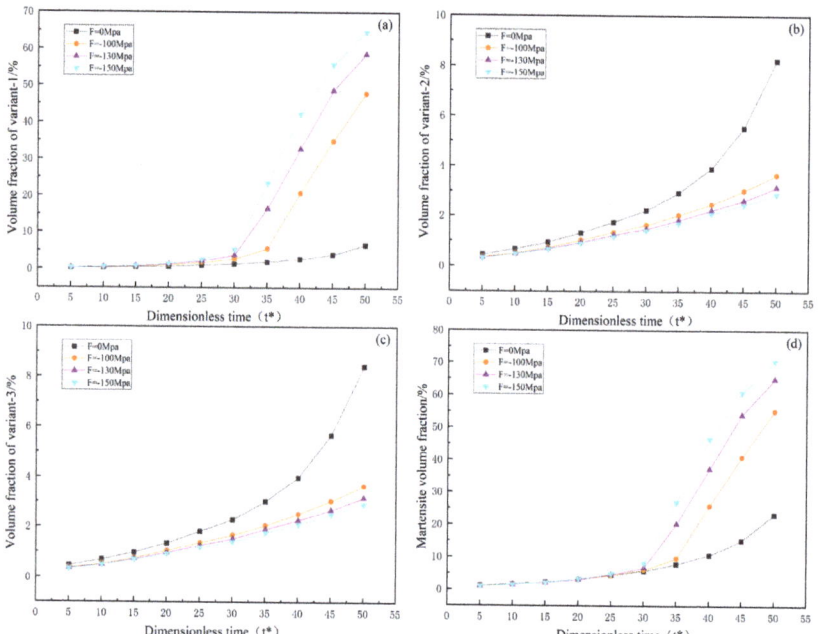

Figure 7. Evolution of martensite volume fraction with respect to time under uniaxial compression: (**a**–**c**) volume fractions of variants 1, 2, and 3; (**d**) total martensite volume fraction.

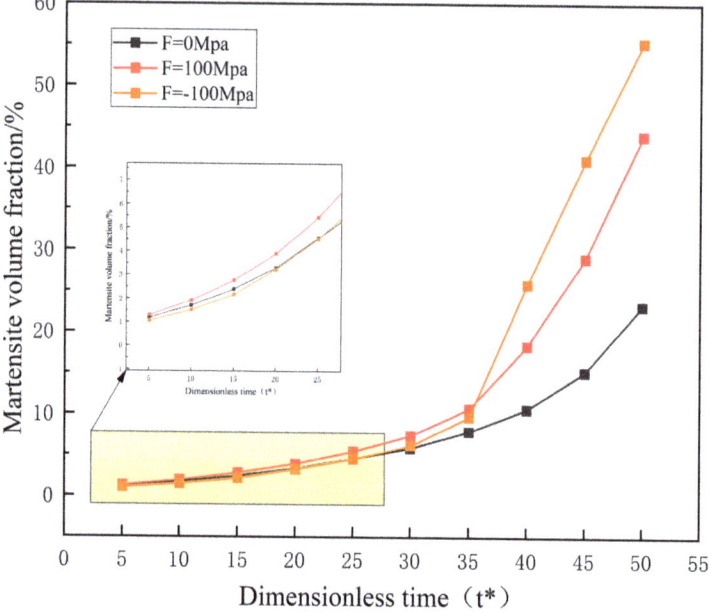

Figure 8. The evolution of martensite volume fraction under different applied uniaxial loads as a function of time.

In order to gain a deeper understanding of the microstructural evolution under a triaxial compression load, Figure 9 depicts the changes in the volume fractions of different martensitic variants and the total martensite with respect to the evolution time. From Figure 9a–c, it can be observed that the volume fractions of martensite and its variants increase as the evolution progresses. However, at the same time, the volume fractions of martensitic variants and the total martensite decrease with an increase in external loading. This simulation result is consistent with the research findings reported by Kakeshita [29] and other researchers. Therefore, under the influence of triaxial compression load, each martensitic variant is subjected to varying degrees of suppression.

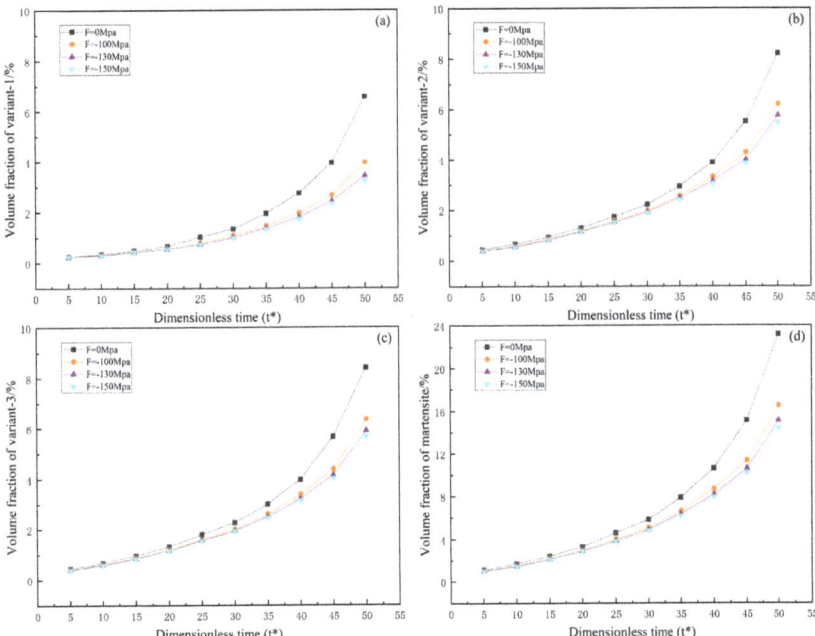

Figure 9. Evolution of martensite volume fraction with respect to time under triaxial compression loading: (**a**–**c**) volume fractions of variants 1, 2, and 3; (**d**) total martensite volume fraction.

To investigate the effects of plasticity and external stress on the elastic strain energy during the phase transformation process, the changes in elastic strain energy density during the martensitic transformation are shown in Figure 10.

In Figure 10, the green and blue lines represent the changes in elastic strain energy density during the microstructural evolution shown in Figures 2 and 3, respectively. The red line represents the changes in elastic strain energy density without plastic accommodation and stress accommodation, where only self-accommodation effects regulate the elastic strain energy density. The results from the curves in Figure 10 indicate that plastic deformation and external tensile stress can reduce the elastic strain energy density, which is consistent with the conclusions obtained in the simulations from the referenced studies [19,30].

External loading and plastic effects both play significant roles in martensitic phase transformation. Figure 11 illustrates the relationship between martensite volume fraction and yield strength.

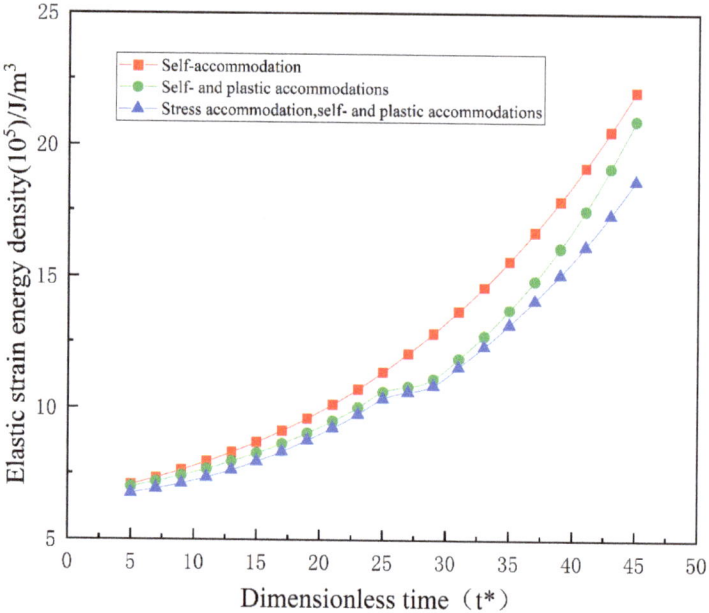

Figure 10. Elastic strain energy density during martensitic phase transformation under different conditions.

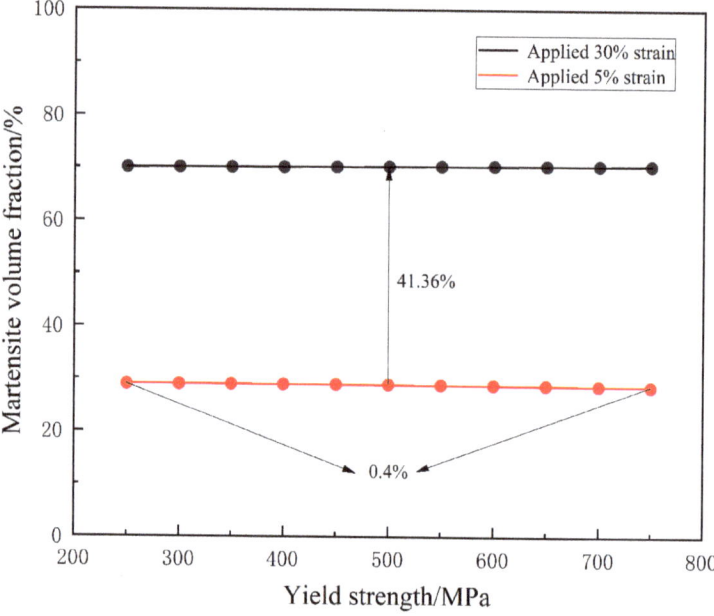

Figure 11. Effect of yield strength on martensite volume fraction under different external loading conditions.

In Figure 11, "applied 30% strain" refers to uniaxial tensile deformation carried out during the simulation at a strain rate of 5.6×10^{-4} s^{-1}, resulting in a final applied strain of 30%. Similarly, "applied 5% strain" indicates that uniaxial tensile deformation was conducted during the simulation at a strain rate of 5.6×10^{-4} s^{-1}, leading to a final applied strain of 5%. During the phase transformation, yield strength is related to plastic deformation. When the yield strength is low, plastic effects are significant. Consequently, the strain energy relaxation due to lattice distortion is more pronounced, leading to a higher martensitic transformation temperature. As shown in Figure 11, the martensite volume fraction is minimally affected by changes in yield strength. In contrast, under the influence of various magnitudes of external loading, the martensite volume fraction exhibits a significant range of variation, spanning two orders of magnitude. This indicates that plastic effects have a minimal influence on the phase transformation process, and they can even be negligibly small under external loading conditions. Under the influence of external loading, the generation of martensite is primarily determined by the external load, with a minimal impact from plastic effects. This simulation result aligns with the experimental findings of Matsuoka and other researchers [31]. Therefore, in martensitic phase transformations under external loading, the external load plays a dominant role and has a greater influence on the transformation process than plastic effects. Previous studies mainly analyzed and discussed experimental results without delving deeply into the underlying physical mechanisms. This research employed a phase field simulation approach to provide theoretical insights into the effects of external loading and plasticity on martensitic phase transformations.

4.4. Prediction of Ms Temperature and TRIP Effect

The change in the martensite transformation temperature M_s with respect to the applied uniaxial load is depicted in Figure 12.

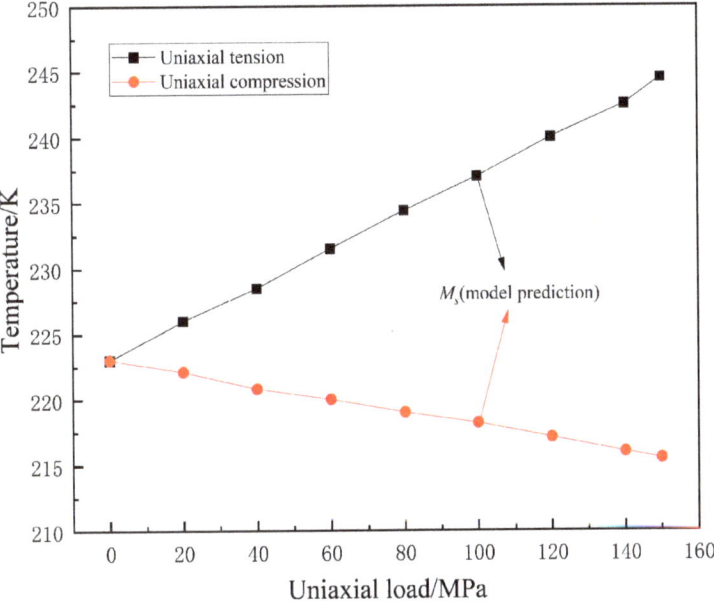

Figure 12. Variation in the martensite start temperature M_s with externally applied uniaxial loading.

From Figure 12, it can be observed that under tensile loading, the M_s temperature increases with the applied uniaxial load. Under compressive loading, the M_s temperature slightly decreases with increasing uniaxial load, with a relatively small effect on the M_s temperature. Overall, in the material studied in this paper, uniaxial tensile loading increases the M_s temperature and promotes martensitic transformation, while uniaxial compressive loading, although not significantly reducing the M_s temperature, accelerates martensitic transformation. This is consistent with the results reported in the referenced studies [15,16]. The phase field simulations in this paper predict the phase transformation temperature, which not only validates the experimental results of previous researchers but also provides an explanation for the influence mechanism of uniaxial loading on martensitic transformation, revealing the connection between external loading and martensitic transformation at a fundamental physical level.

Figure 13 represents the variation in M_s temperature with the magnitude of applied three-dimensional compressive loading. This work references experimental data studied by Patel and other researchers and conducts thermodynamic calculations based on the methods proposed by Patel and Cohen [15]. The red solid line and dashed line represent experimental values and thermodynamic calculation values, while the green scattered points represent the predicted values from the model in this paper. From Figure 13, it can be observed that the M_s temperature decreases with increasing applied load. Based on the changes in volume fraction and M_s temperature mentioned earlier, it can be concluded that three-dimensional compressive loading suppresses martensitic transformation. The prediction of M_s temperature agrees with the results reported in the referenced literature [15,32].

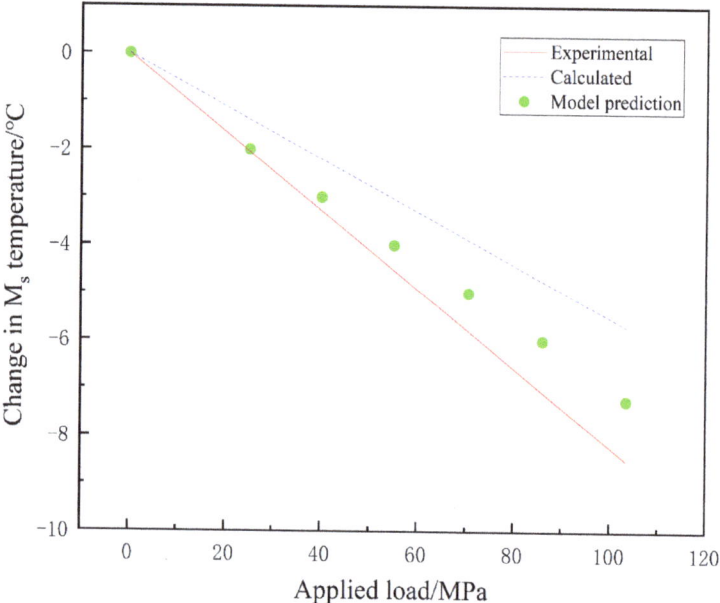

Figure 13. Relationship between the change in M_s temperature and the triaxial compression load.

Metals and alloys exhibit plastic deformation below the yield strength of the parent phase, leading to an increase in plasticity during the phase transformation process. This characteristic is known as Transformation-Induced Plasticity (TRIP) [33]. This study discusses the average equivalent plastic strain during the phase transformation process to investigate this phenomenon. The variation in average equivalent plastic strain with evolution time is shown in Figure 14.

Figure 14 indicates that as evolution progresses, the volume fraction of martensite continuously increases, leading to an increase in plastic strain. The higher degree of plastic strain occurs due to the martensitic phase transformation. Plastic deformation caused by temperature and applied stress changes does not occur during the simulated phase transformation process. Therefore, the martensitic phase transformation induces the plastic strain generated during the evolution. Furthermore, the externally applied stress is much lower than the yield strength of austenite, so the phase transformation process induces plasticity. The local yielding mechanism in this process occurs because the low stress induces the phase transformation, creating substantial internal stresses locally. At this point, the local internal stresses exceed the yield strength of austenite, resulting in local plastic deformation. As the phase transformation continues, the accumulation of local plastic deformation is manifested as macroscopic plastic deformation in the specimen.

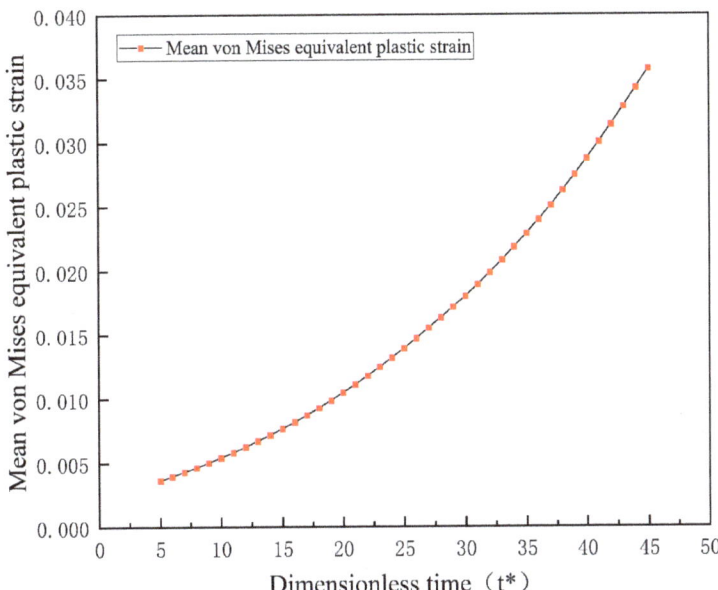

Figure 14. Variation in average equivalent plastic strain under uniaxial tensile load as a function of martensite volume fraction.

5. Conclusions

(1) In the martensitic phase transformation process under external loading conditions, uniaxial tensile loading increases the martensitic phase transformation start temperature M_s and promotes martensitic phase transformation. On the other hand, uniaxial compressive loading, while not significantly lowering M_s, accelerates the transformation of austenite to martensite, resulting in an increase in martensite volume fraction and faster phase transformation kinetics. The acceleration effect of uniaxial compressive loading is superior to that of uniaxial tensile loading. Triaxial compressive loading reduces the M_s temperature and inhibits the martensitic phase transformation. This conclusion aligns with relevant experimental findings.

(2) The implementation of a phase-field model was employed to simulate martensitic phase transformation under the influence of both plastic accommodations and stress accommodations. The simulation results revealed that, in the presence of external loading conditions, the impact of external loading on martensite volume fraction outweighed the influence of plastic effects. Furthermore, external loading and plastic deformation can release the elastic strain energy during the martensitic phase transformation process.

(3) Simulations of microstructure evolution during the martensitic phase transformation in Fe-Ni alloys were performed. The simulation results vividly demonstrate that different variants are favored to transform under corresponding stress conditions. Under uniaxial tensile loading, the growth of martensite variants 2 and 3 is promoted. Under uniaxial compressive loading, martensite variant 1 is favored to grow. Under triaxial compressive loading, the growth of all three martensite variants is inhibited.

Author Contributions: Methodology, G.L. and J.M.; Resources, J.M.; Data curation, G.L. and B.Y.; Writing—original draft, G.L.; Writing—review and editing, G.L., J.M., B.Y., Q.W. and J.W.; All authors have read and agreed to the published version of the manuscript.

Funding: This study was supported by the Tianchi Program for High-level Talents of Xinjiang Uygur Autonomous Region, China (Grant No. 100400028) and the PhD Start-up Fund of Xinjiang University, China (Grant No. 620320011).

Institutional Review Board Statement: Not applicable.

Informed Consent Statement: Not applicable.

Data Availability Statement: The data that support the findings of this study are available from the authors upon reasonable request.

Conflicts of Interest: The authors declare no conflict of interest.

References

1. Mohapatra, S.; Poojari, G.; Marandi, L.; Das, S.; Das, K. A systematic study on microstructure evolution, mechanical stability and the micro-mechanical response of tensile deformed medium manganese steel through interrupted tensile test. *Mater. Charact.* **2023**, *195*, 112562. [CrossRef]
2. Hao, S.; Chen, L.; Wang, Y.; Zhang, W.; She, J.; Jin, M. Influence of loading direction on tensile deformation behavior of a lean duplex stainless steel sheet: The role of martensitic transformation. *Mater. Sci. Eng. A* **2022**, *848*, 143384. [CrossRef]
3. Wang, P.; Zheng, W.; Yu, X.; Wang, Y. Advantageous Implications of Reversed Austenite for the Tensile Properties of Super 13Cr Martensitic Stainless Steel. *Materials* **2022**, *15*, 7697. [CrossRef] [PubMed]
4. Kakeshita, T.; Fukuda, T.; Saburi, T. Time-dependent nature of the athermal martensitic transformations in Fe-Ni-alloys. *Scr. Mater.* **1996**, *34*, 147–150. [CrossRef]
5. Kavousi, S.; Ankudinov, V.; Galenko, P.K.; Zaeem, M. Atomistic-informed kinetic phase-field modeling of non-equilibrium crystal growth during rapid solidification. *Acta Mater.* **2023**, *253*, 118960. [CrossRef]
6. Lv, S.; Wang, S.; Wu, G.; Gao, J.; Yang, X.; Wu, H.; Mao, X. Application of phase-field modeling in solid-state phase transformation of steels. *J. Iron Steel Res. Int.* **2022**, *29*, 867–880. [CrossRef]
7. Tourret, D.; Liu, H.; Llorca, J. Phase-field modeling of microstructure evolution: Recent applications, perspectives and challenges. *Prog. Mater. Sci.* **2022**, *123*, 100810.
8. Qin, R.S.; Bhadeshia, H.K. Phase field method. *Mater. Sci. Technol.* **2010**, *26*, 803–811. [CrossRef]
9. Wang, Y.; Khachaturyan, A.G. Three-dimensional field model and computer modeling of martensitic transformations. *Acta Mater.* **1997**, *45*, 759–773. [CrossRef]
10. Artemev, A.; Wang, Y.; Khachaturyan, A.G. Three-dimensional phase field model and simulation of martensitic transformation in multilayer systems under applied stresses. *Acta Mater.* **2000**, *48*, 2503–2518. [CrossRef]
11. Yamanaka, A.; McReynolds, K.; Voorhees, P.W. Phase field crystal simulation of grain boundary motion, grain rotation and dislocation reactions in a BCC bicrystal. *Acta Mater.* **2017**, *133*, 160–171. [CrossRef]
12. Yeddu, H.K.; Borgenstam, A.; Ågren, J. Stress-assisted martensitic transformations in steels: A 3-D phase-field study. *Acta Mater.* **2013**, *61*, 2595–2606. [CrossRef]
13. Li, Y.; Guan, Q.; He, B. Improving the strength and ductility of medium Mn steel by depleting the stress-assisted martensite. *Scr. Mater.* **2023**, *226*, 115267. [CrossRef]
14. Tian, Y.; Borgenstam, A.; Hedström, P. Comparing the deformation-induced martensitic transformation with the athermal martensitic transformation in Fe-Cr-Ni alloys. *J. Alloys Compd.* **2018**, *766*, 131–139. [CrossRef]
15. Patel, J.R.; Cohen, M. Criterion for the action of applied stress in the martensitic transformation. *Acta Metall.* **1953**, *1*, 531–538. [CrossRef]
16. Hagiwara, I.; Kanazawa, S. Effect of Applied Stress Upon the Martensite Transformation. *J. Jpn. Inst. Met. Mater.* **1961**, *25*, 213–216. [CrossRef]
17. Chen, L.; Shen, J. Applications of semi-implicit Fourier-spectral method to phase field equations. *Comput. Phys. Commun.* **1998**, *108*, 147–158. [CrossRef]
18. Cui, S.; Lei, Z.; Tao, F. Phase-Field Simulation of Thermoelastic Martensitic Transformation in U-Nb Shape Memory Alloys. *Rare Metal Mat. Eng.* **2022**, *51*, 452–460.
19. Heo, T.W.; Chen, L. Phase-field modeling of displacive phase transformations in elastically anisotropic and inhomogeneous polycrystals. *Acta Mater.* **2014**, *76*, 68–81. [CrossRef]
20. Ohmer, D.; Yi, M.; Gutfleisch, O.; Xu, B. Phase-field modelling of paramagnetic austenite–ferromagnetic martensite transformation coupled with mechanics and micromagnetics. *Int. J. Solids Struct.* **2022**, *238*, 111365. [CrossRef]
21. Mamivand, M.; Zaeem, M.A.; El Kadiri, H. Effect of variant strain accommodation on the three-dimensional microstructure formation during martensitic transformation: Application to zirconia. *Acta Mater.* **2015**, *87*, 45–55. [CrossRef]
22. Gao, Y.; Zhou, N.; Wang, D.; Wang, Y. Pattern formation during cubic to orthorhombic martensitic transformations in shape memory alloys. *Acta Mater.* **2014**, *68*, 93–105. [CrossRef]
23. Zhang, X.; Shen, G.; Xu, J.; Gu, J. Analysis of Martensitic Transformation Plasticity Under Various Loadings in a Low-Carbon Steel: An Elastoplastic Phase Field Study. *Metall. Mater. Trans. A* **2020**, *51*, 4853–4867. [CrossRef]

24. Shibata, A.; Murakami, T.; Morito, S.; Furuhara, T.; Maki, T. The origin of midrib in lenticular martensite. *Mater Trans.* **2008**, *49*, 1242–1248. [CrossRef]
25. Liu, H.; Lin, F.X.; Zhao, P.; Moelans, N.; Wang, Y.; Nie, J.F. Formation and autocatalytic nucleation of co-zone {101¯2} deformation twins in polycrystalline Mg: A phase field simulation study. *Acta Mater.* **2018**, *153*, 86–107. [CrossRef]
26. Kubler, R.F.; Berveiller, S.; Bouscaud, D.; Guiheux, R.; Patoor, E.; Puydt, Q. Shot peening of TRIP780 steel: Experimental analysis and numerical simulation. *J. Mater. Process. Technol.* **2019**, *270*, 182–194. [CrossRef]
27. Cluff, S.; Knezevic, M.; Miles, M.P.; Fullwood, D.T.; Mishra, R.K.; Sachdev, A.K.; Brown, T.; Homer, E.R. Coupling kinetic Monte Carlo and finite element methods to model the strain path sensitivity of the isothermal stress-assisted martensite nucleation in TRIP-assisted steels. *Mech. Mater.* **2021**, *154*, 103707. [CrossRef]
28. Yeddu, H.K.; Lookman, T. Phase-field modeling of the beta to omega phase transformation in Zr–Nb alloys. *Mater. Sci. Eng. A* **2015**, *634*, 46–54. [CrossRef]
29. Kakeshita, T.; Shimizu, K.; Akahama, Y.; Endo, S.; Fujita, F. Effect of Hydrostatic Pressure on Martensitic Transformations in Fe–Ni and Fe–Ni–C Alloys. *Trans. JIM* **1988**, *29*, 109–115. [CrossRef]
30. Song, S.J.; Liu, F.; Zhang, Z.H. Analysis of elastic–plastic accommodation due to volume misfit upon solid-state phase transformation. *Acta Mater.* **2014**, *64*, 266–281. [CrossRef]
31. Matsuoka, Y.; Iwasaki, T.; Nakada, N.; Tsuchiyama, T.; Takaki, S. Effect of grain size on thermal and mechanical stability of austenite in metastable austenitic stainless steel. *ISIJ Int.* **2013**, *53*, 1224–1230. [CrossRef]
32. Kakeshita, T.; Saburi, T.; Kindo, K.; Endo, S. Effect of magnetic field and hydrostatic pressure on martensitic transformation and its kinetics. *Jpn. J. Appl. Phys.* **1997**, *36*, 7083. [CrossRef]
33. Soleimani, M.; Kalhor, A.; Mirzadeh, H. Transformation-induced plasticity (TRIP) in advanced steels: A review. *Mater. Sci. Eng. A* **2020**, *795*, 140023. [CrossRef]

Disclaimer/Publisher's Note: The statements, opinions and data contained in all publications are solely those of the individual author(s) and contributor(s) and not of MDPI and/or the editor(s). MDPI and/or the editor(s) disclaim responsibility for any injury to people or property resulting from any ideas, methods, instructions or products referred to in the content.

Review

Constitutive Models for the Strain Strengthening of Austenitic Stainless Steels at Cryogenic Temperatures with a Literature Review

Bingyang He [1], Juan Wang [1,*] and Weipu Xu [2]

[1] Key Laboratory for Liquid-Solid Structural Evolution and Processing of Materials (Ministry of Education), Shandong University, Jinan 250061, China; 18337656719@163.com
[2] Shanghai Institute of Special Equipment Supervision and Inspection Technology, Shanghai 200062, China; neiltwo@126.com
* Correspondence: jwang@sdu.edu.cn

Abstract: Austenitic stainless steels are widely used in cryogenic pressure vessels, liquefied natural gas pipelines, and offshore transportation liquefied petroleum gas storage tanks due to their excellent mechanical properties at cryogenic temperatures. To meet the lightweight and economical requirements, pre-strain of austenitic stainless steels was conducted to improve the strength at cryogenic temperatures. The essence of being strengthened by strain (strain strengthening) and the phase-transformation mechanism of austenitic stainless steels at cryogenic temperatures are reviewed in this work. The mechanical properties and microstructure evolution of austenitic stainless steels under different temperatures, types, and strain rates are compared. The phase-transformation mechanism of austenitic stainless steels during strain at cryogenic temperatures and its influence on strength and microstructure evolution are summarized. The constitutive models of strain strengthening at cryogenic temperatures were set to calculate the volume fraction of strain-induced martensite and to predict the mechanical properties of austenitic stainless steels.

Keywords: austenitic stainless steel; strain strengthening; cryogenic temperature; phase transformation mechanism; mechanical properties; constitutive model

Citation: He, B.; Wang, J.; Xu, W. Constitutive Models for the Strain Strengthening of Austenitic Stainless Steels at Cryogenic Temperatures with a Literature Review. *Metals* **2023**, *13*, 1894. https://doi.org/10.3390/met13111894

Academic Editors: Andrea Di Schino, Damien Fabrègue and Claudio Testani

Received: 16 October 2023
Revised: 7 November 2023
Accepted: 9 November 2023
Published: 15 November 2023

Copyright: © 2023 by the authors. Licensee MDPI, Basel, Switzerland. This article is an open access article distributed under the terms and conditions of the Creative Commons Attribution (CC BY) license (https:// creativecommons.org/licenses/by/ 4.0/).

1. Introduction

The crystal structure of austenitic stainless steel exhibits a face-centered cubic structure (FCC), which has a higher density than that of a body-centered cubic crystal structure (BCC). The corrosion resistance of austenitic stainless steel is significantly good, making it one of the most widely used stainless steels in industrial production [1,2]. The face-centered cubic structure has four groups of slip surfaces and three slip directions on each group of slip surfaces. The 12 slip systems reduce the good plasticity and toughness of austenitic stainless steel. The yield strength of austenitic stainless steels at room temperature is 313.09 MPa, the tensile strength is 804.59 MPa, the yield strength at cryogenic temperature (−196 °C) is 558.91 MPa, and the tensile strength is 1633.52 MPa [3]. Therefore, the application of austenitic stainless steels in cryogenic temperature service equipment, such as cryogenic pressure vessels, liquefied natural gas pipelines, and offshore transportation liquefied petroleum gas storage tanks, often requires a large thickness for safety factors. As a result, the manufacturing cost and weight of the equipment increase, which does not meet the lightweight and economical requirements.

A significant feature of austenitic stainless steel is that it can be strengthened with strain. This phenomenon is called strain strengthening, which is obvious with a decrease in temperature or an increase in deformation during tensile strain. The main reason is the partial phase transformation from austenite to martensite, and the morphology and content of martensite play a key role in strain strengthening at cryogenic temperatures [4].

The thermodynamics required for the phase transformation from austenite γ to martensite α' is shown in Figure 1 [5]. The phase transformation mainly occurs in one of the following three ways: (a) temperatures below the critical temperature M_s, (b) temperatures above M_s while elastic stress provides kinetic phase transformation, or (c) temperatures above M_s while plastic tensile force provides kinetic phase transformation. Under different temperature conditions, there will be different proportions of austenite to martensite during the phase-transformation process. In addition to the macroscopic deformation caused by plastic strain, the process is accompanied by an increase in the phase volume, resulting in microstructural evolution.

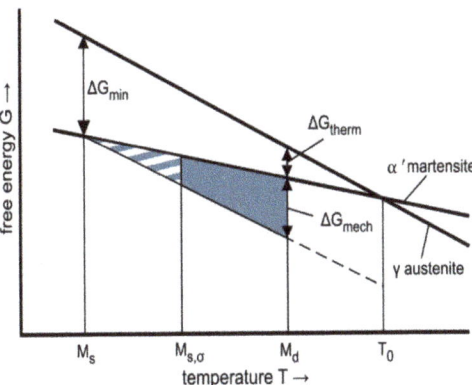

Figure 1. Austenite to martensite phase-transformation free energy (Reprinted from Ref. [5]).

The stress-assisted phase transformation in austenitic stainless steels only occurs when the temperature is close to absolute zero, and the phase transformation is less. The plastic strain-induced phase transformation does not have strict conditions, but it is the main mechanism in austenitic stainless steels [6]. Fewer phase transformations occur when the elastic stress provides kinetic phase transformation because it belongs to the stress-assisted phase transformation process. The strain or deformation-induced phase transformation belongs to the plastic strain-induced phase transformation process. During strain-induced phase transformation, austenite γ (FCC) is not directly transformed into α' (BCC) martensite but into mesophase ε (HCP) austenite [7]. Therefore, the main phase transformation formed during strain strengthening of austenitic stainless steels at cryogenic temperatures is plastic tensile-induced phase transformation.

In this work, recent studies on the strain strengthening of austenitic stainless steels at cryogenic temperatures are reviewed. The influence of strain conditions on mechanical properties at cryogenic temperatures and the phase-transformation mechanism are introduced. The two constitutive models for the simulation of microstructure evolution and mechanical property models for predicting the volumes of phase transformation are summarized.

2. Effect of Strain Conditions on Mechanical Properties

To maximize the mechanical properties of austenitic stainless steels at cryogenic temperatures, research on different aspects has been carried out, including retained austenitic stability [8,9], martensitic content [10–12], grain size of retained austenite and martensite [13,14] and temperatures and methods of strain [15,16]. The effect of cryogenic conditions and strain types on the mechanical properties of austenitic stainless steels is included in this review.

2.1. Effects on the Mechanical Properties of 316 Steel

The effect of strain on the mechanical properties of 316L steel (L-PBF316) prepared by Laser Powder Bed Fusion was studied, and the results showed that the yield strength 594 MPa and tensile strength 689 MPa at 20 °C increased to 751 MPa and 1403 MPa at −196 °C with a decrease in elongation from 49% to 41% [17]. The 316LN steel experienced the pre-strain of 15%, 25%, and 35% at 20 °C, −196 °C, and −268.8 °C, indicating a significant improvement of the yield strength and the tensile strength under a loading rate of 1.0 mm/min (shown in Table 1) [18]. The yield strength increased by 199.6%, 72.5%, and 91.5%. The tensile strength increased by 84.9%, 33.8%, and 34.5%. The yield ratio increased by 61.7%, 28.8%, and 42.3% at 20 °C, −196 °C, and −268.8 °C when the amount of pre-strain increased from 0% to 35%. The influence of pre-strain and temperatures on mechanical properties is shown in Figure 2. Accordingly, the percentage of α' martensite increased sharply at cryogenic temperatures (−196 °C and −268.8 °C) with more dislocations.

Table 1. Tensile properties of 316LN steel under different pre-strain strengthening processes. Reproduced with the permission from [18], [S. Wu et al.], [Cryogenics]; published by Elsevier Ltd., Amsterdam, The Netherlands], [2022].

Amount of Pre-Strain/%	Test Temperature /°C	Yield Strength /MPa	Tensile Strength /MPa	Elongation/%	Yield Ratio
0	20	270	598	68	0.452
15	20	513	799	25	0.642
25	20	636	947	9	0.672
35	20	809	1106	4	0.731
0	−196	672	1311	70	0.513
15	−196	933	1493	55	0.625
25	−196	1030	1615	44	0.638
35	−196	1159	1754	33	0.661
0	−268.8	832	1536	55	0.542
15	−268.8	1063	1740	36	0.611
25	−268.8	1301	1883	29	0.691
35	−268.8	1593	2066	19	0.771

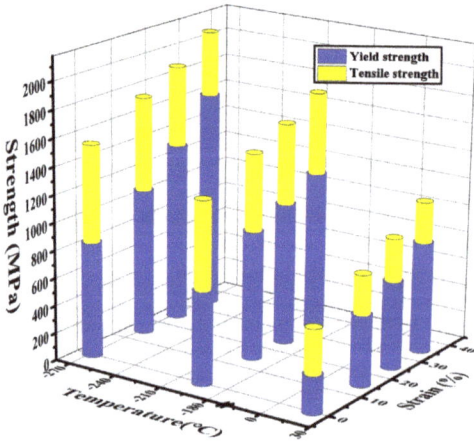

Figure 2. Influence of pre-strain and temperatures on the mechanical properties of 316LN.

The strength of metastable austenitic stainless steels is improved by experiencing cyclic pre-strain at −196 °C while maintaining the high elongation [19]. The phase transformation-induced plastic effect (TRIP) plays a full role in the tensile process. The cyclic plastic strain changes the characteristics of rapid phase transformation after full nucleation, which makes the reinforced microstructure maintain the phase transformation ability. Twin deformation, phase transformation, and serrated yield of 316L steel are difficult to occur when in-situ, small-scale, or high-strain deformation is conducted at −196 °C and −268.8 °C [20]. This is because austenite deforms through twin crystals at low temperatures and low strain by TEM observation. The results of uniaxial tensile and cyclic strain tests on 316L (MASS) at −196 °C showed that the yield strength and tensile strength increased with a decrease in elongation from 65% to 49% [21]. Observed by field emission scanning electron microscopy, the fracture morphology of L-PBF316L steel showed that the tensile strength was improved, and the elongation was preserved 55% at −196 °C [22]. The yield strength and tensile strength of 316 steel increased from 399.6 MPa and 851.2 MPa at −50 °C to 470 MPa and 1160.5 MPa at −130 °C under a loading rate of 2.0 mm/min [23].

2.2. Effect on the Mechanical Properties of 304 Steel

Table 2 shows the mechanical properties of 304 steel after multi-pass cold rolling at different temperatures with a crosshead velocity of 0.5 mm/min [24]. The strength increases with a decrease in temperature or an increase in deformation. In particular, the strength of 304 steel after a multi-pass rolling of 20% at −196 °C is much higher than that of 0 °C, and it is increased by 802 MPa (shown in Figure 3). The yield strength, the tensile strength, and the yield ratio increase with an increase in rolling deformation due to martensitic transformation.

Table 2. Tensile properties of 304 steel under different multi-pass cold rolling processes. Reproduced with the permission from [24], [P. Mallick et al.], [Mater. Charact.]; published by [Elsevier Ltd., Amsterdam, The Netherlands], [2017].

Multi-Pass Cold Rolling/%	Test Temperature /°C	Yield Strength /MPa	Tensile Strength /MPa	Elongation /%	Yield Ratio
/	20	259	675	90	0.384
10	0	703	930	20	0.756
20	0	742	981	30	0.756
30	0	834	1098	40	0.760
40	0	936	1225	10	0.764
10	−196	1061	1306	20	0.812
20	−196	1463	1589	20	0.921

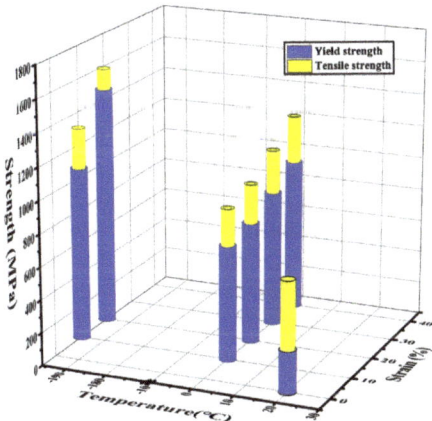

Figure 3. Influence of multi-pass cold rolling on tensile properties of 304 steel.

A total of 304 cylindrical rods with a gradient phase are obtained by torsion at −196 °C [25]. A good combination of strength and plasticity was achieved when austenite gradually decreased, with martensite occurring by the strain gradually increasing from the center of the cylinder to the edge due to the higher fraction of ε/α'. By comparing the stress–strain curve of 25% pre-strain at 0 °C, −20 °C, −40 °C, −80 °C, −120 °C and −196 °C, the yield strength and tensile strength of 304 increased with a decrease in temperatures [26]. However, the tensile experiments of Fe-19Cr-3Mn-4Ni-0.15C-0.17N austenitic steel with different strains at −40 °C and −196 °C indicates that the tensile strength decreases when the strain and temperatures reach a certain value [27]. The tensile strength and impact toughness of 304 steel and its welded joints at cryogenic temperatures increased with an increase in deformation and a decrease in strain temperatures [28]. Strain-induced martensitic transformation plays a significant role in strain strengthening at cryogenic temperatures.

Table 3 shows the effects of strain conditions on the mechanical properties of austenitic stainless steels. The strain was conducted by static or cyclic tensile, torsional, or rolling deformation at low temperatures.

Table 3. Effects of strain conditions on the mechanical properties of austenitic stainless steels.

Strain Conditions	Material Type	Strain Rate /s^{-1}	Test Temperature /°C	Yield Strength /MPa	Tensile Strength /MPa	Elongation /%	Refs.
Static strain	316L	0.00025	−196	751	1403	41	[29]
	316L	—	−196	730	1080	56	[22]
	316L	—	−130	470	1160.5	40	[23]
	316L	—	−268.8	805	1200	28	[22]
Cyclic strain	304	0.005	−196	850	1800	25	[19]
	316L	0.001	−40	681	871	51.5	[29]
	316L	0.001	−80	700	920	72.7	[29]
Torsional strain	304	—	−196	1745	—	23	[25]
	304	—	−196	1147	1357	—	[30]
Rolling strain	304	—	−196	1463	1598	20	[17]
	304	0.0002	−196	2308.4	2165.9	23	[31]

3. Effect of Strain Conditions on Microstructure

Olson and Cohen proposed FCC → HCP, FCC → BCC, and other types of martensitic nucleation mechanisms, including crystal embryos at grain boundaries, sub-grain boundaries, and inclusion particle interfaces [32,33]. Martensitic transformation often arises accompanied by nucleation and growth processes. Nucleation defects are critical for the phase-transformation fraction of martensite [34]. Most martensitic transformations exhibit no thermal characteristics and are sustainable when temperatures decrease, and a small amount of isothermal martensitic transformations do not require cooling [35].

The influence of the original austenitic grain size on the autocatalytic phase transformation was studied with a thermodynamic model [36]. Phase-transformation activation energy decreases with an increase in driving force. The kinetics of the isothermal martensitic transformation of Fe-24Ni-3Mn alloy was analyzed by autocatalytic nucleation method [37]. The nucleation rate first increased and then decreased with the reaction time. The phase-transformation activation energy decreased with a decrease in temperatures. Grain boundaries provide most of the defects [38,39]. It is clarified that potential nucleation locations are required to excite and breed martensitic transformations, and then these fine grains contribute to the occurrence of martensitic transformations.

There are two methods of martensitic transformation. One is the direct transformation of austenite γ to α' martensite, and the other is the transformation of austenite γ to α' mesophase ε (HCP) martensite [32,33]. The $\gamma \rightarrow \alpha'$ is a common martensitic transformation process [40,41]. The phase transformation of $\gamma \to \varepsilon \to \alpha'$ is mostly occurring in Fe-Mn alloy [42] and Fe-Cr-Ni alloy [43] with low stacking fault energy. Regarding the effect of ε on α' formation, it is reported that ε can provide potential nucleation locations to accelerate kinetic phase transformation [44]. However, ε is only a transitional phase [45] and not a prerequisite for the formation of α' [46]. The microstructure evolution of austenite grains after different cryogenic treatment times is shown in Figure 4 [47].

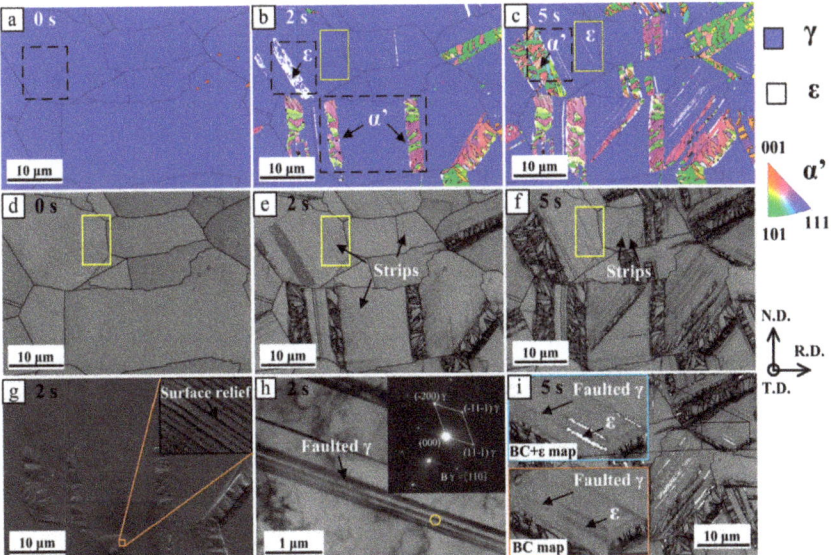

Figure 4. Microstructure evolution of specific regions after cryogenic treatment at 0, 2, and 5 s in liquid nitrogen (−196 °C). (**a–c**) EBSD method to obtain γ, ε phase diagram and α' orientation diagram. (**d–f**) Corresponding background contrast (BC) plot. (**g**) SEM image after 2 s. (**h**) TEM morphological image after 2 s and associated diffraction pattern of the yellow circle. (**i**) Diffraction image of ε phase and BC plot of EBSD after 5 s cryogenic treatment. Reproduced with the permission from [47], [J.L. Wang et al.], [Mater. Charact.]; published by [Elsevier Ltd., Amsterdam, The Netherlands], [2019].

Two transformation processes of $\gamma \to$ faulted $\gamma \to \varepsilon \to \alpha'$ and $\gamma \to$ faulted $\gamma \to$ stacking fault bundles $\to \alpha'$ were experimentally demonstrated in detail from Figure 4. In both, the martensitic transformation was always trigged first by faulted γ and then by ε and stacking fault bundles as transitions, eventually to the final α' transformation. These insights provide direct guidelines for explaining the martensitic transformation characteristics as well as further improving the martensitic transformation kinetic model.

The main phase transformation paths $\gamma \to \alpha$ in Fe-Cr-Ni alloys were only based on the position of α' slatted martensite in the final microstructure evolution after cryogenic treatment and the coexistence of ε martensite [43] or synchrotron diffraction (showing the approximate and predicted spatial positions of γ, ε and α' during in-situ cooling) [44]. Ref. [48] shows that the generation of the strips is directly related to the undulation of the specimen surface. This indicates the local plastic deformation to promote strips formed, and the TEM shows that the crystal structure is still face-centered cubic (FCC), indicating the existence of strips in γ faults. This suggests that ε nucleation in the current system follows a stacking fault mechanism [49]. The formation of stacking fault bundles under an applied stress is schematically illustrated in Figure 5 [50].

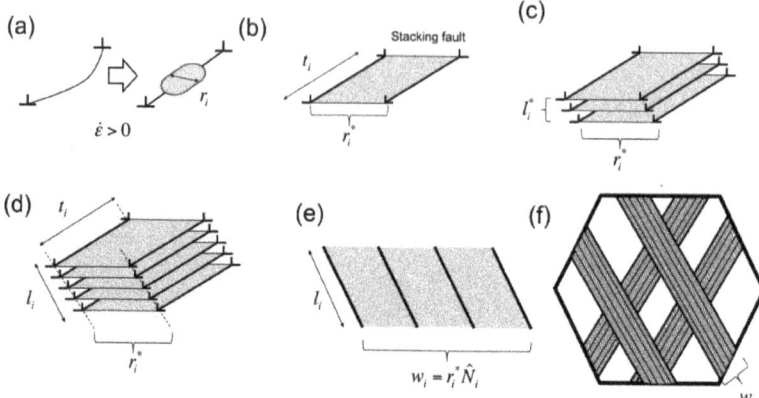

Figure 5. Schematic representation of the proposed mechanism of ε and twin formation [50]. (**a,b**) The size of the stacking fault increases with applied stress (positive strain rate $\dot{\varepsilon} > 0$) until reaching a critical width r_i^*, t_i is the thickness of a stacking fault, r_i is the width of a stacking fault. (**c**) The arrangement for a twin, l_i^* is a critical length. (**d**) The length of the embryo l_i increases and propagates through the grain interiors by subsequent overlapping of stacking faults in adjacent planes. (**e**) The width w_i increases by forming adjacent embryos of constant width r_i^*, $\hat{N}_i$ is the number of embryos in a band. (**f**) The formation of micro-bands. Reproduced with the permission from [50], [E. I. Galindo-Nava et al.], [Acta Mater.]; published by [Elsevier Ltd., Amsterdam, The Netherlands], [2017].

An isolated stacking fault first forms from a perfect dislocation under an applied stress. The size of the stacking fault increases with applied stress until it reaches a critical width. Then, several stacking faults overlap to form an embryo with critical width and length. The length of the embryo increases and propagates through the grain interiors by subsequent overlapping of stacking faults in adjacent planes. The increase in ε/twinning volume fraction increases by the formation and the overlapping of new embryos in various locations of grain, leading to the formation of micro-bands. This process aids in promoting the transformation of martensite [51].

4. Constitutive Models for Simulation of Microstructure Evolution and Mechanical Properties at Cryogenic Temperatures

A macroscopic constitutive model applicable to the plastic behavior of austenitic steels with strain-induced phase transformations was proposed [52], which is based on low elastic mode and large deformation equation. The effects of plastic strain, temperatures, and stress state on martensitic nucleation are mainly used to describe the elastic state and viscosity behavior of austenitic steels. The critical behavior of the alloy was investigated by the second-order phase transformation [53]. The strain-induced phase transformations are decomposed into plastic strain and volumetric strain. The phase transformation caused by plastic strain can be quantified experimentally, whereas the phase transformation caused by volumetric strain cannot be quantified. The experimental data from stress-assisted phase transformation is used to express the volumetric strain-induced phase transformation [52].

A model specific to cryogenic environments was proposed [53] and further enhanced with an anisotropic damage model [54], derived by generalizing Lemaitre's isotropic model to a tensor anisotropy model. It is assumed that the hardening coefficient of the austenitic matrix increases linearly with an increase in martensitic volume fraction in the model. The above models were transformed into a system constitutive model based on irreversible thermodynamics [55], in which dissipation phenomena are coupled through a dissipation potential. Lemaitre's isotropic model was modified to a form that clearly relies on the martensitic volume fraction generated after strain-induced phase transformation at cryogenic temperatures and the material coefficients that rely on the change in martensitic volume fraction during phase transformation, resulting in dramatic material damage [56]. In finite element analysis of other constitutive models of strain-induced phase transformation at cryogenic temperatures [57–60], the results of damage evolution and crack propagation in austenitic stainless steels are also matched well with the results of strain experiments at cryogenic temperatures to maintain reliable accuracy.

Homayounfard et al. proposed a constitutive model of strain-induced phase transformation and material damage intensification at cryogenic temperatures based on continuous damage theory. The damage mechanism is based on large deformation dynamics and hyperelasticity, which combines the existing experimental results, considering the dissipation phenomenon caused by phase transformation and damage propagation during plastic deformation [6]. Damage softening and phase-transformation hardening occur, and they are also induced by strain at cryogenic temperatures. The Von Mises yield criterion (φ) for stainless steels during the strain process at cryogenic temperatures was modified, as shown in Equation (1) [6].

$$\varphi = \frac{\sqrt{3J_2(\tau)}}{1-D} - (K_0 + K_1\xi)r^n - \tau_y^0 \tag{1}$$

- J_2 is the second invariant of partial stress.
- D is the failure parameter.
- τ_y^0 is the initial yield stress.
- ξ is the martensitic volume fraction.
- K_0 is the initial pure austenitic hardening coefficient.
- K_1 is the additional coefficient of ξ.
- r is the hardening variable.

The experimental results show that the phase transformation of austenitic stainless steels begins simultaneously with the generation of damage initiation during strain at cryogenic temperatures, but the increase of martensitic volume fraction inhibits the increase of damage propagation [56]. By modifying the material parameters of the Lemaitre model, the damage potential function (ϕ^D) of austenitic stainless steel during strain-induced phase transformation at cryogenic temperatures is obtained [61], as shown in Equation (2).

$$\phi^D = \frac{S(\xi)}{(s+1)(1-D)}\left(\frac{Y}{S(\xi)}\right)^{s+1} \tag{2}$$

- $S(\xi)$: the material's ability to resist damage growth.
- D is the failure parameter.
- Y is the energy release rate.
- s is the Von Mises stress.

Considering the role of damage-nucleation strain, the damage growth ($\dot{D}$) is derived from the damage dissipation potential function, as shown in Equation (3).

$$\dot{D} = \frac{\dot{\gamma}}{(1-D)}\left(\frac{Y}{S(\xi)}\right)^s H(\varepsilon_P - \varepsilon_D) \tag{3}$$

- $\dot{\gamma}$ is the plastic multiplier.
- H is the step function.
- ε_P is the initial strain of plastic deformation.
- ε_D is the effect of damage-nucleation strain.

Since the dissipation phenomena during strain at cryogenic temperatures in stainless steels include plastic deformation, damage growth, and martensitic transformation, Homayounfard et al. superimposed the plastic variable potential function (ϕ^P), the failure growth potential function (ϕ^D) and the martensitic potential function (ϕ^{tr}) to obtain the dissipation potential (ϕ) in stainless steels after a strain at cryogenic temperatures [6], where the plastic variable potential function (ϕ^p) is replaced by the modified Von Mises yield criterion (φ), as shown in Equation (4).

$$\phi = \left\{ \frac{\sqrt{3 J_2(\tau)}}{1-D} - (K_0 + K_1 \xi) r^n - \tau_y^0 \right\} + \left\{ \frac{S(\xi)}{(s+1)(1-D)} \left(\frac{Y}{S(\xi)} \right)^{s+1} \right\} + \phi^{tr} \quad (4)$$

The evolution of martensitic volume fraction has been described by empirical formulas, and the formula for the martensitic transformation potential function (ϕ^{tr}) has not been developed. It is assumed that the plastic spin is zero in the constitutive model of strain for stainless steels at cryogenic temperatures. The evolution of the plastic deformation gradient ($\overline{d^p}$) is determined by the rotational plastic deformation rate [6], as shown in Equation (5).

$$\overline{d^p} = \dot{\gamma} \frac{\partial \phi^p}{\partial \tau} = \dot{\gamma} N \quad (5)$$

- N represents the flow direction.
- τ represents the yield stress.

The cumulative plastic strain ($\dot{\overline{\varepsilon^p}}$), another important problem during strain of stainless steels at cryogenic temperatures, is also determined according to the rotational plastic deformation rate [56], as shown in Equation (6).

$$\dot{\overline{\varepsilon^p}} = \sqrt{\frac{2}{3} \left\| \overline{d^p} \right\|} = \frac{\dot{\gamma}}{(1-D)} \quad (6)$$

The model parameters of AISI304 steel were determined and corrected according to the results from the tensile test at −196 °C. The calculated results of the strain-induced phase transformation and damage propagation behaviors of AISI304 and AISI316L at −268.8 °C agree with the literature [56,60–63], as shown in Figure 6.

304L and 316L austenitic stainless steels retain 40~50% fracture strain under cryogenic conditions and have good ductility near absolute zero [31,64]. Therefore, it is particularly important to study the phase-transformation behavior induced by fracture strain of austenitic stainless steels at low temperatures. The nonlinear constitutive behavior of austenitic stainless steels 304 and 316 during the ultimate tensile strain at cryogenic temperatures was studied to set up the constitutive model for the cryogenic fracture strain-induced phase transformation of austenitic stainless steels by experiment and simulation [65–67].

Experimental and modeling studies of 316L austenitic stainless steel with symmetrical notches in liquid nitrogen (−196 °C) and liquid helium (−268.8 °C) environments were carried out in ref. [60]. The strain process at cryogenic temperatures is divided into two stages, no-hardening and linear hardening, where the initial flow stresses within each serration in the stress–strain curve of the non-hardening stage are the same, and an ideal plastic model is used, as shown in Equation (7).

$$f(\sigma_{ij}) = \sigma_i - \sigma_0(T) \quad (7)$$

- $f(\sigma_{ij})$ represents the plastic dissipation potential.
- σ_0 represents the flow stress depending on the temperature T.

σ_i can be determined by the bias stress (s_{ij}), as shown in Equation (8).

$$\sigma_i = \sqrt{\frac{3}{2} |s_{ij}|} \quad (8)$$

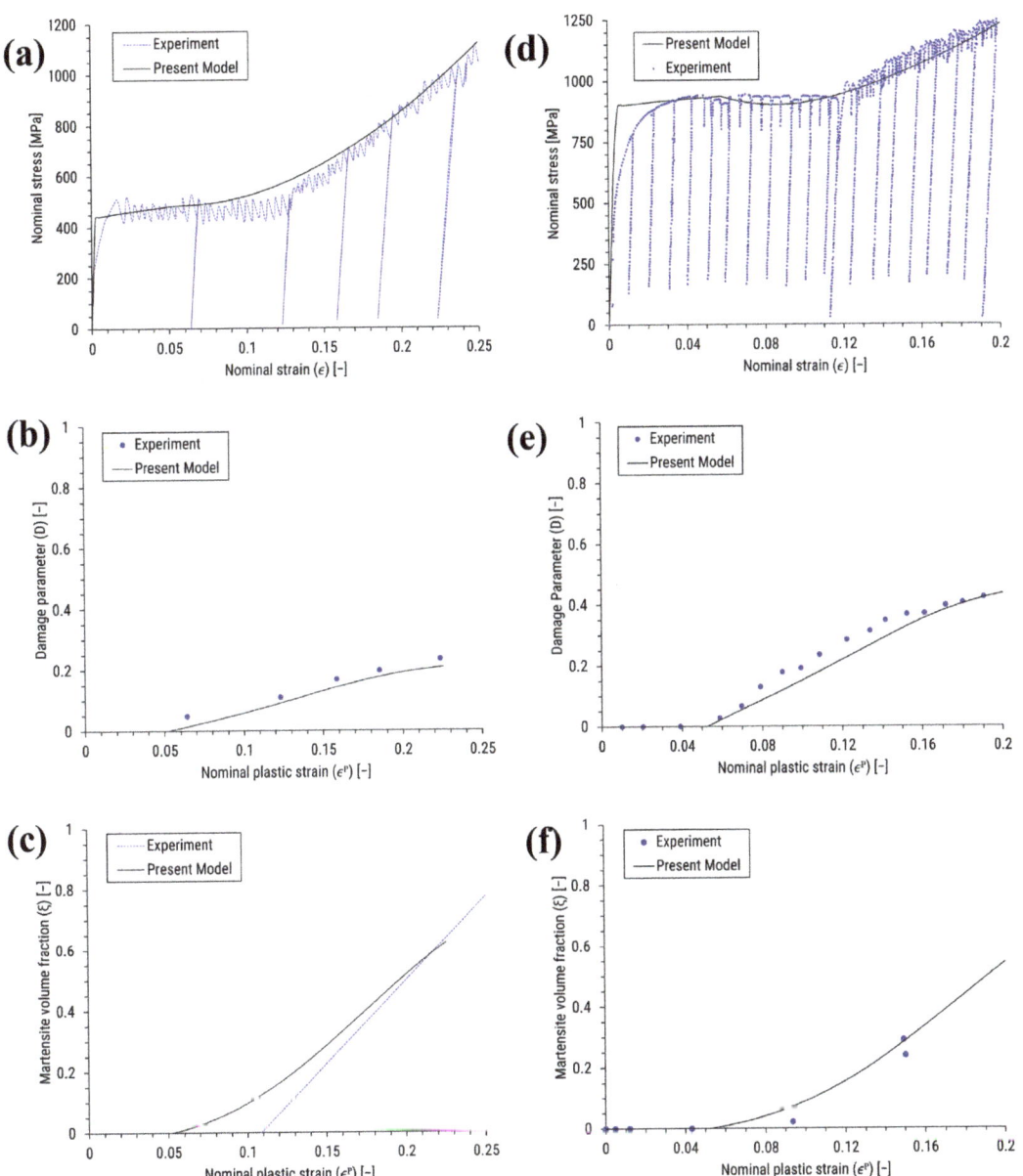

Figure 6. Comparison of constitutive model and experimental results [63] −268.8 °C stainless steel 304: (**a**) nominal stress–strain, (**b**) damage parameters D-ϵ^P, (**c**) martensitic volume fraction ξ-ϵ^P. −268.8 °C stainless steel 316L: (**d**) nominal stress–strain, (**e**) damage parameters D-ϵ^P, (**f**) martensitic volume fraction ξ-ϵ^P. Reproduced with the permission from [6], [M. Homayounfard et al.], [Int. J. Plast.]; published by [Elsevier Ltd., Amsterdam, The Netherlands], [2022].

The linear hardening reaches a critical value, which is almost ideally linear, and the value of the initial flow stress of each serration in the stress–strain curve is approximately linear with hardening, and the elastoplastic constitutive model of linear isotropic hardening is shown in Equation (9).

$$f(\sigma_{ij}) = \sigma_i - \sigma_R(T) \tag{9}$$

- R represents the isotropic hardening parameter.
- C_R represents the hardening modulus, which is limited by Equation (10).

$$dR = C_R dp \tag{10}$$

The results of strain fracture of austenitic stainless steels at cryogenic temperatures simulated in extended finite element (XFEM) are shown in Figure 7.

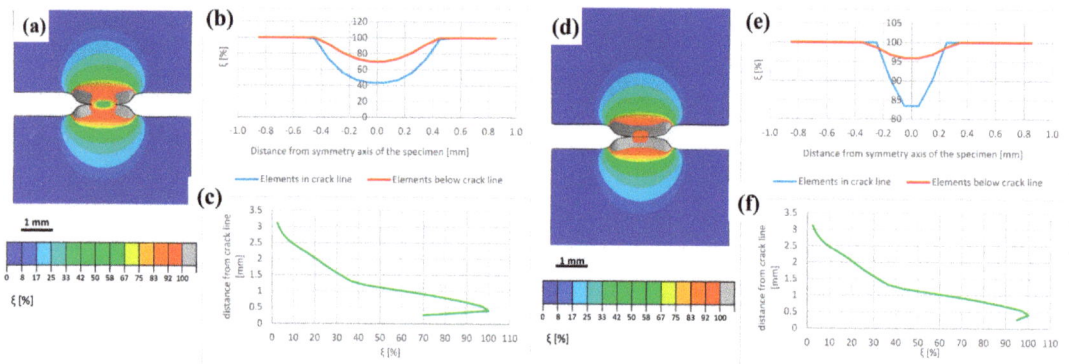

Figure 7. Martensitic volume distribution. (**a**) the strain crack growth stage at cryogenic temperatures, (**b**) the second phase distribution in the crack tips interval at the strain crack growth stage, (**c**) the martensitic distribution on the tensile axis at the strain crack growth stage, (**d**) near fracture stage at cryogenic temperatures, (**e**) the second phase distribution in the crack tips interval near the fracture stage, (**f**) the martensitic distribution on the tensile axis near fracture stage. Reprinted from Ref. [60].

The damage origin and failure of AISI 304L austenitic stainless steel during cold working was studied by a combination of experimental and numerical methods [68]. The numerical prediction of fracture in 304 stainless steel was proposed using a modified Johnson–Cook damage model [69]. A coupled elastoplastic–damage constitutive model for predicting damage in ductile materials was established using the Swift–Voce combinatorial equation as a hardening function to describe the constitutive behavior after necking [66]. The Modified Mohr–Coulomb Criterion (MMC) based on the equivalent plastic strain is used to describe the sudden fracture of 304L stainless steel plates.

There is the largest amount of martensite in the crack tips interval. The closer the distance from the tensile axis is, the less martensite there is. The farther the distance from the cross-section along the tensile axis is, the more martensite there is with an increase in phase transformation. The amount of martensite increases significantly until the tensile specimen is broken completely. At the same time, the evolution of the microstructure near the macroscopic crack growth zone at liquid nitrogen ($-196\,^\circ\text{C}$) and liquid helium ($-268.8\,^\circ\text{C}$) temperatures was determined. Accordingly, the distribution of primary and secondary phase textures at different distances from the fracture surface was experimentally analyzed, which provides a basis for the dynamics of martensitic nucleation and the constitutive relationship.

5. Conclusions

The strain of austenitic stainless steels promotes the phase transformation of austenite γ to martensite α', resulting in the increase of strength at cryogenic temperatures. The strain strengthening occurs in the temperature interval $M_s - M_d$, and the driving force for phase transformation is $\Delta G = \Delta G_{therm} + \Delta G_{mech}$. The plastic tensile-induced phase transformation is the main mechanism.

Studies on the effect of strain on the mechanical properties of 316 steel show that the yield strength, the tensile strength, and the yield ratio increase with a decrease in experimental temperatures and an increase in pre-strain. The strength of austenitic stainless steels is significantly improved with a large elongation maintained after cyclic pre-strain at −196 °C. The effect of strain on the mechanical properties of 304 steel shows that the yield strength, the tensile strength, and the yield ratio increase to a critical value and then decrease with an increase in deformation and a decrease in experimental temperatures. This results from more dislocations interacting with newly formed nanotwins during strain.

The martensitic transformation during strain at cryogenic temperatures occurs in two ways of $\gamma \rightarrow \varepsilon \rightarrow \alpha'$ and $\gamma \rightarrow \alpha'$, where the stacking fault bundles are used as a precursor of martensite α'. In addition to the pre-existing nucleation position, the fault γ is also the nucleation position for the ε phase. The stacking fault bundle is the direct nucleation position for α' martensite. Two constitutive models for the strain strengthening of austenitic stainless steels at cryogenic temperatures are presented. One model is used to describe phase-transformation behavior during strain, and the other is used to predict fracture behavior during strain crack propagation. The calculation of the volume fraction of strain-induced martensite at cryogenic temperatures by two models is well matched with the experimental results.

Author Contributions: Methodology, J.W.; resources, B.H. and W.X.; data curation, B.H. and W.X.; writing-original draft, B.H.; writing—review and editing, J.W.; supervision, J.W. All authors have read and agreed to the published version of the manuscript.

Funding: This research was funded by the National Nature Science Foundation of China (Grant No.52075295 & No.52175305), the Nature Science Foundation of Shandong Province (Grant No. ZR2019MEE117) and Shandong University (Grant No. 2021JCG011).

Data Availability Statement: No new data were created in this study. Data sharing is not applicable to this article.

Conflicts of Interest: The authors declare no conflict of interest.

References

1. Silva, P.M.O.; Filho, M.C.C.; da Cruz, J.A.; Sales, A.J.M.; Sombra, A.S.B.; Tavares, J.M.R.S. Influence on pitting corrosion resistance of AISI 301LN and 316L stainless steels subjected to cold-induced deformation. *Metals* **2023**, *13*, 443. [CrossRef]
2. Borgioli, F. The corrosion behavior in different environments of austenitic stainless steels subjected to thermochemical surface treatments at low temperatures: An overview. *Metals* **2023**, *13*, 776. [CrossRef]
3. Lu, Y.; Hui, H. Investigation on mechanical behaviors of cold stretched and cryogenic stretched austenitic stainless steel pressure vessels. *Procedia Eng.* **2015**, *130*, 628–637. [CrossRef]
4. Li, S.; Xiao, M.; Ye, G.; Zhao, K.; Yang, M. Effects of deep cryogenic treatment on microstructural evolution and alloy phases precipitation of a new low carbon martensitic stainless bearing steel during aging. *Mater. Sci. Eng. A* **2018**, *732*, 167–177. [CrossRef]
5. Hotz, H.; Kirsch, B.; Aurich, J.C. Impact of the thermomechanical load on subsurface phase transformations during cryogenic turning of metastable austenitic steels. *J. Intell. Manuf.* **2020**, *32*, 877–894. [CrossRef]
6. Homayounfard, M.; Ganjiani, M. A large deformation constitutive model for plastic strain-induced phase transformation of stainless steels at cryogenic temperatures. *Int. J. Plast.* **2022**, *156*, 103344. [CrossRef]
7. Mazánová, V.; Heczko, M.; Škorík, V.; Chlupová, A.; Polák, J.; Kruml, T. Microstructure and martensitic transformation in 316L austenitic steel during multiaxial low cycle fatigue at room temperature. *Mater. Sci. Eng. A* **2019**, *767*, 138407. [CrossRef]
8. Zhao, Y.; Cao, Y.; Wen, W.; Lu, Z.; Zhang, J.; Liu, Y.; Chen, P. Effects of Mn content on austenite stability and mechanical properties of low Ni alumina-forming austenitic heat-resistant steel: A first-principles study. *Sci. Rep.* **2023**, *13*, 5769. [CrossRef]
9. Jiang, Y.; Zhou, X.; Li, X.; Lu, K. Stabilizing nanograined austenitic stainless steel with grain boundary relaxation. *Acta Mater.* **2023**, *256*, 119134. [CrossRef]
10. Huang, N.; Tian, Y.; Yang, R.; Xiao, T.; Li, H.; Chen, X. Effect of heat treatment on the cavitation erosion behavior of nanocrystalline surface layer of 304 stainless steel. *Appl. Sci.* **2023**, *13*, 5817. [CrossRef]
11. Fang, T.; Tao, N. Martensitic transformation dominated tensile plastic deformation of nanograins in a gradient nanostructured 316L stainless steel. *Acta Mater.* **2023**, *248*, 118780. [CrossRef]

12. Li, S.; Withers, P.J.; Kabra, S.; Yan, K. The behaviour and deformation mechanisms for 316L stainless steel deformed at cryogenic temperatures. *Mater. Sci. Eng. A* **2023**, *880*, 145279. [CrossRef]
13. Sohrabi, M.J.; Mirzadeh, H.; Sadeghpour, S.; Mahmudi, R. Dependency of work-hardening behavior of a metastable austenitic stainless steel on the nucleation site of deformation-induced martensite. *Mater. Sci. Eng. A* **2023**, *868*, 144600. [CrossRef]
14. Guo, Y.; Zhang, S.; Chen, J.; Fu, B.; Wang, Z.; Pang, L.; Wei, L.; Li, Y.; Ding, Y. The contribution of retained martensite to the high yield strength and sustainable strain hardening of a hierarchical metastable austenitic stainless steel. *Mater. Sci. Eng. A* **2023**, *866*, 144681. [CrossRef]
15. Odnobokova, M.; Belyakov, A.; Dolzhenko, P.; Kostina, M.; Kaibyshev, R. On the strengthening mechanisms of high nitrogen austenitic stainless steels. *Mater. Lett.* **2023**, *331*, 133502. [CrossRef]
16. Odnobokova, M.; Belyakov, A.; Enikeev, N.; Kaibyshev, R.; Valiev, R. Cryogenic impact toughness of a work hardened austenitic stainless steel. *Materialia* **2022**, *23*, 101460. [CrossRef]
17. Mishra, P.; Åkerfeldt, P.; Forouzan, F.; Svahn, F.; Zhong, Y.; Shen, Z.J.; Antti, M.-L. Microstructural Characterization and Mechanical Properties of L-PBF Processed 316 L at Cryogenic Temperature. *Materials* **2021**, *14*, 5856. [CrossRef]
18. Wu, S.; Xin, J.; Xie, W.; Zhang, H.; Huang, C.; Wang, W.; Zhou, Z.; Zhou, Y.; Li, L. Mechanical properties and microstructure evolution of cryogenic pre-strained 316LN stainless steel. *Cryogenics* **2021**, *121*, 103388. [CrossRef]
19. Wang, Z.; Shi, S.; Yu, J.; Li, B.; Li, Y.; Chen, X. Enhanced cryogenic tensile properties through cryogenic cyclic plastic strengthening in a metastable austenitic stainless steel. *Scr. Mater.* **2023**, *222*, 115024. [CrossRef]
20. Han, W.; Liu, Y.; Wan, F.; Liu, P.; Yi, X.; Zhan, Q.; Morrall, D.; Ohnuki, S. Deformation behavior of austenitic stainless steel at deep cryogenic temperatures. *J. Nucl. Mater.* **2018**, *504*, 29–32. [CrossRef]
21. Wolfenden, A.; Suzuki, K.; Fukakura, J.; Kashiwaya, H. Cryogenic fatigue properties of 304L and 316L stainless steels compared to mechanical strength and increasing magnetic permeability. *J. Test. Eval.* **1988**, *16*, 190. [CrossRef]
22. Bidulský, R.; Bidulská, J.; Gobber, F.S.; Kvačkaj, T.; Petroušek, P.; Actis-Grande, M.; Weiss, K.-P.; Manfredi, D. Case Study of the Tensile Fracture Investigation of Additive Manufactured Austenitic Stainless Steels Treated at Cryogenic Conditions. *Materials* **2020**, *13*, 3328. [CrossRef] [PubMed]
23. Lee, K.J.; Chun, M.S.; Kim, M.H.; Lee, J.M. A new constitutive model of austenitic stainless steel for cryogenic applications. *Comput. Mater. Sci.* **2009**, *46*, 1152–1162. [CrossRef]
24. Mallick, P.; Tewary, N.; Ghosh, S.; Chattopadhyay, P. Effect of cryogenic deformation on microstructure and mechanical properties of 304 austenitic stainless steel. *Mater. Charact.* **2017**, *133*, 77–86. [CrossRef]
25. Ma, Z.; Ren, Y.; Li, R.; Wang, Y.-D.; Zhou, L.; Wu, X.; Wei, Y.; Gao, H. Cryogenic temperature toughening and strengthening due to gradient phase structure. *Mater. Sci. Eng. A* **2017**, *712*, 358–364. [CrossRef]
26. Lu, Y.; Hui, H. Investigation on Mechanical Properties of S30403 Austenitic Stainless Steel at Different Temperatures. *J. Press. Vessel. Technol.* **2018**, *140*, 024502. [CrossRef]
27. Hauser, M.; Wendler, M.; Fabrichnaya, O.; Volkova, O.; Mola, J. Anomalous stabilization of austenitic stainless steels at cryogenic temperatures. *Mater. Sci. Eng. A* **2016**, *675*, 415–420. [CrossRef]
28. Ding, H.; Wu, Y.; Lu, Q.; Xu, P.; Zheng, J.; Wei, L. Tensile properties and impact toughness of S30408 stainless steel and its welded joints at cryogenic temperatures. *Cryogenics* **2018**, *92*, 50–59. [CrossRef]
29. Maharaja, H.; Das, B.; Singh, A.; Mishra, S. Comparative assessment of strain-controlled fatigue performance of SS 316L at room and low temperatures. *Int. J. Fatigue* **2023**, *166*, 107251. [CrossRef]
30. Ortwein, R.; Skoczeń, B.; Tock, J. Micromechanics based constitutive modeling of martensitic transformation in metastable materials subjected to torsion at cryogenic temperatures. *Int. J. Plast.* **2014**, *59*, 152–179. [CrossRef]
31. Jiang, W.; Zhu, K.; Li, J.; Qin, W.; Zhou, J.; Li, Z.; Gui, K.; Zhao, Y.; Mao, Q.; Wang, B. Extraordinary strength and ductility of cold-rolled 304L stainless steel at cryogenic temperature. *J. Mater. Res. Technol.* **2023**, *26*, 2001–2008. [CrossRef]
32. Olson, G.B.; Cohen, M. A general mechanism of martensitic nucleation: Part I. General concepts and the FCC → HCP transformation. *Met. Trans. A* **1976**, *7*, 1897–1904. [CrossRef]
33. Olson, G.B.; Cohen, M. A general mechanism of martensitic nucleation: Part II. FCC → BCC and other martensitic transformations. *Met. Trans. A* **1976**, *7*, 1905–1914. [CrossRef]
34. Van Bohemen, S.M.C. Bainite and martensite start temperature calculated with exponential carbon dependence. *Mater. Sci. Technol.* **2012**, *28*, 487–495. [CrossRef]
35. Cui, C.; Gu, K.; Qiu, Y.; Weng, Z.; Zhang, M.; Wang, J. The effects of post-weld aging and cryogenic treatment on self-fusion welded austenitic stainless steel. *J. Mater. Res. Technol.* **2022**, *21*, 648–661. [CrossRef]
36. Li, Y.; Li, J.; Ma, J.; Han, P. Influence of Si on strain-induced martensitic transformation in metastable austenitic stainless steel. *Mater. Today Commun.* **2022**, *31*, 103577. [CrossRef]
37. Pati, S.; Cohen, M. Kinetics of isothermal martensitic transformations in an iron-nickel-manganese alloy. *Acta Met.* **1971**, *19*, 1327–1332. [CrossRef]
38. Mazánová, V.; Heczko, M.; Polák, J. On the mechanism of fatigue crack initiation in high-angle grain boundaries. *Int. J. Fatigue* **2022**, *158*, 106721. [CrossRef]
39. Huang, M.; Wang, L.; Wang, C.; Mogucheva, A.; Xu, W. Characterization of deformation-induced martensite with various AGSs upon Charpy impact loading and correlation with transformation mechanisms. *Mater. Charact.* **2021**, *184*, 111704. [CrossRef]

40. Sohrabi, M.J.; Mirzadeh, H.; Sadeghpour, S.; Geranmayeh, A.R.; Mahmudi, R. Tailoring the strength-ductility balance of a commercial austenitic stainless steel with combined TWIP and TRIP effects. *Arch. Civ. Mech. Eng.* **2023**, *23*, 170. [CrossRef]
41. Injeti, V.; Li, Z.; Yu, B.; Misra, R.; Cai, Z.; Ding, H. Macro to nanoscale deformation of transformation-induced plasticity steels: Impact of aluminum on the microstructure and deformation behavior. *J. Mater. Sci. Technol.* **2018**, *34*, 745–755. [CrossRef]
42. Alves, C.L.M.; Rezende, J.; Senk, D.; Kundin, J. Peritectic phase transformation in the Fe–Mn and Fe–C system utilizing simulations with phase-field method. *J. Mater. Res. Technol.* **2019**, *8*, 233–242. [CrossRef]
43. Tian, Y.; Lienert, U.; Borgenstam, A.; Fischer, T.; Hedström, P. Martensite formation during incremental cooling of Fe-Cr-Ni alloys: An in-situ bulk X-ray study of the grain-averaged and single-grain behavior. *Scr. Mater.* **2017**, *136*, 124–127. [CrossRef]
44. Masumura, T.; Nakada, N.; Tsuchiyama, T.; Takaki, S.; Koyano, T.; Adachi, K. The difference in thermal and mechanical stabilities of austenite between carbon- and nitrogen-added metastable austenitic stainless steels. *Acta Mater.* **2015**, *84*, 330–338. [CrossRef]
45. Ruan, T.; Wang, B.; Li, Y.; Xu, C. Atomistic insight into the solid-solid phase transitions in iron nanotube: A molecular dynamics study. *Mater. Today Commun.* **2021**, *29*, 102833. [CrossRef]
46. Zeng, L.; Song, X.; Chen, N.; Rong, Y.; Zuo, X.; Min, N. A new understanding of transformation induced plasticity (TRIP) effect in austenitic steels. *Mater. Sci. Eng. A* **2022**, *857*, 143742. [CrossRef]
47. Wang, J.; Xi, X.; Li, Y.; Wang, C.; Xu, W. New insights on nucleation and transformation process in temperature-induced martensitic transformation. *Mater. Charact.* **2019**, *151*, 267–272. [CrossRef]
48. Wang, T.; Wei, X.; Zhang, H.; Ren, Z.; Gao, B.; Han, J.; Bian, L. Plastic deformation mechanism transition with solute segregation and precipitation of 304 stainless steel foil induced by pulse current. *Mater. Sci. Eng. A* **2022**, *840*, 142899. [CrossRef]
49. Liu, T.; Liang, L.; Raabe, D.; Dai, L. The martensitic transition pathway in steel. *J. Mater. Sci. Technol.* **2023**, *134*, 244–253. [CrossRef]
50. Galindo-Nava, E.; Rivera-Díaz-Del-Castillo, P. Understanding martensite and twin formation in austenitic steels: A model describing TRIP and TWIP effects. *Acta Mater.* **2017**, *128*, 120–134. [CrossRef]
51. Moallemi, M.; Kermanpur, A.; Najafizadeh, A.; Rezaee, A.; Baghbadorani, H.S.; Nezhadfar, P.D. Deformation-induced martensitic transformation in a 201 austenitic steel: The synergy of stacking fault energy and chemical driving force. *Mater. Sci. Eng. A* **2016**, *653*, 147–152. [CrossRef]
52. Stringfellow, R.; Parks, D.; Olson, G. A constitutive model for transformation plasticity accompanying strain-induced martensitic transformations in metastable austenitic steels. *Acta Metall. Mater.* **1992**, *40*, 1703–1716. [CrossRef]
53. Garion, C.; Skoczen, B. Modeling of plastic strain-induced martensitic transformation for cryogenic applications. *J. Appl. Mech.* **2002**, *69*, 755–762. [CrossRef]
54. Garion, C.; Skoczen, B. Combined model of strain-induced phase transformation and orthotropic damage in ductile materials at cryogenic temperatures. *Int. J. Damage Mech.* **2003**, *12*, 331–356. [CrossRef]
55. Ryś, M.; Egner, H. Energy equivalence based constitutive model of austenitic stainless steel at cryogenic temperatures. *Int. J. Solids Struct.* **2019**, *164*, 52–65. [CrossRef]
56. Homayounfard, M.; Ganjiani, M.; Sasani, F. Damage development during the strain induced phase transformation of austenitic stainless steels at low temperatures. *Model. Simul. Mater. Sci. Eng.* **2021**, *29*, 045004. [CrossRef]
57. Xie, L.; Zhang, H.; Wu, S.; Shen, F.; Xin, J.; Huang, C.; Jiang, M.; Huang, Z.; Wang, W.; Li, L. Development of two-dimensional temperature field solution method based on the stress–strain response of 316LN stainless steel at cryogenic temperatures. *Cryogenics* **2023**, *133*, 103713. [CrossRef]
58. Wei, Y.; Lu, Q.; Kou, Z.; Feng, T.; Lai, Q. Microstructure and strain hardening behavior of the transformable 316L stainless steel processed by cryogenic pre-deformation. *Mater. Sci. Eng. A* **2023**, *862*, 144424. [CrossRef]
59. Chen, L.; Jia, Q.; Hao, S.; Wang, Y.; Peng, C.; Ma, X.; Zou, Z.; Jin, M. The effect of strain-induced martensite transformation on strain partitioning and damage evolution in a duplex stainless steel with metastable austenite. *Mater. Sci. Eng. A* **2021**, *814*, 141173. [CrossRef]
60. Nalepka, K.; Skoczeń, B.; Ciepielowska, M.; Schmidt, R.; Tabin, J.; Schmidt, E.; Zwolińska-Faryj, W.; Chulist, R. Phase transformation in 316L austenitic steel induced by fracture at cryogenic temperatures: Experiment and modelling. *Materials* **2021**, *14*, 127. [CrossRef]
61. Egner, H.; Skoczeń, B.; Ryś, M. Constitutive and numerical modeling of coupled dissipative phenomena in 316L stainless steel at cryogenic temperatures. *Int. J. Plast.* **2015**, *64*, 113–133. [CrossRef]
62. Ryś, M.; Skoczeń, B. Coupled constitutive model of damage affected two-phase continuum. *Mech. Mater.* **2017**, *115*, 1–15. [CrossRef]
63. Tabin, J.; Skoczen, B.; Bielski, J. Damage affected discontinuous plastic flow (DPF). *Mech. Mater.* **2017**, *110*, 44–58. [CrossRef]
64. Li, Y.; Zhong, S.; Luo, H.; Xu, C.; Jia, X. Intermediate stacking fault and twinning induced cooperative strain evolution of dual phase in lean duplex stainless steels with excellent cryogenic strength-ductility combinations. *Mater. Sci. Eng. A* **2022**, *831*, 142347. [CrossRef]
65. Paredes, M.; Grolleau, V.; Wierzbicki, T. On ductile fracture of 316L stainless steels at room and cryogenic temperature level: An engineering approach to determine material parameters. *Materialia* **2020**, *10*, 100624. [CrossRef]
66. Kim, M.-S.; Kim, H.-T.; Choi, Y.-H.; Kim, J.-H.; Kim, S.-K.; Lee, J.-M. A New Computational Method for Predicting Ductile Failure of 304L Stainless Steel. *Metals* **2022**, *12*, 1309. [CrossRef]
67. Tabin, J.; Skoczen, B.; Bielski, J. Discontinuous plastic flow coupled with strain induced fcc–bcc phase transformation at extremely low temperatures. *Mech. Mater.* **2019**, *129*, 23–40. [CrossRef]

68. Ben Othmen, K.; Haddar, N.; Jegat, A.; Manach, P.-Y.; Elleuch, K. Ductile fracture of AISI 304L stainless steel sheet in stretching. *Int. J. Mech. Sci.* **2019**, *172*, 105404. [CrossRef]
69. Pham, H.T.; Iwamoto, T. An evaluation of fracture properties of type-304 austenitic stainless steel at high deformation rate using the small punch test. *Int. J. Mech. Sci.* **2018**, *144*, 249–261. [CrossRef]

Disclaimer/Publisher's Note: The statements, opinions and data contained in all publications are solely those of the individual author(s) and contributor(s) and not of MDPI and/or the editor(s). MDPI and/or the editor(s) disclaim responsibility for any injury to people or property resulting from any ideas, methods, instructions or products referred to in the content.

Article

Effect of Ti/Al Ratio on Precipitation Behavior during Aging of Ni-Cr-Co-Based Superalloys

Dong-Ju Chu [1,2], Chanhee Park [1,2], Joonho Lee [2] and Woo-Sang Jung [1,*]

[1] Energy Materials Research Center, Korea Institute of Science and Technology, Seoul 02792, Republic of Korea; dj.chu@kist.re.kr (D.-J.C.); 218507@kist.re.kr (C.P.)

[2] Department of Materials Science and Engineering, Korea University, Seoul 02841, Republic of Korea; joonholee@korea.ac.kr

* Correspondence: wsjung@kist.re.kr; Tel.: +82-2-958-5429

Abstract: Precipitation behaviors of Ni-Cr-Co-based superalloys with different Ti/Al ratios aged at 750, 800, and 850 °C for up to 10,000 h were investigated using scanning and transmission electron microscopy. The Ti/Al ratio did not significantly affect the diameter of the γ' phase. However, the volume fraction of the γ' phase increased with increasing Ti/Al ratios. The η phase was not observed in alloys with a small Ti/Al ratio, whereas it was precipitated after aging at 850 °C for 1000 h in alloys with a Ti/Al ratio greater than 0.80. Higher aging temperatures and higher Ti/Al ratios led to faster η formation kinetics and accelerated the degradation of alloys. It is thought that the increase in hardness with an increase in the Ti/Al ratio is attributed to the effective inhibition of the γ' phase on dislocation movement due to the increase in the volume fraction of the γ' phase and an increase in the antiphase boundary (APB) energy.

Keywords: nickel-based superalloy; gamma prime; eta; Ti/Al ratio; precipitation; antiphase boundary energy

1. Introduction

Recently, as environmental regulations related to greenhouse gases are strengthened worldwide, research for developing high-performance materials is being conducted to reduce carbon dioxide emissions and improve the economic efficiency of thermal power plants. To meet these regulations and improve economic efficiency, it is important to increase steam temperature and pressure in power plants [1,2]. As a result of this demand, many researchers have conducted studies to develop new alloys with excellent high-temperature strength. Ni-based superalloys are widely used in many extreme engineering applications because of their excellent creep properties, fatigue resistance, and corrosion resistance [2–4]. Many kinds of Ni-based superalloys with excellent mechanical properties have been developed [5–7]. For example, alloys 230 and 282 have been developed by the Haynes International company and alloys 617, 263, 740, and 740 H have been developed by Special Metals Corporation in the USA. Alloy 740H is known to be one of the candidate boiler materials available for steam temperatures higher than 700 °C in thermal power plants. It was developed by increasing the Al content and limiting the contents of Nb, Ti, and B to inhibit the formation and growth of the G phase ((Nb,Ti)$_6$(Ni,Co)$_{16}$Si$_7$) based on alloy 740 [8–10]. Alloy 740H has high creep strength due to precipitation strengthening by the gamma prime (γ') phase, which is an intermetallic compound with an L12 ordered face-centered cubic structure with composition Ni$_3$(Al,Ti) [11–14].

Generally, it is recognized that the mechanical properties of Ni-based superalloys are closely related to the composition, diameter, fraction, morphology, and distribution of the γ' phase. Moreover, the strength of Ni-based superalloys originates from the order strengthening of the γ' phase, which can act as pinning particles to interfere with the movement of dislocation. Shearing by dislocation pairs with the same Burgers vector

is the main deformation mechanism to minimize the total area energy formed by single dislocations in Ni-based superalloys. That is, the movement of trailing dislocation removes the antiphase boundary (APB) formed by the movement of the leading dislocation through the γ' phase. This is called the order strengthening of precipitates cut by dislocations. The amount of order strengthening depends on the APB energy and diameter of the γ' phase. As a result, the γ' phase can inhibit the motion of dislocations as the passing of a perfect dislocation from the γ to the γ' phase can lead to the formation of the APB [15]. Furthermore, many studies have reported on the APB energy of the γ' phase, which has a significant effect on shearing the γ' phase in a larger number of Ni-based alloys. Enomoto and Harada [16] have reported that Ti can interact more strongly with Ni than Al, which can increase the lattice parameter of the γ' phase and lattice misfit as well as the APB energy [16–18]. Also, Dodaran [19], Gorbatov [20], and Chen [21] have calculated the change in the APB energy using an ab initio calculation method. They reported that the APB energy was increased with increasing percentages of Ti. Ti is known by other researchers to be advantageous for high-temperature strength.

However, there is a risk of eta (η) phase precipitation, known as $D0_{24}$ crystal structure (HCP), with composition $Ni_3(Ti,Nb)$, when Ti content exceeds the critical value [22–24]. It has been reported that the η phase is formed in Ni-based alloys with consumption of the γ' phase, which can induce formation of the γ' phase depletion zone [23,25–27]. Consequently, as Ti increases, the possibility of the η phase also increases.

As mentioned above, the strength increases as the amount of Ti is increased. However, it is necessary to appropriately control the amount of Ti and Al to retard the formation of the η phase for improving high-temperature strength. Also, the increase in the APB energy of the γ' phase by adding Ti is expected to make it difficult to shear the γ' phase by dislocations.

In this study, four alloys with different Ti/Al atomic ratios were designed in which the content of Ti + Al was increased to 5.0 at% without any composition change in other alloying elements except for Ti and Al in 740H (4.4 at%). Hardness and microstructural changes of alloys were investigated using scanning electron microscopy (SEM) and transmission electron microscopy (TEM) after exposure to 750 °C, 800 °C, and 850 °C for up to 10,000 h. Also, order strengthening mechanisms by the γ' phase were investigated by comparing the theoretical and experimental results.

2. Experimental Procedure

The volume fraction of the γ' phase was predicted with thermodynamic calculation software (Thermo-calc with TCNI6, Thermo Calc, Stockholm, Sweden). Representative precipitates in Ni-based superalloys included $M_{23}C_6$, MC, γ', and η phases. In designed alloys, the MC phase has less than 0.5%, and the $M_{23}C_6$ phase has less than 0.7% equilibrium volume fraction. Therefore, it was considered that they would not significantly affect materials since these carbides have relatively low volume fractions compared to the γ' phase [28].

Figure 1 shows the equilibrium volume fraction of the γ' phase for Ni-25Cr-20Co-0.5Mo-1.5Nb-0.03C-Ti-Al alloys with different Ti + Al (4.4~6.0 at%) and Ti/Al atomic ratios (0.3~1.1) calculated with thermo Calc software. In Figure 1a, the volume fraction of the γ' phase was increased by increasing the Ti/Al ratio at the same Ti + Al atomic %. The volume fraction of the γ' phase was also increased by increasing the Ti + Al atomic % at the same Ti/Al ratio. As the temperature increased, effects of the Ti/Al ratio and the Ti + Al atomic % on the volume fraction were similar, as shown in Figure 1b,c. Only the equilibrium volume fraction of the γ' phase was decreased.

Figure 2 shows the phase fraction of the γ' phase with change in temperature. The volume fraction of the γ' phase was decreased with increasing temperature. The solvus temperature of the γ' phase was increased with increasing Ti/Al ratio. This means that the thermodynamically stable region of the γ' phase can extend to higher temperatures as the Ti/Al ratio increases. However, the η phase was not calculated with thermo Calc software.

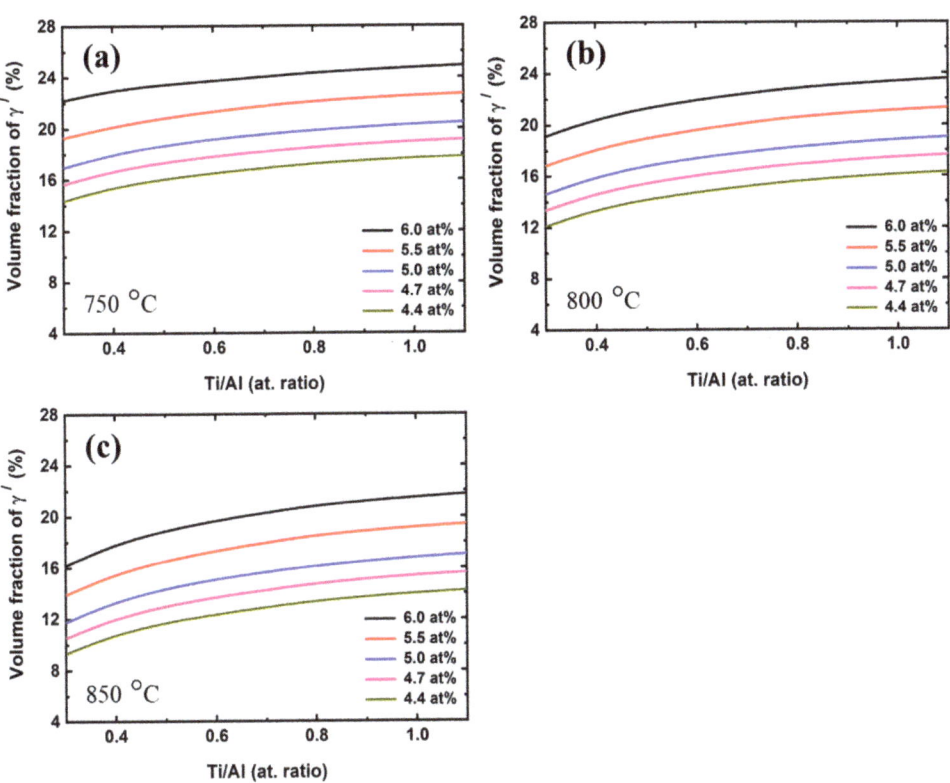

Figure 1. Equilibrium volume fraction of γ' phase for 25Cr-20Co-0.5Mo-1.5Nb-0.03C-Ti-Al with change in Ti + Al atomic % and Ti/Al atomic ratio calculated by Thermo Calc. at (**a**) 750 °C, (**b**) 800 °C, and (**c**) 850 °C.

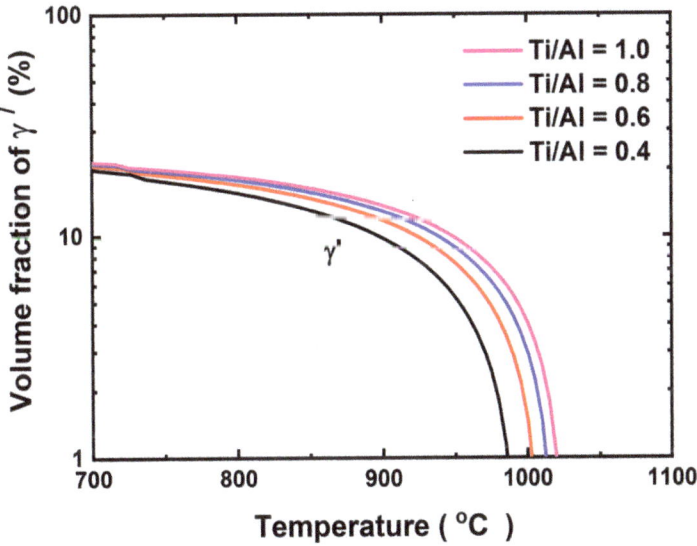

Figure 2. Equilibrium volume fraction of γ' phase for 25Cr-20Co-0.5Mo-1.5Nb-0.03C-Ti-Al with change in the Ti/Al atomic ratio (Ti + Al = 5.0 at%) and aging temperature calculated by Thermo Calc.

Based on calculated results, four alloys (Ti/Al ratio = 0.4, 0.6, 0.8 and 1.0, Ti + Al = 5.0) were designed and prepared by vacuum induction melting (VIM). They were then hot-rolled into 12 mm thick plates at 1150 °C. These rolled plates were heat treated for 30 min at 1150 °C, followed by 16 h at 800 °C.

The chemical compositions of the alloys are shown in Table 1. These alloys were named TA04 (Ti/Al ratio: 0.45), TA06 (Ti/Al ratio: 0.61), TA08 (Ti/Al ratio: 0.80), and TA10 (Ti/Al ratio: 1.09) according to their Ti/Al ratios.

Table 1. Chemical compositions of alloys. (wt%).

Alloy	Ni	Cr	Co	Mo	Nb	Ti	Al	C	Ti/Al (at%)	Ti + Al (at%)
TA04	Bal.	23.1	20.2	0.52	1.5	1.46	1.83	0.03	0.45	5.57
TA06	Bal.	25.2	20.0	0.48	1.4	1.75	1.62	0.03	0.61	5.43
TA08	Bal.	25.2	19.5	0.50	1.4	2.15	1.50	0.03	0.80	5.64
TA10	Bal.	22.8	20.2	0.48	1.5	2.47	1.28	0.03	1.09	5.63

These heat-treated alloys were subjected to long-term aging treatment at 750, 800, and 850 °C for 100, 1000, 3000, 5000, and 10,000 h. Vickers hardness (HV) (HM 122, MITUTOYO, Kawasaki, Japan) at room temperature was measured for each aged specimen under 9.81 N load and 10-s dwell-time conditions. The hardness increments by precipitate were obtained by subtracting the hardness of solution treatment specimens from the hardness of aged specimens. It was assumed that the effects of solid solution strengthening and the grain-size strengthening remained during aging. The microstructure of each aged specimen was observed and analyzed using SEM (INSPECT F50, FEI, Hillsboro, OR, USA). Aged specimens were mechanically polished and etched using 1:10:10-HNO_3:HCl:H_2O solution for 20 min at 50 °C. The fraction of the γ' phase was measured from SEM images with an area of 5 μm × 5 μm using ImageJ software (ImageJ 1.51j8, National Institutes of Health, Bethesda, MD, USA). In addition, the length of the η phase was measured from SEM images.

For TEM (Talos F200X and Tecnai F20 G2, FEI, OR, USA) observations, aged specimens were mechanically thinned to 50 μm or less using an SiC paper followed by twin jet polishing (15% perchloric acid + 85% ethanol at −25 °C, 16 V) with a disk punched at a diameter of 3 mm. The precipitate size and their identification were obtained from bright-field (BF) images taken from the TEM and the γ' phase BF images were observed at the [100] zone axis.

3. Results and Discussion

Figure 3 shows SEM images of the TA10 specimen after heat treatment. In Figure 3a, precipitates were formed at grain boundaries. The grain size was about 50 μm. It was found that the γ' phase was uniformly formed in the grain with an average diameter of 25 nm, as shown in Figure 3b. The grain sizes and diameters of the γ' phases of TA04, TA06, and TA08 after heat treatment were similar to those of the TA10 alloy (data not shown). In addition, the grain sizes of specimens aged up to 10,000 h under each temperature condition did not change significantly.

Figure 4 shows SEM images taken for aged specimens. The spherical γ' phase was uniformly formed in TA10 specimens aged at 800 °C for 3000 h. The average diameter of the γ' phase was about 100 nm, as shown in Figure 4a. The diameter of the γ' phase was significantly increased as aging time and aging temperature increased, as shown in Figure 4b,c. The large-sized γ' phase was also frequently observed in faceted shapes. On the other hand, variation in the Ti/Al ratio had little effect on the diameter or volume fraction of the γ' phase, as shown in Figure 4d.

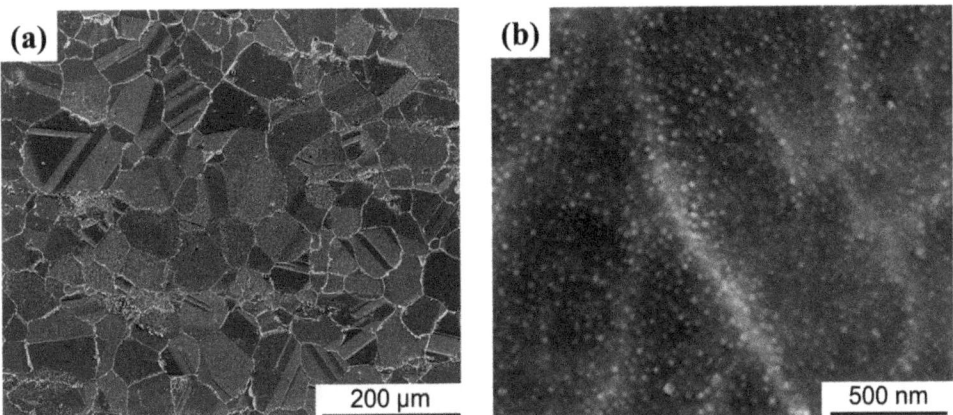

Figure 3. SEM micrographs showing (**a**) grain structure and (**b**) distribution of the γ' phase taken from heat-treated sample for alloy TA10.

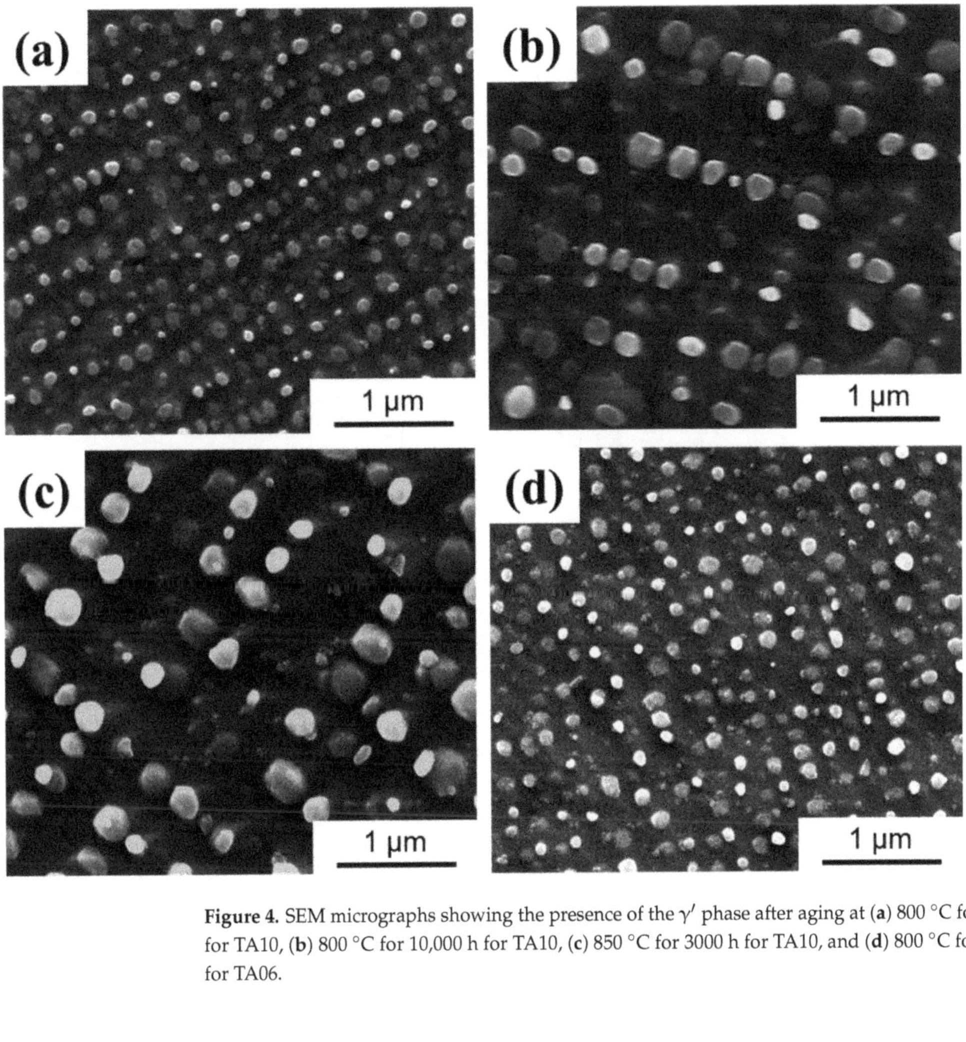

Figure 4. SEM micrographs showing the presence of the γ' phase after aging at (**a**) 800 °C for 3000 h for TA10, (**b**) 800 °C for 10,000 h for TA10, (**c**) 850 °C for 3000 h for TA10, and (**d**) 800 °C for 3000 h for TA06.

Figure 5 shows TEM BF images in the [001] zone axis taken from specimens aged at 750 °C, showing cuboidal shape of the γ' phase. It could be seen that the variation in Ti/Al ratio had little effect on the diameter of the γ' phase. The average diameter of the γ' phase was about 85 nm in specimens aged at 750 °C for 5000 h in TA06 and TA10 alloys and about 110 nm for those aged at 750 °C for 10,000 h in TA06 alloy.

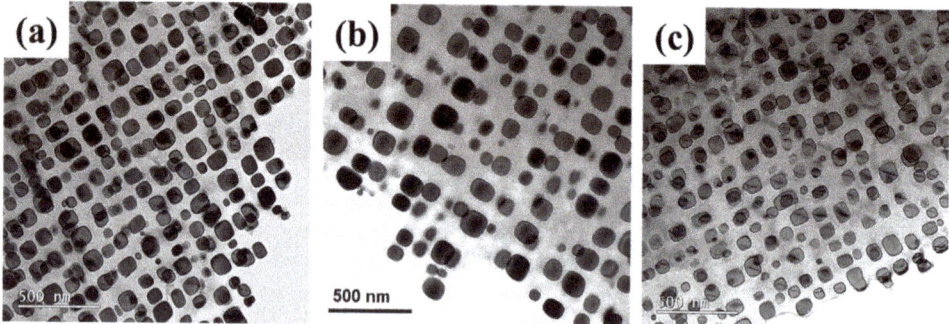

Figure 5. TEM bright-field images showing the presence of γ' phase after aging at 750 °C for (**a**) 5000 h for TA06, (**b**) 10,000 h for TA06, and (**c**) 5000 h for TA10.

The diameter of the γ' phase with aging time is shown in Figure 6. It could be seen that the diameter of the γ' phase increased with increases in aging time and temperature. However, there was little difference in the diameter of the γ' phase among all temperatures even when the Ti/Al ratio was changed.

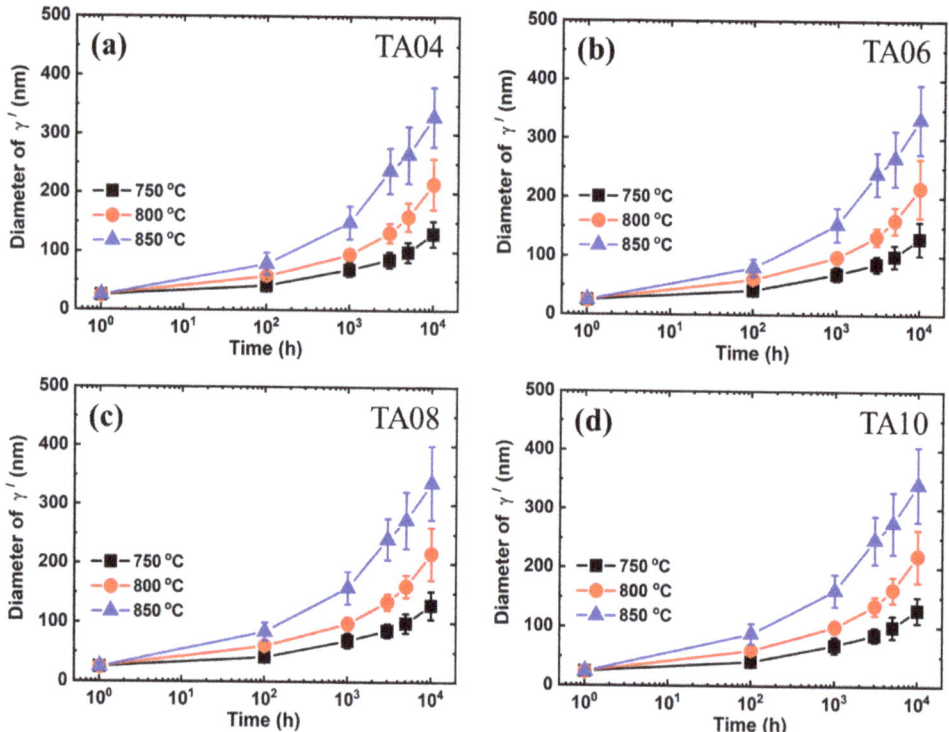

Figure 6. Change in γ' diameter with aging time for (**a**) TA04, (**b**) TA06, (**c**) TA08, and (**d**) TA10 alloys.

The volume fraction of the γ′ phase with aging time at each temperature is depicted in Figure 7. The volume fraction of the γ′ phase except for that for 100 h at 750 °C continuously decreased with the increase in aging time at all temperatures. Since all samples were subjected to heat treatment at 800 °C for 16 h after solution heat treatment followed by aging treatment at each temperature, the reason for the increase in the volume fraction of the γ′ phase in the 100 h aged specimen at 750 °C was thought to be due to additional precipitation of the γ′ phase caused by a difference in solubility. It was also found that the volume fraction of the γ′ phase increased with an increase in Ti/Al ratio at all temperatures (Figure 7a–d).

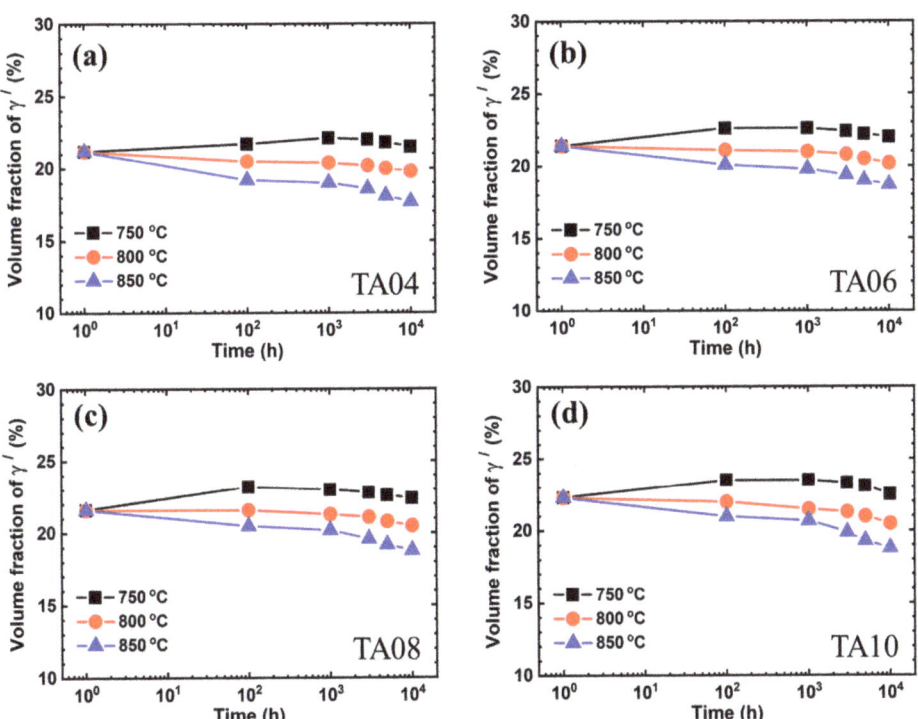

Figure 7. Change in volume fraction of γ′ phase with aging time for (**a**) TA04, (**b**) TA06, (**c**) TA08, and (**d**) TA10 alloys.

Figure 8 presents the Vickers hardness of alloys with aging time. The hardness of alloys slightly increased at the beginning of aging below 100 h at 750 °C. They then gradually decreased as aging time increased. This was closely related to the increase in the volume fraction of the γ′ phase, as shown in Figure 7a. Moreover, as the temperature increased, the hardness gradually decreased during aging (Figure 8b). Additionally, as the temperature further increased, the hardness decreased more rapidly (Figure 8c). Furthermore, it could be seen that the Vickers hardness of the alloys increased with increasing Ti/Al ratio at all temperatures. It was thought that the increase in the γ′ phase volume fraction according to the Ti/Al ratio in Figure 7 was the cause of this increase in hardness in Figure 8.

Figure 9 shows a TEM BF image of the TA10 sample aged for 5000 h at 850 °C. It revealed the existence of needle-like precipitates in addition to the cube-shaped γ′ phase. It was confirmed that this needle-like precipitate was the η phase with ZA = 110 based on careful analysis of the inserted diffraction pattern. One could see that the γ′ phase was absent around the η phase. The formation of the γ′ phase depletion zone was attributed to the formation and growth of the η phase because Ti was shared by both the γ′ and η phases.

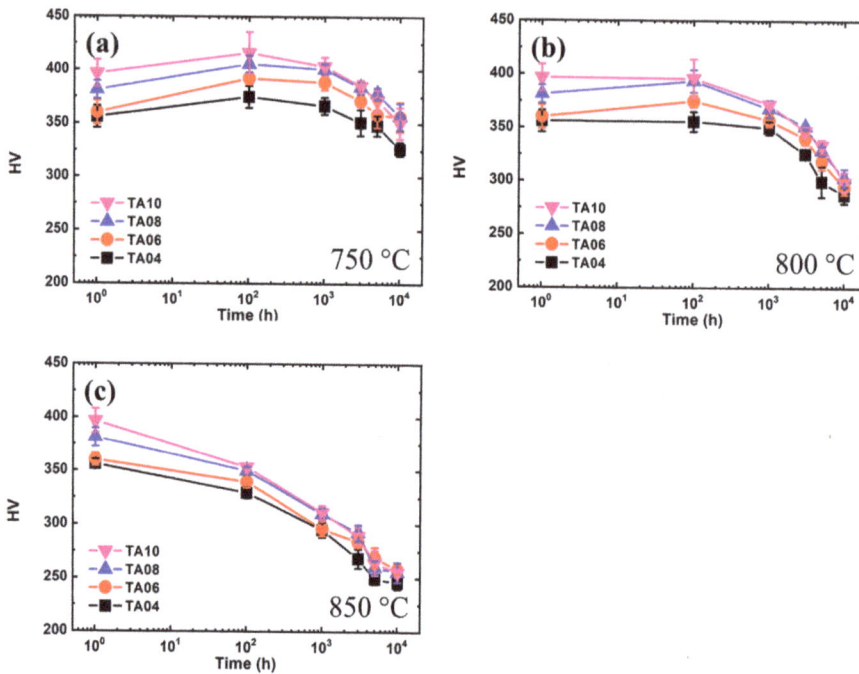

Figure 8. Change in the Vickers hardness with aging time at (**a**) 750 °C, (**b**) 800 °C, and (**c**) 850 °C.

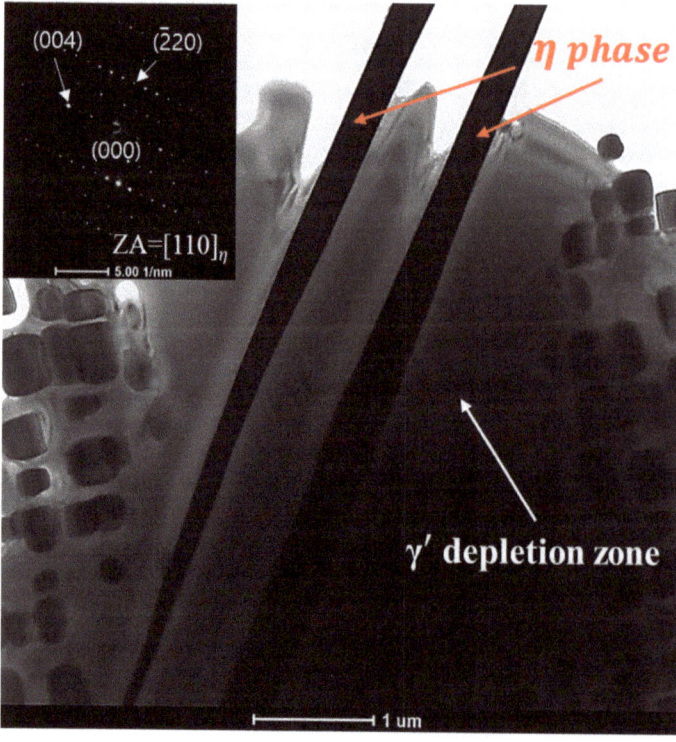

Figure 9. TEM BF image with indicated SADP of alloy TA10 after aging for 5000 h at 850 °C.

SEM micrographs of alloys TA08 and TA10 aged for 10,000 h at 800 °C and 850 °C are presented in Figure 10. One could see the formation of the η phase in samples aged at 800 °C and 850 °C for both TA08 and TA10 alloys. The length of the η phase was longer at higher aging temperatures. In addition, the γ' phase depletion zones were observed around the η phases. The length and number of the η phase also increased with increasing Ti/Al ratio. In addition, as shown in Figure 10a, the η phase was formed with a specific orientation relationship. According to a previous study [29], it was formed in three different directions with angles of 70.5 and 54.7 degrees when observed in the [110] direction of the TEM. Although it was an SEM image, it was possible to infer the approximate orientation relationship. However, it was difficult to quantitatively evaluate the volume fraction of the η phase due to its inhomogeneous distribution at grain boundaries and grain interiors. On the other hand, no evidence of η phase formation was found under any aging conditions for TA04 or TA06 alloys. This means that the η phase is thermodynamically unstable when the Ti/Al ratio is less than 0.61. Furthermore, the existence of the η phase could not be confirmed at low aging temperatures or short aging times, even in the TA08 or TA10 alloys.

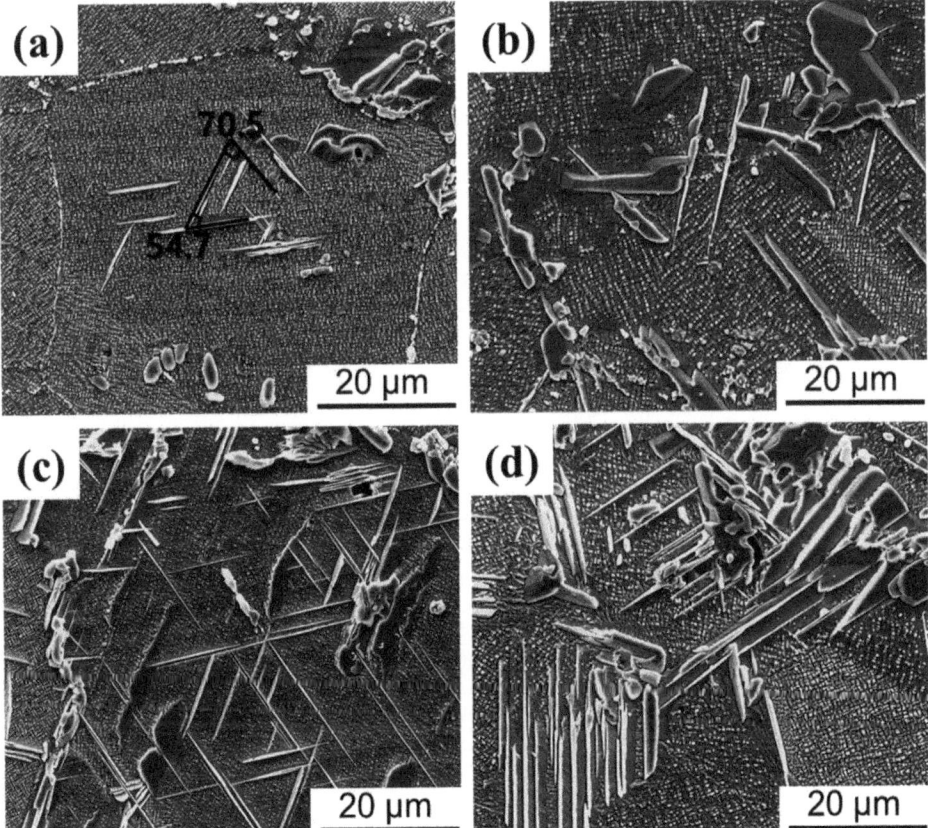

Figure 10. SEM micrographs after 10,000 h aging at (**a**) 800 °C and (**b**) 850 °C for TA08 and at (**c**) 800 °C and (**d**) 850 °C for TA10.

Zener and Frank [30,31] have reported that the length of the needle-like precipitate follows the following equation, assuming the growth of the precipitate is controlled by the lattice diffusion of solute atoms in the matrix:

$$L = k[D(t-\tau)]^{\frac{1}{2}} \qquad (1)$$

where L is the length of the precipitate, k is a dimensionless growth coefficient, D is diffusivity of solute atoms in matrix, and t and τ are isothermal exposure time and incubation time of precipitation, respectively.

Figure 11 shows the length of the η phase with aging time for the TA10 and TA08 alloys under conditions in which η phase formation is confirmed. Despite an inhomogeneous distribution of the η phase, η formation was confirmed in four aging conditions at 850 °C (Figure 11a). It was also found in three conditions at 800 °C for the TA10 alloy (Figure 11a) and 850 °C for the TA08 alloy (Figure 11b). The existence of the η phase was confirmed only in one sample aged for 10,000 h in the TA10 (750 °C) and TA08 (750 °C and 800 °C) alloys. The length of the η phase showed very well linearity with the square root of the aging time at 850 °C for the TA08 and TA10 alloys and 800 °C for the TA10 alloy, respectively. The growth rate of the η phase in TA10 was higher than that in the TA08 alloy. It was thought that the time of the x-intercept of the fitted line in Figure 11 could be considered as the incubation time for η phase formation. Therefore, the incubation time for nucleation of the η phase was thought to be 250 h at 850 °C, 1640 h at 800 °C in TA10, and 1400 h at 850 °C for the TA08 alloy. That is, TA10 had a shorter incubation time for nucleation of the η phase than the TA08 alloy. The shorter incubation time and faster growth rate of TA10 were thought to be due to higher Ti content in the TA10 alloy than in the TA08 alloy.

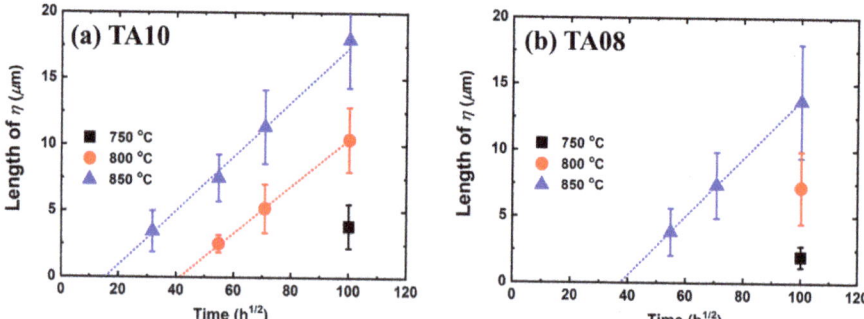

Figure 11. Change in η length with aging time at indicated temperature for (**a**) TA10 and (**b**) TA08 alloys.

It is a well-known phenomenon that the APB formed inside the precipitates suppressed the movement of the dislocation when a precipitate with an ordered structure is sheared by mobile dislocations. This is called order strengthening. Shearing by two dislocations with the same Burgers vector is essential to shear the ordered precipitates. In other words, the shearing process was composed such that the leading dislocation creates an APB during the shearing of the ordered precipitates and the following trailing dislocation removes the formed APB. The order strengthening mechanisms proposed by Reppich [32] are widely accepted in the literature. They proposed the order strengthening increments in two types, assuming that the interaction between leading and trailing dislocations change depending on the size of the precipitates. In the case of small precipitates, they assumed that the trailing dislocation shears the precipitate independently of the leading dislocations and has a straight shape. The trailing dislocations interact with the leading dislocations to minimize the APB area during shearing and have a more complex shape in the case of large precipitates. They called them weakly coupled dislocations (WCD) and strongly coupled dislocations (SCD), respectively.

The order strengthening increments sheared by the WCD and SCD proposed by Reppich [32] are given as follows.

$$\Delta \tau_{ppt.WCD} = \frac{1}{2}\left(\frac{\gamma_{APB}}{b}\right)^{\frac{3}{2}}\left(\frac{bdV}{T}\right)^{\frac{1}{2}} m - \frac{1}{2}\left(\frac{\gamma_{APB}}{b}\right) V \quad (2)$$

$$\Delta\tau_{ppt.SCD} = \left(\frac{\sqrt{3}}{2}\right)\frac{TV^{\frac{1}{2}}w}{bd}\left(1.28\frac{d\gamma_{APB}}{wT}-1\right)^{\frac{1}{2}} \qquad (3)$$

where γ_{APB} is the APB energy of the γ' phase in the {111} plane, b is the Burgers vector of dislocation in the matrix, d is the diameter of the γ' phase, V is the volume fraction of the γ' phase, m is numerical factor determined by the the γ' phase morphology, w is a constant that explains the elastic repulsion between SCD, and T is the line tension of the dislocation. The line tension was estimated as $T = 0.5Gb^2$ and G is the shear modulus.

Figure 12 shows the shear stress increments ($\Delta\tau_{ppt}$) resulting from the order strengthening of the γ' phase calculated using Equations (2) and (3), plotted according to the diameter of the γ' phase for all conditions. The values used in the calculation for b, m, w, G, and γ_{APB} were 0.254 nm, 0.72, 1, 80 GPa [32], and 0.25 Jm^{-2} [33], respectively. The volume fraction (V) and diameter (d) of the γ' phase were the values used in Figures 6 and 7.

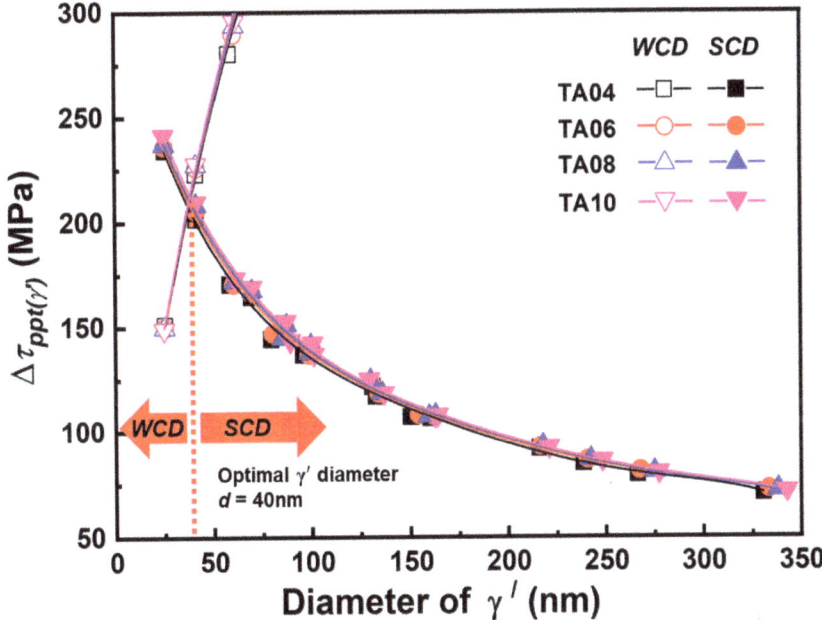

Figure 12. Shear stress increments by order strengthening against the diameter of γ' phase calculated using WCD and SCD models at 750, 800, and 850 °C.

The $\Delta\tau_{ppt}$ was increased with the diameter of the γ' phase in the WCD model. On the contrary, the $\Delta\tau_{ppt}$ was decreased with the diameter of the γ' phase in the SCD model. The shearing of the γ' phase is controlled by lower values of the WCD and SCD at given diameters of the γ' phase. One can see that the transition diameter from the WCD to SCD takes place at 40 nm. The transition diameter and the $\Delta\tau_{ppt}$ are shown to be nearly the same values despite the change in the Ti/Al ratio. The transition diameter between WCD and SCD changes greatly depending on the APB energy. It means that the transition diameter decreases as the APB energy increases. This is because of the balance between the total energy generated by the APB and the repulsive force of the leading–trailing dislocations with the same Burgers vector.

In order to compare the experimentally obtained precipitation strengthening increments with the theoretical order strengthening increments of the γ' phase ($\Delta\sigma_{order.cal.}$), the $\Delta\tau_{ppt}$ were converted to $\Delta\sigma_{order.cal.}$ by multiplying the Taylor factor M, which is taken as 3 [34–36].

Tensile tests were not conducted in all aged conditions, so σ_{YS} was obtained by converting from the hardness values in the calculation of strengthening. The conversion factor between yield strength and hardness was measured to 2.3 from the standard heat treatment specimens, as shown in Table 2. It is thought that this relationship is reasonable compared to the values of 2.34 and 2.46 presented by Wu et al. [37] and Osada et al. [38], respectively. The increment in precipitation strengthening ($\Delta\sigma_{ppt.exp.}$) was obtained by subtracting the hardness of the solution-treated specimens from the hardness of the aged specimens.

Table 2. Yield strength (YS) and Vickers hardness (HV) values in standard heat treatment specimens.

Alloy	YS	Hv	YS/Hv
TA04	827.0	356.0	2.32
TA06	840.9	360.0	2.33
TA08	853.5	381.3	2.24
TA10	927.8	396.6	2.34

The relationship between $\Delta\sigma_{order.cal.}$ and $\Delta\sigma_{ppt.exp.}$ in all aged conditions is shown in Figure 13. One can see that $\Delta\sigma_{order.cal.}$ is linearly proportional to $\Delta\sigma_{ppt.exp.}$.

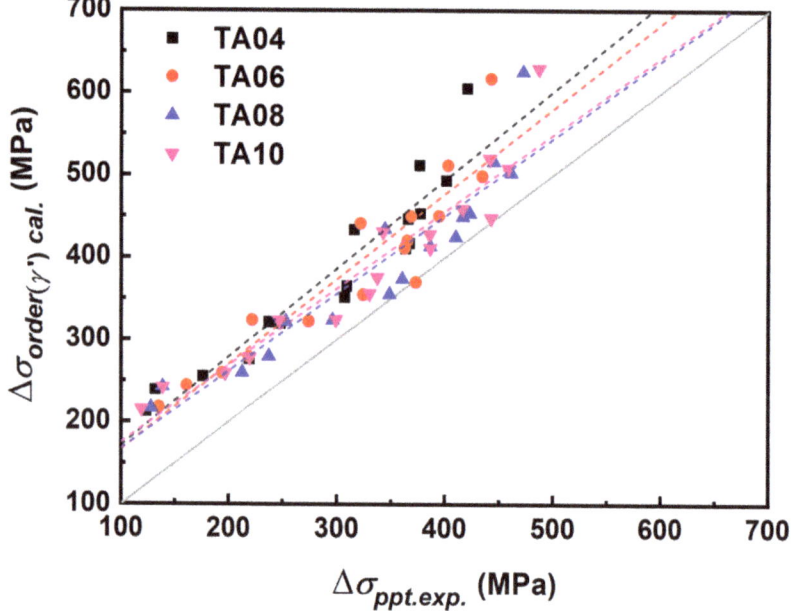

Figure 13. Comparison of measured precipitation strengthening increments and calculated order strengthening increments.

Also, it can be confirmed that the differences between $\Delta\sigma_{order.cal.}$ and $\Delta\sigma_{ppt.exp.}$ decrease as the Ti/Al ratio increases. It is a well-known fact that the order strengthening is controlled by the diameter, volume fraction, and APB energy of the γ' phase.

Therefore, the difference in order strengthening increments with the Ti/Al ratio is mainly thought to originate from the difference in APB energy considering that there is no significant difference between the diameter and volume fraction of the γ' phase, as shown in Figures 6 and 7. It can be inferred that the APB energy increases with the Ti/Al ratio and it was found that the APB energy increased by approximately 27% per atomic % of Ti. This value is higher than the 9% and 10% values reported by Dodaran et al. [19] and Gorbatov et al. [20], respectively. However, it is consistent with the fact that Ti increases the APB energy of the γ' phase.

4. Conclusions

In this study, precipitation behavior was investigated according to the Ti/Al ratio during aging in Ni-Cr-Co-based superalloys. The diameter of the γ' phase did not change significantly with an increasing Ti/Al ratio, although the volume fraction of the γ' phase increased. The diameter of the γ' phase was increased with aging times and aging temperatures. Additionally, the η phase did not form during aging for 10,000 h at 850 °C when the Ti/Al ratio was less than 0.61. However, the η phase formed when the Ti/Al ratio was equal to or higher than 0.80. As the temperature increased, η formed quickly and the growth rate was also fast. As a result, it was found that the TA06 alloy showed high strength without the η phase formation. Also, it was found that the APB energy was increased by approximately 27% per atomic % of Ti from the consideration of order strengthening.

Author Contributions: Conceptualization, D.-J.C. and W.-S.J.; Data curation, D.-J.C.; Formal analysis, D.-J.C.; Investigation, D.-J.C. and C.P.; Methodology, D.-J.C.; Project administration, W.-S.J.; Supervision, W.-S.J.; Validation, J.L.; Visualization, J.L.; Writing—original draft, D.-J.C.; Writing—review & editing, J.L. and W.-S.J. All authors have read and agreed to the published version of the manuscript.

Funding: The authors gratefully acknowledge the financial support from the Energy Technology Development Program of [Korea Institute of Energy Technology Evaluation and Planning] grant number [20181110100410] and HRD Program for Industrial Innovation of [Korea Institute for Advancement of Technology] grant number [P0023676] in the Republic of Korea.

Data Availability Statement: The data presented in this study are available in this article.

Conflicts of Interest: The authors declare no conflict of interest.

References

1. Smith, T.M.; Esser, B.D.; Antolin, N.; Carlsson, A.; Williams, R.E.A.; Wessman, A.; Hanlon, T.; Fraser, H.L.; Windl, W.; McComb, D.W.; et al. Phase Transformation Strengthening of High-Temperature Superalloys. *Nat. Commun.* **2016**, *7*, 13434. [CrossRef]
2. Wu, Z.; Chen, X.; Fan, Z.; Zhou, Y.; Dong, J. Studies of High-Temperature Fatigue Behavior and Mechanism for Nickel-Based Superalloy Inconel 625. *Metals* **2022**, *12*, 755. [CrossRef]
3. Shin, G.S.; Yun, J.Y.; Chul Park, M.; Kim, S.J. Effect of Mo on the Thermal Stability of Γ' Precipitate in Inconel 740 Alloy. *Mater. Charact.* **2014**, *95*, 180–186. [CrossRef]
4. Li, H.; Feng, W.; Zhuang, W.; Hua, L. Microstructure Analysis and Segmented Constitutive Model for Ni-Cr-Co-Based Superalloy during Hot Deformation. *Metals* **2022**, *12*, 357. [CrossRef]
5. Ma, Y.W.; Lee, K.W.; Kong, B.O.; Hong, H.U.; Lee, Y.S. Evaluation of Weld Joint Strength Reduction Factor Due to Creep in Alloy 740H to P92 Dissimilar Metal Weld Joint. *Met. Mater. Int.* **2021**, *27*, 4408–4417. [CrossRef]
6. Xie, X.; Chi, C.; Zhao, S.; Dong, J.; Lin, F. Superalloys and the Development of Advanced Ultra-Supercritical Power Plants. *Mater. Sci. Forum* **2013**, *748*, 594–603. [CrossRef]
7. Hanning, F.; Khan, A.K.; Steffenburg-Nordenström, J.; Ojo, O.; Andersson, J. Investigation of the Effect of Short Exposure in the Temperature Range of 750–950 °C on the Ductility of Haynes® 282® by Advanced Microstructural Characterization. *Metals* **2019**, *9*, 1357. [CrossRef]
8. Wei, L.; Wang, S.; Yang, Q.; Cheng, Y.; Tan, S. Investigation on Precipitation Phenomena and Mechanical Properties of Ni-25Cr-20Co Alloys Aged at High Temperature. *J. Mater. Res.* **2018**, *33*, 3479–3489. [CrossRef]
9. Wang, X.; Liu, Y.; Song, Y.; Li, H.; Hu, X.; Ji, Y. Application of Neural Network in Micromechanical Deformation Behaviors of Inconel 740H Alloy. *Int. J. Adv. Manuf. Technol.* **2023**, *125*, 2339–2348. [CrossRef]
10. Kim, D.M.; Kim, C.; Yang, C.H.; Park, J.U.; Jeong, H.W.; Yim, K.H.; Hong, H.U. Heat Treatment Design of Inconel 740H Superalloy for Microstructure Stability and Enhanced Creep Properties. *J. Alloys Compd.* **2023**, *946*, 169341. [CrossRef]
11. Yan, C.; Zhengdong, L.; Godfrey, A.; Wei, L.; Yuqing, W. Microstructure Evolution and Mechanical Properties of Inconel 740H during Aging at 750 °C. *Mater. Sci. Eng. A* **2014**, *589*, 153–164. [CrossRef]
12. Zhao, S.; Xie, X.; Smith, G.D.; Patel, S.J. Gamma Prime Coarsening and Age-Hardening Behaviors in a New Nickel Base Superalloy. *Mater. Lett.* **2004**, *58*, 1784–1787. [CrossRef]
13. Mehdizadeh, M.; Farhangi, H. Investigation of Microstructural Evolution and Mechanical Properties of IN617 Superalloy During Long-Term Operating at High Temperature. *Met. Mater. Int.* **2022**, *28*, 2719–2734. [CrossRef]
14. Liang, X.; Wu, J.; Zhao, Y. Creep Fracture Mechanism of a Single Crystal Nickel Base Alloy Under High Temperature and Low Stress. *Met. Mater. Int.* **2022**, *28*, 841–847. [CrossRef]
15. Reed, R. *The Superalloys: Fundamentals and Applications*; Cambridge University Press: Cambridge, UK, 2006; Volume 1999, ISBN 9780521859042.

16. Enomoto, M.; Harada, H. Analysis of g′/g Equilibrium in Nickel-Aluminum-X Alloys by the Cluster Variation Method with the Lennard-Jones Potential. *Metall. Trans. A Phys. Metall. Mater. Sci.* **1989**, *20A*, 649–664. [CrossRef]
17. Xu, Y.; Zhang, L.; Li, J.; Xiao, X.; Cao, X.; Jia, G.; Shen, Z. Relationship between Ti/Al Ratio and Stress-Rupture Properties in Nickel-Based Superalloy. *Mater. Sci. Eng. A* **2012**, *544*, 48–53. [CrossRef]
18. Ou, M.; Ma, Y.; Ge, H.; Xing, W.; Zhou, Y.; Zheng, S.; Liu, K. Microstructure Evolution and Mechanical Properties of a New Cast Ni-Base Superalloy with Various Ti Contents. *J. Alloys Compd.* **2018**, *735*, 193–201. [CrossRef]
19. Dodaran, M.; Ettefagh, A.H.; Guo, S.M.; Khonsari, M.M.; Meng, W.J.; Shamsaei, N.; Shao, S. Effect of Alloying Elements on the γ′ Antiphase Boundary Energy in Ni-Base Superalloys. *Intermetallics* **2020**, *117*, 106670. [CrossRef]
20. Gorbatov, O.I.; Lomaev, I.L.; Gornostyrev, Y.N.; Ruban, A.V.; Furrer, D.; Venkatesh, V.; Novikov, D.L.; Burlatsky, S.F. Effect of Composition on Antiphase Boundary Energy in Ni3Al Based Alloys: Ab Initio Calculations. *Phys. Rev. B* **2016**, *93*, 224106. [CrossRef]
21. Chen, E.; Tamm, A.; Wang, T.; Epler, M.E.; Asta, M.; Frolov, T. Modeling Antiphase Boundary Energies of Ni3Al-Based Alloys Using Automated Density Functional Theory and Machine Learning. *npj Comput. Mater.* **2022**, *8*, 80. [CrossRef]
22. Cui, C.Y.; Gu, Y.F.; Ping, D.H.; Harada, H.; Fukuda, T. The Evolution of η Phase in Ni-Co Base Superalloys. *Mater. Sci. Eng. A* **2008**, *485*, 651–656. [CrossRef]
23. Shingledecker, J.P.; Pharr, G.M. The Role of Eta Phase Formation on the Creep Strength and Ductility of Inconel Alloy 740 at 1023 k (750 °C). *Metall. Mater. Trans. A Phys. Metall. Mater. Sci.* **2012**, *43*, 1902–1910. [CrossRef]
24. Li, J.; Wu, Y.; Liu, Y.; Li, C.; Ma, Z.; Yu, L.; Li, H.; Liu, C.; Guo, Q. Enhancing Tensile Properties of Wrought Ni-Based Superalloy ATI 718Plus at Elevated Temperature via Morphology Control of η Phase. *Mater. Charact.* **2020**, *169*, 110547. [CrossRef]
25. Detrois, M.; Jablonski, P.D.; Hawk, J.A. The Effect of η Phase Precipitates on the Creep Behavior of Alloy 263 and Variants. *Mater. Sci. Eng. A* **2021**, *799*, 140337. [CrossRef]
26. Liu, G.; Kong, L.; Xiao, X.; Birosca, S. Microstructure Evolution and Phase Transformation in a Nickel-Based Superalloy with Varying Ti/Al Ratios: Part 1—Microstructure Evolution. *Mater. Sci. Eng. A* **2022**, *831*, 142228. [CrossRef]
27. Liu, G.; Xiao, X.; Véron, M.; Birosca, S. The Nucleation and Growth of η Phase in Nickel-Based Superalloy during Long-Term Thermal Exposure. *Acta Mater.* **2020**, *185*, 493–506. [CrossRef]
28. Oh, J.H.; Yoo, B.G.; Choi, I.C.; Santella, M.L.; Jang, J. Il Influence of Thermo-Mechanical Treatment on the Precipitation Strengthening Behavior of Inconel 740, a Ni-Based Superalloy. *J. Mater. Res.* **2011**, *26*, 1253–1259. [CrossRef]
29. Asgari, S.; Sharghi-Moshtaghin, R.; Sadeghahmadi, M.; Pirouz, P. On Phase Transformations in a Ni-Based Superalloy. *Philos. Mag.* **2013**, *93*, 1351–1370. [CrossRef]
30. Zener, C. Theory of Growth of Spherical Precipitates from Solid Solution. *J. Appl. Phys.* **1949**, *20*, 950–953. [CrossRef]
31. Frank, F.C. Radially Symmetric Phase Growth Controlled by Diffusion. *Proc. R. Soc. London. Ser. A. Math. Phys. Sci.* **1950**, *201*, 586–599. [CrossRef]
32. Reppich, B. Some New Aspects Concerning Particle Hardening Mechanisms in γ′ Precipitating Ni-Base Alloys-I. Theoretical Concept. *Acta Metall.* **1982**, *30*, 87–94. [CrossRef]
33. Dimiduk, D.M.; Thompson, A.W.; Williams, J.C. The Compositional Dependence of Antiphase-Boundary Energies and the Mechanism of Anomalous Flow in Ni3 Al Alloys. *Philos. Mag. A Phys. Condens. Matter. Struct. Defects Mech. Prop.* **1993**, *67*, 675–698. [CrossRef]
34. Zhou, D.; Ye, X.; Teng, J.; Li, C.; Li, Y. Effect of Nb on Microstructure and Mechanical Property of Novel Powder Metallurgy Superalloys during Long-Term Thermal Exposure. *Materials* **2021**, *14*, 656. [CrossRef]
35. Galindo-Nava, E.I.; Connor, L.D.; Rae, C.M.F. On the Prediction of the Yield Stress of Unimodal and Multimodal γ′ Nickel-Base Superalloys. *Acta Mater.* **2015**, *98*, 377–390. [CrossRef]
36. Goodfellow, A.J.; Galindo-Nava, E.I.; Schwalbe, C.; Stone, H.J. The Role of Composition on the Extent of Individual Strengthening Mechanisms in Polycrystalline Ni-Based Superalloys. *Mater. Des.* **2019**, *173*, 107760. [CrossRef]
37. Wu, H.; Zhuang, X.; Nie, Y.; Li, Y.; Jiang, L. Effect of Heat Treatment on Mechanical Property and Microstructure of a Powder Metallurgy Nickel-Based Superalloy. *Mater. Sci. Eng. A* **2019**, *754*, 29–37. [CrossRef]
38. Osada, T.; Gu, Y.; Nagashima, N.; Yuan, Y.; Yokokawa, T.; Harada, H. Optimum Microstructure Combination for Maximizing Tensile Strength in a Polycrystalline Superalloy with a Two-Phase Structure. *Acta Mater.* **2013**, *61*, 1820–1829. [CrossRef]

Disclaimer/Publisher's Note: The statements, opinions and data contained in all publications are solely those of the individual author(s) and contributor(s) and not of MDPI and/or the editor(s). MDPI and/or the editor(s) disclaim responsibility for any injury to people or property resulting from any ideas, methods, instructions or products referred to in the content.

Microstructure Optimization for Design of Porous Tantalum Scaffolds Based on Mechanical Properties and Permeability

Yikai Wang [1,2], Xiao Qin [1,2], Naixin Lv [1,2], Lin Gao [1,2,*], Changning Sun [1,2,*], Zhiqiang Tong [1,2] and Dichen Li [1,2]

1 State Key Laboratory for Manufacturing System Engineering, School of Mechanical Engineering, Xi'an Jiaotong University, Xi'an 710054, China; wangyikai@stu.xjtu.edu.cn (Y.W.); qinxiao@stu.xjtu.edu.cn (X.Q.); lnx9228@stu.xjtu.edu.cn (N.L.); zhiqiang.tong@xjtu.edu.cn (Z.T.); dcli@mail.xjtu.edu.cn (D.L.)
2 National Medical Products Administration (NMPA), Key Laboratory for Research and Evaluation of Additive Manufacturing Medical Devices, Xi'an Jiaotong University, Xi'an 710054, China
* Correspondence: gaolin2013@mail.xjtu.edu.cn (L.G.); sun.cn@xjtu.edu.cn (C.S.)

Abstract: Porous tantalum (Ta) implants have important clinical application prospects due to their appropriate elastic modulus, and their excellent bone growth and bone conduction ability. However, porous Ta microstructure designs generally mimic titanium (Ti) implants commonly used in the clinic, and there is a lack of research on the influence of the microstructure on the mechanical properties and penetration characteristics, which will greatly affect bone integration performance. This study explored the effects of different microstructure parameters, including the fillet radius of the middle plane and top planes, on the mechanics and permeability properties of porous Ta diamond cells through simulation, and put forward an optimization design with a 0.5 mm midplane fillet radius and 0.3 mm top-plane fillet radius in order to significantly decrease the stress concentration effect and improve permeability. On this basis, the porous Ta structures were prepared by Laser Powder Bed Fusion (LPBF) technology and evaluated before and after microstructural optimization. The elastic modulus and the yield strength were increased by 2.31% and 10.39%, respectively. At the same time, the permeability of the optimized structure was also increased by 8.25%. The optimized microstructure design of porous Ta has important medical application value.

Keywords: porous Ta; microstructure optimization; mechanical properties; permeability

1. Introduction

Tantalum and its compounds have been widely used in aerospace, electronics, and especially medical fields due to their excellent properties such as high hardness, high toughness, high ductility, and corrosion resistance [1–3]. Porous Ta is generally considered a promising implantable biomaterial due to its biocompatibility, non-cytotoxicity, and suitability for the attachment and growth of osteoblasts [1,4]. In recent years, more and more clinical orthopedic implants made of porous Ta materials have been used, including artificial humerus femoral prostheses [5], ankle and hip prostheses [6] and acetabular pelvic prostheses [7]. In addition, compared with other porous metals, the friction coefficient between porous Ta implants and human bone is 40~80% higher [8,9], which is conducive to its binding with the host bone interface and reduces loosening of the implant in the initial stage of implantation [10–12]. Thus, porous Ta has broad application prospects in the field of bone defect repair.

The cellular structure of porous Ta implants has a direct impact on its mechanical properties, and also plays a very important role in the bonding between the implant and the host bone. The effects of different cellular microstructure characteristics of porous Ta implants on the mechanical properties, biological properties, bone integration ability, and bone formation need to be further studied. Huang et al. chose diamond as the cell structure, and found that the microstructure can be selected according to the biological performance requirements or mechanical requirements of the application environments [13]. Gao et al. studied the compressive mechanical behavior and failure mechanism of porous Ta, and found that the structural failure sites were mostly located in vertical struts and their connections [14]. Wang et al. designed and manufactured a new three-dimensional, multi-scale interconnected porous Ta scaffold, and biological experiments in vivo and in vitro showed that the microstructure characteristics of the new porous Ta scaffold were similar to bone trabeculae, and it had potential for bone tissue engineering applications [15]. Wauthle et al. tested the mechanical properties of porous Ta scaffolds with 80% porosity prepared with a regular dodecahedral cell structure, and the results showed that it was conducive to reducing the stress shielding effect within the range of human canceller bone [1]. Yang et al. used LPBF technology to fabricate porous Ta structures, and their microstructure showed a rough surface, with promising application prospects in bone filling and reconstruction [16]. It should be noted that Ta-related nanostructured materials can be obtained by simpler process methods, such as electro-chemical anodizing [17]. Additionally, nano-structured Ta oxide is more corrosion-resistant than pure metal, with great value for application in the medical field. In addition, the morphological characteristics of internal channels in porous structures also affect the mechanical properties, bone integration, and bone growth ability of implants [18–20]. Thus, the excellent bone integration properties of porous Ta scaffolds are closely related to their micro-mechanical properties and permeability properties. However, the existing microstructure design of porous Ta scaffolds follows the design scheme of Ti alloy, and the influence of microstructure on its mechanical properties and permeability is still unclear, which could lead to a bottleneck of porous Ta applications. Due to the superior mechanical and biological properties of porous Ta, previous studies on porous optimization schemes based on porous Ti cannot work well for porous Ta scaffolds. Therefore, it is urgent to design a microstructure optimization method suitable for porous Ta scaffolds.

In this paper, the influence of microstructures of porous Ta with diamond crystal cells on the mechanical properties and permeability was systematically investigated, and an optimized microstructure design was proposed. On this basis, the porous Ta structures were prepared by 3D printing and evaluated before and after microstructural optimization. This study could provide a reference for the optimal design of the cellular structure of clinical porous Ta implants.

2. Materials and Methods

2.1. Optimal Design Strategy of Porous Ta Microstructure

In this study, a diamond crystal cell was selected for porous Ta structures. The mechanical properties of the structures were simulated by ANSYS v2022 software. According to the analysis of a stress cloud diagram of the unicellular structure (Figure 1a), the stress concentration effect mainly occurred at the intersection of two struts on the top surface (Figure 1b, region 2), followed by the intersection of the struts on the middle surface of the unicellular structure (Figure 1b, regions 1 and 3). Based on this, the single-cell structure optimization areas were divided as shown in Figure 1b. Due to the isotropy of the structure, regions 1 and 3 were regarded as the top surface, and region 2 as the middle surface. The forming of the single-cell strut was achieved by rotating along the central axis. This research investigated the optimization design parameters with variable fillets on the top and middle surfaces. The maximum stress, coefficient of variation, kurtosis, and skewness were used as criteria to analyze and evaluate the mechanical properties. Different optimization schemes are shown in Table 1. In order to ensure the unity of the structures, the porosities of the porous structures were kept at 80%, which is commonly applied in bioimplants, by changing the strut diameters.

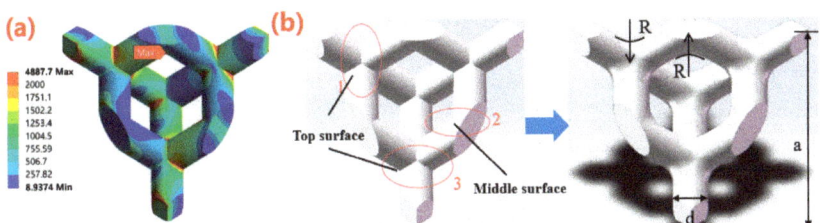

Figure 1. Schematic diagram of optimization of microstructural single cells. (**a**) Stress nephogram of a unit diamond cell structure; (**b**) optimized region division and unit cell fillet optimization design. The numbers 1 and 3 refer to the top corner optimization area, and the number 2 refers to the middle corner optimization area. The letter R refers to the radius of the rounded corner, and the letters a and d refer to the height and diameter of the cell, respectively.

Table 1. Single-cell structure optimization design parameters.

Parameters	Radius
Middle-plane fillets	0.1 mm~1.0 mm (interval 0.1 mm)
Top-plane fillets	0.1 mm~0.5 mm (interval 0.1 mm)

2.2. Simulation of Mechanical Properties and Permeability for Optimization

2.2.1. Mechanical Properties Simulation

A static mechanical simulation was carried out to study the influence of different optimization parameters on the mechanical properties of the porous Ta structures. Two substrates were added on the top and bottom of the porous structures to match the stress environment during the compression experiments. The properties of the substrates were set as rigid bodies, the thickness of the substrates was 0.2 mm, and the side length was equal to the length of the porous structures. A schematic diagram of the overall structures is shown in Figure 2a. Considering the irregular structures in this study, the mesh type selected was a tetrahedral mesh. A downward displacement load was applied to the upper substrate as 1% of the total support height with a fixed lower substrate. The equivalent elastic modulus and stress concentration effect were analyzed. The stress concentration effect can be directly observed through the stress–strain cloud map, and the equivalent elastic modulus was mainly solved by the following formula [21]:

$$E = \frac{\sigma}{\varepsilon} = \frac{R_Y \cdot L}{\Delta L \cdot A}$$

R_Y—axial branch reaction of fixed constraint (N); A—nominal contact area of porous structure (mm^2); ΔL—deformation in the loading direction of the support (mm); L—length of support (mm).

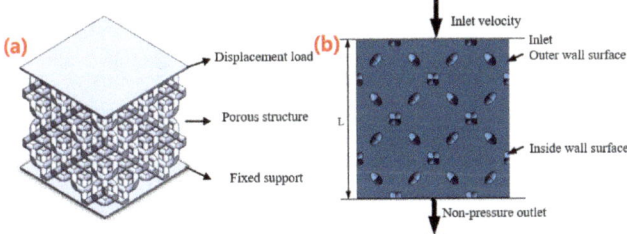

Figure 2. Simulation diagram of porous structure performance. (**a**) Loading conditions for mechanical simulation of porous structures; (**b**) fluid domain model and boundary condition setting of porous structures.

2.2.2. Permeability Simulation

A fluid simulation was carried out to study the influence of optimization parameters on the permeability of the porous structures. The fluid would enter from the upper surface, flow through the inner and outer walls in the microstructure, and exit from the lower surface as shown in Figure 2b. The liquid selected was water to simulate the flow trend of blood in the human body, with a density of 1000 kg/m^3, a dynamic viscosity of 0.001 Pa/s, an inlet flow rate of 0.001 m/s, an and outlet pressure of 0 Pa. The permeability of porous structures was calculated according to Darcy's law [22]:

$$k = \frac{\nu u l}{\Delta p}$$

k—permeability of porous structure (m); ν—inlet velocity of porous structure (m·s^{-1}); l—total height of porous structure model (m); Δp—the pressure gradient of the fluid domain, that is, the pressure difference between the upper and lower surfaces of the fluid domain (Pa).

Wall shear stress (WSS) is defined as the force per unit area exerted by the wall upward of the fluid in the local tangent plane [23]. Previous studies showed that it is conducive to cell growth when the wall shear stress ranges from 5×10^{-5} Pa to 2.5×10^{-2} Pa [24]. Otherwise, it would inhibit cell growth. So, the proportion of suitable areas for cell survival was also calculated as the ratio of the area with appropriate shear stress to the whole area.

2.3. Manufacturing of Porous Ta

The manufacturing equipment in this study was an LPBF (Laser Powder Bed Fusion) additive manufacturing system independently developed by Xi'an Jiaotong University (Figure 3a). The apparatus was equipped with a YLR-500-WC fiber laser with a rated power of 500 W, a spot diameter of 45 μm, and a wavelength of 1070.94 nm. The spherical Ta powder was provided by Suzhou Yaoyi New Material Technology Co., Ltd., Suzhou, China. The purity was 99.99% and the particle size was 15–53 μm. Porous structure models of $12 \times 12 \times 12$ mm^3 before and after optimization were established for preparation (Figure 3b). After the samples was fabricated, they were firstly sandblasted to remove the unmelted powder adhering to the surface of the structs, and then, cut from the substrate using a wire cutting machine (WGM4, Bossage CNC Equipment Manufacturing Co., Ltd., Suzhou, China). Finally, the samples were cleaned in an ultrasonic cleaning machine with anhydrous ethanol for 1 h. The optimal process parameters (P: 150 W, v: 270 mm/s, t: 0.05 mm, h: 0.07 mm) were selected to manufacture the structures with 80% porosity [25]. A total of 10 samples were prepared with 5 samples with and without optimization, respectively.

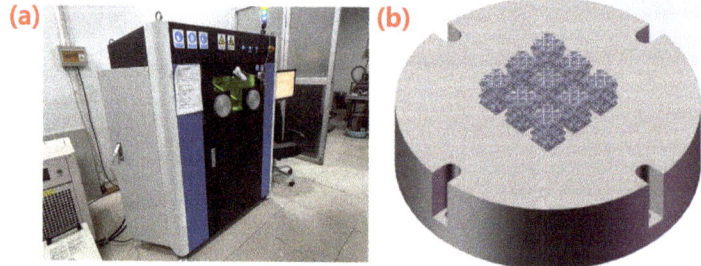

Figure 3. Porous Ta manufacturing platform. (**a**) LPBF additive manufacturing system; (**b**) sample preparation diagram.

2.4. Evaluation of Properties of Porous Ta Optimization

2.4.1. Microstructure Observation

A scanning electron microscope (Gemini500, ZEISS Microscopy, Jena, Germany) was used to observe the microstructure of the porous structures, and Image J software was used to measure the diameters of the struts. Ten different positions in each sample were selected for measurements of the average values. The porosities of porous Ta samples after post-treatment were measured using the weighing method.

2.4.2. Mechanics Performance Testing

The samples were compressed along the forming and horizontal directions at a constant loading rate of 1 mm/min until failure by an electronic universal material testing machine (CMT4304, Max. 35 kN, Xian Letry Testing Machinese Co., Ltd., Xi'an, China). The compression experiment was set according to the national standard ISO 13314:2011 [26] (mechanical testing of metals–ductility testing–compression test for porous and cellular metals). The strain values of the porous structures during compression were measured by a force–displacement (stress–strain) sensor on the test machine, and the elastic modulus and yield strength of the porous structures were calculated according to the stress–strain curves.

2.4.3. Permeability Testing

In the permeability tests, sealing tapes were used to seal four sides of the porous Ta samples in the fixture to ensure that the sample would not leak during the test. The flow rate at the inlet of the pipe was controlled to keep the height of the horizontal plane unchanged. During the test, the time required for each sample to discharge 500 mL water was measured. The test was repeated 5 times for each sample to calculate the average penetrating time.

2.5. Statistical Analysis Method

The statistical analysis was performed by one-way ANOVA and Student's t-test using SPSS 25.0 (IBM Corporation, Armonk, NY, USA). The maximum stress, coefficient of variation, kurtosis, and skewness were calculated as criteria to evaluate the mechanical properties. The maximum stress reflected the influence of different optimization schemes on the stress concentration effect. The coefficient of variation represented the stress distribution in the analysis area. The skewness demonstrated the asymmetry of the mechanical properties data. All skewness values greater than 0 indicated that the stress distribution is a positive skewness distribution, that is, arithmetic mean > median > mode. The kurtosis represented the steepness of the data distribution. The kurtosis value of data that completely follow a normal distribution is 0. When the kurtosis value is greater than 0, it indicates that the data distribution is in a steep peak state and has a heavier tail than the normal distribution; on the contrary, when the kurtosis value is less than 0, it indicates that the distribution has a lighter tail and belongs to the flat peak state. The absolute value of kurtosis indicates how close the distribution is to the normal distribution [27–29].

3. Results and Discussion

3.1. Simulation Analysis Results

3.1.1. Simulation Results of Mechanical Properties

Figure 4a shows the simulation results of the equivalent elastic modulus after optimization with different radii of the middle-plane fillets. With an increase in the radius, the elastic modulus gradually grew from 5.83 GPa to 6.57 GPa, about 14.66% higher than the original modulus (5.73 GPa). The maximum stress, coefficient of variation, skewness, and kurtosis change curves after optimization are shown in Figure 5. The maximum stress gradually decreased with an increase in the radius of middle-plane fillets, from 4338.80 MPa to 2662.30 MPa, with a maximum reduction of 50.66% compared with that of the original porous structure (5395.70 MPa). The coefficient of variation of all optimized structures was greater than 0, indicating that the stress distribution was not uniform. The structure with a radius of 0.50 mm showed the smallest coefficient of variation (0.56), indicating a relatively more uniform stress distribution in the structures. As shown in Figure 5c, the skewness value was 0.51 with a 0.50 mm radius of the middle-plane fillets, indicating that the positive skewness of the structure was more obvious at this time. Most of the stress values were small, demonstrating that the stress concentration effect was significantly improved. The kurtosis values of all structures were less than 0, demonstrating that the stress distribution of the optimized structures belonged to the flat peak state, with a minimum value of −1.36 when the radius was 0.10 mm, and a maximum value of −0.48 when the radius equaled 0.50 mm. The absolute value of kurtosis reached its smallest when the optimization radius was 0.50 mm, with a more uniform stress distribution, as shown in Figure 5 Thus, this study selected a midplane fillet radius of 0.50 mm.

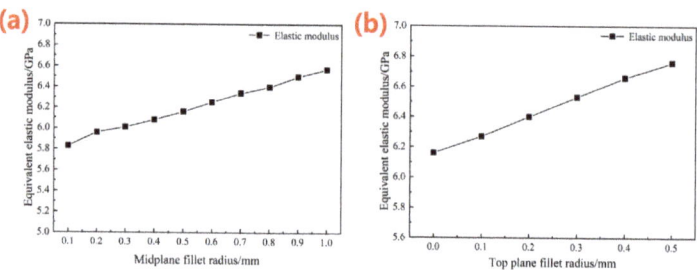

Figure 4. Simulation results of equivalent elastic modulus of different microstructures with an increase in the middle-plane fillet radius (**a**) and top-plane fillet radius (**b**).

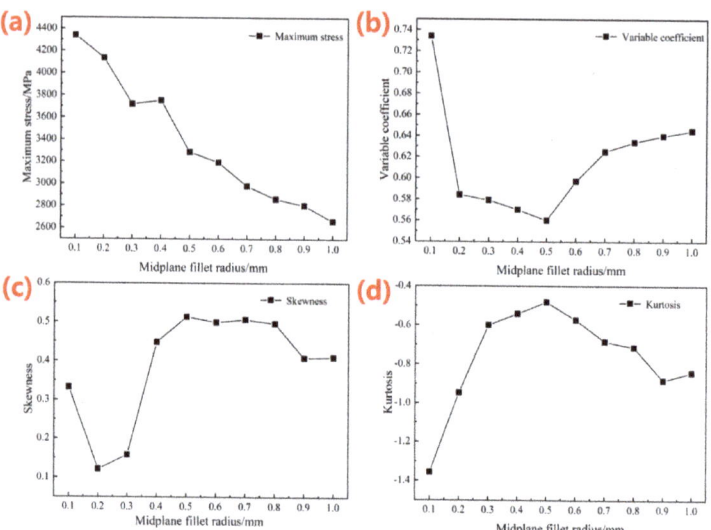

Figure 5. Simulation results of (**a**) maximum stress, (**b**) coefficient, (**c**) skewness change curve, and (**d**) kurtosis during optimization for the radius of middle-plane fillets.

Figure 4b shows the simulation results of the equivalent elastic modulus with radius optimization of the top-plane fillets. The modulus of the optimized structure was further improved by 9.74% with an increase in fillet radius, from 6.16 GPa to 6.76 Gpa. The maximum stress, coefficient of variation, skewness, and kurtosis curves of the structures after the top fillet optimization are shown in Figure 6. All skewness values greater than 0 indicated that the stress distribution after the top fillet optimization was also a positive skewness distribution; all kurtosis values greater than 0 revealed that the stress distribution curve was a steep peak state. And the smaller the absolute value of kurtosis value was, the more concentrated the stress distribution was around the arithmetic mean value. With the increase in the optimized radius, the maximum stress decreased gradually, and a minimum value of 3282.10 MPa was obtained when the radius was 0.5 mm, with a decrease of 13.64%. The coefficient of variation, kurtosis, and skewness obtained the minimum values at a radius of 0.30 mm, which were 0.65, 0.70, and 0.05, respectively. With an optimization radius of 0.3 mm, the small coefficient of variation indicated a more uniform stress distribution, and the minimum skewness and kurtosis values demonstrated a smaller stress distribution range.

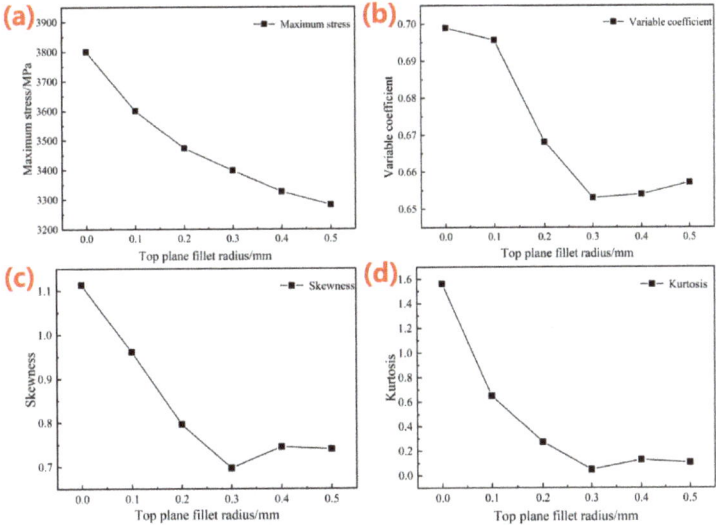

Figure 6. Simulation results of (**a**) maximum stress, (**b**) coefficient, (**c**) skewness change curve, and (**d**) kurtosis with the increase in the radius of top-plane fillets.

Combined with the simulation results of fillet optimization of the midplane and top plane, the equivalent elastic modulus of the 80%-porosity cell structure increased by 13.96% from 5.73 GPa to 6.53 GPa (midplane fillet radius: 0.5 mm; top-plane fillet radius: 0.3 mm). The maximum stress was reduced by 39.17% from 5395.70 MPa to 3282.10 MPa, the stress distribution range was smaller and more uniform, and the stress concentration effect was significantly improved, as shown in Figure 7.

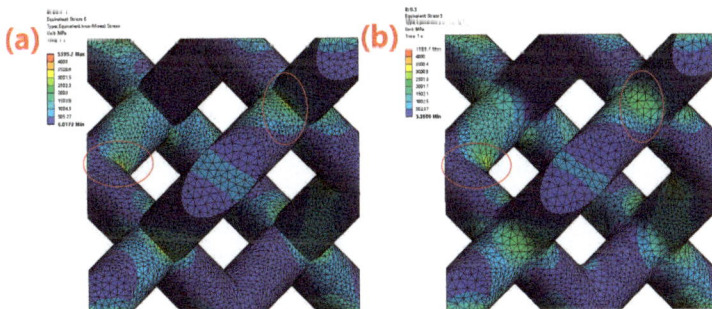

Figure 7. Stress nephogram comparison between (**a**) the original cell structure and (**b**) the optimized cell structure. The red circle areas represent the rounded corner optimization area.

3.1.2. Simulation Results of Permeability Performance

Figure 8a,b show the results of the permeability and the proportion of suitable areas for cell survival with the radius optimization of top-plane fillets. As the radius became larger, the permeability of the porous structures decreased at the beginning, and then, increased. When the radius was 0.3 mm, the maximum permeability was 25.67×10^{-9} m^2, with an increase of 2.81%. Similarly, the proportion of suitable areas for cell survival also reached the maximum value of 93.61% at this point (Figure 8b). Thus, the optimized results of the radius of the top-plane fillets based on permeability were consistent with the results from the mechanical properties, and a radius of 0.30 mm was selected for the top-plane fillet radius after optimization.

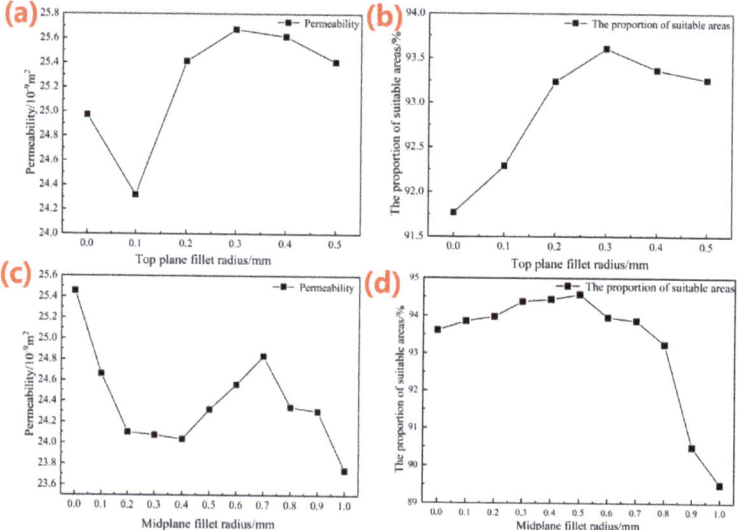

Figure 8. Simulation results of permeability and the proportion of biologically suitable areas with radius increase of top-plane fillets (**a**,**b**) and midplane fillets (**c**,**d**), respectively.

Figure 8c,d shows the permeability and the proportion of suitable areas for cell survival results with midplane fillet radius optimization. The proportion of the biologically suitable areas firstly decreased when the radius was less than 0.7 mm, and then, increased with the radius change. On the other hand, the proportion of suitable areas for cell survival remained stable when the radius was smaller than 0.7 mm, and reduced suddenly with a larger radius (Figure 8d). Therefore, this research chose a midplane fillet radius of 0.50 mm to keep consistence with the mechanical properties analysis.

Figure 9 shows the comparison of the WSS distribution of the single-cell structure before and after fillet radius optimization of the top plane and midplane, and the WSS range was set to 5.0×10^{-5} to 2.5×10^{-2} Pa. It can be seen that the WSS distribution of the optimized structure was more uniform, which is more conducive to cell adhesion. Figure 9b compared the WSS distribution cloud map before and after the optimization of the fillets. It can be seen from the figure that the structure optimization effectively improved the phenomenon of the sudden change at the angle of the struct, and made the WSS distribution more uniform. According to the simulation results of different fillet radius designs, the permeability and proportion of the biologically suitable areas of 80%-porosity porous structures with optimized fillet radii of 0.50 mm on the top plane and 0.30 mm on the midplane are 24.318×10^{-9} m^2 and 94.57%, respectively, which is consistent with the permeability of human bone structure.

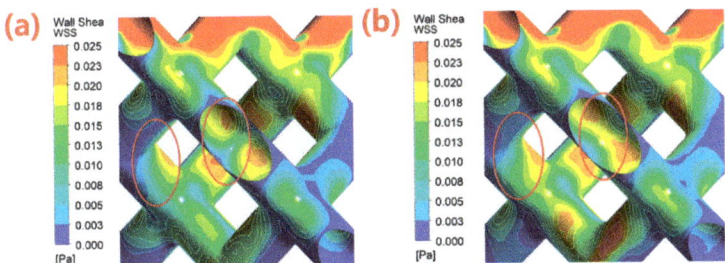

Figure 9. Comparison of WSS distribution cloud image between (**a**) original single cell structure and (**b**) cell structure with optimization of fillet radius (top-plane fillet radius: 0.3 mm, midplane fillet radius: 0.5 mm). The red circle areas represent the rounded corner optimization area.

3.2. Porous Model after Microstructure Optimization for Manufacturing

A suitable porous microstructure design can not only improve the carrying capacity and long-term stability of porous structures and reduce the stress concentration effect inside the structure, but also improve the permeability of porous structures, which is more conducive to the transport of nutrients and cell adhesion in the porous structure after implantation. Based on the results of the microstructure optimization of mechanical properties and permeability properties, a porous structure with fillet radius optimization was established (Figure 10). The radius of fillets of the midplane was 0.50 mm, and the radius of the top surface fillet was 0.30 mm.

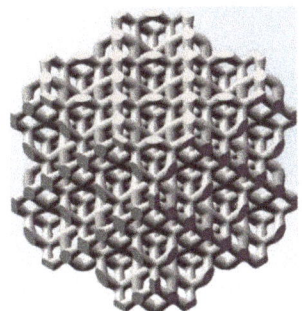

Figure 10. Porous structure with fillet optimization.

3.3. Morphology of Porous Ta with Optimized Microstructure

The prepared porous Ta samples are shown in Figure 11 [25]. Compared with the microstructure of the un-optimized structure, it can be seen that the optimization of the fillets could effectively improve the non-uniformity of the connection transition between the struts (Figure 12), which helped reduce the stress concentration effect inside the structure and improve its mechanical properties. The results showed that the strut diameters of the porous structure after optimization in the horizontal direction and the forming direction were 0.86 ± 0.01 mm and 0.80 ± 0.01 mm, respectively. The difference in the strut diameter before and after the optimization was not significant. Due to the existence of fabrication error, the porosities of the porous structures before and after optimization were still smaller than the design value of 80%.

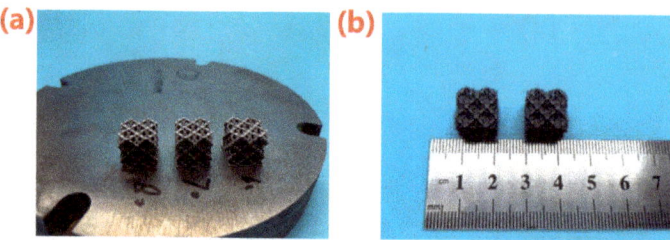

Figure 11. Macroscopic photographs of porous Ta scaffold samples. (**a**) Macroscopic photographs; (**b**) macroscopic geometry size [25].

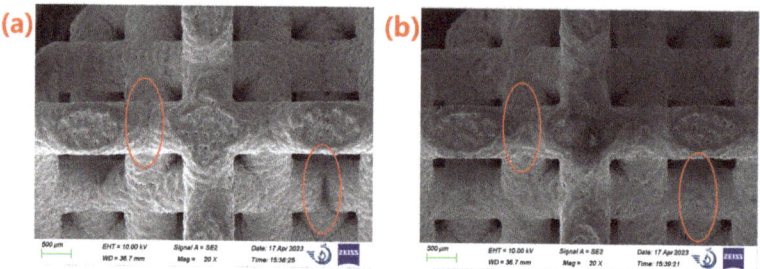

Figure 12. Electron microscopic photos of microstructure of porous Ta. (**a**) The original microstructure; (**b**) the fillet-optimized microstructure. The red circle areas represent the rounded corner optimization area.

3.4. Mechanical Test Results of Porous Ta with Optimized Microstructure

The compressive stress–strain curves of the fillet-optimized structure are shown in Figure 13. The elastic modulus and yield strength of different porous structures were measured, and the results are shown in Table 2. The results showed that the optimized design of a porous structure can effectively improve the yield strength, while the elastic modulus increased slightly after optimization. This might be due to the fact that the optimized design could improve the stress concentration effect during loading with a relatively smooth transition between the struts of porous structures.

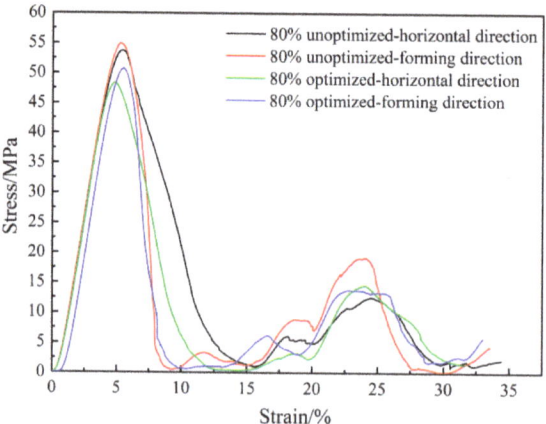

Figure 13. Compressive stress–strain curve results of porous Ta with and without fillet optimization scheme.

Table 2. Test results of mechanical properties between the porous Ta with original and optimized structures.

Structure Type	Elasticity Modulus (GPa)		Yield Strength (MPa)	
	Horizontal Direction	Forming Direction	Horizontal Direction	Forming Direction
Original structure	1.352 ± 0.007	1.298 ± 0.006	53.981 ± 0.124	49.771 ± 0.091
Optimized structure	1.384 ± 0.053	1.328 ± 0.009	59.590 ± 0.053	54.226 ± 0.010

3.5. Permeability Test Results of Porous Ta with Optimized Microstructure

The results demonstrated that the permeability of the 80%-porosity porous structure increased by 8.25% from $6.63 \pm 0.16 \times 10^{-9}$ m^2 to $7.18 \pm 0.18 \times 10^{-9}$ m^2 after optimization. This might be due to the fact that the transition of the struts in the porous structure after the fillet optimization was more uniform, and the permeability was increased by improving the pore distribution inside the structures. Both of the permeability test results and the simulation results show the same change trend, indicating that the established permeability calculation model had good reliability. The test results were smaller than the simulation values. This may be result from two reasons: the strut diameter of the fabricated porous structure was too large, causing the shape of the fluid domain to be different from the simulation model; the manufactured struts had a certain degree of roughness, and some liquid droplets would be adsorbed on the struts when the liquid flowed through the surfaces of the struts, thus affecting the permeability. Therefore, fillet optimization can improve the mechanical properties and permeability of porous structures, which is consistent with the simulation results.

4. Conclusions

This paper created a fillet optimization design with a porous cell structure, obtained the optimal design parameters through the simulation of mechanical and penetration properties, and verified the results experimentally using porous Ta. We determined the following based on the above simulation and experimental analysis results:

(1) The optimized parameters of the porous diamond structure are as follows: the fillet radius of the top plane is 0.50 mm, and the radius of the midplane fillet is 0.30 mm. The equivalent elastic modulus of the 80%-porosity cell structure is increased by 13.96%, the maximum stress is reduced by 39.17%, and the proportion of the suitable area for cell growth is increased by 1.15%.

(2) The elastic modulus in the forming and horizontal directions of the porous Ta before and after fillet optimization are increased from 1.298 ± 0.006 GPa and 1.352 ± 0.007 GPa to 1.328 ± 0.002 GPa and 1.384 ± 0.006 GPa, representing increases of 2.31% and 2.37%, respectively. At the same time, the permeability is increased by 8.25% from $6.63 \pm 0.16 \times 10^{-9}$ m^2 to $7.18 \pm 0.18 \times 10^{-9}$ m^2.

(3) The yield strength of the optimized porous Ta samples prepared by LPBF has a relatively obvious increase of 10.39%. This might be due to the fact that the porous Ta after the fillet optimization of can effectively reduce the stress concentration effect.

Compared with previous studies on the structural design optimization of porous scaffolds [13,14,16,18–20], the microstructure optimization of porous Ta in this study can maximize the mechanical properties and permeability based on the excellent mechanical and biological properties of porous Ta. Although this fillet optimization scheme can also be applied to porous Ti scaffolds, the porous Ta scaffold could demonstrate superior performance with better innate properties as bioimplants. Furthermore, the optimization method in this study combined with the mechanical properties of porous Ta can enable the preparation of porous structures with higher porosity and permeability that meet the mechanical requirements of bioimplants, which are incomparable to porous Ti scaffolds. This has important significance for the research and application of porous Ta scaffolds with high porosity.

Author Contributions: Conceptualization, L.G.; Methodology, Y.W. and X.Q.; Software, Y.W.; Formal analysis, X.Q. and C.S.; Investigation, X.Q. and L.G.; Resources, Z.T.; Data curation, N.L.; Writing—original draft, Y.W. and N.L.; Writing—review & editing, L.G. and C.S.; Visualization, N.L., L.G. and Z.T.; Supervision, C.S. and D.L.; Project administration, Z.T. and D.L.; Funding acquisition, D.L. All authors have read and agreed to the published version of the manuscript.

Funding: This work was supported in part by the Key R&D Program of Ningxia Province (grant number 2020BCH01001), National Natural Science Foundation of China (grant number 52175276), and Program for Innovation Team of Shaanxi Province (2023-CX-TD-17).

Institutional Review Board Statement: This study did not involve human or animal studies.

Informed Consent Statement: This study did not involve humans.

Data Availability Statement: The data presented in this study are available on request from the corresponding author.

Conflicts of Interest: The authors declare no conflict of interest.

References

1. Wauthle, R.; van der Stok, J.; Yavari, S.A.; Van Humbeeck, J.; Kruth, J.-P.; Zadpoor, A.A.; Weinans, H.; Mulier, M.; Schrooten, J. Additively manufactured porous tantalum implants. *Acta Biomater.* **2015**, *14*, 217–225. [CrossRef] [PubMed]
2. Gao, H.; Jin, X.; Yang, J.; Zhang, D.; Zhang, S.; Zhang, F.; Chen, H. Porous structure and compressive failure mechanism of additively manufactured cubic-lattice tantalum scaffolds. *Mater. Today Adv.* **2021**, *12*, 100183. [CrossRef]
3. Dong, C.; Bi, X.; Yu, J.; Liu, R.; Zhang, Q. Microstructural evolution and sintering kinetics during spark plasma sintering of pure tantalum powder. *J. Alloys Compd.* **2019**, *781*, 84–92. [CrossRef]
4. Balla, V.K.; Bodhak, S.; Bose, S.; Bandyopadhyay, A. Porous tantalum structures for bone implants: Fabrication, mechanical and in vitro biological properties. *Acta Biomater.* **2010**, *6*, 3349–3359. [CrossRef] [PubMed]
5. Bobyn, J.D.; Stackpool, G.J.; Hacking, S.A.; Tanzer, M.; Krygier, J.J. Characteristics of bone ingrowth and interface mechanics of a new porous tantalum biomaterial. *J. Bone Jt. Surg. Br. Vol.* **1999**, *81*, 907–914. [CrossRef]
6. Gruen, T.A.; Poggie, R.A.; Lewallen, D.G.; Hanssen, A.D.; Lewis, R.J.; O'Keefe, T.J.; Stulberg, S.D.; Sutherland, C.J. Radiographic evaluation of a monoblock acetabular component: A multicenter study with 2-to 5-year results. *J. Arthroplast.* **2005**, *20*, 369–378. [CrossRef] [PubMed]
7. Moore, R.E.; Austin, M.S. Use of porous tantalum cones in revision total knee arthroplasty. *Oper. Tech. Orthop.* **2012**, *22*, 209–221. [CrossRef]
8. Levine, B.R.; Sporer, S.; Poggie, R.A.; Della Valle, C.J.; Jacobs, J.J. Experimental and clinical performance of porous tantalum in orthopedic surgery. *Biomaterials* **2006**, *27*, 4671–4681. [CrossRef]
9. Nasser, S.; Poggie, R.A. Revision and salvage patellar arthroplasty using a porous tantalum implant. *J. Arthroplast.* **2004**, *19*, 562–572. [CrossRef]
10. Zhao, D.; Li, J. Preparation of porous Ta and its application as bone implant materials. *Acta Met. Sin.* **2017**, *53*, 1303–1310. (In Chinese)
11. Asri, R.I.M.; Harun, W.S.W.; Samykano, M.; Lah, N.A.C.; Ghani, S.A.C.; Tarlochan, F.; Raza, M.R. Corrosion and surface modification on biocompatible metals: A review. *Mater. Sci. Eng. C* **2017**, *77*, 1261–1274. [CrossRef] [PubMed]
12. Gao, H.; Yang, J.; Jin, X.; Qu, X.; Zhang, F.; Zhang, D.; Chen, H.; Wei, H.; Zhang, S.; Jia, W.; et al. Porous tantalum scaffolds: Fabrication, structure, properties, and orthopedic applications. *Mater. Des.* **2021**, *210*, 110095. [CrossRef]
13. Huang, G.; Pan, S.; Qiu, J. The osteogenic effects of porous Tantalum and Titanium alloy scaffolds with different unit cell structure. *Colloids Surf. B Biointerfaces* **2022**, *210*, 112229. [CrossRef] [PubMed]
14. Gao, R.; Xiong, Y.; Zhang, H.; Dong, L.; Li, J.; Li, X. Mechanical properties and biocompatibility of radial gradient porous titanium/tantalum prepared by SLM. *Rare Met. Mater. Eng.* **2021**, *50*, 249–254. (In Chinese)
15. Wang, X.; Zhu, Z.; Xiao, H.; Luo, C.; Luo, X.; Lv, F.; Liao, J.; Huang, W. Three-Dimensional, MultiScale, and Interconnected Trabecular Bone Mimic Porous Tantalum Scaffold for Bone Tissue Engineering. *ACS Omega* **2020**, *5*, 22520–22528. [CrossRef] [PubMed]
16. Yang, J.; Jin, X.; Gao, H.; Zhang, D.; Chen, H.; Zhang, S.; Li, X. Additive manufacturing of trabecular tantalum scaffolds by laser powder bed fusion: Mechanical property evaluation and porous structure characterization. *Mater. Charact.* **2020**, *170*, 110694. [CrossRef]
17. Pligovka, A. Reflectant photonic crystals produced via porous-alumina-assisted-anodizing of Al/Nb and Al/Ta systems. *Surf. Rev. Lett.* **2021**, *28*, 2150055. [CrossRef]
18. Chen, Z.; Yan, X.; Yin, S.; Liu, L.; Liu, X.; Zhao, G.; Ma, W.; Qi, W.; Ren, Z.; Liao, H.; et al. Influence of the pore size and porosity of selective laser melted Ti6Al4V ELI porous scaffold on cell proliferation, osteogenesis and bone ingrowth. *Mater. Sci. Eng. C* **2020**, *106*, 110289. [CrossRef]

19. Tsuruga, E.; Takita, H.; Itoh, H.; Wakisaka, Y.; Kuboki, Y. Pore size of porous hydroxyapatite as the cell-substratum controls BMP-induced osteogenesis. *J. Biochem.* **1997**, *121*, 317–324. [CrossRef]
20. Lv, J.; Jia, Z.; Li, J.; Wang, Y.; Yang, J.; Xiu, P.; Zhang, K.; Cai, H.; Liu, Z. Electron Beam Melting Fabrication of Porous Ti6Al4V Scaffolds: Cytocompatibility and Osteogenesis. *Adv. Eng. Mater.* **2015**, *17*, 1391–1398. [CrossRef]
21. Gao, X.; Zhao, Y.; Wang, M.; Liu, Z.; Liu, C. Parametric design of hip implant with gradient porous structure. *Front. Bioeng. Biotechnol.* **2022**, *10*, 850184. [CrossRef] [PubMed]
22. Chao, L.; Jiao, C.; Liang, H.; Xie, D.; Shen, L.; Liu, Z. Analysis of mechanical properties and permeability of trabecular-like porous scaffold by additive manufacturing. *Front. Bioeng. Biotechnol.* **2021**, *9*, 779854. [CrossRef] [PubMed]
23. Urschel, K.; Tauchi, M.; Achenbach, S.; Dietel, B. Investigation of wall shear stress in cardiovascular research and in clinical practice—from bench to bedside. *Int. J. Mol. Sci.* **2021**, *22*, 5635. [CrossRef] [PubMed]
24. Li, J.P.; Habibovic, P.; van den Doel, M.; Wilson, C.E.; de Wijn, J.R.; van Blitterswijk, C.A.; de Groot, K. Bone ingrowth in porous titanium implants produced by 3D fiber deposition. *Biomaterials* **2007**, *28*, 2810–2820. [CrossRef] [PubMed]
25. Gao, L.; Wang, Y.; Qin, X.; Lv, N.; Tong, Z.; Sun, C.; Li, D. Optimization of Laser Powder Bed Fusion Process for Forming Porous Ta Scaffold. *Metals* **2023**, *13*, 1764. [CrossRef]
26. *ISO 13314:2011*; Mechanical testing of metals—Ductility testing—Compression test for porous and cellular metals. ISO: Geneva, Switzerland, 2011.
27. Cianetti, F.; Palmieri, M.; Braccesi, C.; Morettini, G. Correction formula approach to evaluate fatigue damage induced by non-Gaussian stress state. *Procedia Struct. Integr.* **2018**, *8*, 390–398. [CrossRef]
28. Yan, X.L.; Wang, X.L.; Zhang, Y.Y. Influence of roughness parameters skewness and kurtosis on fatigue life under mixed elastohydrodynamic lubrication point contacts. *J. Tribol.* **2014**, *136*, 031503. [CrossRef]
29. Niesłony, A.; Böhm, M.; Owsiński, R. Crest factor and kurtosis parameter under vibrational random loading. *Int. J. Fatigue* **2021**, *147*, 106179. [CrossRef]

Disclaimer/Publisher's Note: The statements, opinions and data contained in all publications are solely those of the individual author(s) and contributor(s) and not of MDPI and/or the editor(s). MDPI and/or the editor(s) disclaim responsibility for any injury to people or property resulting from any ideas, methods, instructions or products referred to in the content.

Article

Evaluation of Austenitic Stainless Steel ER308 Coating on H13 Tool Steel by Robotic GMAW Process

Jorge Eduardo Hernandez-Flores [1,2], Bryan Ramiro Rodriguez-Vargas [3], Giulia Stornelli [3], Argelia Fabiola Miranda Pérez [4], Felipe de Jesús García-Vázquez [1], Josué Gómez-Casas [1] and Andrea Di Schino [3,*]

1. Faculty of Engineering, Universidad Autonoma de Coahuila, Ciudad Universitaria, Arteaga 25350, Mexico; flores_jorge@uadec.edu.mx (J.E.H.-F.); felipegarcia@uadec.edu.mx (F.d.J.G.-V.); jogomezc@uadec.edu.mx (J.G.-C.)
2. Corporacion Mexicana de Investigación en Materiales S.A., Ciencia y Tecnología 790, Saltillo 400, Saltillo 25290, Mexico
3. Dipartimento di Ingegneria, Università degli Studi di Perugia, Via G. Duranti 93, 06125 Perugia, Italy; bryanramiro.rodriguezvargas@studenti.unipg.it (B.R.R.-V.); giulia.stornelli@unipg.it (G.S.)
4. Engineering Department, Mechatronics, Bionics and Aerospace, Universidad Popular Autonoma del Estado de Puebla, 17 Sur, 901, Barrio de Santiago, Puebla 72800, Mexico; argeliafabiola.miranda@upaep.mx
* Correspondence: andrea.dischino@unipg.it

Abstract: Within the drilling, petrochemical, construction, and related industries, coatings are used to recover components that failed during service or to prevent potential failures. Due to high stresses, such as wear and corrosion, which the materials are subjected to, industries require the application of coating between dissimilar materials, such as carbon steels and stainless steels, through arc welding processes. In this work, an austenitic stainless steel (ER308) coating was applied to an H13 tool steel substrate using the gas metal arc welding (GMAW) robotic process. The heat input during the process was calculated to establish a relationship between the geometry obtained in the coating and its dilution percentage. Furthermore, the evolution of the microstructure of the coating, interface, and substrate was evaluated using XRD and SEM techniques. Notably, the presence of martensite at the interface was observed. The mechanical behavior of the welded assembly was analyzed through Vickers microhardness, and a pin-on-disk wear test was employed to assess its wear resistance. It was found that the dilution percentage is around 18% at high heat input (0.813 kJ/mm) but decreases to about 14% with reduced heat input. Microhardness tests revealed that at the interface, the maximum value is reached at about 625 HV due to the presence of quenched martensite. Moreover, increasing the heat input favors wear resistance.

Keywords: coating; robotic GMAW; H13 tool steel; 308 stainless steel

Citation: Hernandez-Flores, J.E.; Rodriguez-Vargas, B.R.; Stornelli, G.; Pérez, A.F.M.; García-Vázquez, F.d.J.; Gómez-Casas, J.; Di Schino, A. Evaluation of Austenitic Stainless Steel ER308 Coating on H13 Tool Steel by Robotic GMAW Process. *Metals* **2024**, *14*, 43. https://doi.org/10.3390/met14010043

Academic Editor: Hong Shen

Received: 22 November 2023
Revised: 19 December 2023
Accepted: 26 December 2023
Published: 29 December 2023

Copyright: © 2023 by the authors. Licensee MDPI, Basel, Switzerland. This article is an open access article distributed under the terms and conditions of the Creative Commons Attribution (CC BY) license (https://creativecommons.org/licenses/by/4.0/).

1. Introduction

The use of coatings as a surface modification technique plays an essential role in the restoration of components that have experienced failures during their time in production and in the prevention of potential future failures. This is particularly crucial in applications where these components are subjected to aggressive conditions, such as mechanical stresses that may weaken the properties of the original material, as well as exposure to corrosive environments [1–7].

Surface modification techniques exploit filler metals, either in powder form or in solid state, with chemical compositions dependent on the substrate. This enables the adaptation of components to high corrosive environments and high mechanical stresses [8]. An example of this is the use of filler materials with a high nickel content, such as Inconel, to enhance high-temperature resistance [2,9,10]. In applications exposed to aggressive environments where corrosion resistance is critical, stainless steel or high chromium alloy filler materials are employed. Austenitic stainless steels are the most common due to their high corrosion resistance and relatively lower cost compared to other stainless alloys [11–14].

The use of coatings in dissimilar joints between carbon steels and austenitic stainless steels offers economic benefits in various applications where materials with high corrosion resistance are needed. This approach eliminates the use of expensive all-austenitic stainless steel plates and replaces them with low-cost coated components, resulting in significant cost-effectiveness [15]. However, it is essential to consider several factors, such as the process to use, weldability of the parts, their chemical composition, and the microstructural development that occurs, since this combination of features significantly influences the corrosion resistance and mechanical properties of the resulting joint [16]. Previous research has emphasized the importance of analyzing the interface between the substrate and stainless steel coatings. This is because the thermal gradients generated in the joint, due to the heat input, promote the diffusion and loss of some alloying elements from the weld pool to the substrate, which can affect the material properties and lead to susceptibility to corrosion [17,18].

Various surface modification processes are used, depending on the specific end application. These processes may include welding arc technologies such as gas metal arc welding (GMAW) [19], gas tungsten arc welding (GTAW) [20], cold metal transfer (CMT) [21,22], plasma transferred arc (PTA) [22,23], and high-energy processes such as laser technologies [24–26] or additive manufacturing (AM) [27,28].

The GMAW process is one of the most widely used processes in the industrial sector due to its low cost and ease of automation. However, a critical parameter to determine is the metal transfer mode, since it influences the deposition rate of the filler material, the geometry of the weld beads, and consequently, the dilution percentages [17,29].

Spray mode transfer is generally used in high-thickness materials, as it involves continuous fusion of the filler metal and allows for rapid deposition on the substrate using high voltage and current parameters. However, this metal transfer tends to generate a challenging-to-control weld pool, resulting in high dilution and significant heat input [30,31]. In contrast, the short-circuit mode transfer (GMAW-S) relies on an arc that briefly extinguishes, and that is achieved by using lower welding voltages and current, combined with alternating current. This reduces heat input and ensures excellent thermal stability, enabling the production of high-quality coatings, even on thinner substrates [32,33].

The combination of unique features resulting from surface modification through welding processes in dissimilar alloys is of significant interest for tool steel, such as H13 steel, which is used in the manufacturing of extrusion dies, forming applications, forging, and die-casting of aluminum alloys [34,35]. For such types of applications, improving the corrosion and erosion resistance of H13 steel is imperative, as its use in high-temperature processes with aggressive lubricants/coolants leads to rapid degradation of the material [36,37]. In aluminum extrusion dies, the combination of high temperature and the aggressiveness of molten aluminum in terms of corrosion can lead to pitting and the formation of intermetallic layers. This shortens the lifespan of these components, necessitating replacement and sometimes the substitution of H13 steel with more expensive materials [38–41]. In this context, the main novelty of the research reported in this paper is in defining different material selection strategies. In particular, the use of anticorrosive coatings with alloys exhibiting high corrosion resistance, such as austenitic stainless steels, can be exploited to improve the durability and performance of these components [42].

The present research evaluates the feasibility of the deposition of an ER308 austenitic stainless steel coating on an H13 tool-grade steel substrate through the variation of key operating parameters such as welding current, which allow for evaluation of the influence of heat input on the dilution percentage, microstructural development, hardness, and wear, using the robotic system GMAW-S welding process. We tested a limited number of process parameters, which allowed us to maintain precise control and greater reproducibility across experimental conditions.

2. Materials and Methods

In this work, an H13 tool steel plate with a thickness of 10 mm and ER308 with a high nickel content were used, respectively, as the substrate and filler metal. The initial state of H13 steel is a tempered martensite, following austenitization heat treatment at 1010 °C for 30 min, cooling in air, and tempering at 650 °C for 2 h. The chemical compositions and mechanical properties of these materials are detailed in Tables 1 and 2.

Table 1. Chemical composition of substrate and filler metal (wt. %).

Material	Chemical Composition									
	C	Cr	Cu	Mn	Mo	Ni	Si	V	Al	Fe
H13	0.45	4.95	0.071	0.39	1.26	0.16	0.93	0.47	0.05	Bal.
ER308	0.11	16.80	0.14	0.48	0.23	11.49	0.59	0.17	0.07	Bal.

Table 2. Mechanical properties of H13 tool steel and ER308 filler metal.

Material	Mechanical Properties				
	Hardness (HRC)	Tensile Strength Ultimate (MPa)	Tensile Strength Yield (MPa)	Modulus of Elasticity (GPa)	Elongation after Fracture (%)
H13	25–28	1100	820	215	9.0
ER308	-	>600	-	-	>30.0

Bead-on-weld deposits were made on the H13 steel plate using a GMAW robotic process with short-circuit mode transfer. A KUKA KR16-2 robot (Kuka, Augsburg, Germany) was used, connected to a Lincoln POWERWAVE 455 m power source (Lincoln Electric, Cleveland, OH, USA) with an 80%Ar—20%CO_2 (10 L/min) mixture as the shielding atmosphere. The H13 steel plate was preheated to 210 °C to prevent residual stresses and reduce the risk of cracks in the heat-affected zone. The process parameters used are displayed in Table 3.

Table 3. Parameters used for coating deposition.

Designation	Welding Parameters				
	Current (A)	Voltage (V)	Welding Speed (mm/s)	Wire Feed Speed (mm/s)	Heat Input (kJ/mm)
C1	254.3	20.0	5	5	0.813
C2	252.0	19.8			0.798
C3	229.5	19.6			0.719
C4	208.3	19.4			0.646

Electrode extension: 10 mm
Arc length: 1 mm

2.1. Heat Input Calculation

The heat input (HI) was calculated based on Equation (1). According to the information provided by the welding process, the efficiency (η) is assumed to be 80%.

$$HI = \frac{V * I}{S} * \eta \quad (1)$$

where HI is the heat input in kJ/mm, V is the welding voltage in V, I is the welding current in A, and S is the welding speed in mm/s.

2.2. Dilution Percent Calculation

The dilution percentage of each coating was calculated using Equation (2) in accordance with Figure 1.

$$\text{Dilution}(\%) = \frac{A_m}{A_c + A_m} * 100 \tag{2}$$

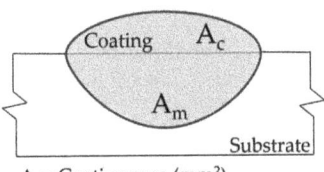

A_c = Coating area (mm^2)
A_m = Molten area (mm^2)

Figure 1. Scheme of the cross-section of a coating (dilution percentage measurement).

It is important to highlight that this parameter is of interest since it strongly depends on the process parameters and its values tend to demonstrate the union and adhesion of the filler metal on the substrate [43,44]. Generally, in coating or hard banding, dilution plays an outstanding role in the economic part, since values below 10% increase the lack of adhesion integrity, while values above 20% can increase the cost of filler metal [45,46].

2.3. Macro- and Microstructural Characterization

The preparation of both the substrate and the coating was metallographic prepared in accordance with standardized and conventional procedures. To reveal the microstructure of the substrate, a Nital 5% solution was used for 5 s. Visualizing the coating microstructure involved an electrolytic etching process using oxalic acid (10 g $C_2H_2O_4$ + 100 mL H_2O) at 6 V for 60 s.

Macrostructural evaluation was conducted using a Nikon SMZ 745T (Nikon Corp., Tokyo, Japan) stereoscope, and to observe the microstructural development, a Nikon Eclipse MA200 optical microscope and a Tescan MIRA3 (Tescan Analytics, Brno, Czech Republic) scanning electron microscope (SEM) were used. The SEM was equipped with an energy-dispersive X-ray spectroscopy (EDS) detector to perform a semi-quantitative analysis of the chemical composition in different regions of the coating. Additionally, the phases in the coating, interface, and substrate were analyzed using X-ray diffraction Phillips XPert 3040 (Philips, Amsterdam, Netherlands with the following parameters: anode excitation voltage of 45 kV, 30 mA current, scanning angle from 35° to 100° (2θ), scanning speed of 0.02° (2θ)/s, and Cu Kα monochromatic radiation.

2.4. Microhardness

Vickers microhardness evaluation was conducted in accordance with ASTM E384 using a Wilson Hardness Tukon 2500 (Buehler, Lake Bluff, IL, USA) microhardness tester with a 500 g$_F$ load. This test was performed both longitudinally and transversely on the coating. In the longitudinal test, microhardness was assessed along all the deposited beads. Meanwhile, the transverse test examined microhardness through the coating, interface, and substrate.

2.5. Pin-on-Disk Test

Dry sliding wear tests were conducted on the substrate and coating using a 100Cr6 pin with a diameter of 6 mm and ~63 HRC (hardness of the pin greater than the hardness of the samples [47]). The test conditions to which the samples were subjected are shown in Table 4.

Table 4. Parameters used in the pin-on-disk test.

Parameters	Value
Normal force	3 N
Rotating speed	10 cm/s
Test radius	4.5 mm
Sliding distance	170 m
Sphere radius	3 mm
Environment	Air
Temperature nominal	28 °C
Specimen dimensions	1 mm × 1 mm

Subsequently, the width of the wear tracks was measured using SEM and the volume loss was calculated using Equation (3), with reference to standard ASTM G99 [48].

$$V = \frac{\pi \cdot R \cdot d^3}{6r} \quad (3)$$

where V is the volume loss (mm^3), R is the radius of the wear track on the sample (mm), r is the radius of the pin (mm), and d is the average wear track diameter (mm).

Equation (4) was used to calculate the wear rate:

$$k = \frac{V}{F_N \cdot L} \quad (4)$$

where k represents the specific wear rate (mm^3/N × m), V is the volume loss (mm^3), F_N is the applied normal force (N), and L is the sliding distance (m).

3. Results and Discussion

3.1. Macrostructure of Welds and Dilution Percentage

The macrostructural evaluation (Figure 2) confirms the absence of pores and cracks in all samples and demonstrates proper fusion between the coatings and the substrate.

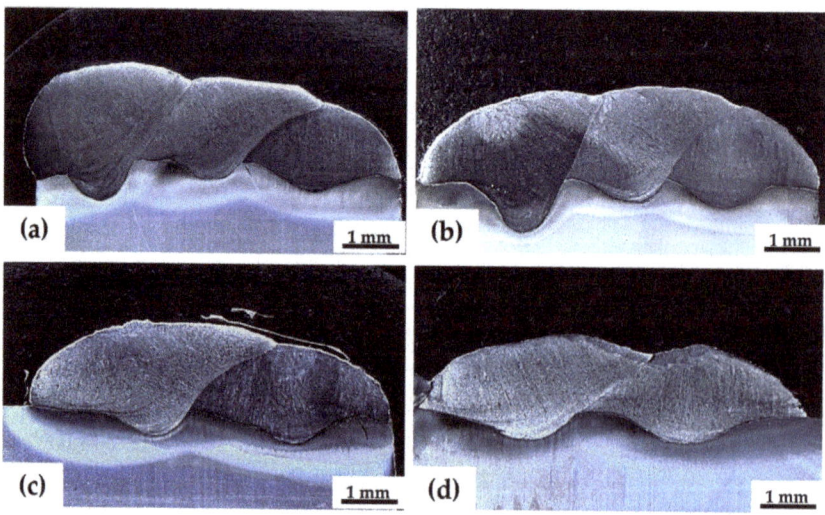

Figure 2. Macrograph of the coatings obtained with (**a**) 0.81 kJ/mm, (**b**) 0.79 kJ/mm, (**c**) 0.71 kJ/mm, and (**d**) 0.64 kJ/mm.

Figure 3a illustrates the relationship between heat input and its influence on the maximum and minimum thickness (Figure 3b) of each coating. It is observed that when heat input reaches a higher value, 0.81 kJ/mm, a significant increase in coating thickness occurs, with a maximum value of 7.49 mm and a minimum value of 4.91 mm. As the heat input decreases, the coating thickness decreases until it reaches a minimum value of 3.83 mm for the heat input of 0.64 kJ/mm. This behavior is related to the higher temperature reached between coating and substrate due to a greater heat input, which promotes substrate melting and a greater coating thickness [49]. Moreover, when significantly more thermal energy is used during the welding process, the resulting cooling rate is slower [50–52] and, consequently, the percentage dilution increases (Figure 4), reaching values of 18.3% and 18.1% at 0.81 kJ/mm and 0.79 kJ/mm, respectively. In contrast, at low energy input (0.65 kJ/mm), the dilution percentage decreases to 14.37%. Figure 2a,b shows how a higher heat input allows for non-uniform dilution throughout the coating as the cooling rate increases [53,54]. These results help understand how the process parameters used influence coating thickness and provide a base for effectively tuning these parameters.

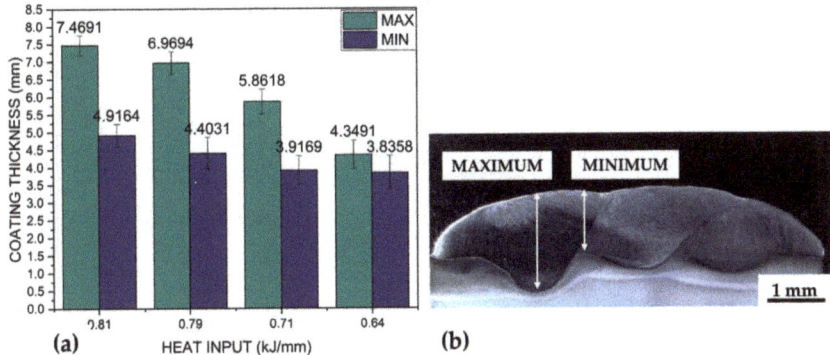

Figure 3. (**a**) Coating thickness as a function of the heat input. (**b**) Measurement of maximum and minimum coating thickness.

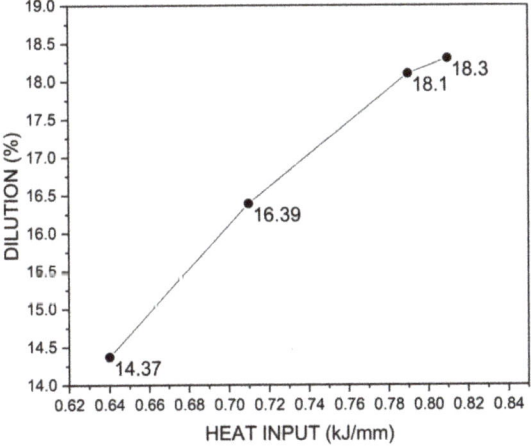

Figure 4. Influence of the heat input on the dilution percentage.

3.2. Microstructure and XRD Measurements

Figure 5 shows the microstructural evolution of each coating, analyzed in different zones (top and middle parts of the coating and the substrate interface). It is observed that, regardless of the heat input, all samples have a similar microstructure. The top zone consists of a large number of columnar dendrites (Figure 5a–d), while towards the middle part of the coating, a columnar growth is evident (Figure 5e–h). This microstructural behavior is attributed to the thermal gradient effect, which influences the cooling rate of each coating. When the filler material is deposited on the substrate, it forms a weld pool, which, as the torch advances, starts various solidification processes. In the internal zone of the weld pool, where the highest temperature is reached, the low cooling rate allows adequate diffusion of chemical elements and favors the columnar austenitic grain formation that grows in the material deposition direction. Instead, at the top of the coating, different solidification processes are observed due to the greater cooling rate reached when the part comes in contact with the atmosphere. This prevents the correct diffusion of chemical elements, thus hindering the formation of austenitic grains and promoting dendritic growth [55–57].

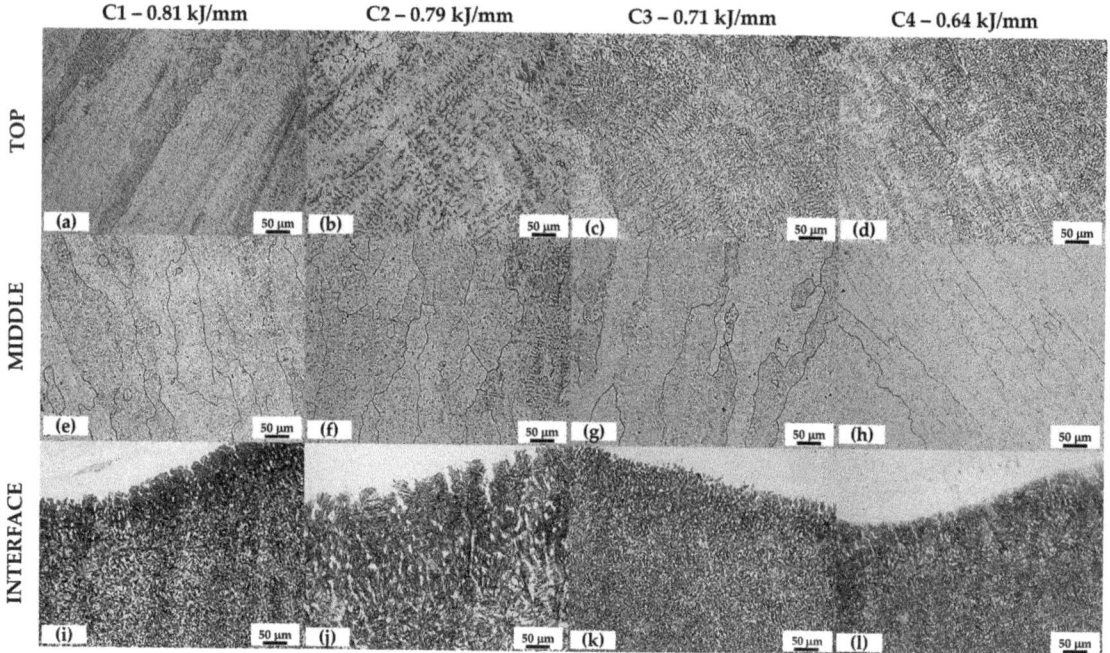

Figure 5. Evolution of microstructure across coating for each set of process parameters: (**a**) C1—0.81 kJ/mm top zone (columnar dendrites), (**b**) C2—0.79 kJ/mm top zone (columnar dendrites), (**c**) C3—0.71 kJ/mm top zone (columnar dendrites), (**d**) C4—0.64 kJ/mm top zone (columnar dendrites), (**e**) C1—0.81 kJ/mm middle zone (columnar grains), (**f**) C2—0.79 kJ/mm middle zone (columnar grains), (**g**) C3—0.71 kJ/mm middle zone (columnar grains), (**h**) C4—0.64 kJ/mm middle zone (columnar grains), (**i**) C1—0.81 kJ/mm interface zone, (**j**) C2—0.79 kJ/mm interface zone, (**k**) C3—0.71 kJ/mm interface zone, and (**l**) C4—0.64 kJ/mm interface zone.

Furthermore, as shown by the XRD results for sample C4 (0.64 kJ/mm) in Figure 6, the microstructures reported in Figure 5 result in a solidification mode of the coating in the form of ferrite-austenite, in agreement with Creq and Nieq [46,58]. The XRD analysis reveals, for the various zones (coating, interface, and substrate), the presence of intense peaks at angles of 44° and 84° identified as peaks corresponding to a body-centered cubic (BCC) crystalline structure. Moreover, in the spectra relating to coating and interface, peaks with more intense diffraction angles at 43° and 75° can be observed, identifying the presence of a face-centered cubic (FCC) phase. Therefore, the combination of both ferrite and austenite phases are present in the areas of the coating and at the interface with the substrate. The presence of ferrite in an austenitic steel coating is due to the high cooling rates to which the materials are subjected. The rapid and inhomogeneous solidification prevents δ-ferrite from completely transforming into austenite, causing it to remain a residual ferrite at room temperature [46,59,60].

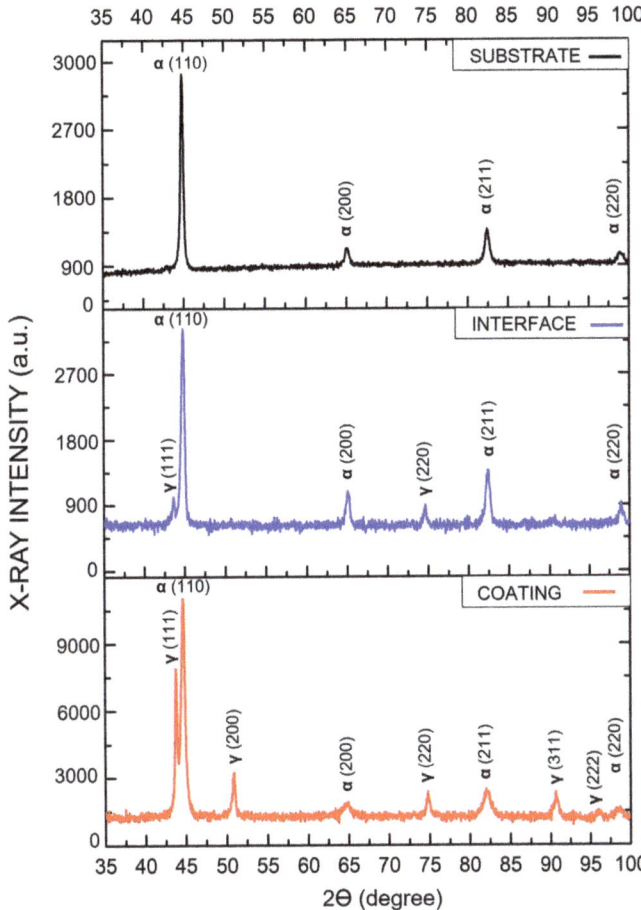

Figure 6. XRD patterns collected from the coating, interface, and substrate of sample C4 (0.64 kJ/mm).

An important aspect of microstructural analysis is the morphology and quantity of residual ferrite in the coating. At the top of each coating (Figure 5a–d), a lathy ferrite is observed (Figure 7), and as the heat input decreases, the amount of ferrite tends to increase. Previous research has shown that if the cooling rate is slow, the growth of austenite is favored by the diffusion phenomenon, and the predominant residual ferrite presents a vermicular morphology. On the other hand, if the cooling rates are high, diffusion is limited and the formation of lathy ferrite is promoted. Furthermore, it has been demonstrated that the amount of ferrite formed during solidification also depends on the cooling rate, as well as on the heat input, chemical composition, and heat treatment methods. Therefore, with higher cooling rates, and especially in the areas in contact with the ambient temperature, the ferrite content appears to be higher [61,62].

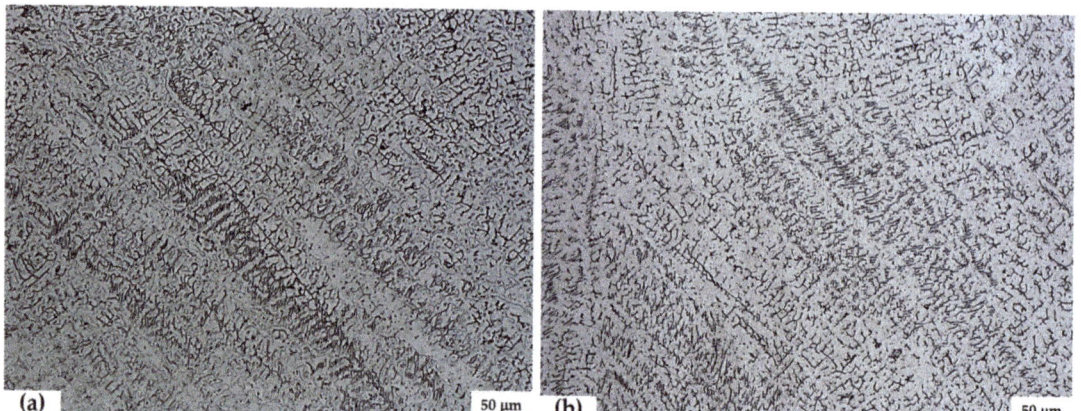

Figure 7. Columnar dendritic microstructure (lathy ferrite) on top of the coating: (**a**) 0.79 kJ/mm, (**b**) 0.64 kJ/mm.

As mentioned above, the interface in all coatings exhibits a similar microstructure. However, for a more detailed analysis, Figure 8a shows the microstructure at the interface for coating C1, highlighting the presence of martensite and austenite. Prior research has found that high Ni and Mo content stabilizes austenite during solidification, leading to the formation of austenite [63]. Martensite formation at the interface zone is attributed to factors such as the cooling rate during solidification of the weld pool [64–66]. Additionally, the high number of microalloying elements in the chemical composition of H13 steel and the presence of tempered martensite in the substrate (see Figure 8b) lead to the formation of untempered martensite during the rapid cooling of the welding process [67,68].

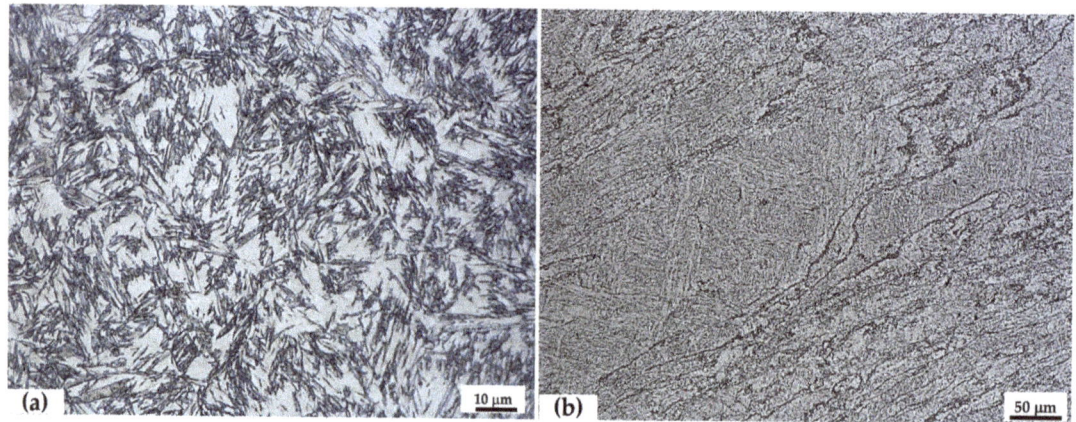

Figure 8. Microstructure by OM of (**a**) the interface of sample C1 (0.81 kJ/mm) and (**b**) substrate H13.

3.3. Microhardness Test

Figure 9 depicts the microhardness profiles of the coatings for different heat input conditions. Two profiles run transversely from the top of the coating to the substrate, and one horizontally across the various weld beads deposited on the substrate (Figure 9e).

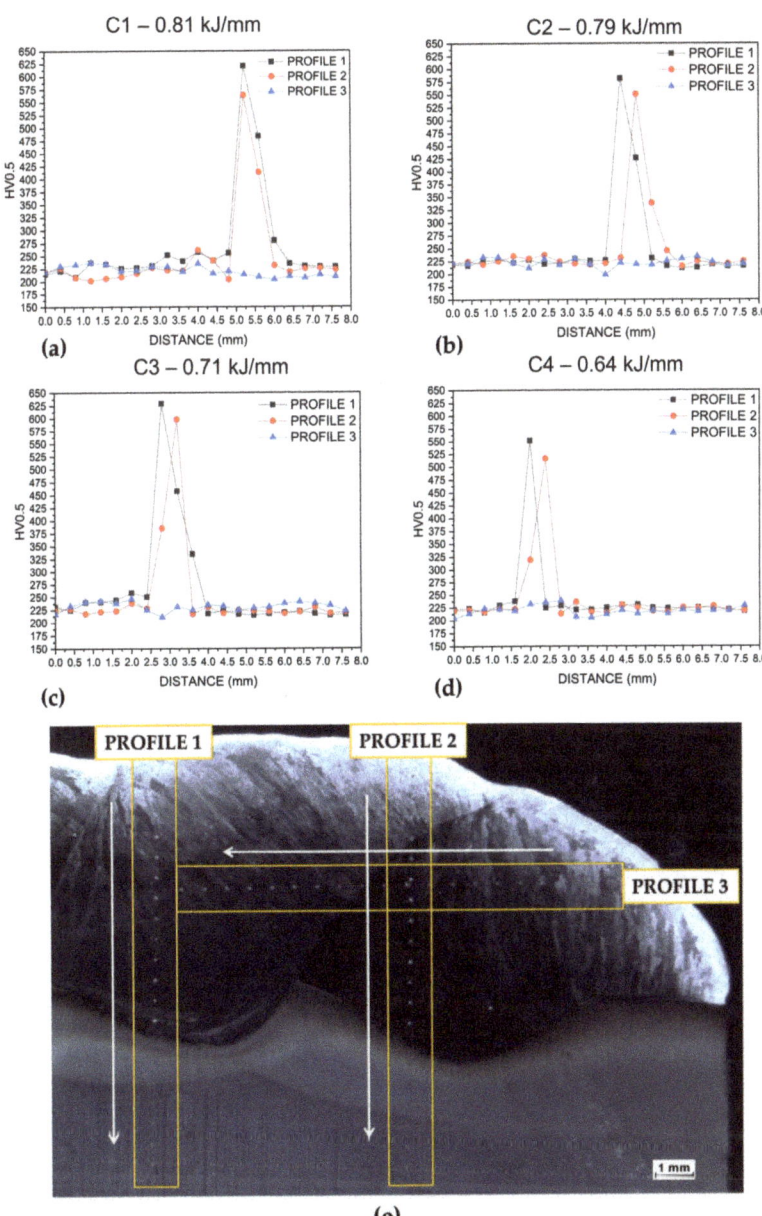

Figure 9. Vickers microhardness profiles. (**a**) C1—0.81 kJ/mm; (**b**) C2—0.79 kJ/mm; (**c**) C3—0.71 kJ/mm; (**d**) C4—0.64 kJ/mm; (**e**) diagram of the direction of microhardness evaluation carried out on each sample.

The hardness behavior is similar in all samples, despite different heat inputs. The coating and the substrate exhibit hardness values of about 225 HV, while a significantly higher increase to over 600 HV is observed when evaluating the interface and the area of the substrate immediately close to the interface. This behavior is associated with the presence of martensite in these areas, which produces a substantial increase in microhardness, making the interface less ductile [69–71].

The results highlight the importance of paying special attention to the interface in applications involving dynamic stresses, as this region can be mechanically vulnerable.

3.4. Wear Test

Figure 10 illustrates the volume loss and wear rate for both the substrate and the coatings. Samples with high heat input (0.81 and 0.79 kJ/mm) exhibit greater wear resistance (with a volume loss of 0.085 and 0.087 mm^3, respectively) compared to those with lower thermal input, showing a slight improvement compared to the substrate (0.095 mm^3). Likewise, the wear rate remains higher in the low heat input samples, with values of 3.64 and 3.72 mm^3/N·m $\times 10^{-3}$ in coatings C3 (0.71 kJ/mm) and C4 (0.64 kJ/mm), respectively. This suggests that the coatings have a lower mechanical strength when subjected to abrasive wear, and this behavior might be related to the variation in the content of residual austenite and ferrite in the coating and their morphology [72].

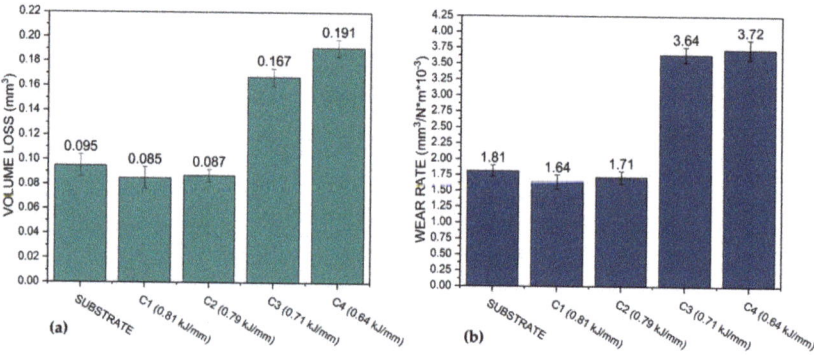

Figure 10. Tribological results: (**a**) volume loss; (**b**) wear rate.

Figure 11 shows that the substrate exhibits a friction coefficient of about 0.55, whereas the samples with a heat input of 0.81 kJ/mm and 0.79 kJ/mm have higher friction coefficients of 0.75 and 0.72, respectively. Moreover, these values are also greater than samples with a lower heat input, which have friction coefficients of 0.68 (C3—0.71 kJ/mm) and 0.61 (C4—0.64 kJ/mm). To understand the behavior of the peaks in the friction coefficient graphs, it is crucial to consider the process occurring during tribology testing. When the pin in the tribology test starts to rub against the sample, the static friction coefficient is obtained, represented by the initial peak on the left side of the graph. As the test continues, the dynamic friction coefficient is calculated and the pin starts wearing the sample surface, resulting in the release of particles from the tested material (in this case, the coatings) [73]. These released particles disperse over the wear zone, and due to the temperature generated during the dry pin-on-disk test, these particles can re-adhere to the surface. This leads to what is known as adhesive wear. Once this point is reached, the test exhibits a stable friction behavior, known as the stick–slip effect [74].

The morphology of the worn surfaces was analyzed using SEM and is presented in Figure 12. The images reveal evidence of abrasion wear in all the wear tracks caused by the detachment of oxide particles formed in the coating, which start to slide along the wear track [73,75]. Delamination flakes and a smooth wear surface can also be observed in the samples. Note that a reduced wear track width is evident in the samples with a high heat input, consistent with the volume loss results (see Figure 8).

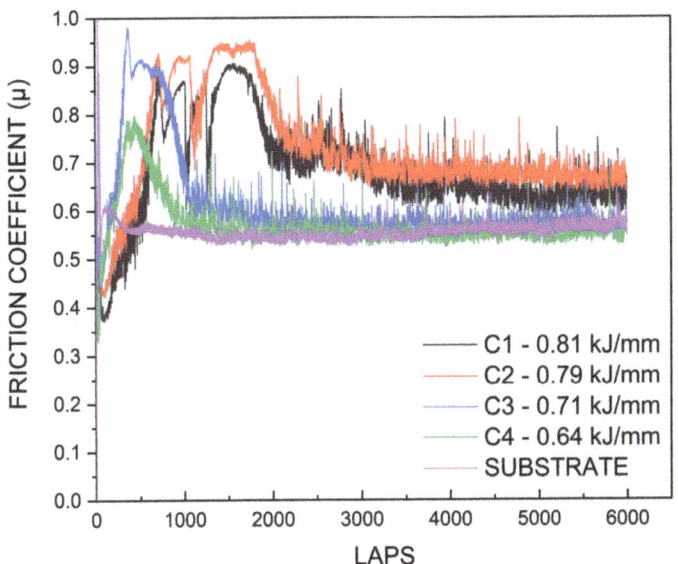

Figure 11. Graph of friction coefficients.

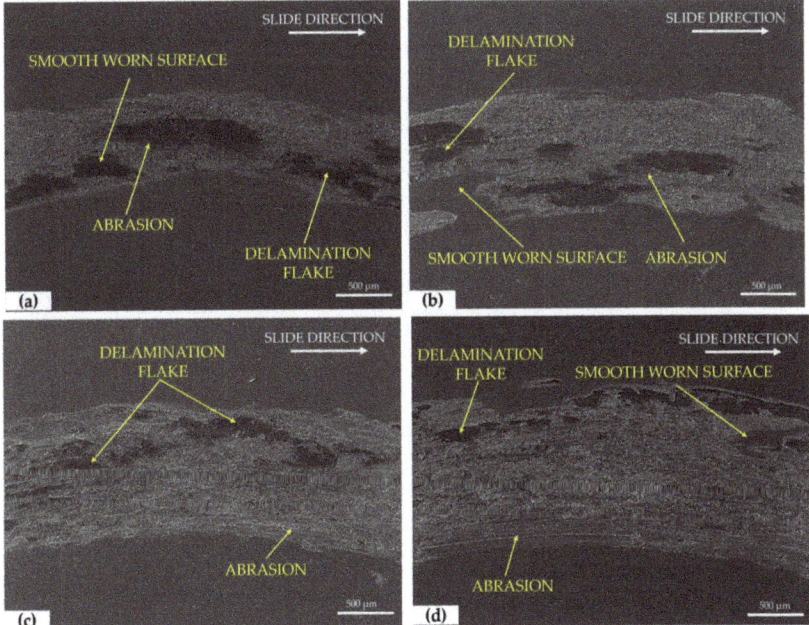

Figure 12. SEM micrographs of the wear tracks. (**a**) C1—0.81 kJ/mm; (**b**) C2—0.79 kJ/mm; (**c**) C3—0.71 kJ/mm; (**d**) C4—0.64 kJ/mm.

Chemical analysis using EDS was performed on the wear tracks (Figure 13), evaluating different areas. The results of this chemical analysis indicate the presence of oxygen, chromium, and iron in all samples. The high oxygen content is related to the fact that since the test was conducted under dry conditions, the coating surface heats up and the particles released during the test enter an oxidation state [76,77]. These oxides promote the formation of a protective layer on the surface, reducing friction and wear [78].

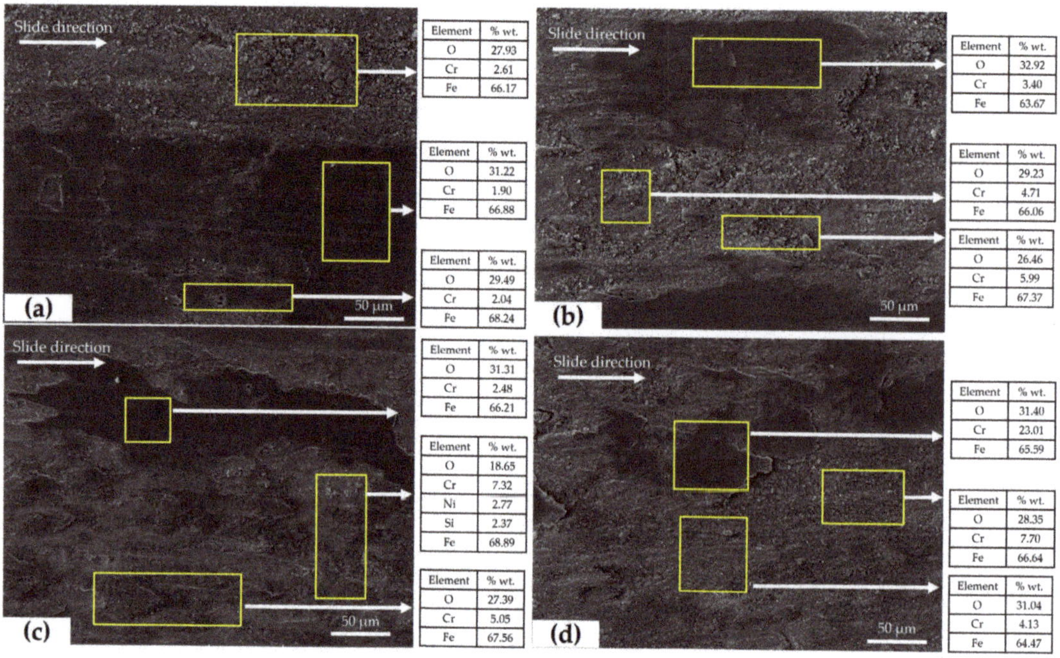

Figure 13. EDS of the wear tracks. (**a**) C1—0.81 kJ/mm; (**b**) C2—0.79 kJ/mm; (**c**) C3—0.71 kJ/mm; (**d**) C4—0.64 kJ/mm.

4. Conclusions

In this study, coatings of austenitic stainless steel ER308 were applied on an H13 tool-grade steel substrate using the S-GMAW process. Their macro-/microstructural and mechanical evolution was analyzed using various characterization techniques. The main conclusions of the study are as follows:

1. No visible discontinuities such as pores, cracks, or lack of fusion were observed in any of the coatings.
2. The thickness of the coating and its dilution are directly related to the heat input. For a high heat input (0.81 kJ/mm), the dilution percentage and the maximum coating thickness reach higher values (18.3% and 7.46 mm, respectively) compared to the values obtained with lower heat input 0.64 kJ/mm (14.3%—4.34 mm).
3. In all samples, the coating mainly consists of austenite, with lathy ferrite at the top of the coating and columnar austenitic grains in the center. The interface shows the presence of austenite and quenched martensite.
4. Microhardness tests revealed that the coating and the substrate have similar hardness values, which increase significantly at the interface, reaching a maximum of 625 HV due to the presence of martensite in this area.
5. Samples with a higher heat input (0.81, 0.79 kJ/mm) exhibited greater wear resistance.

The conclusions of the present study allow us to define the feasibility of joining dissimilar materials by adequate dilution of the filler material on the substrate. In subsequent work, the corrosion resistance of the welded assembly must be evaluated and analyzed to complete the application of this investigation.

Author Contributions: Conceptualization, J.E.H.-F. and B.R.R.-V.; methodology, J.E.H.-F., F.d.J.G.-V. and J.G.-C.; formal analysis, J.E.H.-F., G.S. and B.R.R.-V.; investigation, J.E.H.-F. and J.G.-C.; writing—original draft preparation, J.E.H.-F.; writing—review and editing, B.R.R.-V., G.S., A.F.M.P. and A.D.S.; supervision, F.d.J.G.-V. and A.D.S. All authors have read and agreed to the published version of the manuscript.

Funding: This research was funded (Scholarship number 21632109-2023) by the National Council of Humanities, Sciences and Technologies (CONAHCYT), Mexico.

Data Availability Statement: The data presented in this study are available on request from the corresponding author. The data are not publicly available due to privacy.

Acknowledgments: The authors thank the engineer Ramiro Rodríguez Rosales for his suggestions and contributions to this research.

Conflicts of Interest: The authors declare no conflict of interest.

References

1. Chang, S.-H.; Tsai, B.-C.; Huang, K.-T.; Yang, C.-H. Investigation of the Wear and Corrosion Behaviors of CrN Films onto Oxynitriding-Treated AISI H13 Alloy Steel by the Direct Current Magnetron Sputtering Process. *Thin Solid Films* **2022**, *758*, 139434. [CrossRef]
2. Kotarska, A.; Poloczek, T.; Janicki, D. Characterization of the Structure, Mechanical Properties and Erosive Resistance of the Laser Cladded Inconel 625-Based Coatings Reinforced by TiC Particles. *Materials* **2021**, *14*, 2225. [CrossRef] [PubMed]
3. Gonzalez-Carmona, J.M.; Mambuscay, C.L.; Ortega-Portilla, C.; Hurtado-Macias, A.; Piamba, J.F. TiNbN Hard Coating Deposited at Varied Substrate Temperature by Cathodic Arc: Tribological Performance under Simulated Cutting Conditions. *Materials* **2023**, *16*, 4531. [CrossRef] [PubMed]
4. Grossi, E.; Baroni, E.; Aprile, A.; Fortini, A.; Zerbin, M.; Merlin, M. Tribological Behavior of Structural Steel with Different Surface Finishing and Treatments for a Novel Seismic Damper. *Coatings* **2023**, *13*, 135. [CrossRef]
5. Di Schino, A.; Porcu, G.; Scoppio, L.; Longobardo, M.; Turconi, G.L. Metallurgical Design and Development of C125 Grade for Mild Sour Service Application. In Proceedings of the NACE CORROSION CONFERENCE SERIES, San Diego, CA, USA, 12–16 March 2006; pp. 061251–0612514. Available online: https://onepetro.org/NACECORR/proceedings-abstract/CORR06/All-CORR06/NACE-06125/118115 (accessed on 19 December 2023).
6. Di Schino, A.; Valentini, L.; Kenny, J.M.; Gerbig, Y.; Ahmed, I.; Haefke, H. Wear resistance of a high-nitrogen austenitic stainless steel coated with nitrogenated amorphous carbon films. *Surf. Coat. Technol.* **2002**, *161*, 224–231. [CrossRef]
7. Stornelli, G.; Gaggiotti, M.; Mancini, S.; Napoli, G.; Rocchi, C.; Tirasso, C.; Di Schino, A. Recrystallization and Grain Growth of AISI 904L Super-Austenitic Stainless Steel: A Multivariate Regression Approach. *Metals* **2022**, *12*, 200. [CrossRef]
8. Ranjan, R.; Kumar Das, A. Protection from Corrosion and Wear by Different Weld Cladding Techniques: A Review. *Mater. Today Proc.* **2022**, *57*, 1687–1693. [CrossRef]
9. Liu, T.; Chang, M.; Cheng, X.; Zeng, X.; Shao, H.; Liu, F. Characteristics of WC Reinforced Ni-Based Alloy Coatings Prepared by PTA + PMI Method. *Surf. Coat. Technol.* **2020**, *383*, 125232. [CrossRef]
10. Magarò, P.; Marino, A.L.; Di Schino, A.; Furgiuele, F.; Maletta, C.; Pileggi, R.; Sgambitterra, E.; Testani, C.; Tului, M. Effect of Process Parameters on the Properties of Stellite-6 Coatings Deposited by Cold Gas Dynamic Spray. *Surf. Coat. Technol.* **2019**, *377*, 124934. [CrossRef]
11. Aslam, M.; Sahoo, C.K. Development of Hard and Wear-Resistant SiC-AISI304 Stainless Steel Clad Layer on Low Carbon Steel by GMAW Process. *Mater. Today Commun.* **2023**, *36*, 106444. [CrossRef]
12. Kumar, V.; Ranjan Sahu, D.; Mandal, A. Parametric Study and Optimization of GMAW Based AM Process for Multi-Layer Bead Deposition. *Mater. Today Proc.* **2022**, *62*, 255–261. [CrossRef]
13. Di Schino, A.; Kenny, J.M.; Barteri, M. High temperature resistance of a high nitrogen and low nickel austenitic stainless steel. *J. Mat. Sci. Lett.* **2003**, *22*, 691–693. [CrossRef]
14. Di Schino, A.; Kenny, J.M.; Salvatori, I.; Abbruzzese, G. Modelling primary recrystallization and grain growth in a low nickel austenitic stainless steel. *J. Mater. Sci.* **2001**, *36*, 593–601. [CrossRef]
15. Kang, K.; Kawahito, Y.; Gao, M.; Zeng, X. Effects of Laser-Arc Distance on Corrosion Behavior of Single-Pass Hybrid Welded Stainless Clad Steel Plate. *Mater. Des.* **2017**, *123*, 80–88. [CrossRef]
16. Li, C.; Qin, G.; Tang, Y.; Zhang, B.; Lin, S.; Geng, P. Microstructures and Mechanical Properties of Stainless Steel Clad Plate Joint with Diverse Filler Metals. *J. Mater. Res. Technol.* **2020**, *9*, 2522–2534. [CrossRef]
17. Vlasov, I.V.; Gordienko, A.I.; Eremin, A.V.; Semenchuk, V.M.; Kuznetsova, A.E. Structure and Mechanical Behavior of Heat-Resistant Steel Manufactured by Multilayer Arc Deposition. *Metals* **2023**, *13*, 1375. [CrossRef]
18. Valiente Bermejo, M.A.; Eyzop, D.; Hurtig, K.; Karlsson, L. Welding of Large Thickness Super Duplex Stainless Steel: Microstructure and Properties. *Metals* **2021**, *11*, 1184. [CrossRef]
19. Singh Singhal, T.; Kumar Jain, J. GMAW Cladding on Metals to Impart Anti-Corrosiveness: Machine, Processes and Materials. *Mater. Today Proc.* **2020**, *26*, 2432–2441. [CrossRef]

20. Zhu, L.; Cui, Y.; Cao, J.; Tian, R.; Cai, Y.; Xu, C.; Han, J.; Tian, Y. Effect of TIG Remelting on Microstructure, Corrosion and Wear Resistance of Coating on Surface of 4Cr5MoSiV1 (AISI H13). *Surf. Coat. Technol.* **2021**, *405*, 126547. [CrossRef]
21. Tang, X.; Zhang, S.; Cui, X.; Zhang, C.; Liu, Y.; Zhang, J. Tribological and Cavitation Erosion Behaviors of Nickel-Based and Iron-Based Coatings Deposited on AISI 304 Stainless Steel by Cold Metal Transfer. *J. Mater. Res. Technol.* **2020**, *9*, 6665–6681. [CrossRef]
22. Rajeev, G.P.; Kamaraj, M.; Bakshi, S.R. Comparison of Microstructure, Dilution and Wear Behavior of Stellite 21 Hardfacing on H13 Steel Using Cold Metal Transfer and Plasma Transferred Arc Welding Processes. *Surf. Coat. Technol.* **2019**, *375*, 383–394. [CrossRef]
23. Yuan, Y.; Wu, H.; You, M.; Li, Z.; Zhang, Y. Improving Wear Resistance and Friction Stability of FeNi Matrix Coating by In-Situ Multi-Carbide WC-TiC via PTA Metallurgical Reaction. *Surf. Coat. Technol.* **2019**, *378*, 124957. [CrossRef]
24. Morales, C.; Fortini, A.; Soffritti, C.; Merlin, M. Effect of Post-Fabrication Heat Treatments on the Microstructure of WC-12Co Direct Energy Depositions. *Coatings* **2023**, *13*, 1459. [CrossRef]
25. Hao, J.; Niu, Q.; Ji, H.; Liu, H. Effect of Ultrasonic Rolling on the Organization and Properties of a High-Speed Laser Cladding IN 718 Superalloy Coating. *Crystals* **2023**, *13*, 1214. [CrossRef]
26. Bailey, N.S.; Katinas, C.; Shin, Y.C. Laser Direct Deposition of AISI H13 Tool Steel Powder with Numerical Modeling of Solid Phase Transformation, Hardness, and Residual Stresses. *J. Mater. Process. Technol.* **2017**, *247*, 223–233. [CrossRef]
27. Lizzul, L.; Sorgato, M.; Bertolini, R.; Ghiotti, A.; Bruschi, S.; Fabbro, F.; Rech, S. On the Influence of Laser Cladding Parameters and Number of Deposited Layers on As-Built and Machined AISI H13 Tool Steel Multilayered Claddings. *CIRP J. Manuf. Sci. Technol.* **2021**, *35*, 361–370. [CrossRef]
28. Le, V.T.; Mai, D.S. Microstructural and Mechanical Characteristics of 308L Stainless Steel Manufactured by Gas Metal Arc Welding-Based Additive Manufacturing. *Mater. Lett.* **2020**, *271*, 127791. [CrossRef]
29. Miranda-Pérez, A.F.; Rodríguez-Vargas, B.R.; Calliari, I.; Pezzato, L. Corrosion Resistance of GMAW Duplex Stainless Steels Welds. *Materials* **2023**, *16*, 1847. [CrossRef]
30. Stinson, H.; Ward, R.; Quinn, J.; McGarrigle, C. Comparison of Properties and Bead Geometry in MIG and CMT Single Layer Samples for WAAM Applications. *Metals* **2021**, *11*, 1530. [CrossRef]
31. Amushahi, M.H.; Ashrafizadeh, F.; Shamanian, M. Characterization of Boride-Rich Hardfacing on Carbon Steel by Arc Spray and GMAW Processes. *Surf. Coat. Technol.* **2010**, *204*, 2723–2728. [CrossRef]
32. Chen, T.; Xue, S.; Wang, B.; Zhai, P.; Long, W. Study on Short-Circuiting GMAW Pool Behavior and Microstructure of the Weld with Different Waveform Control Methods. *Metals* **2019**, *9*, 1326. [CrossRef]
33. Das, S.; Vora, J.J.; Patel, V.; Li, W.; Andersson, J.; Pimenov, D.Y.; Giasin, K.; Wojciechowski, S. Experimental Investigation on Welding of 2.25 Cr-1.0 Mo Steel with Regulated Metal Deposition and GMAW Technique Incorporating Metal-Cored Wires. *J. Mater. Res. Technol.* **2021**, *15*, 1007–1016. [CrossRef]
34. Chhabra, P.; Kaur, M.; Singh, S. High Temperature Tribological Performance of Atmospheric Plasma Sprayed Cr3C2-NiCr Coating on H13 Tool Steel. *Mater. Today Proc.* **2020**, *33*, 1518–1530. [CrossRef]
35. Telasang, G.; Dutta Majumdar, J.; Padmanabham, G.; Manna, I. Wear and Corrosion Behavior of Laser Surface Engineered AISI H13 Hot Working Tool Steel. *Surf. Coat. Technol.* **2015**, *261*, 69–78. [CrossRef]
36. Hadidi, H.; Saminathan, R.; Zouli, N. Enhancement of Diffusive Oxidation Resistance and Topographical Corrosion Resistance of Hot Work Tool Steels by Nickel Multifunctional Coatings. *J. Mater. Res. Technol.* **2024**, *28*, 1227–1232. [CrossRef]
37. Calvo-García, E.; Valverde-Pérez, S.; Riveiro, A.; Álvarez, D.; Román, M.; Magdalena, C.; Badaoui, A.; Moreira, P.; Comesaña, R. An Experimental Analysis of the High-Cycle Fatigue Fracture of H13 Hot Forging Tool Steels. *Materials* **2022**, *15*, 7411. [CrossRef] [PubMed]
38. Shinde, T.; Pruncu, C.; Dhokey, N.B.; Parau, A.C.; Vladescu, A. Effect of Deep Cryogenic Treatment on Corrosion Behavior of AISI H13 Die Steel. *Materials* **2021**, *14*, 7863. [CrossRef] [PubMed]
39. Xu, G.; Wang, K.; Dong, X.; Yang, L.; Ebrahimi, M.; Jiang, H.; Wang, Q.; Ding, W. Review on Corrosion Resistance of Mild Steels in Liquid Aluminum. *J. Mater. Sci. Technol.* **2021**, *71*, 12–22. [CrossRef]
40. Lou, D.C.; Akselsen, O.M.; Onsøien, M.I.; Solberg, J.K.; Berget, J. Surface Modification of Steel and Cast Iron to Improve Corrosion Resistance in Molten Aluminium. *Surf. Coat. Technol.* **2006**, *200*, 5282–5288. [CrossRef]
41. Chen, G.; Xue, L.; Wang, J.; Tang, Z.; Li, X.; Dong, H. Investigation of Surface Modifications for Combating the Molten Aluminum Corrosion of AISI H13 Steel. *Corros. Sci.* **2020**, *174*, 108836. [CrossRef]
42. Ostolaza, M.; Arrizubieta, J.I.; Lamikiz, A.; Cortina, M. Functionally Graded AISI 316L and AISI H13 Manufactured by L-DED for Die and Mould Applications. *Appl. Sci.* **2021**, *11*, 771. [CrossRef]
43. Dass, A.; Moridi, A. State of the Art in Directed Energy Deposition: From Additive Manufacturing to Materials Design. *Coatings* **2019**, *9*, 418. [CrossRef]
44. De Oliveira, U.; Ocelík, V.; De Hosson, J.T.M. Analysis of Coaxial Laser Cladding Processing Conditions. *Surf. Coat. Technol.* **2005**, *197*, 127–136. [CrossRef]
45. Davis, J.R. Hardfacing, Weld Cladding, and Dissimilar Metal Joining. In *Welding, Brazing, and Soldering*; ASM International: Tokyo, Japan, 1993; pp. 789–829.
46. Lippold, J.C.; Kotecki, D.J. *Welding Metallurgy and Weldability of Stainless Steels*; John Wiley & Sons, Inc.: Hoboken, NJ, USA, 2005; ISBN 978-0-471-47379-4.

47. Arsić, D.; Lazić, V.; Mitrović, S.; Džunić, D.; Aleksandrović, S.; Djordjević, M.; Nedeljković, B. Tribological Behavior of Four Types of Filler Metals for Hard Facing under Dry Conditions. *Ind. Lubr. Tribol.* **2016**, *68*, 729–736. [CrossRef]
48. ASTM G99; ASTM E384 International Standard Test Method for Wear Testing with a Pin-on-Disk Apparatus G99–17. ASTM International: Conshohocken, PA, USA, 2022.
49. Luchtenberg, P.; de Campos, P.T.; Soares, P.; Laurindo, C.A.H.; Maranho, O.; Torres, R.D. Effect of Welding Energy on the Corrosion and Tribological Properties of Duplex Stainless Steel Weld Overlay Deposited by GMAW/CMT Process. *Surf. Coat. Technol.* **2019**, *375*, 688–693. [CrossRef]
50. Chen, L.; Tan, H.; Wang, Z.; Li, J.; Jiang, Y. Influence of Cooling Rate on Microstructure Evolution and Pitting Corrosion Resistance in the Simulated Heat-Affected Zone of 2304 Duplex Stainless Steels. *Corros. Sci.* **2012**, *58*, 168–174. [CrossRef]
51. Rodriguez, B.R.; Miranda, A.; Gonzalez, D.; Praga, R.; Hurtado, E. Maintenance of the Austenite/Ferrite Ratio Balance in GTAW DSS Joints Through Process Parameters Optimization. *Materials* **2020**, *13*, 780. [CrossRef]
52. Jiang, Z.; Chen, X.; Yu, K.; Lei, Z.; Chen, Y.; Wu, S.; Li, Z. Improving Fusion Zone Microstructure Inhomogeneity in Dissimilar-Metal Welding by Laser Welding with Oscillation. *Mater. Lett.* **2019**, *2019*, 126995. [CrossRef]
53. Bunaziv, I.; Ren, X.; Hagen, A.B.; Hovig, E.W.; Jevremovic, I.; Gulbrandsen-Dahl, S. Laser Beam Remelting of Stainless Steel Plate for Cladding and Comparison with Conventional CMT Process. *Int. J. Adv. Manuf. Technol.* **2023**, *127*, 911–934. [CrossRef]
54. Wu, D.; An, Q.; Zhao, G.; Zhang, Y.; Zou, Y. Corrosion Resistance of Stainless Steel Layer Prepared by Twin-Wire Indirect Arc Surfacing Welding. *Vacuum* **2020**, *177*, 109348. [CrossRef]
55. Li, S.; Eliniyaz, Z.; Zhang, L.; Sun, F.; Shen, Y.; Shan, A. Microstructural Evolution of Delta Ferrite in SAVE12 Steel under Heat Treatment and Short-Term Creep. *Mater. Charact.* **2012**, *73*, 144–152. [CrossRef]
56. Kim, Y.-G.; Lee, D.J.; Byun, J.; Jung, K.H.; Kim, J.; Lee, H.; Taek Shin, Y.; Kim, S.; Lee, H.W. The Effect of Sigma Phases Formation Depending on Cr/Ni Equivalent Ratio in AISI 316L Weldments. *Mater. Des.* **2011**, *32*, 330–336. [CrossRef]
57. Fu, J.W.; Yang, Y.S.; Guo, J.J.; Ma, J.C.; Tong, W.H. Microstructure Formation in Rapidly Solidified AISI 304 Stainless Steel Strip. *Ironmak. Steelmak.* **2009**, *36*, 230–233. [CrossRef]
58. Lee, D.J.; Byun, J.C.; Sung, J.H.; Lee, H.W. The Dependence of Crack Properties on the Cr/Ni Equivalent Ratio in AISI 304L Austenitic Stainless Steel Weld Metals. *Mater. Sci. Eng. A* **2009**, *513–514*, 154–159. [CrossRef]
59. Tate, S.B.; Liu, S. Solidification Behaviour of Laser Welded Type 21Cr–6Ni–9Mn Stainless Steel. *Sci. Technol. Weld. Join.* **2014**, *19*, 310–317. [CrossRef]
60. Weidong, H. Microstructure Evolution of 316L Stainless Steel During Laser Rapid Forming. *Acta Metall. Sin.* **2006**, *42*, 361–368. Available online: https://www.ams.org.cn/EN/abstract/abstract5556.shtml (accessed on 19 December 2023).
61. Li, K.; Li, D.; Liu, D.; Pei, G.; Sun, L. Microstructure Evolution and Mechanical Properties of Multiple-Layer Laser Cladding Coating of 308L Stainless Steel. *Appl. Surf. Sci.* **2015**, *340*, 143–150. [CrossRef]
62. Li, D.; Lu, S.; Li, D.; Li, Y. Investigation of the microstructure and impact properties of the high nitrogen stainless steel weld. *Acta Metall. Sin.* **2013**, *49*, 129. [CrossRef]
63. Zhu, Y.; Chen, J.; Li, X. Numerical Simulation of Thermal Field and Performance Study on H13 Die Steel-Based Wire Arc Additive Manufacturing. *Metals* **2023**, *13*, 1484. [CrossRef]
64. Wang, Y.; Sebeck, K.; Tess, M.; Gingrich, E.; Feng, Z.; Haynes, J.A.; Lance, M.J.; Muralidharan, G.; Marchel, R.; Kirste, T.; et al. Interfacial Microstructure and Mechanical Properties of Rotary Inertia Friction Welded Dissimilar 422 Martensitic Stainless Steel to 4140 Low Alloy Steel Joints. *Mater. Sci. Eng. A* **2023**, *885*, 145607. [CrossRef]
65. Hamada, A.; Ghosh, S.; Ali, M.; Jaskari, M.; Järvenpää, A. Studying the Strengthening Mechanisms and Mechanical Properties of Dissimilar Laser-Welded Butt Joints of Medium-Mn Stainless Steel and Automotive High-Strength Carbon Steel. *Mater. Sci. Eng. A* **2022**, *856*, 143936. [CrossRef]
66. Ansari, R.; Movahedi, M.; Pouranvari, M. Microstructural Features and Mechanical Behavior of Duplex Stainless Steel/Low Carbon Steel Friction Stir Dissimilar Weld. *J. Mater. Res. Technol.* **2023**, *25*, 5352–5371. [CrossRef]
67. Katlhe, P.; Paul, S.; Singh, R.; Yan, W. Experimental Characterization of Laser Cladding of CPM 9V on H13 Tool Steel for Die Repair Applications. *J. Manuf. Process.* **2015**, *20*, 492–499. [CrossRef]
68. Li, S.; Li, J.; Sun, G.; Deng, D. Modeling of Welding Residual Stress in a Dissimilar Metal Butt Welded Joint between P92 Ferritic Steel and SUS304 Austenitic Stainless Steel. *J. Mater. Res. Technol.* **2023**, *23*, 4938–4954. [CrossRef]
69. Guan, H.; Chai, L.; Wang, Y.; Xiang, K.; Wu, L.; Pan, H.; Yang, M.; Teng, C.; Zhang, W. Microstructure and Hardness of NbTiZr and NbTaTiZr Refractory Medium-Entropy Alloy Coatings on Zr Alloy by Laser Cladding. *Appl. Surf. Sci.* **2021**, *549*, 149338. [CrossRef]
70. Li, X.; Jiang, P.; Nie, M.; Liu, Z.; Liu, M.; Qiu, Y.; Chen, Z.; Zhang, Z. Enhanced Strength-Ductility Synergy of Laser Additive Manufactured Stainless Steel/Ni-Based Superalloy Dissimilar Materials Characterized by Bionic Mechanical Interlocking Structures. *J. Mater. Res. Technol.* **2023**, *26*, 4770–4783. [CrossRef]
71. Wang, X.; Lei, L.; Yu, H. A Review on Microstructural Features and Mechanical Properties of Wheels/Rails Cladded by Laser Cladding. *Micromachines* **2021**, *12*, 152. [CrossRef] [PubMed]
72. Di Schino, A.; Testani, C. Corrosion Behavior and Mechanical Properties of AISI 316 Stainless Steel Clad Q235 Plate. *Metals* **2020**, *10*, 552. [CrossRef]
73. Zhang, M.; Li, M.; Chi, J.; Wang, S.; Ren, L.; Fang, M. Microstructure and Tribology Properties of In-Situ MC(M:Ti,Nb) Coatings Prepared via PTA Technology. *Vacuum* **2019**, *160*, 264–271. [CrossRef]

74. Luiz, V.D.; dos Santos, A.J.; Câmara, M.A.; Rodrigues, P.C.d.M. Influence of Different Contact Conditions on Friction Properties of AISI 430 Steel Sheet with Deep Drawing Quality. *Coatings* **2023**, *13*, 771. [CrossRef]
75. Zhao, C.; Xing, X.; Guo, J.; Shi, Z.; Zhou, Y.; Ren, X.; Yang, Q. Microstructure and Wear Resistance of (Nb,Ti)C Carbide Reinforced Fe Matrix Coating with Different Ti Contents and Interfacial Properties of (Nb,Ti)C/α-Fe. *Appl. Surf. Sci.* **2019**, *494*, 600–609. [CrossRef]
76. Fox-Rabinovich, G.S.; Gershman, I.S.; Endrino, J.L. Accelerated Tribo-Films Formation in Complex Adaptive Surface-Engineered Systems under the Extreme Tribological Conditions of Ultra-High-Performance Machining. *Lubricants* **2023**, *11*, 221. [CrossRef]
77. Morales, C.; Merlin, M.; Fortini, A.; Fortunato, A. Direct Energy Depositions of a 17-4 PH Stainless Steel: Geometrical and Microstructural Characterizations. *Coatings* **2023**, *13*, 636. [CrossRef]
78. Ferreira Filho, D.; Souza, D.; Gonçalves Júnior, J.L.; Reis, R.P.; Da Silva Junior, W.M.; Tavares, A.F. Influence of Substrate on the Tribological Behavior of Inconel 625 GMAW Overlays. *Coatings* **2023**, *13*, 1454. [CrossRef]

Disclaimer/Publisher's Note: The statements, opinions and data contained in all publications are solely those of the individual author(s) and contributor(s) and not of MDPI and/or the editor(s). MDPI and/or the editor(s) disclaim responsibility for any injury to people or property resulting from any ideas, methods, instructions or products referred to in the content.

Article

Regulated Phase Separation in Al–Ti–Cu–Co Alloys through Spark Plasma Sintering Process

Seulgee Lee [1], Chayanaphat Chokradjaroen [2], Yasuyuki Sawada [3], Sungmin Yoon [4] and Nagahiro Saito [1,2,3,5,6,*]

1. Department of Chemical Systems Engineering, Graduate School of Engineering, Nagoya University, Furo-cho, Chikusa-ku, Nagoya 464-8603, Japan; sg@sp.material.nagoya-u.ac.jp
2. Department of International Collaborative Program in Sustainable Materials and Technology for Industries between Nagoya University and Chulalongkorn University, Graduate School of Engineering, Nagoya University, Furo-cho, Chikusa-ku, Nagoya 464-8603, Japan; eig@sp.material.nagoya-u.ac.jp
3. Institute of Innovation for Future Society, Nagoya University, Nagoya 464-8601, Japan; ysawada@sp.material.nagoya-u.ac.jp
4. Department of Micro-Nano Mechanical Science and Engineering, Nagoya University, Furo-cho, Chikusa-ku, Nagoya 464-8603, Japan; yoon.sungmin.v8@f.mail.nagoya-u.ac.jp
5. Conjoint Research Laboratory in Nagoya University, Shinshu University, Furo-cho, Chikusa-ku, Nagoya 464-8603, Japan
6. Japan Science and Technology Corporation (JST), Open Innovation Platform with Enterprises, Research Institute and Academia (OPERA), Furo-cho, Chikusa-ku, Nagoya 464-8603, Japan

* Correspondence: hiro@sp.material.nagoya-u.ac.jp; Tel.: +81-52-7893259

Abstract: With the goal of developing lightweight Al-Ti-containing multicomponent alloys with excellent mechanical strength, an Al–Ti–Cu–Co alloy with a phase-separated microstructure was prepared. The granulometry of metal particles was reduced using planetary ball milling. The particle size of the metal powders decreased as the ball milling time increased from 5, 7, to 15 h (i.e., 6.6 ± 6.4, 5.1 ± 4.3, and 3.2 ± 2.1 μm, respectively). The reduction in particle size and the dispersion of metal powders promoted enhanced diffusion during the spark plasma sintering process. This led to the micro-phase separation of the $(Cu, Co)_2AlTi$ ($L2_1$) phase, and the formation of a Cu-rich phase with embedded nanoscale Ti-rich (B2) precipitates. The Al–Ti–Cu–Co alloys prepared using powder metallurgy through the spark plasma sintering exhibited different hardnesses of 684, 710, and 791 HV, respectively, while maintaining a relatively low density of 5.8–5.9 g/cm^3 (<6 g/cm^3). The mechanical properties were improved due to a decrease in particle size achieved through increased ball milling time, leading to a finer grain size. The $L2_1$ phase, consisting of $(Cu, Co)_2AlTi$, is the site of basic hardness performance, and the Cu-rich phase is the mechanical buffer layer between the $L2_1$ and B2 phases. The finer network structure of the Cu-rich phase also suppresses brittle fracture.

Keywords: Al-Ti-containing multicomponent alloys; phase separation; hardness; spark plasma sintering; powder metallurgy

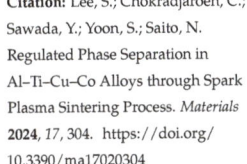

Citation: Lee, S.; Chokradjaroen, C.; Sawada, Y.; Yoon, S.; Saito, N. Regulated Phase Separation in Al–Ti–Cu–Co Alloys through Spark Plasma Sintering Process. *Materials* **2024**, *17*, 304. https://doi.org/10.3390/ma17020304

Academic Editors: Andrea Di Schino and Claudio Testani

Received: 14 December 2023
Revised: 2 January 2024
Accepted: 5 January 2024
Published: 7 January 2024

Copyright: © 2024 by the authors. Licensee MDPI, Basel, Switzerland. This article is an open access article distributed under the terms and conditions of the Creative Commons Attribution (CC BY) license (https://creativecommons.org/licenses/by/4.0/).

1. Introduction

A lightweight design for components, such as reducing the weight of parts in automobiles and aircraft, can largely improve energy efficiency and reduce CO_2 emission. Al and Ti alloys have been widely used, due to their lightweight and ease of processing [1]. However, the improvement in their mechanical properties is still ongoing, aiming to expand their application area. Recently, the design of Al-Ti-containing multicomponent alloys to improve the properties of lightweight materials has been proposed. For example, Huang et al. reported that the AlCrTiV alloy exhibited a hardness of 710 HV with a density of 4.5 g/cm^3 [2]. Additionally, Liao et al. investigated a series of $Ti_x(AlCrNb)_{100-x}$ alloys with hardness values of 290–516 HV and a density of 4.85–5.25 g/cm^3 [3]. X–Al–Ti alloy systems also have been developed to improve mechanical properties through ordered body-centered cubic (BCC) (B2) phase-strengthened alloy [4,5]. While previous studies have emphasized the high hardness of the B2 phase, it is imperative to recognize that the presence of the B2 phase within the metallurgical structure does not universally enhance the material's mechanical properties. To solve this problem, there has been growing attention on the examination of the $L2_1$ phase as another ordered BCC phase. This phase is characterized by a higher degree of ordering than the BCC and B2 phases, following the sequence: BCC < B2 < $L2_1$ [6]. Specifically, the $L2_1$ phase exhibits a restricted slip system and a better resistance to creep than the B2 and BCC phases [7]. Consequently, $L2_1$ has been proposed to improve strength and ductility significantly [8]. In the Co–Al–Ti systems, the $L2_1$ and B2 phases are known to exhibit a phase-separated structure [9], and Cu stabilizes the $L2_1$ phase [10]. Therefore, in this study, the addition of Cu into the Co–Al–Ti system was anticipated to elevate the ratio of the $L2_1$ phase and enhance the mechanical properties. Moreover, concerning weight density, the suitable range for lightweight materials is below 7.00 g/cm^3 [11,12]. In the case of the X–Al–Ti (X = Cu, Co) system, the weight density is approximately 6.8 g/cm^3, so the addition of Cu does not compromise lightweight characteristics.

Various methods are used to fabricate alloys, such as arc melting [13], mechanical alloying [14], and hot isostatic pressing [15]. Among these methods, powder metallurgy based on spark plasma sintering (SPS) has advantages due to its rapid heating and cooling process. Compared to conventional sintering, SPS is recognized for producing nanocrystalline alloys through high densification achieved in a short duration [16]. The heat treatment through SPS enables a material design that effectively preserves local structures, referred to as non-equilibrium states, making it an excellent process for optimizing metal microstructures and achieving nanoscale crystallization. In addition, the morphology of powder, particularly its dimension, could significantly influence the final microstructure in powder metallurgy using SPS. A reduced particle size generally results in finer grain size, contributing to the enhancement of mechanical properties [17]. Another aspect of improving these mechanical properties involves optimizing phase-separated microstructures. For optimized phase-separated microstructures, it is necessary to implement a non-equilibrium process in which no bonding occurs between the phase regions.

Considering the aforementioned perspective, this study systematically varied the particle size of Al, Ti, Co, and Cu powder mixtures through the powder metallurgy process with SPS to optimize the phase-separated microstructure featuring the refinement of $L2_1$ and its effect on hardness was investigated. Planetary ball milling was conducted under three duration conditions: 5, 7, and 15 h, as a process parameter for particle size reduction. The alloys were then fabricated through SPS, utilizing metal powders prepared by ball milling. The resulting samples were carefully evaluated using scanning electron microscopy (SEM), energy dispersive spectroscopy (EDS), X-ray diffractometry (XRD), and high-resolution transmission electron microscopy (HRTEM). Hardness measurements were carried out using a Vickers hardness tester. Finally, the correlation between the phase-separated microstructure and the resultant mechanical properties was elucidated.

2. Materials and Methods

The initial powders of Al (<30 μm), Ti (<44 μm), Cu (<100 μm), and Co (<43 μm) (The Nilaco Co., Ltd., Tokyo, Japan) with high purity (>99.5 wt.%) were used, and stearic acid was added as the process control agent (PCA) for lubrication. Each elemental powder with 25 at. % was used and mechanically milled using a high-energy planetary ball mill (Pulverisette 7, Fritsch Japan Co. Ltd., Tokyo, Japan) with 5 mm diameter zirconia balls and zirconia vials (45 mL) under a high-purity argon atmosphere. The milling times were 5, 7, and 15 h, respectively. In addition, 0.5 wt.% of PCA was added during the 7 h of milling, followed by an additional 0.5 wt.% PCA for continuous milling up to 15 h. The ball-to-powder ratio was 10:1, and the rotational speed was 300 rpm. The ball milling was stopped for 10 min every 30 min to prevent overheating. The ball-milled powder mixtures were compacted under vacuum conditions (about 5 Pa) using SPS equipment (SPS-211LX,

Fuji Electronics Industry Co., Osaka, Japan). The hypothetical melting point of the alloy (T_m) is calculated as follows:

$$T_m = \sum_{i=1}^{n} c_i (T_m)_i \quad (1)$$

where c_i is the atomic percentage of the ith component element, and $(T_m)_i$ is the melting point of the ith component element. The T_m of Al–Ti–Cu–Co alloy is 1227 °C, and the necessary sintering temperature level is generally between 0.6 and 0.8 T_m [18]. Considering this, along with our preliminary experiments, we set the optimal sintering temperature at 900 °C. During the sintering process, a pressure of 30 MPa was applied, with a holding time of 3 min, as illustrated in the heating/cooling profile provided in Figure S1. The obtained cylindrical samples with approximately 10 mm diameter and 2 mm thickness were polished with SiC paper from 320 to 2000 grit, followed by final mirror polishing using 1 μm diamond powder. The samples obtained in this study were classified according to the milling time; samples sintered at 900 °C using powders for 5, 7, and 15 h ball-milled powders were designated as 5 h–900, 7 h–900, and 15 h–900, respectively.

The unmilled and ball-milled powder mixtures and sintered samples were characterized using an XRD (Smartlab, Rigaku Co., Tokyo, Japan) equipped with monochromatic Cu Kα radiation (λ = 1.5418 Å) X-rays source with scanning at 2θ angle from 20 to 100 degrees at a scan rate of 3 deg/min. Subsequently, the microstructures were examined using SEM (S-4800, Hitachi, Tokyo, Japan) equipped with EDS at an accelerating voltage of 15 kV. Electron backscatter diffraction (EBSD) analysis was also conducted to reveal the phase orientation of the sintered sample using a cold-field emission SEM (SU-8230, Hitachi, Tokyo, Japan) equipped with an HR EBSD detector. Before EBSD scanning, the final polishing of the samples was conducted using an ion milling system (IM4000, Hitachi, Tokyo, Japan). EBSD detector interfaced with the scan region of 100 μm × 150 μm and scan step of 0.5 μm. TEM specimens were prepared using a focused ion beam (FIB, FB-2100, Hitachi, Tokyo, Japan) for cross-section, and then analyzed with HRTEM (JEM-2100, JEOL Ltd., Tokyo, Japan) and field emission SEM (FE-SEM, JSM-7610 F, JEOL Ltd., Tokyo, Japan) at an accelerating voltage of 200 kV. The average particle size distribution was assessed based on SEM images (Figure S2) using ImageJ software (version 1.54d, National Institutes of Health, Bethesda, MD, USA) at least 800 spots for each sample. The density of the sintered sample was measured by an electro-densimeter (MDS300, Alfa Mirage Co., Ltd., Osaka, Japan) based on Archimedes' principle and automatically calculated from sample weights in air and water. Vickers hardness values were obtained on the polished surface of the samples using a Vickers hardness tester (FM-700, Future-tech Co., Tokyo, Japan). The applied load was 1 kgf with a dwell time of 15 s. The hardness test was carried out 10 times on the sample, and then the average of the resulting values was determined.

3. Results and Discussion

3.1. Characterization of Ball-Milled Powders

The morphology, alloying tendency, and average particle size of ball-milled powders were confirmed with SEM and EDS (Figure 1). In Figure 1a, 5 h ball-milled powders exhibited a partial alloying of ductile Al and Cu with copper segregation. Additionally, Ti and Co were not alloyed with other elements and were unevenly mixed. With the extension of ball milling time at 7 and 15 h (Figure 1b,c), the particle size reduction and alloying further progressed, leading to the homogeneous distribution of Ti and Co. Finally, the 15 h ball-milled particles showed uniform mixing with the finest particle size. Moreover, some degree of copper segregation observed at 5 and 7 h (arrow mark) became more refined and evenly distributed at 15 h. As depicted in Figure 1d–f, the average particle sizes of powders subjected to 5, 7, and 15 h of ball milling were 6.6 ± 6.4, 5.1 ± 4.3, and 3.2 ± 2.1 μm, respectively. Both the average particle size and size deviation exhibited a decrease with increasing ball milling time. In this study, ball milling was terminated at 15 h, when it was confirmed that equiaxed morphology with fine size was consistently achieved. The extension of ball milling time can lead to the intense cold welding and agglomeration of powders. Normally, an additional PCA is required to prevent these phenomena; however, the addition of PCA could influence the contamination level of final products [19]. For example, Anas et al. reported that PCA promoted powder oxidation and subsequently increased oxide contamination in bulk samples [20]. Furthermore, prolonged ball milling time could increase impurities from the ball milling medium (e.g., container and ball) [21].

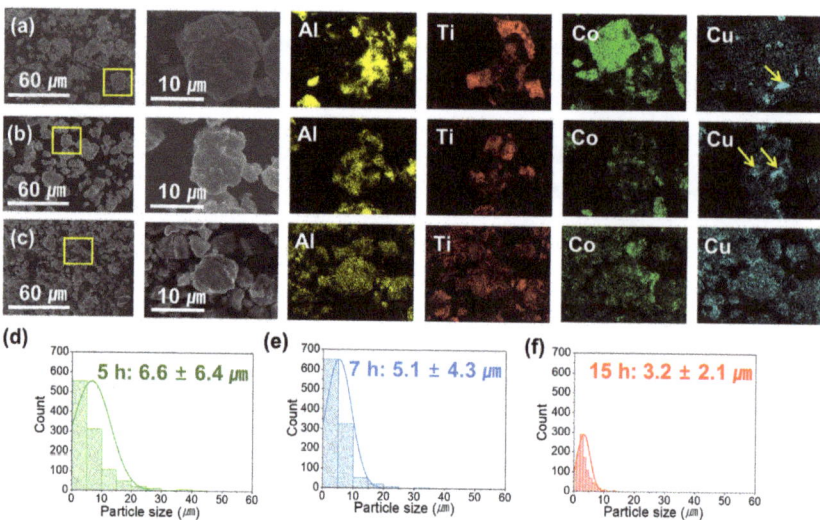

Figure 1. SEM/EDS results of the ball-milled Al–Ti–Cu–Co powders at (**a**) 5 h, (**b**) 7 h, and (**c**) 15 h, each yellow box in (**a**–**c**) indicates the magnified region on the right SEM image, respectively and the particle size distribution of the ball-milled Al–Ti–Cu–Co powders at (**d**) 5 h, (**e**) 7 h, and (**f**) 15 h.

The XRD pattern of the unmilled and the ball-milled powders after 5, 7, and 15 h are shown in Figure 2. The diffraction peaks of unmilled powders clearly showed the pattern attributed to pure elemental powder. The disappearance of diffraction peaks attributed to Al after 5 h of ball milling indicates that Al alloyed with other elements (Ti, Co, Cu). As the ball milling time at 5 and 7 h was prolonged, the decrease and broadening of the peaks corresponding to Ti and Co were observed, referring to a severe lattice distortion and grain size reduction induced by the high-impact energy generated during planetary ball milling [22]. Finally, the primary FCC peaks attributed to Cu and the minor HCP peaks corresponding to Co and Ti were observed at 15 h ball-milled powders. Additionally, the reduction in the overall peak intensity occurred, implying that a severe plastic deformation was induced by the ball milling process that contributed to fine grain sizes and increased internal stresses [23]. The primary objective at this stage is the reduction in particle size and the uniform mixing of powders, and the observed alloying reactions did not significantly impact the subsequent process.

3.2. Phase-Separated Microstructure of the Al-Ti-Cu-Co Alloys Obtained with SPS

Figure 3 shows the backscattered electron (BSE) image with EDS mapping of the Al–Ti–Cu–Co alloys (designated as 5 h–900, 7 h–900, and 15 h–900, respectively), which were sintered at 900 °C using ball-milled powders with durations of 5, 7, and 15 h. The BSE mode provided composition contrast based on atomic number, which was used to distinguish different phases in the obtained Al–Ti–Cu–Co alloys. EDS results for each phase representing the different contrasts observed in this alloy are presented in Table 1. Along with the EDS mapping and elemental composition, it could be suggested that the predominant phase corresponding to the dark gray contrast (mark 1) represented the Al–Ti–Cu–Co phase region with similar atomic ratios corresponding to the $(Cu, Co)_2AlTi$ phase. The precipitated phase with white contrast (mark 2) had a smaller volume fraction and corresponded to a Cu-rich phase. The coarse Cu-rich phase became finer and more homogeneously distributed with increasing milling time, as shown in Figure 3a–c. The EDS composition results also revealed the homogenization of Cu within the predominant (Cu, Co)$_2$AlTi phase as milling time was prolonged, i.e., 5 h (24.4 at.%) < 7 h (27.4 at.%) < 15 h (28.4 at.%). According to the previous results of Figure 1, the segregated Cu in the ball-milled powders at 5 and 7 h could affect the segregation of the Cu-rich phase during sintering. Furthermore, a finer and more uniform powder distribution without Cu segregation was finally obtained when the milling time was increased. This affected the homogeneous distribution of nano-sized Cu-rich precipitates during the SPS process.

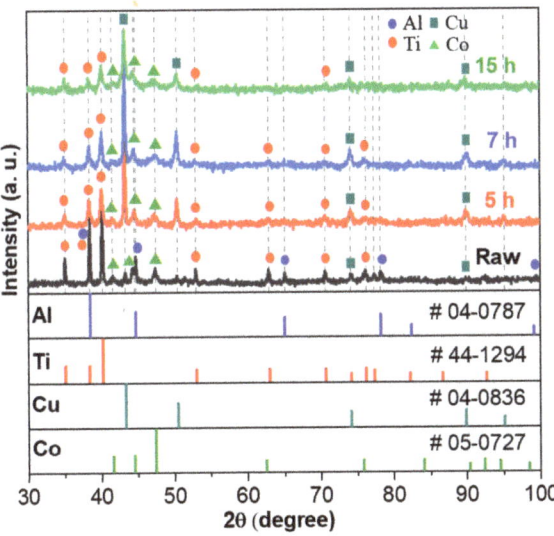

Figure 2. XRD pattern of unmilled Al–Ti–Cu–Co powders, and ball-milled Al–Ti–Cu–Co powders at different ball milling times at 5, 7, and 15 h (#: powder diffraction files (PDF) card no.).

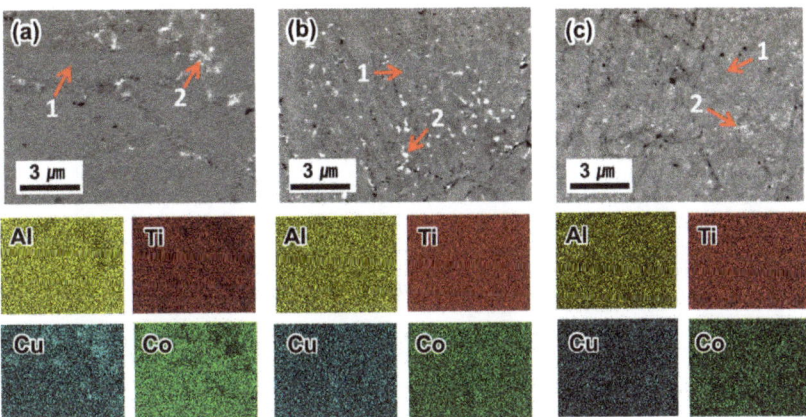

Figure 3. BSE images with EDS mapping of (**a**) 5 h–900, (**b**) 7 h–900, and (**c**) 15 h–900.

In addition, a higher concentration of Al in the Cu-rich phase was observed in EDS point analysis, indicating preferential Cu-Al bonding. Additionally, the presence of Cu in the (Cu, Co)$_2$AlTi phase suggests that Cu is also actively combined with other elements. This could be explained by the dominant role of diffusion rate during the sintering process. In general, the diffusion rate is closely related to the melting point of the elements. Elements with lower melting points exhibit weaker atomic bonds and higher diffusion coefficients [24]. The alloying sequences of the Al–Ti–Cu–Co systems could be Al → Cu → Co → Ti; therefore, Al and Cu preferentially diffuse and combine. During sintering, the molten Al (Al(*l*)) covers the surface of solid Cu (Cu(*s*)), initiating a chemical reaction on the Cu surface, ultimately leading to the formation of a Cu-rich phase. Subsequently, the remaining Al rapidly combined with other elements to form (Cu, Co)$_2$AlTi phase. Finally, Cu in the Cu-rich phase diffused and moved into the (Cu, Co)$_2$AlTi phase, leading to a decrease in the fraction of the Cu-rich phase and an increase in the Cu content of the (Cu, Co)$_2$AlTi phase.

Table 1. Chemical compositions of different phases in the 5 h–900, 7 h–900, and 15 h–900.

Sample	Region	Al (at. %)	Ti (at. %)	Co (at. %)	Cu (at. %)
5 h–900	Overall	23.45	21.85	23.69	31.01
	Dark gray (1)	22.99	25.24	27.36	24.41
	White (2) *	18.84	12.48	10.88	57.79
7 h–900	Overall	23.06	23.49	23.47	29.98
	Dark gray (1)	23.20	24.69	24.73	27.38
	White (2) *	20.42	14.16	16.03	49.38
15 h–900	Overall	23.19	23.39	23.93	29.48
	Dark gray (1)	22.77	23.22	25.57	28.44
	White (2) *	23.79	19.83	14.84	41.55

* Note: due to the limitations of EDS resolution, the Cu composition may be underestimated by the interaction volume of the nano-sized white phase (2) and dark gray phase (1).

Figure 4 shows the XRD patterns of the 5 h–900, 7 h–900, and 15 h–900 samples. Compared to the XRD patterns of the ball-milled powder, it showed that phase transformation occurred after the SPS. The microstructure of the Al–Ti–Cu–Co alloys dominantly exhibited diffraction peaks corresponding to the Heusler L2$_1$ phase (AlCu$_2$Mn prototype, Fm3m, cF16). The L2$_1$ phase is a highly ordered BCC-based structure that consists of eight smaller BCC lattices [25]. The Cu-rich phase was not detected due to its very low volume fraction (<2 vol.%) and nanocrystalline property [26]. In multi-component alloys, the crystal structure can be significantly distorted because of the randomness with which different-sized atoms occupy lattice points. This intrinsic lattice distortion can increase the scattering effect, ultimately weakening the detectable diffraction signal [27]. Combined with BSE and EDS results, the dominant phase is (Cu, Co)$_2$AlTi, represented by the dark gray phase with most of the volume fraction (Figure 3).

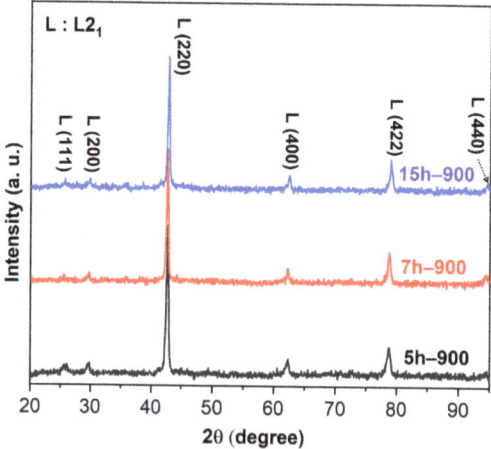

Figure 4. XRD results of the Al–Ti–Cu–Co alloys corresponding to 5 h–900, 7 h–900, and 15 h–900.

Figure 5 depicts the inverse pole figure (IPF) of the 5 h–900, 7 h–900, and 15 h–900, respectively. The Al–Ti–Cu–Co alloy showed a random crystalline orientation. It is noteworthy that the grain size became finer from 3.1 ± 1.5 µm, 1.8 ± 1.0 µm to 1.0 ± 0.5 µm, depending on the increased ball milling time of the powders. The grain morphology exhibited a more equiaxed shape. The fine grain size achieved after ball milling significantly influenced the small grain size reduction by SPS; the SPS method is known for its rapid heating rate and short holding time, which promotes alloy densification and controlled grain growth [28]. The results suggested that the fine particle size induced by ball milling led to the fine grain size of the L2$_1$ phase and the Cu-rich phase due to the fast heating and cooling rates of SPS.

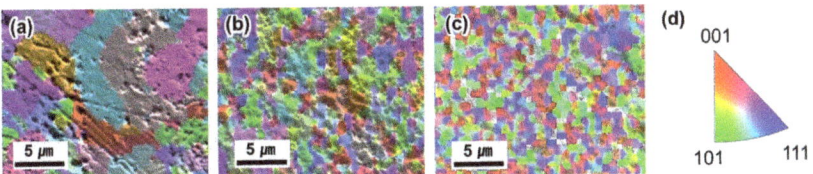

Figure 5. IPF orientation colored grain map overlaid with EBSD band contrast image of (**a**) 5 h–900, (**b**) 7 h–900, (**c**) 15 h–900, and (**d**) standard stereographic triangle.

L2$_1$ and B2 are indistinguishable in XRD peaks because they have essentially the same crystal structure. Hence, the HRTEM/EDS analysis was performed on the cross-section of the 7 h–900 sample for the additional clarification of the phases. As depicted in Figure 6a–f, a nanoscale Ti-rich phase with a light gray contrast (marked in a blue circle) is distributed within the Cu-rich phase region with a white contrast. The dark spots on the sample indicate aluminum oxide inclusions produced by small amounts of oxygen adsorbed on the powder during the powder-handling process. Figure 6g–i depicts the ordered L2$_1$ phase structure, and the Cu-rich phase revealed heterogeneous structures including precipitates. In the Al–Ti–Cu–Co alloy system, a phase separation into the B2 and L2$_1$ phases has been previously reported at 1173 K (900 °C), and the presence of the B2 phase can be observed in the Ti-rich portion [29]. Therefore, in this study, the Ti-rich phase within the Cu-rich phase can be presumed to form the B2 phase. Consequently, HRTEM/EDS results further confirmed a phase-separated microstructure. Furthermore, the Cu-rich phase region was observed as a network phase that interconnects the L2$_1$ and B2 phases.

3.3. Mechanical Properties

Figure 7a shows the hardness and density graph of the Al-Ti–Cu–Co alloys and the previously reported Al-Ti-containing multicomponent alloys as a comparative material. The hardness values of 5 h–900, 7 h–900, and 15 h–900 samples were 684 ± 15, 710 ± 9, and 791 ± 9 HV, respectively, which increased with prolonged ball milling time. Especially, a notable sharp rise was observed in the 15 h–900 sample. The Al–Ti–Cu–Co alloys exhibited a prominent combination of density and hardness compared to previously reported Al-Ti-containing multicomponent alloys. In addition, the indentation mark was distinguishable and had no cracking (Figure 7b), implying that the material could absorb enough energy to arrest cracking without embrittlement [30]. Furthermore, Table 2 presents the calculated strength and measured density of the Al–Ti–Cu–Co alloys compared with conventional metallic alloys. The Al–Ti–Cu–Co alloys exhibited relatively high specific strength (419–489 MPa cm^{3}g^{-1}) compared to conventional materials, such as Inconel 718 (1050 MPa cm^{3}g^{-1}) [10] and ultra-high tensile strength steel (1500 MPa cm^{3}g^{-1}) [31]. Finally, the Al–Ti–Cu–Co alloys could be considered promising candidate materials for advanced applications that are required for lightweight and high performance.

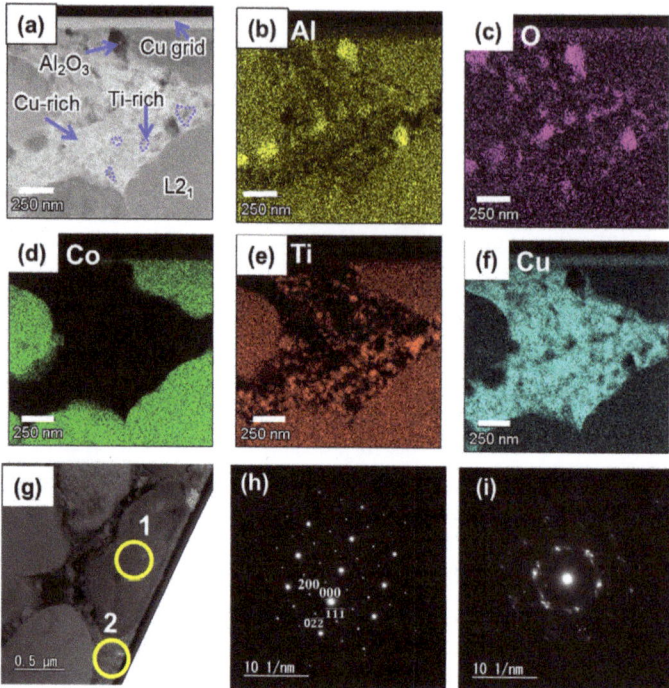

Figure 6. (**a**) TEM image of the 7 h-900, and TEM-EDS mapping elemental distribution of (**b**) Al, (**c**) O, (**d**) Co, (**e**) Ti, and (**f**) Cu, (**g**) bright-field TEM image, (**h**) diffraction pattern taken under the [110] zone axis from L2$_1$ phase (marked 1 in (**g**)), and (**i**) diffraction pattern taken under the [111] zone axis from Cu-rich network phase (marked 2 in (**g**)).

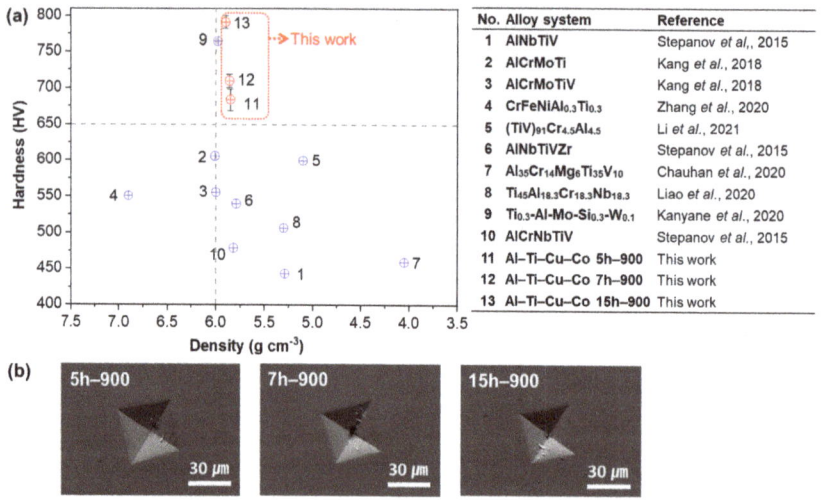

Figure 7. (**a**) Hardness and density graph of the Al–Ti–Cu–Co alloys compared to Al-Ti-containing multicomponent alloys [3,12,26,32–37]. (**b**) Indentation mark obtained after Vickers hardness test.

Table 2. The density, hardness, and strength of Al–Ti–Cu–Co alloys, Inconel 718, and ultra-high tensile strength steel (HTSS).

Alloys	Measured Density (g cm^{-3})	Hardness (HV)	Strength (MPa) *	Specific Strength (MPa cm^{3}g^{-1})
5 h–900	5.85 ± 0.02	684 ± 15	2451	419
7 h–900	5.86 ± 0.02	710 ± 9	2557	436
15 h–900	5.90 ± 0.02	791 ± 9	2887	489
Inconel 718 [12]	8.18	355	1050	128
HTSS [31]	7.90	540	1500	190

* Calculated using the relationship. HV ≈ $3\sigma_y$.

The strengthening mechanism of the Al–Ti–Cu–Co alloys, which exhibit excellent specific strength, can be explained as follows. Firstly, the high hardness of the Al–Ti–Cu–Co alloys is attributed to the presence of the L2$_1$ phase. Zhang et al. reported that the BCC and ordered BCC phases in Al-Ti-containing multicomponent alloys increase both hardness and strength [33]. As a result, the formation of the highly ordered L2$_1$ phase as the major phase contributed significantly to the high hardness of Al–Ti–Cu–Co alloys. Secondly, the particle size of the starting powder decreased as the ball milling time increased, resulting in a finer phase-separated structure after SPS. The reduced particle size and increased lattice strain generated by ball milling can accelerate the short-distance diffusion of constituent elements during SPS by providing more interfaces for phase formation [38]. This resulted in grain size reduction and uniform dispersion of the L2$_1$ and Cu-rich phases, as evidenced by the increase in Cu concentration in the L2$_1$ phase due to the facilitated elemental diffusion during extended ball milling time. In addition, the facilitated short-distance diffusion during SPS can result in the formation of non-equilibrium phases [39]. In conventional sintering methods like HIP, slow and extended heating and cooling rates allow sufficient time for the material to attain a state of thermodynamic equilibrium, leading to the formation of equilibrium phases through long-distance diffusion. In contrast, the rapid heating and short-time sintering in SPS facilitated the formation of optimized non-equilibrium phases through localized heating of the powder surface. As a result, the regulation of non-equilibrium phases yields distinctive material properties not achievable through conventional sintering processes. Lastly, precipitation hardening was achieved through the formation of Cu-rich phases, along with the B2 Ti-rich nanoprecipitate phase. The nanoscale size and uniform distribution of precipitates played a crucial role in precipitation strengthening. Nano-precipitates have been reported to act as a strengthening phase, impeding dislocation movement and enhancing the hardness [40]. Previous studies have demonstrated that the presence of L2$_1$ phases, coupled with other precipitates, contributes to the enhancement of mechanical properties in multi-component alloys [41]. Therefore, nano-precipitates could significantly enhance the strength of the material through the precipitation strengthening effect. Consequently, phase-separated microstructure consisting of the finer grain L2$_1$ phase and homogeneous distribution of Cu-rich phase led to a notable improvement in hardness. Finally, the Cu-rich phase was reinforced with a nanoscale Ti-rich B2 phase and served as a network phase connecting the hard L2$_1$ and B2 phases, thereby mitigating embrittlement.

4. Conclusions

In this study, the phase separation microstructure of the Al–Ti–Cu–Co alloy was regulated by powder metallurgy processes and SPS using metal powder with varying particle sizes. Efforts were made to optimize the phase separation microstructure of the Al–Ti–Cu–Co alloys, and it was shown that the hardness increased, and brittle property was also suppressed, albeit at the Vickers level. Microstructure control of the L2$_1$ phase of (Cu, Co)$_2$AlTi and the Cu-rich phase incorporating the nanoscale Ti-rich B2 phase was achieved by increasing the ball milling time. The grain size reduction after powder metallurgy processes using SPS resulted in an enhancement in hardness (i.e., the hardness values of

5 h–900, 7 h–900, and 15 h–900 were 684 ± 15, 710 ± 9, and 791 ± 9 HV, respectively). This microstructure optimization can be achieved due to the rapid temperature rise and drop rate of the SPS process, which is a spatially non-equilibrium process. The materials obtained in this study, the Al–Ti–Cu–Co alloys, showed a good combination of hardness and density properties of 684–791 HV and 5.8–5.9 g/cm^3 (<6 g/cm^3) compared to other Al-Ti-multicomponent alloys and conventional materials (e.g., Inconel 718 and ultra-high strength steels).

Supplementary Materials: The following supporting information can be downloaded at: https://www.mdpi.com/article/10.3390/ma17020304/s1. Figure S1: Heating and cooling profile during SPS. Figure S2: SEM images of ball-milled Al–Ti–Cu–Co powders at (a–c) 5 h, (d–f) 7 h, (g–i) 15 h.

Author Contributions: S.L.: Conceptualization, Methodology, Investigation, and Writing—Original Draft. C.C.: Project administration and Writing—Review and Editing. Y.S.: Resources and Supervision. S.Y.: Formal analysis. N.S.: Supervision and Writing—Review and Editing. All authors have read and agreed to the published version of the manuscript.

Funding: This research is partially supported by "Knowledge Hub Aichi", Priority Research Project from the Aichi Prefectural Government. This research was financially supported by Open Innovation Platform with Enterprises, Research Institute, and Academia (OPERA, Grant No: JPMJOP1843) and Strategic International Collaborative Research Program (SICORP, Grant No: JPMJSC18H1) of Japan Science and Technology Agency.

Institutional Review Board Statement: Not applicable.

Informed Consent Statement: Not applicable.

Data Availability Statement: Data are contained within the article and supplementary materials.

Acknowledgments: We would like to thank Takeshi Hagio, Masao Tabuchi, Yoshio Watanabe, and Eiji Abe, for their valuable advice on characterization.

Conflicts of Interest: The authors declare no conflict of interest.

References

1. Baqer, Y.M.; Ramesh, S.; Yusof, F.; Manladan, S. Challenges and advances in laser welding of dissimilar light alloys: Al/Mg, Al/Ti, and Mg/Ti alloys. *Int. J. Adv. Manuf. Technol.* **2018**, *95*, 4353–4369. [CrossRef]
2. Huang, X.; Miao, J.; Luo, A.A. Lightweight AlCrTiV high-entropy alloys with dual-phase microstructure via microalloying. *J. Mater. Sci.* **2019**, *54*, 2271–2277. [CrossRef]
3. Liao, Y.; Li, T.; Tsai, P.; Jang, J.; Hsieh, K.; Chen, C.; Huang, J.; Wu, H.-J.; Lo, Y.-C.; Huang, C. Designing novel lightweight, high-strength and high-plasticity Ti$_x$(AlCrNb)$_{100-x}$ medium-entropy alloys. *Intermetallics* **2020**, *117*, 106673. [CrossRef]
4. Yen, S.-Y.; Murakami, H.; Lin, S.-K. Low-density CoAlTi-B2 strengthened Al-Co-Cr-Mo-Ti bcc refractory high-entropy superalloy designed with the assistance of high-throughput CALPHAD method. *J. Alloys Compd.* **2023**, *952*, 170027. [CrossRef]
5. Oh, M.C.; Sharma, A.; Lee, H.; Ahn, B. Phase separation and mechanical behavior of AlCoCrFeNi-X (X = Cu, Mn, Ti) high entropy alloys processed via powder metallurgy. *Intermetallics* **2021**, *139*, 107369. [CrossRef]
6. Zhou, L.; Miller, M.K.; Lu, P.; Ke, L.; Skomski, R.; Dillon, H.; Xing, Q.; Palasyuk, A.; McCartney, M.; Smith, D. Architecture and magnetism of alnico. *Acta Mater.* **2014**, *74*, 224–233. [CrossRef]
7. Feng, R.; Gao, M.C.; Zhang, C.; Guo, W.; Poplawsky, J.D.; Zhang, F.; Hawk, J.A.; Neuefeind, J.C.; Ren, Y.; Liaw, P.K. Phase stability and transformation in a light-weight high-entropy alloy. *Acta Mater.* **2018**, *146*, 280–293. [CrossRef]
8. Palm, M.; Sauthoff, G. Deformation behaviour and oxidation resistance of single-phase and two-phase L21-ordered Fe–Al–Ti alloys. *Intermetallics* **2004**, *12*, 1345–1359. [CrossRef]
9. Yamada, R.; Mohri, T. Origin of the phase separation into B2 and L21 ordered phases in X–Al–Ti (X: Fe, Co, and Ni) alloys based on the first-principles cluster variation method. *J. Phys. Condens. Matter* **2020**, *32*, 174002. [CrossRef]
10. Ishikawa, K.; Kainuma, R.; Ohnuma, I.; Aoki, K.; Ishida, K. Phase stability of the X$_2$AlTi (X: Fe, Co, Ni and Cu) Heusler and B2-type intermetallic compounds. *Acta Mater.* **2002**, *50*, 2233–2243. [CrossRef]
11. Zhang, Y.; Zuo, T.T.; Tang, Z.; Gao, M.C.; Dahmen, K.A.; Liaw, P.K.; Lu, Z.P. Microstructures and properties of high-entropy alloys. *Prog. Mater Sci.* **2014**, *61*, 1–93. [CrossRef]
12. Chauhan, P.; Yebaji, S.; Nadakuduru, V.N.; Shanmugasundaram, T. Development of a novel light weight Al$_{35}$Cr$_{14}$Mg$_6$Ti$_{35}$V$_{10}$ high entropy alloy using mechanical alloying and spark plasma sintering. *J. Alloys Compd.* **2020**, *820*, 153367. [CrossRef]

13. Montero, J.; Zlotea, C.; Ek, G.; Crivello, J.-C.; Laversenne, L.; Sahlberg, M. TiVZrNb multi-principal-element alloy: Synthesis optimization, structural, and hydrogen sorption properties. *Molecules* **2019**, *24*, 2799. [CrossRef] [PubMed]
14. Murali, M.; Babu, S.K.; Krishna, B.J.; Vallimanalan, A. Synthesis and characterization of AlCoCrCuFeZnx high-entropy alloy by mechanical alloying. *Prog. Nat. Sci. Mater. Int.* **2016**, *26*, 380–384. [CrossRef]
15. Tang, Z.; Senkov, O.N.; Parish, C.M.; Zhang, C.; Zhang, F.; Santodonato, L.J.; Wang, G.; Zhao, G.; Yang, F.; Liaw, P.K. Tensile ductility of an AlCoCrFeNi multi-phase high-entropy alloy through hot isostatic pressing (HIP) and homogenization. *Mater. Sci. Eng. A* **2015**, *647*, 229–240. [CrossRef]
16. Praveen, S.; Murty, B.; Kottada, R.S. Phase evolution and densification behavior of nanocrystalline multicomponent high entropy alloys during spark plasma sintering. *JOM* **2013**, *65*, 1797–1804. [CrossRef]
17. Ahmed, H.M.; Ahmed, H.A.; Hefni, M.; Moustafa, E.B. Effect of grain refinement on the dynamic, mechanical properties, and corrosion behaviour of Al-Mg alloy. *Metals* **2021**, *11*, 1825. [CrossRef]
18. Babalola, B.J.; Shongwe, M.B.; Obadele, B.A.; Olubambi, P.A. Densification, microstructure and mechanical properties of spark plasma sintered Ni-17% Cr binary alloys. *Int. J. Adv. Manuf. Technol.* **2019**, *101*, 1573–1581. [CrossRef]
19. Kleiner, S.; Bertocco, F.; Khalid, F.; Beffort, O. Decomposition of process control agent during mechanical milling and its influence on displacement reactions in the Al–TiO2 system. *Mater. Chem. Phys.* **2005**, *89*, 362–366. [CrossRef]
20. Anas, N.; Ramakrishna, M.; Dash, R.; Rao, T.N.; Vijay, R. Influence of process control agents on microstructure and mechanical properties of Al alloy produced by mechanical alloying. *Mater. Sci. Eng. A* **2019**, *751*, 171–182. [CrossRef]
21. Canakci, A.; Erdemir, F.; Varol, T.; Patir, A. Determining the effect of process parameters on particle size in mechanical milling using the Taguchi method: Measurement and analysis. *Measurement* **2013**, *46*, 3532–3540. [CrossRef]
22. Chermahini, M.D.; Sharafi, S.; Shokrollahi, H.; Zandrahimi, M.; Shafyei, A. The evolution of heating rate on the microstructural and magnetic properties of milled nanostructured $Fe_{1-x}Co_x$ (x = 0.2, 0.3, 0.4, 0.5 and 0.7) powders. *J. Alloys Compd.* **2009**, *484*, 54–58. [CrossRef]
23. Basariya, M.R.; Srivastava, V.; Mukhopadhyay, N. Effect of milling time on structural evolution and mechanical properties of garnet reinforced EN AW6082 composites. *Metall. Mater. Trans. A* **2015**, *46*, 1360–1373. [CrossRef]
24. Chen, Y.-L.; Hu, Y.-H.; Hsieh, C.-A.; Yeh, J.-W.; Chen, S.-K. Competition between elements during mechanical alloying in an octonary multi-principal-element alloy system. *J. Alloys Compd.* **2009**, *481*, 768–775. [CrossRef]
25. Qian, F.; Sharp, J.; Rainforth, W.M. Characterisation of L21-ordered Ni2TiAl precipitates in FeMn maraging steels. *Mater. Charact.* **2016**, *118*, 199–205. [CrossRef]
26. Kang, M.; Lim, K.R.; Won, J.W.; Lee, K.S.; Na, Y.S. Al-ti-containing lightweight high-entropy alloys for intermediate temperature applications. *Entropy* **2018**, *20*, 355. [CrossRef]
27. Yeh, J.-W.; Chang, S.-Y.; Hong, Y.-D.; Chen, S.-K.; Lin, S.-J. Anomalous decrease in X-ray diffraction intensities of Cu–Ni–Al–Co–Cr–Fe–Si alloy systems with multi-principal elements. *Mater. Chem. Phys.* **2007**, *103*, 41–46. [CrossRef]
28. Keller, C.; Tabalaiev, K.; Marnier, G.; Noudem, J.; Sauvage, X.; Hug, E. Influence of spark plasma sintering conditions on the sintering and functional properties of an ultra-fine grained 316L stainless steel obtained from ball-milled powder. *Mater. Sci. Eng. A* **2016**, *665*, 125–134. [CrossRef]
29. Kainuma, R.; Ohnuma, I.; Ishikawa, K.; Ishida, K. Stability of B2 ordered phase in the Ti-rich portion of Ti–Al–Cr and Ti–Al–Fe ternary systems. *Intermetallics* **2000**, *8*, 869–875. [CrossRef]
30. Keryvin, V.; Hoang, V.; Shen, J. Hardness, toughness, brittleness and cracking systems in an iron-based bulk metallic glass by indentation. *Intermetallics* **2009**, *17*, 211–217. [CrossRef]
31. Mori, K.-I.; Bariani, P.; Behrens, B.-A.; Brosius, A.; Bruschi, S.; Maeno, T.; Merklein, M.; Yanagimoto, J. Hot stamping of ultra-high strength steel parts. *Cirp Ann.* **2017**, *66*, 755–777. [CrossRef]
32. Stepanov, N.; Shaysultanov, D.; Salishchev, G.; Tikhonovsky, M. Structure and mechanical properties of a light-weight AlNbTiV high entropy alloy. *Mater. Lett.* **2015**, *142*, 153–155. [CrossRef]
33. Zhang, M.; Ma, Y.; Dong, W.; Liu, X.; Lu, Y.; Zhang, Y.; Li, R.; Wang, Y.; Yu, P.; Gao, Y. Phase evolution, microstructure, and mechanical behaviors of the CrFeNiAlxTiy medium-entropy alloys. *Mater. Sci. Eng. A* **2020**, *771*, 138566. [CrossRef]
34. Li, D.; Dong, Y.; Zhang, Z.; Zhang, Q.; Chen, S.; Jia, N.; Wang, H.; Wang, B.; Jin, K.; Xue, Y. An as cast Ti-V-Cr-Al light-weight medium entropy alloy with outstanding tensile properties. *J. Alloys Compd.* **2021**, *877*, 160199. [CrossRef]
35. Stepanov, N.; Yurchenko, N.Y.; Shaysultanov, D.; Salishchev, G.; Tikhonovsky, M. Effect of Al on structure and mechanical properties of Al_xNbTiVZr (x = 0, 0.5, 1, 1.5) high entropy alloys. *Mater. Sci. Technol.* **2015**, *31*, 1184–1193. [CrossRef]
36. Kanyane, L.; Popoola, A.; Malatji, N.; Sibisi, P. Synthesis and characterization of Ti_xAlSi$_x$MoW light-weight high entropy alloys. *Mater. Today Proc.* **2020**, *28*, 1231–1238. [CrossRef]
37. Stepanov, N.; Yurchenko, N.Y.; Skibin, D.; Tikhonovsky, M.; Salishchev, G. Structure and mechanical properties of the AlCr$_x$NbTiV (x = 0, 0.5, 1, 1.5) high entropy alloys. *J. Alloys Compd.* **2015**, *652*, 266–280. [CrossRef]
38. Zhang, K.; Fu, Z.; Zhang, J.; Shi, J.; Wang, W.; Wang, H.; Wang, Y.; Zhang, Q. Nanocrystalline CoCrFeNiCuAl high-entropy solid solution synthesized by mechanical alloying. *J. Alloys Compd.* **2009**, *485*, L31–L34. [CrossRef]
39. Cavaliere, P. *Spark Plasma Sintering of Materials: Advances in Processing and Applications*; Springer: Berlin/Heidelberg, Germany, 2019.

40. Gwalani, B.; Soni, V.; Lee, M.; Mantri, S.; Ren, Y.; Banerjee, R. Optimizing the coupled effects of Hall-Petch and precipitation strengthening in a $Al_{0.3}CoCrFeNi$ high entropy alloy. *Mater. Des.* **2017**, *121*, 254–260. [CrossRef]
41. Wang, L.; Wang, J.; Niu, H.; Yang, G.; Yang, L.; Xu, M.; Yi, J. BCC+ L21 dual-phase $Cu_{40}Al_{20}Ti_{20}V_{20}$ near-eutectic high-entropy alloy with a combination of strength and plasticity. *J. Alloys Compd.* **2022**, *908*, 164683. [CrossRef]

Disclaimer/Publisher's Note: The statements, opinions and data contained in all publications are solely those of the individual author(s) and contributor(s) and not of MDPI and/or the editor(s). MDPI and/or the editor(s) disclaim responsibility for any injury to people or property resulting from any ideas, methods, instructions or products referred to in the content.

Article

Microstructure and Properties Variation of High-Performance Grey Cast Iron via Small Boron Additions

Grega Klančnik [1,*], Jaka Burja [2], Urška Klančnik [3], Barbara Šetina Batič [2], Luka Krajnc [1] and Andrej Resnik [4]

[1] Pro Labor, d.o.o., Podvin 20, SI-3310 Žalec, Slovenia
[2] Institute of Metals and Technology, Lepi pot 11, SI-1000 Ljubljana, Slovenia
[3] Valji d.o.o., Železarska Cesta 3, SI-3220 Štore, Slovenia
[4] Omco Metals Slovenia d.o.o. Cesta Žalskega Tabora 10, SI-3310 Žalec, Slovenia
* Correspondence: grega.klancnik@prolabor.si

Abstract: A study was undertaken to investigate the effects of small boron additions on the solidification and microstructure of hypo-eutectic alloyed grey cast iron. The characteristic temperatures upon crystallisation of the treated metal melt were recorded, specifically those concerning small boron addition by using thermal analysis with the ATAS system. Additionally, a standardised wedge test was set to observe any changes in chill performance. The microstructures of thermal analysis samples were analysed using a light optical microscope (LOM) and field emission scanning electron microscopy (FE-SEM) equipped with energy dispersive spectroscopy (EDS), which reveal variations in graphite count number with the addition of boron within observed random and undercooled flake graphite. The effect of boron was estimated by the classical analytical and statistical approach. The solidification behaviour under equilibrium conditions was predicted by a thermodynamic approach using Thermo-Calc. Based on all gathered data, a response model was set with boron for given melt quality and melt treatment using the experimentally determined data. The study reveals that boron as a ferrite and carbide-promoting element under the experimental set shows weak nucleation potential in synergy with other heterogenic nuclei at increased solidification rates, but no considerable changes were observed by the TA samples solidified at slower cooling rates, indicating the loss of the overall inoculation effect. The potential presence of boron nitride as an inoculator for graphite precipitation for a given melt composition and melt treatment was not confirmed in this study. It seems that boron at increased solidification rates can contribute to overall inoculation, but at slower cooling rates these effects are gradually lost. In the last solidification range, an increased boron content could have a carbide forming nature, as is usually expected. The study suggests that boron in traces could affect the microstructure and properties of hypo-eutectic alloyed grey cast iron.

Keywords: boron; cast iron; solidification; thermal analysis; Thermo-Calc

1. Introduction

The nucleation and growth of graphite is complex. Based on Stefanescu et al. [1], the graphite aggregates in iron–carbon melts crystallise via hexagonal faceted graphite platelets, and impurities (O, S, etc.) inside the metal melt may change the graphite growth considerably by affecting the thickening mechanism of the graphene sheets. The impurities and formed nuclei can affect graphite morphology and shape considerably. Oxides (MgO, AlO); nitrides and carbo–nitrides ((Ti, Zr)(C, N), AlN, (Mg, Si, Al)N, TiN etc.); sulfides (MgS, (Mg, Ca)S, MnS, etc.); phosphides ((Mg, La, Ce)P); sulpho-oxides (Mg, Ca(SO), etc.); or complexes between all of them are all considered for potential nuclei [1–4]. Based on Sommerfeld et al., [3] MnS precipitates are preferred sites for graphite nucleation in iron cast melts based on investigations done on EN-GJL-200, but the temperature for MnS formation changes in relation to the melt chemical composition, for example with the addition of Ti and resulting Ti–carbo–sulfides, which deplete sulphur in the melt needed for MnS formation at higher temperatures. Based on Stefan et al. [5], complex second compound nucleates on manganese sulfides (Mn, X)S are found for graphite nucleants, where X = Mg, Ca, Sr, Al; all in sizes under 10 μm. It is notable that oxy-forming elements, such as Mg, improve the (Mn, X)S graphite nucleation by forming a thin superficial layer as oxide or oxy-sulfide. This is presumably due to a better crystallographic compatibility compared to purely sulfide with cubic system. According to Campbell [6], mostly liquid oxide films (bi-films) of different compositions based on SiO_2 act as substrates for nucleation of oxy-sulfide nuclei and the growth of flake and nodular graphite. In the case of boron, which is commonly used for white irons, liquid borates are expected to form. Based on Campbell [6] liquid iron–borate can form, preventing oxide formation substrates for graphite growth. Which exact type of M_XB_Y will form is related to the melt composition. However, boron is an element with an affinity to oxygen, potentially forming oxy-sulfides as well as nitrides. It seems that B_2O_3, if formed, is not considered to be a highly stable oxide based on a Richardson–Ellingham diagram, as shown later in the paper. Based on [7–9], boron, in combination with Cr, affected carbide and boride formation significantly in white cast iron and high-speed alloy steels. Boron in traces inside different systems is considered as infinite dilution [10]. It can also be considered as an impurity in some cases. Based on the same assumption, relationships that are close to linear are expected and therefore also observed in this study.

Data in the literature regarding the effects of boron in cast irons identifies boron as an influential and strong carbide stabiliser and ferrite-promoting element; therefore, it is an element that has the potential power for affecting the matrix hardness (with boric pearlite, etc.) and overall hardness. The formed boron complexes are commonly described as α-Fe + Fe_3C + Fe_2B. It is also recognised that certain elements, such as Sr and Sb can enhance the hardness of pearlite; therefore, the effects of boron can be expected to be balanced with certain inoculation/elemental additions. It is recognised that time—and an additional quantity of inoculants—effects the microstructure and hardness in lamellar graphite cast irons [11]. For example, by the microalloying in the grey lamellar iron with simultaneous phosphorous and boron (P + B), the micro-abrasive wear was studied where increasing hardness of pearlitic matrix and refined interlamellar spacing was noticed [12]; however, it is not clear how much this was contributed by the boron addition.

A hardness increase in chill is possible due to the formation of a carbide network. The potential hardness drop can also be expected with the addition of boron due to ferrite promotion, but it is possible to counteract with N and Ti on high-strength GJS [13,14]. These responses are usually not comparable overall, despite the fact that an overall average hardness increase is mostly expected with boron for grey iron. However, it was also shown that hardness decrease can also appear after exceeding a certain boron limit. This limit of different material behaviour seems to vary considerably with different iron grades. This is potentially related to synergistic effects with other active-forming elements, such as the aforementioned Ti and N. The boron effect is potentially sensible also to CEQ. As boron is recognised as a ferrite stabiliser—and this could be unwanted for pearlitic grades but beneficial for grades which are aimed to be mainly ferritic in the final condition (as ductile GJS types). However, chill with an excessive presence of carbide network should be avoided. In the case of Keivan, et al. [15] no considerable change in graphite size with boron was noticed up to 131 ppm of added B for GJS type on different thickness sections, but the effect was visible as a change in ferrite content. Similarly, no change in distribution and number density of graphite nodules was noticed by Bugten, et al. [16] up to 130 ppm B. In the case of certain ductile irons, the boron seems to degrade the graphite sphericity towards compacted (vermicular) graphite and degrades the needed mechanical properties compared to nodular/ductile irons. The degradation of mechanical properties is also partially the result of primary carbide precipitation. This also seems to be valid for thicker sections. However, it seems that this boron response is not always consistent, as shown in the patent and in [13,16,17], where boron in ductile iron minimises the dissolved nitrogen in the molten iron by estimating that BN were formed and are potentially working as new heterogeneous nuclei for graphite precipitation, consequently increasing the nodule count, promoting the ferrite content, diminishing the nitrogen effect on carbide stability and affecting the stability of pearlite during annealing. This was achieved by setting the optimal boron content [15,17].

In some cases, excessive boron (>200 ppm B) for plain grey cast iron presumably promotes B-type and D-type graphitic structures, therefore decreasing tensile strength over certain boron limits. The effect of boron on annealing is quite contradictory, as boron can work as a graphite promoter in the case of malleable irons by promoting temper carbon from carbides, and it also works as a carbide stabiliser in white cast irons for increasing wear and—in some cases, such as for nodular—improves the carbide's annealing stability. Based on shorter diffusion paths due to higher nucleation graphite count, the annealing ability could also be improved [18–21].

It is recognised that boron narrows the eutectic region between stable (EG) and metastable solidification (EC); $\Delta T_E = T_{EG} - T_{EC}$, along with V, Cr, Mn, Ti, and Mo. The intensity of the ΔT_E change is also related to the distribution coefficient of elements between cementite and austenite, as with other elements, and is therefore dictated by the actual solidification rate [22]. Interestingly, a controlled addition of boron and nitrogen can also have a positive effect on the control of heavy grey iron castings by the potential formation of boron-based nuclei for graphite growth. However, the actual presence of BN has not yet been confirmed [23].

Based on the above, boron appears to be a sensible element, and sensibility rises with boron content. When added intentionally, the elemental yield varies, but when boron is already in the scrap or iron, the boron is fully preserved (high yield). When boron is entrapped in the metal melt it can have an important influence on the quality of produced as-cast irons and on annealed grey irons [19,20,24]. The literature shows that boron can promote heterogeneous graphite distribution, variations in mechanical properties after it exceeds certain limits and that interactional effects with other elements are also possible.

In this study, an alloyed grey cast iron is considered instead of plain cast iron for testing the interactions with intentionally added boron during solidification. The boron in this study is highly limited and the limitation is set from the measurements established under industrial environments where boron is considered as contaminant.

2. Materials and Methods

The chemical composition was controlled by using the starting non-boron (unmodified) cast iron used for remelting to have repeatable behaviour during solidification. This would also help to achieve repeatable chemical composition (which was taken as the reference material further in this paper). The tight elemental control is important for controlling the starting eutectic solidification interval (freezing) and to observe the main target factor considered as a single factorial test, namely the boron influence. Additionally, the temperature profile, furnace reheating, maximum allowable temperatures, time of melting and casting temperatures were carefully controlled for maximising the repeatability of the tests. The melting was carried out in an open-air induction furnace equipped with a furnace isolative lid for Ar surface purging to minimise slag formation. The inner wall of the high-frequency furnace was made of MgO-based refractory. The unmodified material was cut into smaller pieces within similar melting and tapping times and surface-to-volume ratios for comparable oxide entrapment via surface contaminants that potentially also work as nuclei (affecting active carbon equivalent, ACEL). Secondary remelting of the starting non-boron grey iron can help improve the chemical homogeneity of prepared final melts. For sufficient and repeated inoculation, FeSi(low Al) was used in all cases (approx. 0.4 wt.%) to promote repeatable starting heterogeneous nucleation sites. The time interval between FeSi addition and first tapping is 5 min. The primary melt treatment was already carried out in the induction furnace just before tapping into the well-preheated graphite ladle. It was then cast into the Thermal Analysis cell (TA) and wedge (chill) cells with and without boron, in this order. In the case of TA, all pourings are made without the addition of strong carbide forming tellurium, as this was done before the test to observe the suitability of the equations used in this study for white solidification temperature prediction ($T_{EC} = T_{Metastable}$). No additional chill was allowed for TA or the wedge. As a tool for analysing the recorded solidification nature, ATAS MetStar tool (Nova Cast, Ronneby, Sweden) was used. Microstructural observation on prepared metallographic samples was done by ImageJ (National Institutes of Health, USA) [25] (https://imagej.net/). The maximum field for graphite count detection per image was 1027 µm × 1369 µm. The FeB additions were introduced as a fine powder by small additions in both mould cavities of the TA cell and wedge cell to maximise the boron yield. The yield was not constant; therefore, all boron values are represented only as measured ones. The experimental setup is presented in a way that observes the short-time inoculation effect (last moment "in-mold" inoculation), if it exists with the addition of boron, as suggested in the literature. In the case of TA, the samples were extracted just under the thermocouple junction to have representative values coupled with the TA data.

The wedge cavity was prepared with dry silica sand and by using a special 3D-printed core, as shown in Figure 1a. For TA analysis, in Figure 1b, a regular TA cup (Nova Cast, Ronneby, Sweden) was used equipped with a K-type thermocouple calibrated within ±1.2 °C at 1000 °C [26].

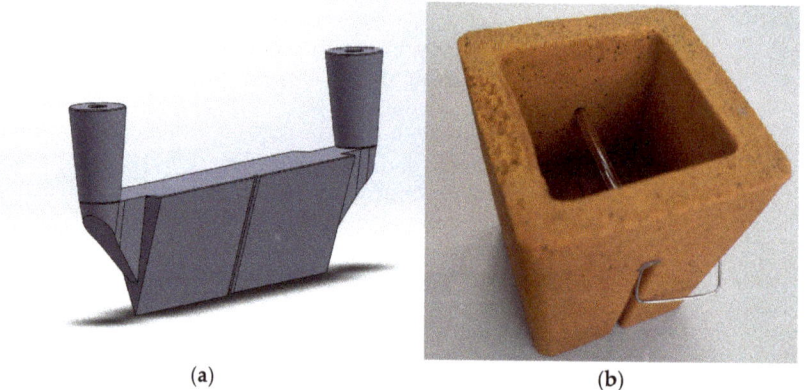

(a) (b)

Figure 1. An example of (**a**) the 3D model for wedge test with the feeder design used for the 3D printed single core for sand mould preparation to be used for chill tendencies observations after solidification. (**b**) ATAS cup.

The hardness of the TA analysed samples was measured using the Brinell method, with a 2.5 mm tungsten carbide indenter and a 187.5 kg load (HBW 2.5/187.5) (based on ISO 6506-1 [27]), measured on metallographically prepared samples. The microstructure was observed using metallographic samples in unetched conditions using a Light optical microscope (Olympus DP70 LOM, Tokyo, Japan) and FE-SEM (Zeiss CrossBeam 550, EDAX Octane elite EDS detector, Oberkochen, Germany). Using FE-SEM, the focus was on both the inclusions and boron in the inclusions. The microstructure using LOM was evaluated by regular standards, such as ISO 945 [28].

The thermodynamic prediction was made by Thermo-Calc (Thermo-Calc 2017a, TCFE8, Sweden). For the estimation of physical properties, JMatPro 6.1 (Sente Software Ltd., Guildford, UK) was used. The input data for the calculations was performed with a chemical composition based on optical emission spectrometry (ARL 3460, ThermoFisher Scientific, Waltham, MA, USA) and LECO for sulphur and carbon determination (LECO CS-600, Leco Corporation, St. Joseph, MI, USA). For nitrogen and oxygen LECO TC436 (Leco Corporation, St. Joseph, MI, USA) was also used. The boron measuring limit was set under 5 ppm.

Figure 1a, above, depicts the 3D model of the wedge test specimen, which adheres to the geometry prescribed in standard test method A, W4 [29], while Figure 1b shows a photograph of the measuring ATAS cup (cell) from Nova Cast [26].

The results of the wedge test (W4) and TA were primarily correlated with the measured boron content based on linearity. The search for a correlation between chill and pouring temperatures and related superheat (difference between pouring temperature and liquidus temperature) was included to observe if temperature control was sufficient for the observation of boron influence as a single factorial test. For this purpose, Pearson's correlation coefficient (r) was calculated where +1 or −1 reveals a perfect positive and negative correlation between the measured variable (y) and, for example, the measured boron variable (x). If r is under the absolute value of ±0.30 (a rule of thumb) or close to 0, no linear correlation is observed. The data of linear and nonlinear regression for the observed population were evaluated by the R^2 coefficient (coefficient of determination), and if R^2 is 1, then the fit is perfect. The statistical significance of r (assessed by the calculated p-value) was also calculated for each potential linear relationship between variables. A statistical significance was established with 0.05. When the p-value was less than 0.05 (critical level of significance), then the null hypothesis (H_0) was rejected (no linear relationship between the two variables with $r = 0$), and an alternate hypothesis was accepted (there is a relationship between variables), based on the model presented in [30–32]. If the p-value is under 0.01, then the response is considered very significant.

3. Results and Discussion

3.1. Chemical Composition

The measured chemical composition is given in Table 1. The basic composition is set for flake graphite formation. The flake structure is intended for the studied special grey iron at elevated temperatures. A higher content of carbide stabilisers can also be recognised in Table 1 (Cr, Mo, V, Ti) for increased wear and temperature stability. For this study, boron as a contaminant is also included. The chemical composition was kept stable for a stable carbon equivalent control (CEQ), and its counterpart carbon equivalent was also evaluated through liquidus (CEL, ACEL). The carbon content was relatively low to provide evidence of the potential chill effect with the boron additions. The CEQ was preserved as a fixed value, as the boron effect can vary with CEQ due to the changed graphitisation ability, according to [18]. The Si value represents the final inoculated value, as the final chemical composition is treated with FeSi before boron interactions with the melt and performed TA. The nitrogen, possibly also affecting chill as a carbide stabiliser [17], is controlled in the range 85–100 ppm range and is regarded as reasonable. The nitrogen pickup from scrap, alloys and the atmosphere was under control [33].

Table 1. The chemical composition of melts is based on the extracted TA samples. The values are gathered and referred to as OES and LECO (in wt.%) values. Boron, nitrogen and total oxygen are in ppm.

	C	Si	Mn	Cr + Mo	Ti + V	Ni + Cu + Co	[O]$_{tot}$	B	N	CEQ *
								0	94	
								19	85	
	3.22	2.22	0.57	0.60	0.13	0.78	33–71	30	100	3.77
								37	97	
St.dev.	0.01	0.05	0.01	0.01	0.01	0.07	15	/	6.48	0.02

* CEQ = CEL = %C + %Si/4 + %P/2.

In Table 1 standard deviation of basic composition between heats are given, with boron as an exception.

3.2. Thermodynamic Approaches

A thermodynamic equilibrium prediction of the temperature-related phase evolution, using Thermo-Calc, was established to clarify the potential synergy of the boron with other elements under equilibrium conditions. The Diamond_FCC_A4 and Cementite were suspended for stable prediction of phase evolution versus temperature. The gas phase was allowed for the same calculations. The typical diagram of phase evolution concerning temperature is shown in Figure 2. Based on such diagrams, characteristic temperatures were gathered and presented in Table 2. When adding boron, a change is predicted in decreased solidus temperature ($T_{solidus}$), as shown in Table 2.

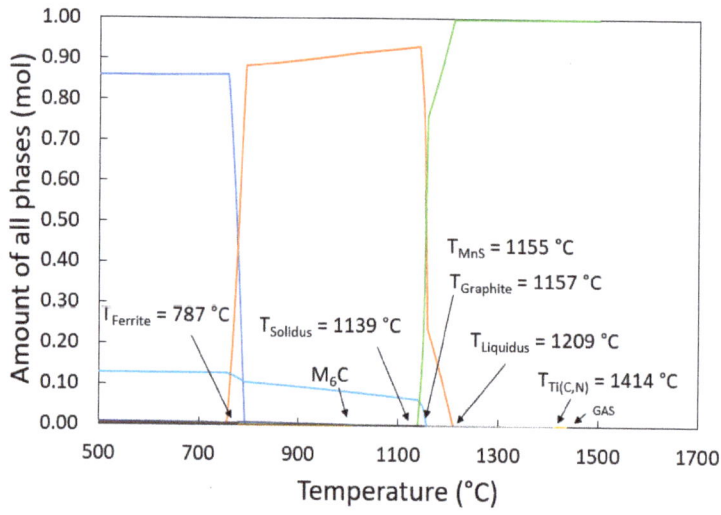

Figure 2. An example of equilibrium phase evolution in relation to temperature. A reference alloy without boron (untreated sample: 0 ppm B). When boron is added M_3B_2 is stable (Table 2). Calculation: Thermo-Calc (TCFE8).

Table 2. Predicted characteristic temperatures determined by the Thermo-Calc (TCFE-8) calculation (in °C) by fixed nitrogen content of 100 ppm and 150 ppm.

B (ppm)	N (ppm)	Ti(C, N)	Austenite	Graphite	MnS	$T_{solidus}$	M_3B_2	M_6C	MC	Ferrite	MC_Eta	BN
0	100	1414	1209	1157	1155	1139		1014	1005	787	774	No
19	100	1414	1208	1157	1155	1087	1087	1013	1008	787	774	No
30	100	1414	1208	1157	1155	1087	1088	1010	1008	787	774	No
37	100	1414	1207	1157	1155	1087	1088	1008	1008	787	774	No
37	150	1414	1207	1157	1156	1088	1087	1013	995	786	774	1125

Based on Table 2, there is no expected precipitation of BN for given compositions (0–37 ppm B and 100 ppm N) and, therefore, boron is not expected to work as a potential inoculator based solely on the presumption of formed BN. The only nitride or carbo–nitride that are present before the primary and eutectic austenite and eutectic graphite start to precipitate from the liquid is Ti(C, N) or co-precipitated (Ti, V, Fe)$_1$(N, C, Va)$_1$. Based on these calculations, when nitrogen is sufficiently raised, BN seems to be thermodynamically stable, but the formation of BN is predicted to occur after graphite precipitation. Therefore, up to 150 ppm of raised nitrogen (N) the BN seems not to be present as a major active nucleator for graphite precipitation. However, the precipitation of BN is still predicted above the solidus temperature, which gives it the possibility of being a weak inoculator in the last solidification region. The BN is formed at 1125 °C at 37 ppm B at raised nitrogen, and is also potentially in the form of solute enrichment in the last liquid region. According to the calculations, when boron is included, it is integrated into the graphite phase from start at approx. 1157 °C, and boron content rises to approx. 1086 °C. This is the temperature where M_3B_2 starts to become stable. Manganese sulfides (MnS), as non-metallic inclusions, are also predicted and expected in the last solidification region. The potential complexes formed between iron–molybdenum based M_3B_2 and MnS are not excluded due to their thermodynamic co-existence, and this is also the case for Ti(C, N), which is formed in the early stage of solidification. Based on the predicted equilibrium solidification, carbides (M_6C, MC is also interpretated as M(C, N) type, etc.) are not formed from the liquid phase; however, under actual non-equilibrium conditions, primary carbides can be found, as shown later in this study.

Based on the measured chemical composition, shown in Table 1, the activities of carbon in the liquid phase versus boron can also be calculated. The carbon activity predictions are carried out by first calculating the activity coefficient of carbon (f_C). The calculation is made with the presumption of dilute solutions by first incorporating the Henryan activities (using 1 wt.% alternative reference state) in a multi-component liquid system to describe the thermodynamic behaviour of element (C) by first-order interaction parameters (coefficients) of carbon (e_i^j, where i = C and solute j = Si, Mn, B etc.). The protocol and interaction values for this calculation are found elsewhere [34]. The carbon activity using the Raoultian reference scale, $a_{C\ (R)}$, in liquid is preferred and recalculated from the Henryan scale. The carbon activity values reveal practically no obvious activity changes with such a small boron content, toward the ability to graphite/carbide formation; $a_{C(R)}^{o\ ppm\ B}$ = 0.356 and $a_{C(R)}^{100\ ppm\ B}$ = 0.358. The calculated values using the CALPHAD approach reveal similar values ($a_{C(R)}^{0-37\ ppm\ B}$ = 0.381) concerning boron for carbon activity in liquid, and these values are similar to the values for stable pseudo-Fe-C system (ternary Fe-C-Si) for given carbon and silicon content [35–37]. However, by increasing the boron content, the carbon activity gradually increases. An increase in the activity of carbon in liquid reveals an improved ability for carbon to separate as graphite in the Fe-C system [38]. However, based on the diagram below, Figure 3, the temperature difference between stable and metastable eutectic intervals is barely noticeable for a given boron interval, but a higher boron content reveals a weak tendency toward mottled or white solidification if the content of boron is sufficient. This is possible when elemental enrichment appears in the last solidification front. In this study a low superheat was obtained to limit any elemental enrichment. However, in the case of added nitride forming Ti, as in our case, a graphitising influence can also be promoted among carbide promotion, as it is related to the nitrogen levels achieved in the melt [38]. This indicates a complex graphitisation behaviour for a given composition.

3.3. Wedge Test

After pouring the melt into the wedge cell and letting it cool to ambient temperature, the test chill specimen was broken in half to observe the fractured topography, as seen in Figure 4. An obvious difference in chill is observed within the heats. The measured data is given in Table 3. For the wedge test, according to test method A, the evaluation is carried out according to standard procedure [29]. As it is recognised by the standard A367, the amount of chill is strongly affected by the pouring temperature. The recorded pouring temperature was also controlled at its best by having the same melting loads (melt enthalpy capacities) and similar temperature profiles from melting to pouring. The maximum melt temperature before pouring was up to 1310 °C. A linear correlation was observed between the measured maximum pouring temperatures with measured clear chill depths measured by the width, W, ($r = -0.92$). However, a linear regression of the data indicates that the pouring temperature is outside measurable significance (p-value = 0.08, R^2 = 0.845), especially when compared with the boron measured values with W values. A close-to-linear trend was established for the measured boron values on wedge samples with measured clear chill (with p-value = 0.038, R^2 = 0.924). However, according to statistics, the change due to boron is regarded as significant instead of very significant. It should be emphasised that, based on the data, synergistic effects could be possible between boron and the pouring temperature (and related superheat), but in the current study, multilinearity for the same observed parameter was not observed.

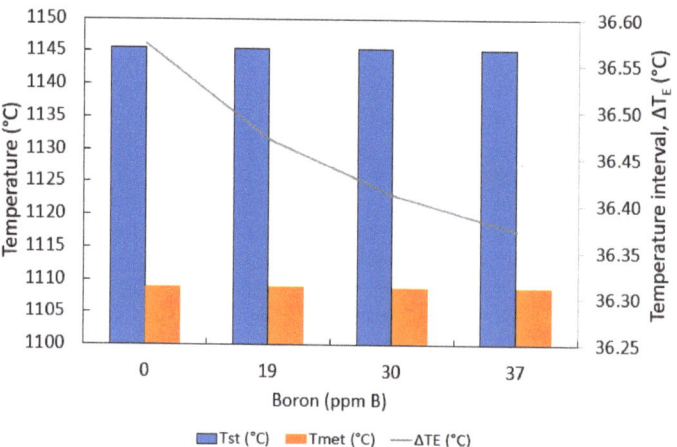

Figure 3. Effect of boron on eutectic solidification region defined by $T_{EG} - T_{EC}$ and calculated based on equations [22,38,39] reveals that boron is in function of a weak carbide stabiliser decreasing T_{Stable} and $T_{Metastable}$. The intensity is related to actual boron distribution in the last solidification front. The theoretic temperature difference (ΔT_E) between both eutectics for the studied melts is practically negligible.

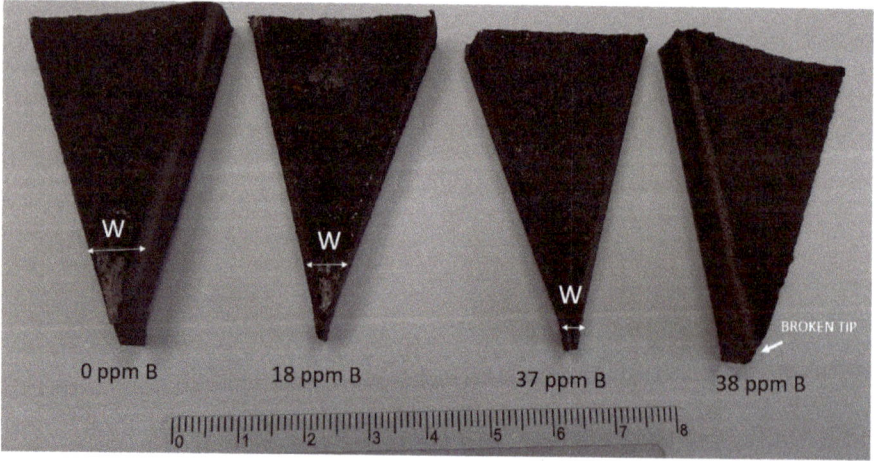

Figure 4. Decreased chill (W) with the addition of boron is obvious. The dark grey appearance of the majority of the sample's cross-section reveals the overall effective melt inoculation treatment (by also including FeSi). No macro-porosity was observed in the cross-sections.

Table 3. Clear chill boundary measured from the apex to the starting of the second (mottled) zone (in mm) with the measured boron content. More in [29].

Boron (ppm B)	0	18	37	38
W	10.98	9.60	3.45	<3.45 *

* The tip edge is broken therefore the W is not reliably measured.

3.4. Thermal Analysis (TA)

Based on the recorded (temperature–time) cooling curves, several parameters were extracted to evaluate basic information about the melt behaviour [39]. The basic parameters are T_{Liq}, T_{ELow}, T_{EHigh}, T_{ES}, T_{Sol}, Rec, UQ, ACEL, S1 and GRF2, and these are interpreted as liquidus temperature; the temperature of eutectic undercooling; the temperature of eutectic recalescence; eutectic start; solidus temperature (a solidified structure is represented with grey in Figure 5); recalescence (defined as $T_{EHigh} - T_{ELow}$, the rise of temperature due to graphite precipitation); undercooling quotient; active carbon equivalent (calculated from T_{Liq}); relative amount of primary austenite precipitation (yellow region in Figure 5); and first derivate of the cooling curve at solidification finish (graphite factor number 2), respectively. GRF2 is also a measure for inverted heat conductivity in a solid, being important for the purposes of graphite quantity and shape [40]. In this paper, the sign Rec is used for recalescence and the small letter r for the Pearson correlation coefficient, to avoid confusion.

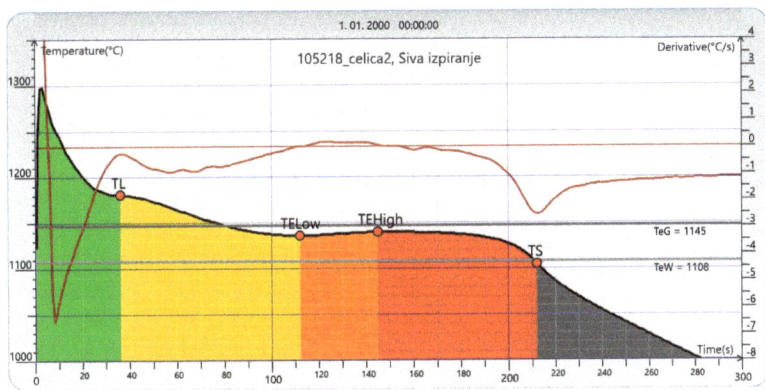

Figure 5. An example of an ATAS analysed cooling curve with the first derivate, revealing the hypothetic solidification nature of the investigated alloyed grey iron. Sample: 37 ppm B. Color explanation: Green—Liquid phase, so-called superheat (between $T_{pouring}$ and $T_L = T_{Liq}$), Yellow—S1 (between T_{Liq} and T_{ELow}; precipitated as pro-eutectic austenite), Orange—S2 (between T_{E*Low} and T_{EHigh}; Eutectic region No.1), Red-S3 (between T_{EHigh} and $T_S = T_{Sol}$; Eutectic region No.2) and Grey (< T_{Sol}; solidified as-cast structure).

An automated ATAS system was used for analysing the recorded solidification curves of the non-boron and boron-treated samples. Instead of using a TA cup with Te seed promoting white solidification, an equation based on reference [22] is used instead, which also includes the (soluble) boron influence. Similar equations are valid for stable (graphite) solidification. Based on the temperature interval between stable (grey) and metastable (white) eutectics, the basic melt characteristics were observed using the ATAS system, as shown in Table 4. All samples are hypo-eutectic based on the calculated constant CEQ, with 3.77 and active carbon equivalent (ACEL) changes from batch to batch, as observed by ACEL [25]. The evidenced primary austenite precipitation suggests the possibility of forming D and/or E graphite distribution (according to EN ISO 945 [28]). In the case of a higher ACEL, a transition towards A distribution is possible with potential locally thicker flakes due to high eutectic temperatures obtained by TA. The ACEL increases on average with boron-contaminated samples (approaching the eutectic region with C ≈ 4.3%) but with no obvious correlation. The relevant austenite precipitation index (S1 region, the yellow area in Figure 5) is estimated to be rather similar, and with the highest values for the non-boron heat equivalent, with 43.68% of the primary austenite compared to lowest values of 41.71% within boron treated heat. The overall increase of eutectic graphite (based on ACEL) is observed on boron-treated samples, but the trend is non-linear as the data are scattered. The results reveal that boron contamination with values up to 37 ppm B and its effects on chill or graphitisation are not straightforward. The data also reveal the usefulness of using TA during the melt quality check (like *ACEL*) when preparing the melt with scrap, instead of simply relying on the chemical composition and related CEQ.

Table 4. Measured parameters by using ATAS. The boron values are the measured boron contents in TA samples. $T_{Stable} = 1145\ °C$ and $T_{Metastable} = 1108\ °C$ are calculated by [22,39] and ACEL by [38,41].

Internal Name	ppm B	T_{Liq}	T_{ELow}	T_{EHigh}	T_{ES}	T_{Sol}	Rec	UQ	ACEL	S1	GRF2
102130_1	0	1186	1127	1133	1166	1089	6	1.0	3.89	43.68	58
105218_1	19	1181	1136	1141	1163	1108	5	1.2	3.94	42.86	46
102130_2	30	1184	1130	1134	1169	1095	4	1.5	3.91	41.71	58
105218_2	37	1181	1136	1140	1168	1105	4	1.5	3.94	43.44	52

ACEL = 14.45 − 0.0089 · T_{liq} [38,41].

There is one measurable relation with boron independent of *ACEL* (related to T_{Liq}), and this is recalescence, *Rec* (*Rec* = $T_{EHigh} - T_{ELow}$). It is agreed that inoculation reduces eutectic undercooling [41], and that this can be observed by the undercooling quotient here described as *UQ* (Equation (1)). The $Rec_{Reference}$ in Equation (1) stands for recalescence of untreated (0 ppm B) heat and Rec_{Boron} related eutectic undercooling measured for boron-added melts:

$$UQ = Rec_{Reference} / Rec_{Boron} \qquad (1)$$

Based on the *UQ* values in Table 4, where the *UQ* increase indicates possibly the eutectic cell number change, it is clear that the change regarding boron can be described as having a weak inoculation phenomenon only. This is also true when considering that T_{sol} is higher for boron-treated melts (contrary to predictions, Figure 2). This reveals that boron is an important element, even in traces, during the production, as well as for special grey iron melts. It is notable that, for a given laboratory set, the recalescence or, more importantly, *UQ*, stops at values of 30 ppm B, and could indicate an achieved maximum allowable inoculation for a given laboratory set. It should be remembered that this inoculation must be a synergy of boron complexes with already present nucleation sites (from scrap, added FeSi, etc.) or, as shown by Thermo-Calc, partially related to the composition itself.

The liquidus temperature somehow decreases with the boron content for special cast iron grades and, therefore, the added boron did not act as a nucleus for primary austenite crystallisation, according to measured data. Additionally, the graphite eutectic and solidus temperatures are higher than the reference indicating small changes in the solidification paths. Overall increased solidus reveals that, compared to the reference (of non-boron) sample, the tendency to segregations and related chill was decreased. This is similarly the case for T_{ELow}. The measured effects using TA are the opposite of those found in Pawaskar's study [18], and this is consistent with the observed wedge specimen depth change with boron in this study. The closing of the solidification interval between liquidus and solidus is interesting, and similar to observations for plain cast irons in reference [18,19]. A higher nucleation potential for given heats, compared to plain cast grey irons, can be expected, and this is present due to a probable nucleus formation based on pre-existing nitrides. It is also interesting that the non-calibrated eutectic cell count (ECC) was followed under slow solidification rates, and no obvious ECC correlation with boron was observed, as it was not obvious with the measured graphite count. This is despite the higher counts observed for the highest boron content, as shown in Figure 6. This reveals that an obvious graphitisation effect was not achieved and confirmed during slow cooling rates, as it was significant for the wedge test.

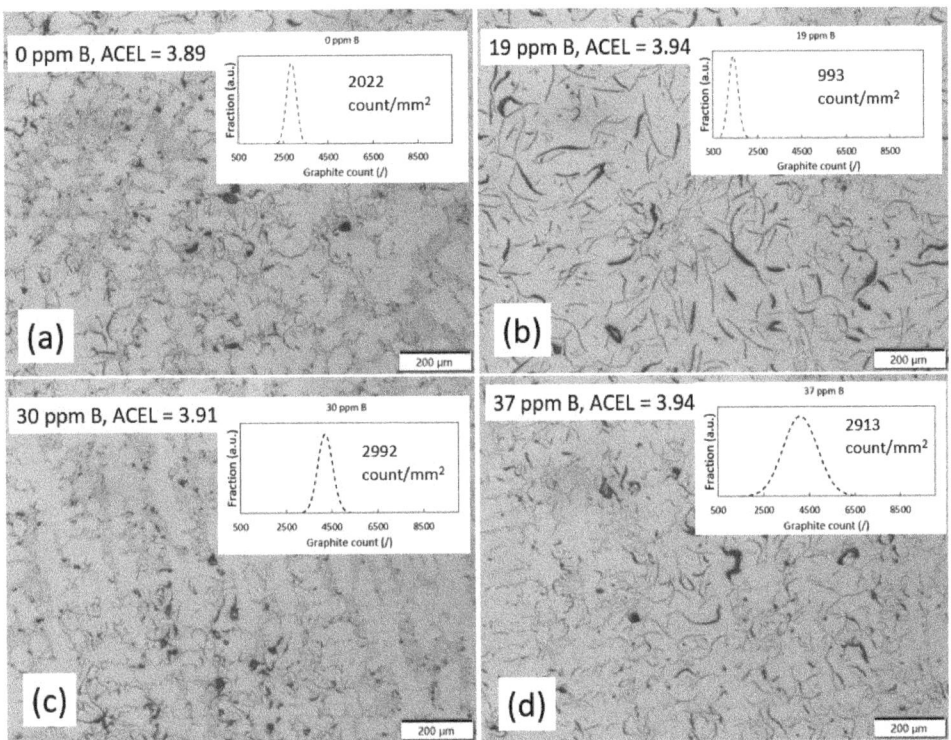

Figure 6. An example of the as-cast structure of TA samples and related graphite count distribution (**a**) 0 ppm B with finer size D/E graphite as reference (1 sigma, min–max = 2630–3058); (**b**) 19 ppm B with D/E graphite with thicker graphite flakes inside the eutectic cells of A as the majority (1 sigma, min–max = 1186–1606); (**c**) 30 ppm B with D/E with finer and local coarse size inside the eutectic cells (1 sigma, min–max = 3884–4531); (**d**) 37 ppm B with D/E with thicker eutectic flakes of A as the majority (1 sigma, min–max = 2995–5197). Calculated *ACEL* [41] from the obtained T_{Liq} is between 3.89 and 3.94. Enlargement: 50×. State: Polished.

3.5. Thermal Conductivity

Thermal conductivity is an important physical property for elevated temperature use. The increased fractions of transformed austenitic primary crystals and branched D-type graphite lower the ability to transfer heat and are related to achieved cooling rates, inoculation efficiency, etc., meaning that it is not simply related to basic nominal composition [40]. A higher content of A-type is usually desired where high thermal conductivity and low friction losses are needed [23]. Thermal conductivity was not directly measured at room temperature. However, it is estimated by using the indirect method by observing the end of the solidification curve during TA, which can indicate the graphite shape. In this case, the angle of the derivate marked as GRF2 was used, as shown in Table 4. According to Stefanescu et al. [39], low angles can correspond to increased thermal conductivity. Compared to the analysed boron content and GRF2, no linear trend with thermal conductivity was observed ($r = -0.23$, $R^2 = 0.05$). The lowest angles (highest conductivity) were determined with the sample with the highest *ACEL*, regardless of the boron content (19 and 37 ppm B). This complements the improved graphite distribution uniformity observed in Figure 6. In the case of the sample with the lowest *ACEL*, the undercooled dendritic graphite indicates lower heat transfer ability, even if graphite count is increased (compared to 0 and 30 ppm B with GRF2 and graphite count). Carbides have an influence on the conductivity, but the fractions are comparable with the TA samples, as shown under FE-SEM investigation later.

An example of estimation of thermal conductivity for a stable reference sample with ferritic–pearlitic matrix and lamellar type of graphite is between 48 and 35 W/mK for 25 and 500 °C, respectively. Boron within the given limits is considered, based on GRF2, as non-significant to thermal conductivity values. Based on prediction (using the JMatPro 6.1, internal database: General physical properties,) the thermal conductivity is close to linear with the increased working temperature.

3.6. Microstructure

3.6.1. LOM

Despite having the same carbon equivalent (CEQ), the microstructures of the TA extracted samples (taken from the centre of the measuring cup) differ considerably in terms of their slow solidification rate, as shown in Figure 6. According to ATAS, the calculated *ACEL* from measured T_{Liq} identify the actual changes observed from batch to batch and goes well with the formed microstructure. A higher *ACEL* brings about a combination of A and D graphite types, with a majority of A (compare sample with 19 ppm B and 30 ppm B), instead of mostly D/E type of (undercooled/dendritic) graphite. This shows that actual achieved *ACEL*, as a measure for the achieved melt quality, prevails for the given samples compared to small boron content for microstructure evolution. This microstructure evolution stands for the absolute cooling rate (°C/s), taken from the eutectic start (T_{ES}, being between the T_{Liq} and T_{ELow}) at 1 °C/s.

The most homogeneous microstructure is found to be the one with the highest T_{ELow} identified for 19 and 37 ppm B, as shown in Figure 6b,d. This was achieved despite a variation in the graphite count.

Figure 6 shows the highest measured graphite count obtained for the final two highest boron-containing melts but also the lowest measured value for smallest boron addition. Based on graphite count distribution, the graphite count appears to be more dispersed (higher standard deviation, SD), indicating a considerable graphite size variation. The higher SD is related to smaller flakes present locally and due to overall scatter by image analyses. In the case of 37 ppm B, this could partially be a result of the change from thin and fine D/E towards thicker A, which is also due to a higher *ACEL* value. Based solely on the achieved *ACEL*, it is assumed that inoculation efficiency or combined inoculation rate between the melts are not repeatable compared to the unmodified sample (0 ppm B).

3.6.2. FE-SEM

Using the backscattered electron detector (BSD) detector shown in Figure 7 and the undercooled FE-SEM microstructures, it is possible to observe the presence of primary carbo nitrides (white) composed of Ti, V and Mo, as well as their positions. In the last solidification region, carbides are more enhanced, confirming their non-equilibrium solidification nature. The dark phases are graphite inside the matrix (grey), showing mainly undercooled flake graphite (type D) with local interdendritic flake graphite (type E). Visually, carbides appear to increase with boron. However, measurements reveal that for boron values 0, 30 and 37 ppm B, there is a 0.64, 0.61 and 1.27 areal %, respectively. Therefore, there is no obvious difference in carbides, with an exception of those with the highest boron content.

An example of the location of the last solidification region is shown in Figure 8, where complex interactions with carbonitrides and non-metallic inclusions (MnS) are observed. Both complex carbonitrides and MnS are predicted by Thermo-Calc. The boron was not observed. Additionally, the pearlite (not shown here) did not reveal the presence of any boron.

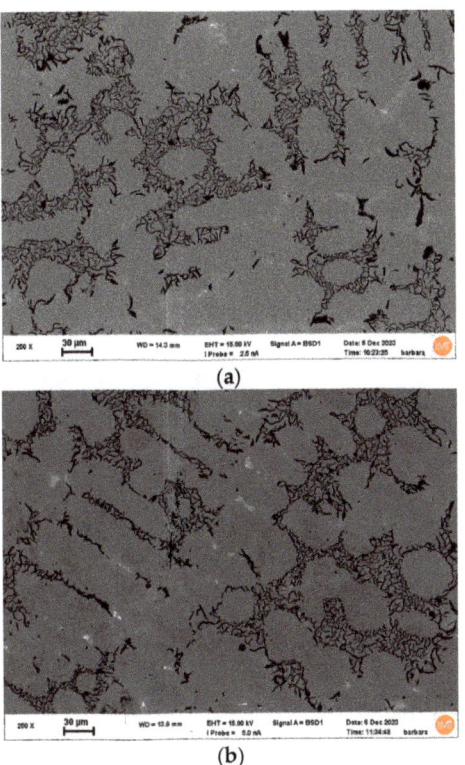

Figure 7. FE-SEM-BSD: an example of (**a**) reference sample with 0 ppm B; (**b**) 30 ppm B.

Based on the executed FE-SEM, it cannot be established whether or not boron-based nuclei were present in the last solidification region. It is recognised that BN should be ideal for graphite growth due to its hexagonality, but it is also recognised that boron has a limited oxygen affinity [13,23]. For example, boron, under highly oxidising conditions, can react with dissolved oxygen or by reducing less stable oxides than boron to form B_2O_3 type of oxides, while potentially also working as a nuclei for graphite, as it is hexagonal [42]. If considering B_2O_3, SiO_2, MgO and others (TiO, Ti_2O_3) as potential nuclei, B_2O_3 with approx. −320 kJ/mol (and others with <−320 kJ/mol) appears to be less stable based, on the Gibbs energy of formation per mole [O] in the desired temperature interval of 1145 °C and 1108 °C. However, the stability of B_2O_3 is close to SiO_2. It should also be added that B_2O_3 can thermodynamically interact (based on the Gibbs energy of reaction, ΔG_R) with other more stable oxides and form new (double) oxide products. For example, Ca is a commonplace additive for inoculation inside complex ferroalloys (i.e., FeSi) [43]. By interacting with oxygen, it forms a basic oxide, and by interacting with acids, boron oxide can potentially form CaB_xO_y. These or similar complexes in theory should at least partially remove boron from the residual melt and affect the nucleation potential as a cumulative effect. It is also recognised that oxide slags saturated with B_2O_3 can easily be reduced in steel containing Si [44], indicating that boron-related nucleation phenomena, if present, could fade with time. Limited boron saturation is expected inside oxides under a sufficiently high mass transfer rate of boron inside the melt into the boron-free oxide system (nuclei) and with sufficient time for saturation. The intensity of boron saturation in oxide-based nuclei and other oxides depends on the actual achieved distribution ratio of boron between oxides (nuclei) and the remaining melt (defined by L_B = wt.% B_{Oxide}/wt.% B_{Liquid}). This was not investigated in this study. However, the content of boron should remain practically unchanged during regular grey cast iron production, as melt refining in a metallurgical sense is unwanted for grey cast iron to preserve sufficiently high nuclei density. The variation in distribution of boron between oxides and residual melt is found in some of the literature data [10,44].

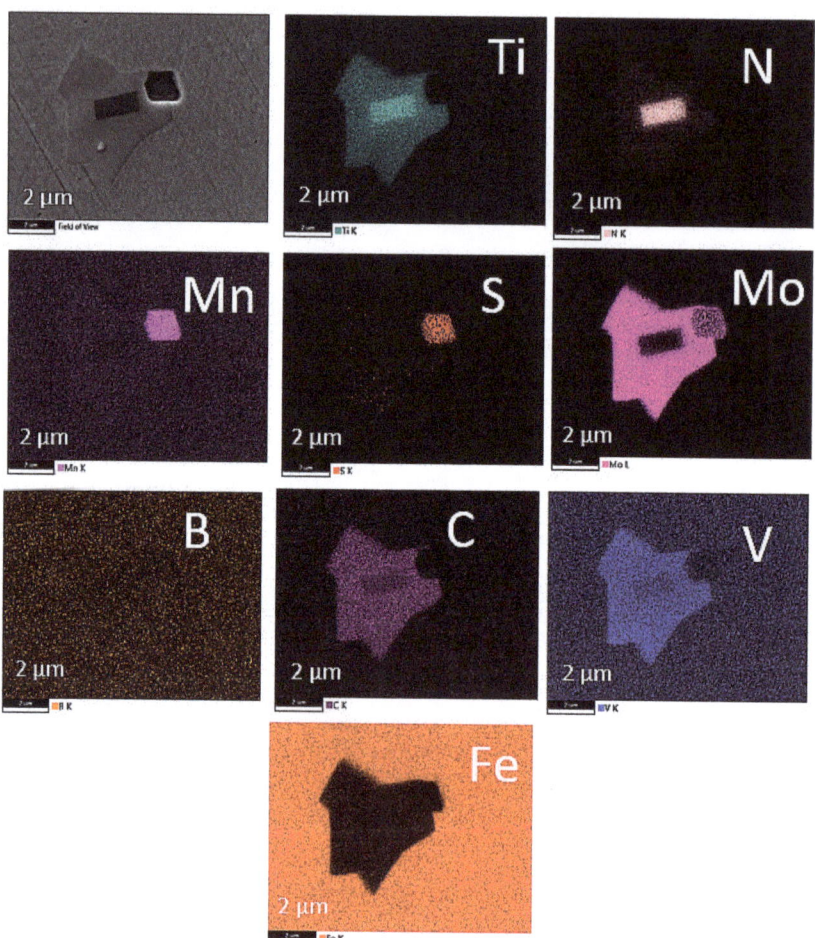

Figure 8. FE-SEM-EDS elemental distribution maps: these maps are calculated to show the actual element content (ZAF wt%), and not just signal intensity to avoid overlaps of B and Mo X-ray peaks. Sample: 37 ppm B.

3.7. Hardness at Slow Cooling Rates

Hardness as a function of the change of mechanical properties was tested for slow cooling rates, to evaluate whether or not there is a significant difference between the measured HB values of the TA samples (extracted adjacent to thermocouples) concerning boron by considering linearity. No linearity was observed ($r = -0.38$, $R^2 = 0.14$). However, by having a polynomial fit (of the second order), the regression model shows improved correlation (adjusted $R^2 = 0.68$), but is still rather weak in relation to R^2, where more than half of the data is represented well with the model, see Figure 9. However, the model is well-adjusted according to the p-value for dependent coefficients (B_i), by using test statistics with the ANOVA approach. This means that regression is sufficiently explained by the given model. The coefficients are given and explained further in the paper. The hardness scatter could be related to material homogeneity/heterogeneity observed after solidification and the related presence of graphite-free regions and changes in carbide percentage (areal.

%), as shown in Figures 6 and 7. This means that simple changes in the graphite count with boron is not the main cause of hardness variation but rather the achieved distribution of the soft ferritic regions. Based on [23,24], the excess of boron could stabilise pearlite and give the matrix a harder response. Increased primary carbides at slow cooling rates are potentially recognised in the last solidification front (according to FE-SEM) and the hardness jump for the highest values of boron could be the result of local coarse carbide formation. Based on the hardness response and the fact that boron is a ferrite stabiliser, an increased soft ferrite matrix could be the reason for the starting hardness drop [21]. A correlation between the addition of boron and the graphite count affecting the hardness was not correlated, as shown in Table 5.

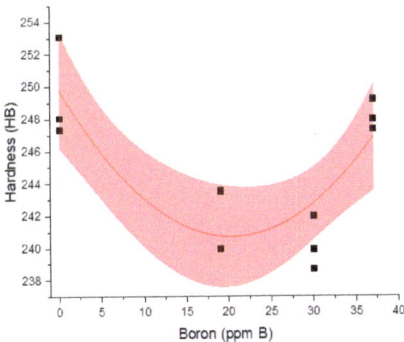

Figure 9. The hardness of TA extracted samples to evaluate the changes with the addition of boron with a 95% confidence band (red interval). The hardness is valid for 1 °C/s of solidification rate. No additional heat treatment (ferritisation) was performed.

Table 5. Pearson correlation coefficient, r, for boron sensibility test by actual towards predicted for correlation versus boron.

Correlation	Pearson Correlation Coefficient, r	p-Value (Level)	R^2	Significant
Liquidus (T_{Liq}) versus B	−0.69	0.3084	0.48	No
ACEL versus B	0.69	0.3084	0.48	No
ACEL versus GRF2	−0.85	0.1472	0.73	No
T_{ELow} versus B	0.65	0.3458	0.43	No
T_{EHigh} versus B	0.49	0.5091	0.24	No
T_{EE} versus B	0.50	0.5003	0.25	No
T_{sol} versus B	0.59	0.4101	0.35	No
Rec versus B	−0.98	0.0182	0.96	Yes
UQ versus B	0.97	0.0301	0.94	Yes
GRF2 versus B	−0.23	0.7734	0.05	No
S1 versus B	−0.43	0.5670	0.19	No
W versus B	−0,96	0.0387	0.92	Yes
W versus Rec	0.96	0.0443	0.91	Yes
HB versus B	−0.38	0.6228	0.14	No
$T_{Pouring}$ versus W4	−0.92	0.0806	0.85	No
$T_{Pouring}$ versus HB	0.06	0.9403	0.00	No
Graph. count versus B	0.54	0.4558	0.29	No

Based on the small boron values, the presence of the measuring error for the boron also contributes to the scattered hardness relation with the boron for slow cooling rates of TA samples. Therefore, this relation of boron with hardness remains unclear. However, based on the overall results, a hardness drop was achieved compared to the reference (0 ppm B), despite local hardness jumps.

The equation for polynomial hardness fit with boron for slow cooling at 1 °C/s was fit using the equation:

$$y = Intercept + \sum B_i \cdot x^i; i = 1, 2, \ldots \quad (2)$$

where y is the predicted hardness, x represents the boron value in ppm and B_1 and B_2 are dependent and fitted coefficients. The starting hardness as an independent value is presented with $HB_{0\_ppm\_B}$. The final and simplest quadratic relation is given by:

$$HB = HB_{0\_ppm\_B} - 0.8941 \cdot boron \text{ (ppm)} + 0.02204 \cdot boron^2 \text{ (ppm)} \quad (3)$$

The potential boron effect was considered using a factor W when considerable cooling rates are present. It is recognised that boron had a certain effect on Rec, as well as on W, which indicates the weak inoculation phenomena if the potential boron is not completely lost with time, as in the case of TA samples with slow melt cooling.

3.8. Correlation Probability

For all variables, the Pearson correlation coefficient, r, was calculated. The strength of relationships of measured values with boron varies from weak to strong linear correlation. Based on the results, it can be concluded that a strong linear relationship between two variables exists for Rec, related UQ and W with boron. This also goes with the coefficient of determination, R^2. Strong linearity was also observed between $T_{Pouring}$ and $W4$ for intense cooling, but practically close to zero correlation for $T_{Pouring}$ and measured hardness for slow cooling (TA cell). Additionally, inversed heat conductivity, GRF2, reveals a stronger linear correlation with ACEL than with boron, indicating the non-significance of the boron contribution to thermal conductivity change in this study.

The statistical significance of the Pearson correlation coefficient was tested where Rec and related UQ, as well as W, reveal that there is a statistically significant linear relationship. In the case of W versus Rec, the significance could also be a result of a third parameter. All other factors, including ACEL and T_{sol}, have a moderate or weak linear correlation; therefore, it can be concluded that the boron influence is rather weak for the studied concentrations.

Based on all analysed values, boron performs as a weak inoculant only under short hold and solidification times (wedge test). Based on linear regression an equation is set:

$$W4 = W4_{0\_ppm\_B} + (-0.21273) \cdot boron \text{ (ppm)} \quad (4)$$

where a negative slope indicates a decreasing chill tendency for intense cooling rates with added boron, according to the laboratory settings. The influence of boron on recalescence is also described with a linear model, where $Rec_{0\_ppm\_B}$ represents recalescence measured for reference melt, without boron:

$$Boron\ influence\ on\ recalescence = Rec = Rec_{0\_ppm\_B} + (-0.058 \cdot boron \text{ (ppm)}) \quad (5)$$

Both Rec and $W4$ are well correlated, as shown in Table 5. It should be emphasised, that Rec changes with time, as its potential inoculation fades with time. Therefore, the equation above is set for the cooling/solidification rate of approx. 1 °C/s with the direct addition of boron. As the noise is rather high for Equation (5), the inoculation and/or change in recalescence phenomena cannot be considered solely as a part of boron, but it is assumed that other heterogeneous nuclei also contribute to this phenomenon.

Based on Table 5, that ACEL versus B and liquidus temperature versus B have the same Pearson correlation coefficient, as expected. This is due to calculation of ACEL directly from measured T_{Liq}.

4. Conclusions

Based on the conducted research, it was concluded that no very significant linear trend was observed at slow cooling rates of approx. 1 °C/s with the addition of boron for any of the measured parameters. The exception is that boron shows a significant decreasing effect in the recalescence (Rec) during eutectic graphite solidification on the temperature-time cooling curve. This is contrary to expectations. This indicates that boron could potentially affect graphite growth as short-term inoculation effect, meaning that boron can work as low-effective inoculator only. It is important to note that, based on the ATAS system, the boron effect was noticed even when it was almost negligible.

Graphite transformations from poor inoculated D to well inoculated A-type with boron at slow cooling rates was non-repeatable—the effect of boron is easily lost, and overall inoculation is dictated by ACEL.

At slow cooling rates, the effect of added boron concerning thermal conductivity observed indirectly by the GRF2 factor was recognised as non-significant under the condition of weak linear correlation. The inverted heat conductivity, GRF2, was more affected by the achieved ACEL than boron itself, because ACEL has a direct influence on the formed as-cast microstructure. The hardness at lower cooling rates was correlated with boron under the parabolic model showing the complex behaviour of the melts. An overall hardness decrease was observed with boron.

If bonded boron is present, a limited influence on the temperature interval between the stable and metastable eutectic is expected. There is no measurable effect of boron on nucleation of primary austenite.

The chill test for the increased solidification rates confirmed that the influence of boron was significant but not very significant. In the current study, the minor increasing graphitisation trend with boron is observed by the chill test only (short solidification times, fast cooling rates and the preservation of nucleation sites). This nucleation trend is also indicated by T_{sol} using TA at slower cooling rates, although it showed no distinctive trend.

Both TA and wedge chill test results reveal that boron, even in trace amounts, is influential and has very complex behaviour. The overall melt condition seems to be of higher importance than the actual achieved boron content, within the studied boron limits. None of the observed factors had a very significant response with boron up to 38 ppm B when using a linear regression model.

Boron can work as inoculator, which is consistent with some observations made and tests carried out by other authors.

Author Contributions: Conceptualisation, G.K. and J.B.; methodology, G.K. and J.B.; software, G.K. and U.K.; validation, G.K., U.K.; formal analysis, U.K, B.Š.B. and G.K.; investigation, G.K., U.K., B.Š.B. and J.D., resources, G.K., J.B., B.Š.B., U.K. and A.R.; data curation, U.K.; writing—original draft preparation, G.K.; writing—review and editing, G.K., J.B., U.K., L.K., B.Š.B. and A.R.; visualisation, G.K.; All authors have read and agreed to the published version of the manuscript.

Funding: This research was funded by Republic of Slovenia and the European Union from the Recovery and Resilience Mechanism, NextGenerationEU.

Data Availability Statement: The data is unavailable due to privacy.

Acknowledgments: The authors would like to acknowledge industrial partner OMCO Metals Slovenia d.o.o. and Andrej Kump from Nova Cast, Sweden. The authors would also like to thank Luka Snoj, Anže Bajželj and Nejc Velikajne for their help with the experimental part. The authors acknowledge Financial resources for the project DigitKrom provided by the Republic of Slovenia (Spodbude za raziskovalno razvojne projekte NOO, 303-2-00109/2021/4) and the European Union from the Recovery and Resilience Mechanism, NextGenerationEU.

Conflicts of Interest: Author Andrej Resnik was employed by the company Omco Metals Slovenia d.o.o. The remaining authors declare that the research was conducted in the absence of any commercial or financial relationships that could be construed as a potential conflict of interest.

References

1. Stefanescu, D.M.; Alonso, G.; Larrañaga, P.; De la Fuente, E.; Suarez, R. Reexamination of crystal growth theory of graphite in iron-carbon alloys. *Acta Mater.* **2017**, *139*, 109–121. [CrossRef]
2. Tonkovič, M.P.; Mrvar, P.; Vončina, M.; Donik, Č.; Godec, M.; Petrič, M. Analysis and thermodynamic stability of nuclei in spheroidal graphite in Fe–C–Si alloys. *Mater. Tech.* **2021**, *55*, 533–539.
3. Sommerfeld, A.; Böttger, B.; Tonn, B. Graphite Nucleation in Cast Iron Melts Based on Solidification Experiments and Microstructure Simulation. *J. Mater. Sci. Technol.* **2008**, *24*, 3.
4. Alonso, G.; Larrañaga, P.; De la Fuente, E.; Stefanescu, D.M.; Natxiondo, A.; Suarez, R. Kinetics of nucleation and growth of graphite at different stages of solidification for spheroidal graphite iron. *Int. J. Met.* **2017**, *11*, 14–26. [CrossRef]
5. Stefan, E.; Chisamera, M.; Riposan, I.; Stan, S. Graphite nucleation sites in commercial grey cast irons. *Mater. Today Proc.* **2021**, *45*, 4091–4095. [CrossRef]
6. Campbell, J. The Structure of Cast Irons. *Mater. Sci. Forum* **2018**, *925*, 86–89. [CrossRef]
7. Tasgin, Y.; Kaplan, M.; Yaz, M. Investigation of effects of boron additives and heat treatment on carbides and phase transition of highly alloyed duplex cast iron. *Mater. Des.* **2009**, *30*, 3174–3179. [CrossRef]
8. Zhang, J.; Yang, P.; Wang, R. Investigation of the Microstructures and Properties of B-Bearing High-Speed Alloy Steel. *Coatings* **2022**, *12*, 1650. [CrossRef]
9. Kolokoltsev, V.M.; Petrochenko, E.V.; Molochkova, O.S. Influence of boron modification and cooling conditions during solidification on structural and phase state of heat- and wear-resistant white cast iron. *CIS Iron Steel Rev.* **2018**, *15*, 11–15. [CrossRef]
10. Jakobsson, L.K. Distribution of Boron between Silicon and CaO-SiO$_2$, MgO-SiO$_2$, CaO-MgO-SiO$_2$ and CaO-Al$_2$O$_3$-SiO$_2$ Slags at 1600 °C. Ph.D. Thesis, Norwegian University of Science and Technology, Trondheim, Norway, November 2013.
11. Çolak, M.; Uslu, E.; Teke, Ç.; Şafak, F.; Erol, Ö.; Erol, Y.; Çoban, Y.; Yavuz, M. Investigation of the Effect of Solidification Time and Addition Amount of Inoculation on Microstructure and Hardness in Lamellar Graphite Cast Iron. *Arch. Foundry Eng.* **2022**, *22*, 24–33. [CrossRef]
12. Mendas, M.; Benayoun, S. Comparative study of abrasion via microindentation and microscratch tests of reinforced and unreinforced lamellar cast iron. *Friction* **2019**, *7*, 457–465. [CrossRef]
13. Zou, Y.; Ogawa, M.; Nakae, H. Interaction of Boron with Copper and Its Influence on Matrix of Spheroidal Graphite Cast Iron. *ISIJ Int.* **2012**, *52*, 505–509. [CrossRef]
14. Ha, J.S.; Hong, J.W.; Kim, J.W.; Han, S.B.; Choi, C.Y.; Song, H.J.; Jang, J.S.; Kim, D.Y.; Ko, D.C.; Yi, S.H.; et al. The Effect of Boron (B) on the Microstructure and Graphite Morphology of Spheroidal Graphite Cast Iron. *Materials* **2023**, *16*, 4225. [CrossRef] [PubMed]
15. Kasvayee, K.A.; Ciavatta, M.; Ghassemali, E.; Svensson, I.L.; Jarfors, A.E.W. Effect of Boron and Cross-Section Thickness on Microstructure and Mechanical Properties of Ductile Iron. *Mater. Sci. Forum* **2018**, *925*, 249–256. [CrossRef]
16. Bugten, A.V.; Michels, L.; Brurok, R.B.; Hartung, C.; Ott, E.; Vines, L.; Li, Y.; Arnberg, L.; Di Sabatino, M. The Role of Boron in Low Copper Spheroidal Graphite Irons. *Met. Mater. Trans. A* **2023**, *54*, 2023–2539. [CrossRef]
17. Eppich, R. Cast Iron Alloy Containing Boron. Patent: WO 2006/133355 A2, 8 June 2005.
18. Pawaskar, S.D. Effect of Boron in Cast Iron. Master's Thesis, Missouri University of Science and Technology, Rolla, MO, USA, September 2022. Available online: https://scholarsmine.mst.edu/masters_theses/8109 (accessed on 10 October 2023).
19. Krynitsky, A.I.; Stern, H. Effect of Boron on the Structure and Some Physical Properties of Plain Cast Irons. *J. Res. Natl. Bur. Stand.* **1949**, *42*, 465–479. [CrossRef]
20. Ankamma, K. Effect of Trace Elements (Boron and Lead) on the Properties of Gray Cast Iron. *J. Inst. Eng. Ser. D* **2014**, *95*, 19–26. [CrossRef]
21. Chen, X.; Li, Y.; Zhang, H. Microstructure and mechanical properties of high boron white cast iron with about 4 wt% chromium. *J. Mater. Sci.* **2011**, *46*, 957–963. [CrossRef]
22. Kanno, T.; You, Y.; Morinaka, M.; Nakae, H. Effect of Alloying Elements on Graphite and Cementite Eutectic Temperature of Cast Iron. *ISIJ Int.* **1998**, *70*, 465–470. [CrossRef]
23. Strande, K.; Tiedje, N.S.; Chen, M. A Contribution to the Understanding of the Combined Effect of Nitrogen and Boron in Grey Cast Iron. *Int. J. Met.* **2017**, *11*, 61–70. [CrossRef]
24. Röhrig, K. Einfluss von Legierungselementen auf die Eigenschaften von Gusseisen (G) Legiertes Gusseisen, Gusseisenwerkstoffe. *Gesserei-Prax.* **2019**, *4*, 21–26.
25. Persson, P.E.; Udroiu, A.; Vomacka, P.; Xiaojing, W.; Sjögren, T. ATAS as a tool for analyzing, stabilizing and optimizing the graphite precipitation in grey cast iron. In Proceedings of the 69th World Foundry Congress, FICMES, Hangzhou, China, 16–20 October 2010.
26. Thermal Analysis Cup. Available online: https://www.novacast.se/product/thermal-analysis-cup (accessed on 20 December 2023).
27. ISO 6506-1:2005; Metallic Materials-Brinell Hardness Test, Method 1: Test Method. ISO: Geneve, Switzerland, 2007.
28. DIN ISO 945-1:2008; Graphite Classification by Visual Analysisi, Part-1. ISO: Berlin, Germany, 2008.
29. A367-11; Standard Test Methods of Chill Testing of Cast Iron. ASTM International: Singapore, 2011.

30. Montgomery, D.C. *Design and Analysis of Experiments*, 8th ed.; Wiley: Hoboken, NJ, USA, 2013.
31. Pearson Correlation Coefficient: Formula, Examples. Available online: https://vitalflux.com/pearson-correlation-coefficient-statistical-significance/ (accessed on 20 December 2023).
32. Timelli, G.; Bonollo, F. Fluidity of aluminium die castings alloys. *Int. J. Cast. Met. Res.* **2007**, *20*, 304–311. [CrossRef]
33. Satir-Kolorz, A.H.; Feichtinger, H.K. On the solubility of nitrogen in liquid iron and steel alloys using elevated pressure. *Int. J. Mat. Res.* **1991**, *82*, 689–697. [CrossRef]
34. Sigworth, G.K.; Elliot, J.F. The Thermodynamics of Liquid Dilute Iron Alloys. *Met. Sci.* **1974**, *8*, 298–310. [CrossRef]
35. Hultgren, R.; Desai, P.D.; Hawkins, D.T.; Gleiser, M.; Kelley, K.K. *Selected Values of the Thermodynamic Properties of Binary Alloys*; American Society for Metals: Metals Park, OH, USA, 1975.
36. Naraghi, R.; Selleby, M.; Ågren, J. Thermodynamic of stable and metastable structures in Fe-C system. *Calphad* **2014**, *46*, 148–158. [CrossRef]
37. Banya, S.; Matoba, S. Activity of carbon and oxygen in liquid iron. *Tetsu-Hagane Overseas* **1963**, *3*, 21–28. [CrossRef]
38. Stefanescu, D.M. *Thermodynamic Properties of Iron-Base Alloys*; ASM Handbook Committee: 1988; ASM International: Metals Park, OH, USA, 1988; Volume 15.
39. Stefanescu, D.M.; Suarez, R.; Kim, S.B. 90 years of thermal analysisi as a control tool in the melting of cast iron. *China Foundry* **2020**, *17*, 69–84. [CrossRef]
40. Holmgren, D.; Diószegi, A.; Svensson, I.L. Effects of Inoculation and Solidification Rate on the Thermal Conductivity of Grey Cast Iron. *Giessereiforschung* **2006**, *58*, 12–17.
41. Thermal Analysis of Cast Iron, Heraeus Electro-Ni. Available online: https://www.heraeus.com/media/media/hen/media_hen/products_hen/iron/thermal_analysis_of_cast_iron.pdf (accessed on 15 December 2023).
42. Sato, A.; Aragane, G.; Ogata, S.; Yamada, K.; Yoshimatsu, S. A method of Recovery of Boron from Fig Iron and Boron Oxide from Slag. *Trans. Iron Steel Inst. Jpn.* **1986**, *26*, 949–954. [CrossRef]
43. TDR, Product Line Overview. Available online: http://www.tdrlegure.si/files/pdf/Katalog-2019-ARR.pdf (accessed on 15 June 2023).
44. Sychev, A.V.; Salina, V.A.; Babenko, A.A.; Zhuchkov, V.I. Distribution of Boron between Oxide Slag and Steel. *Steel Transl.* **2017**, *47*, 105–107. [CrossRef]

Disclaimer/Publisher's Note: The statements, opinions and data contained in all publications are solely those of the individual author(s) and contributor(s) and not of MDPI and/or the editor(s). MDPI and/or the editor(s) disclaim responsibility for any injury to people or property resulting from any ideas, methods, instructions or products referred to in the content.

Article

Molecular Dynamics Simulation Research on Fe Atom Precipitation Behaviour of Cu-Fe Alloys during the Rapid Solidification Processes

Xufeng Wang, Xufeng Gao, Yaxuan Jin, Zhenhao Zhang, Zhibo Lai, Hanyu Zhang and Yungang Li *

College of Metallurgy and Energy, North China University of Science and Technology, Tangshan 063210, China; wangxf@ncst.edu.cn (X.W.); ruoyichen2021@163.com (X.G.); j1137025165@163.com (Y.J.); 13053334352@163.com (Z.Z.); 247537441@163.com (Z.L.); 15631867824@163.com (H.Z.)
* Correspondence: lyg@ncst.edu.cn

Abstract: To explore the crystalline arrangement of the alloy and the processes involving iron (Fe) precipitation, we employed molecular dynamics simulation with a cooling rate of 2×10^{10} for $Cu_{100-X}Fe_X$ (where X represents 1%, 3%, 5%, and 10%) alloy. The results reveal that when the Fe content was 1%, Fe atoms consistently remained uniformly distributed as the temperature of the alloy decreased. Further, there was no Fe atom aggregation phenomenon. The crystal structure was identified as an FCC-based Cu crystal, and Fe atoms existed in the matrix in solid solution form. When the Fe content was 3%, Fe atoms tended to aggregate with the decreasing temperature of the alloy. Moreover, the proportion of BCC crystal structure exhibited no obvious changes, and the crystal structure remained FCC-based Cu crystal. When the Fe content was between 5% and 10%, the Fe atoms exhibited obvious aggregation with the decreasing temperature of the alloy. At the same time, the aggregation phenomenon was found to be more significant with a higher Fe content. Fe atom precipitation behaviour can be delineated into three distinct stages. The initial stage involves the gradual accumulation of Fe clusters, characterised by a progressively stable cluster size. This phenomenon arises due to the interplay between atomic attraction and the thermal motion of Fe-Fe atoms. In the second stage, small Fe clusters undergo amalgamation and growth. This growth is facilitated by non-diffusive local structural rearrangements of atoms within the alloy. The third and final stage represents a phase of equilibrium where both the size and quantity of Fe clusters remain essentially constant following the crystallisation of the alloy.

Keywords: Cu-Fe alloy; Fe atom precipitation; molecular dynamics simulation

Citation: Wang, X.; Gao, X.; Jin, Y.; Zhang, Z.; Lai, Z.; Zhang, H.; Li, Y. Molecular Dynamics Simulation Research on Fe Atom Precipitation Behaviour of Cu-Fe Alloys during the Rapid Solidification Processes. *Materials* **2024**, *17*, 719. https://doi.org/10.3390/ma17030719

Academic Editors: Andrea Di Schino and Claudio Testani

Received: 19 December 2023
Revised: 29 January 2024
Accepted: 30 January 2024
Published: 2 February 2024

Copyright: © 2024 by the authors. Licensee MDPI, Basel, Switzerland. This article is an open access article distributed under the terms and conditions of the Creative Commons Attribution (CC BY) license (https://creativecommons.org/licenses/by/4.0/).

1. Introduction

Cu-Fe alloys are significant amorphous alloy materials. Owing to their thermal stability and mechanical properties [1–3], such alloys have been extensively adopted in electronics, aerospace, automobiles, and other fields. As is well known, changes in the Fe content of Cu-Fe alloys have a significant influence on the alloy structure and properties. Elevating the Fe content within the Cu-Fe lattice serves to augment both the stability of the lattice and the material's hardness and strength. Nevertheless, this enhancement comes at the expense of reduced material toughness and plasticity, resulting in a more brittle fracture surface. Simultaneously, the wear resistance of Cu-Fe alloys demonstrates an increase proportional to the rising Fe content [4,5]. However, with the increasing Fe content, the resistivity of Cu-Fe alloys will gradually increase, causing the alloy's conductivity to decrease. This phenomenon can be attributed to the higher resistivity of Fe crystals compared to Cu crystals. Additionally, the formation of Fe cluster structures disrupts the integrity of the crystal structure, further contributing to the increased resistivity and reduced conductivity of the alloy [6–8]. Currently, the utilisation of rapid solidification technology for researching and producing typical immiscible alloys like Cu-Fe alloys holds great potential. The crystal structure of Cu-Fe alloys will change with increasing Fe content during rapid cooling. Cu-Fe alloys with low Fe content (Fe content is less than 3%) present a solid solution structure, which renders the formation of complex polycrystalline structures difficult. With a high Fe content (Fe content is greater than 3%), a phase separation structure is exhibited, and the Cu and Fe form separate crystal regions [9,10]. In the high-temperature crystalline state, Cu-Fe alloys exhibit a polycrystalline structure and a serialised microstructure. However, when subjected to rapid cooling, they transform into an amorphous structure with local order. The change in cooling rate can directly result in a significant alteration in the characteristics of the Fe-rich phase. During heat treatment, the morphology of the Fe-rich phase of a Cu-Fe alloy in a water-cooled solidified state will change from nanometer-sized spheres to micron-sized multi-branch petal shapes [11,12]. Currently, most experimental studies focus on the morphology and mechanical properties of Cu-Fe alloys, but there is limited research on the precipitation mechanism of Fe clusters. The precipitation of Fe clusters is crucial for the performance of alloys. Molecular dynamics simulation is a highly valuable method for investigating the microscopic properties of crystalline materials and is extensively utilised in the examination of metal clusters and metal nanoparticles [13,14]. Starting from the atomic scale, Xu et al. conducted a thorough analysis of the effects of various contents and cooling rates on solidification nucleation during rapid solidification. They further gained insight into the nucleation mechanism at the atomic scale [15]. J. Phys et al. investigated the structural evolution mechanism of Cu clusters and Fe clusters during the solidification process of a Cu-Fe alloy using molecular dynamics simulation. The nanoclusters were characterised using adaptive adjacency relationships, radial density distribution, and potential energy. They conducted comprehensive research on the structural units and bond energy of Cu-Fe alloys and presented their evolution process and nanoparticle morphology. The research results showed that Fe and Cu are immiscible. Solidification is closely related to solid volume and other factors [16]. However, the influence of the Fe doping concentration in the Cu matrix has not been thoroughly studied. Currently, there is a substantial body of literature examining the growth mechanism of Fe-rich phases and clusters, as well as their influence on alloys. However, there are few articles that systematically describe the formation process of Fe clusters and the impact of changes in Fe content on Fe cluster formation in low-Fe environments. Using molecular dynamics simulation, an investigation was conducted into the distribution and morphological changes of Fe atoms in Cu-Fe alloys with different Fe contents (1%, 3%, 5%, and 10%) under rapid cooling (2×10^{10} K/s). With the analysis of the structural changes of Cu-Fe alloys during rapid cooling, the formation process and precipitation mechanism of Fe atomic clusters with different Fe contents were investigated. Additionally, the structural changes of the alloys were analysed through the relationships between statistical average potential energy, radial distribution function, coordination number, and mean square displacement, as well as the changes in the alloy temperature. At the same time, by means of the atomic structure visualisation technology, the size and number changes of Fe clusters were statistically and analytically analysed, so as to explore the influence mechanism of Fe atom precipitation during the rapid cooling process of Cu-Fe alloys with different Fe contents. Investigating the distribution and segregation mechanisms of Fe atoms within alloys holds significant importance. It provides essential guidance for optimising the manufacturing process and enhancing the performance of Cu-Fe alloys.

2. Simulation Method

Based on existing research, the $Cu_{100-X}Fe_X$ (where X represents 1%, 3%, 5%, and 10%) alloy model was established using Atomsk (Pierre Hirel 2010—Version 0.11) (The Swiss-army knife of atomic simulations) software [17]. Specifically, the unit cell of the Cu single crystal was constructed first and then copied to generate a $30 \times 30 \times 12$ supercell. There were 43,200 Cu atoms in total in the simulation box, of which 432, 1296, 2160, and 4320 were replaced randomly by Fe atoms to create the alloy model needed for simulation. Subsequently, MD simulation analysis was conducted using the large-scale atomic/molecular massively parallel simulator (LAMMPS) designed and developed by Sandia National Laboratory, Albuquerque, NM, USA [18]. During the simulation process, the separation of Fe atoms from Cu-Fe alloys was modelled using the interaction potential function introduced by Bonny et al. [19], which has been commonly applied to investigate the phase separation phenomenon in Fe-Cu alloys over the past several years. The NPT ensemble was implemented in all MD simulation steps, and a time step of 2 fs was selected. The system pressure was fixed at 0 bar, and the Nose-Hoover algorithm was employed to regulate both the pressure and temperature of the system. The simulation was conducted under periodic boundary conditions. The model first ran 300,000 time steps at 2200 K so as to allow for the atoms to diffuse fully and the alloy system to reach the equilibrium state. The alloy was then cooled to 300 K at a rate of 2×10^{10} K/s. The rapid cooling process was achieved through the specification of a total number of time steps for the simulation, which was set to 4.75×10^7 in this study. The structural information and kinetic and thermodynamic parameters of the alloy system were recorded in the simulated cooling process. Finally, OVITO (version 3.8.3.) [20] software was used to visualise the simulation results, and common neighbour analysis and cluster analysis were performed to examine the crystal microstructure, formation, and growth patterns of Fe clusters in the simulation process.

3. Results and Discussion

3.1. The Influence of the Fe Content on the $Cu_{95}Fe_5$ Alloy Structure

3.1.1. Average Atomic Potential Energy

Curves depicting the variation in APE (average atomic potential energy) with temperature serve as an accurate representation of microstructural changes that occur during rapid cooling. These curves offer a clear and straightforward means of analysing the phase transition process within the system [21,22]. Generally, if a curve veers away from a straight line without any sudden shifts, it suggests that the system is amorphous. However, if the curve veers away from a straight line and there are sudden shifts, it suggests that the system is crystalline [23,24]. The plot in Figure 1 shows the average potential energy curves for Fe contents of 1%, 3%, 5%, and 10% as a function of temperature.

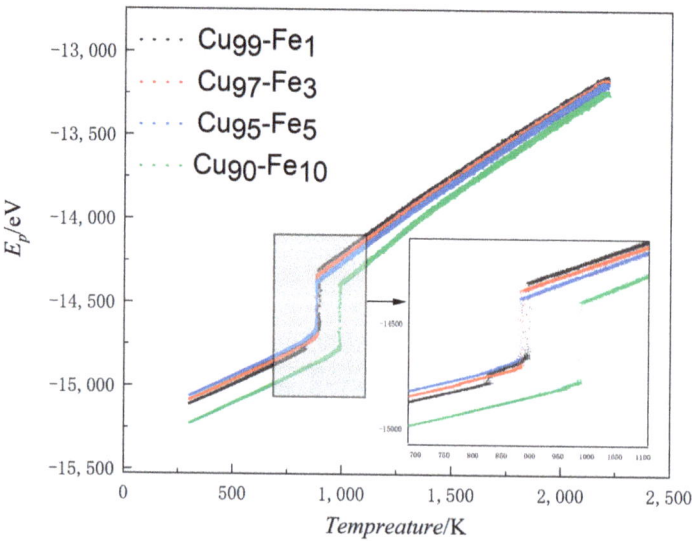

Figure 1. Average atomic potential energy-temperature variation line (APE-Temp variation line) for Cu-Fe alloys.

During the simulation process, as the system's temperature decreases, the intensity of the thermal motion of atoms in the system also decreases. The gradual decrease in speed facilitates the combination of atoms, leading them to combine with each other, thus forming ordered structures. The formation of this ordered structure results in a decrease in system volume. When the volume of the system decreases to a certain point, the local cluster structure formed in the alloy will rapidly disintegrate in order to minimise the energy of the system. The atoms will then recombine into a more stable structure, leading to crystallization. This process is accompanied by a sudden decrease in the system's energy. Manifested as a sudden change in the average atomic potential temperature curve. At Fe content levels of 1%, 3%, 5%, and 10%, the APE-Temp change lines exhibited temperatures of 894 K, 887 K, 887 K, and 994 K, respectively, indicating the formation of crystals in the system. The APE-Temp variation line suggests that as the Fe content in the alloy increased, the average atomic potential energy decreased, resulting in a more stable system structure.

3.1.2. Analysis of Crystal Structure

To further investigate the effect of Fe content on the alloy structure during solidification, the crystal configuration was analysed as shown in Figure 2. The various crystal structures in the system are determined using common neighbour analysis (CNA) in OVITO software (version 3.8.3). The CNA is an algorithm to compute a fingerprint for pairs of atoms that is designed to characterise the local structural environment. Typically, the CNA is used as an effective filtering method to classify atoms in crystalline systems, with the goal of getting a precise understanding of which atoms are associated with which phases and which are associated with defects. In this paper, the modifier selected is adaptive CNA (with variable cutoff), which determines the optimal cutoff radius automatically for each individual particle.

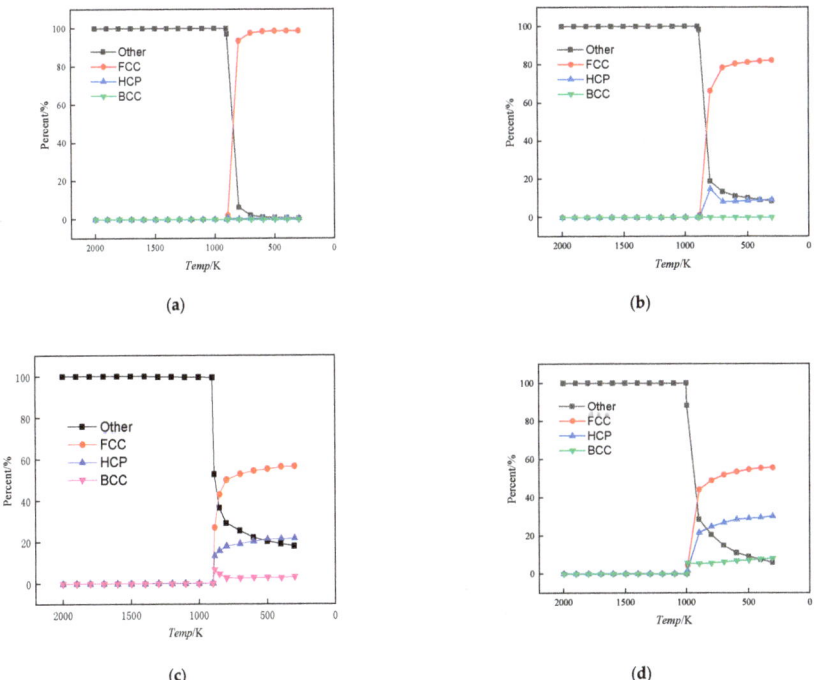

Figure 2. Crystal type ratio versus alloy temperature change curve. (a) $Cu_{99}Fe_1$, (b) $Cu_{97}Fe_3$, (c) $Cu_{95}Fe_5$, (d) $Cu_{90}Fe_{10}$.

As shown, when the system temperature was in the interval from 2000 K to the phase transition temperature (the phase transition temperatures of $Cu_{99}Fe_1$, $Cu_{97}Fe_3$, $Cu_{95}Fe_5$, and $Cu_{90}Fe_{10}$ were 894 K, 887 K, 887 K, and 994 K, respectively), the system existed in a liquid state, devoid of any crystalline structures. The Other category represents atoms that do not belong to any specific crystalline structure and are generally considered defects, occupying positions outside the lattice sites. At the phase transition temperature, there was a rapid decline in the proportion of other type atoms, concurrent with a substantial increase in the proportion of the FCC crystal structure. Figure 2 shows that at a temperature of 300 K within the alloy, the proportion of the FCC crystal structure diminished as the Fe content in the alloy increased. The crystal structure of Cu in this system was FCC, and the proportion of FCC crystal structure decreased, indicating a decrease in the percentage of Cu crystals or a decline in the proportion of Cu solid solution. Similarly, the crystal structure of Fe in this system was BCC, and there was almost no BCC crystal in the alloy when the Fe content was between 1% and 3%, indicating that Fe atoms may not have been clustered or clusters of smaller Fe atoms were involved in the formation of HCP and FCC crystal structures. The observation of the BCC crystal structure at 5% Fe content, as depicted in the temperature change curve, reveals that this structure initially increased and then decreased with the phase transition temperature. This phenomenon suggests that clusters formed by Fe atoms exhibited an initial increase followed by a decrease. Considering the subsequent process of Fe cluster formation, it can be concluded that the abrupt decrease in the BCC crystal structure at the onset of crystallisation was due to the continuous formation of Fe clusters within the alloy before crystallisation occurred. Before the alloy crystallised, it contained both large and small Fe clusters within the system. As the alloy reached the crystallisation temperature, some of the smaller clusters decomposed into individual atoms or condensed as a whole. These condensed clusters adopted the FCC and HCP crystal structures, with Cu atoms serving as the basis for atom arrangement to achieve greater compactness. Ultimately, this process led to the formation of the Cu matrix with embedded Fe atoms, primarily in the FCC and HCP crystal types. The BCC crystal structure of the alloy with 10% Fe content increased gradually as it cooled down from the phase transition temperature, unlike the alloy with 5% Fe content. This slow increase was caused by a few iron (Fe) atoms near the iron clusters combining to form larger iron clusters.

3.1.3. Radial Distribution Function Analysis

The radial distribution function, expressed as g(r), is a statistical parameter reflecting the distribution characteristics of system atoms and $g_{\alpha-\beta}(r)$ represents the ratio of the probability of finding an atom β at a fixed atomic distance α from radius r to the conditional probability [25–27]. The radial distribution function serves as a valuable tool for distinguishing between liquid, crystalline, and amorphous states in a material. It offers structural insights, including atomic radius, inter-atomic spacing, and coordination number. Through the analysis of peaks in the radial distribution function curve, the strength of inter-atomic interactions can be assessed, and the extent of short- and mid-range ordering in the material's atomic arrangement can be determined [28,29]. The radial distribution functions of $Cu_{99}Fe_1$, $Cu_{97}Fe_3$, $Cu_{95}Fe_5$, and $Cu_{90}Fe_{10}$ alloys were calculated at 2000 K–300 K with a cooling rate of 2×10^{10} K/s, and the results are shown in Figure 3.

The results of the radial distribution function analysis, g(r), conducted in the present study on Cu-Fe alloy systems with varying Fe contents, cooled at a rate of 2×10^{10}, reveal specific temperature intervals. When the Fe content was 1%, 3%, 5%, and 10%, corresponding to temperature ranges of 2000 K–894 K, 2000 K–887 K, 2000 K–887 K, and 2000 K–994 K, respectively, the first peak in the radial distribution function exhibited a distinct non-spiky pattern with clearly resolved symmetric peaks. These characteristics are indicative of the system being in a liquid state within these temperature intervals [30]. Upon further analysis, it was evident that this conclusion aligns closely with other physical properties and demonstrates high experimental reliability. In the subsequent temperature intervals, the shape of the first peak in the radial distribution function for systems with varying Fe contents all displayed symmetric, sharp peaks. These peaks were notably more pronounced than the other peaks, which strongly suggests that the system transitioned into a crystalline state [31,32]. In light of the aforementioned analyses, the phase transition temperature for each alloy system could be reliably derived.

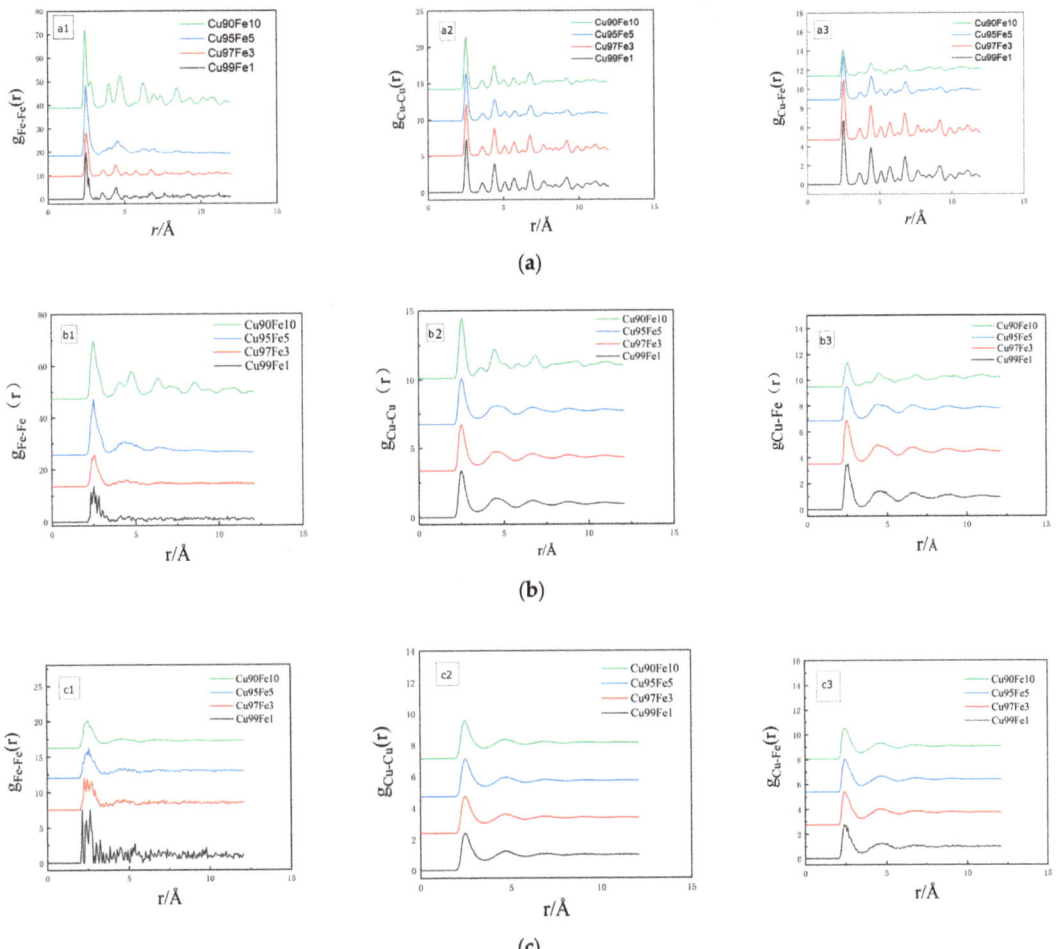

Figure 3. Plot of radial distribution function of alloy at different temperatures, (**a**) 300 K, (**b**) crystallisation temperature, (**c**) 2000 K.

Upon analysing the radial distribution function g(r), the first peak values of $g_{Fe-Fe}(r)$ were found to be higher than those of $g_{Cu-Cu}(r)$ and $g_{Cu-Fe}(r)$ for alloys with varying Fe contents and temperatures. As the temperature of the alloy decreased, the difference between the Fe-Fe atomic pairs of the first peak and the Cu-Fe and Cu-Cu atomic pairs of the first peak became increasingly larger, and the Cu-Cu atomic pairs of the first peak gradually exceeded the Cu-Fe atomic pairs of the first peak. At an alloy temperature of 300 K, it was observed that the higher the Fe content in the alloy, the greater the first peak value for Fe-Fe atomic pairs in the radial distribution function. These findings indicate a substantial attraction between Fe-Fe atom pairs, in contrast to the relatively weaker attraction between Cu-Fe and Cu-Cu atom pairs. The increase in the number of Fe atoms within the alloy plays a significant role in facilitating the formation of Fe clusters. The analysis of the value of the second peak of the radial distribution function provided information about the changes in the structure and properties of the system with temperature [33]. In the $Cu_{95}Fe_5$ and $Cu_{90}Fe_{10}$ alloy systems, it is clear that as the temperature decreased, the value of the second peak of the $g_{Fe-Fe}(r)$ curve increased. After the phase transition temperature, the values of the $g_{Fe-Fe}(r)$ curve were consistently above 1, indicating that most of the first and second nearest neighbours of Fe atoms were occupied by Fe atoms at this stage. The second peaks of $g_{Cu-Fe}(r)$ and $g_{Cu-Cu}(r)$ decreased with the decrease in temperature until dropping below 1 at 300 K, indicating that at this time, the probability of Cu atoms and Fe atoms appearing in the second nearest-neighbour layer of Cu atoms was the same as the conditional probability, and the interatomic correlation was low. In summary, the following conclusions can be drawn from the analysis: The interaction force between Fe-Fe atoms was notably stronger than that between other atom pairs; the alloy's Fe-aggregation tendency increased with the increase in the number of Fe atoms; and the aggregation of Fe atoms was an active behaviour. When the temperature of the alloy decreased, the probability of the appearance of Fe atoms around the Fe atoms increased gradually, which could be attributed to the fact that a closer interaction could be established between Fe atoms with equal charges. In this process, the proportion of the first and second nearest neighbours of Fe atoms occupied by Fe atoms also increased, thereby promoting the formation of Fe atoms into clusters.

3.1.4. Allotropic Analysis

The coordination number (CN) indicates how many molecules are found in the range of each coordination sphere. Integrating g(r) in spherical coordinates to the first minimum of the RDF will give the coordination number of a molecule. The coordination number is a parameter that provides insight into the local atomic bonding within alloys. It is indicative of the short-range ordering of atomic arrangements and describes the proximity or closeness of atoms to a central atom. The coordination number is a statistical method for exploring the arrangement pattern of atoms' nearest neighbours, and an increase in the coordination number indicates an increase in the local packing density of atoms [34,35]. Figure 4 shows the variation of atomic coordination number with temperature under simulated conditions.

When the alloy contained 1% Fe, the coordination numbers of Fe-Fe and Cu-Fe atoms remained relatively constant as the temperature decreased. However, once the temperature reached the phase change point, the coordination number of Cu-Cu atoms suddenly decreased. This indicates that the atomic arrangement became denser and the alloy underwent crystallisation. At this point, the crystal structure of Cu dominated the alloy, and Fe clusters did not form. When the alloy contained 3% Fe and reached the phase transition temperature, the coordination number of Fe-Fe atoms increased. This signifies that Fe-Fe atoms tend to aggregate at this specific temperature and composition. However, the coordination number of Cu-Fe atoms notably remained largely unchanged during this phase transition. Given the phenomenon that the proportion of BCC crystal structure was basically unchanged with the decrease in temperature in Figure 2b, an observation can be made that when the Fe content was 3%, although the Fe-Fe atoms had the tendency to aggregate with each other, the Cu atoms accounted for a larger proportion of the Fe atoms. Moreover, the clusters formed by Fe atom aggregation were solidly dissolved in the Cu matrix of the FCC crystal structure, and the structure of the alloy was still dominated by the Cu crystal structure, which is the reason for the basically unchanged coordination number of the Cu-Fe atoms. This is also the reason why the Cu-Fe atomic coordination number was basically unchanged [36]. When the Fe content was between 5% and 10%, the trends of Fe-Fe, Cu-Fe, and Cu-Cu atomic coordination numbers with temperature change were essentially the same, and the change was more obvious for the Fe content of 10% than that of 5%. When the temperature of the alloy reached the phase transition temperature, the coordination number of Cu-Cu atom pairs suddenly decreased, and crystals were formed. Before reaching the phase transition temperature, the coordination number of Fe-Fe atoms steadily increased as the temperature decreased. In contrast, the coordination number of Cu-Fe atoms decreased gradually before the phase transition temperature. These trends suggest that in the liquid state, Cu-Fe atoms were dispersed, while Fe-Fe atoms began to exhibit increasing attraction and aggregation. This process continued until the phase transition temperature was reached. After the phase transition temperature, the coordination numbers of these atoms remained relatively stable, indicating a distinct change in the state of the alloy.

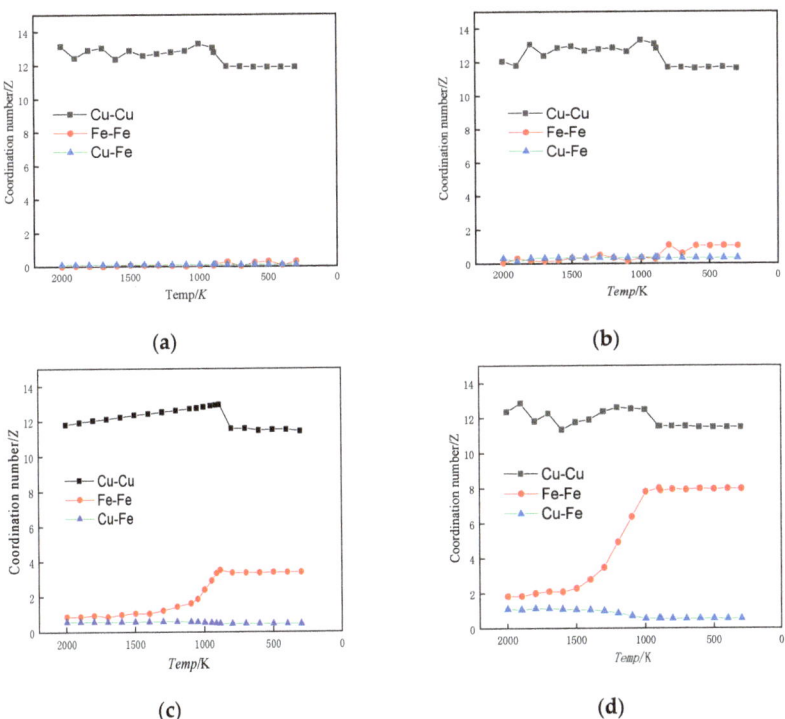

Figure 4. Coordination number versus temperature profile. (**a**) $Cu_{99}Fe_1$, (**b**) $Cu_{97}Fe_3$, (**c**) $Cu_{95}Fe_5$, (**d**) $Cu_{90}Fe_{10}$.

3.1.5. Mean-Square Displacement Analysis

Atomic diffusion is the process by which atoms move from one of their positions to another across lattice vacancies or other defective sites during thermal movement within a crystal [37]. The atomic diffusion coefficient is a physical quantity that describes the rate of atomic diffusion motion [38]. In molecular dynamics simulations of alloy systems, one common method to calculate the diffusion coefficients of atoms is by monitoring the mean square displacements (MSD) of atoms throughout the simulation [39].

$$D^* = \frac{<r^2(t)>}{2Nt} \tag{1}$$

where N—the dimension of the simulation system, N = 3 for this system; t—the simulation time, ps; r(t), r(0)—the position of the atom at time t and the initial position of the atom, respectively.

$$MSD = <r^2(t)> = \frac{1}{N}\sum_i^N <|r_i(t) - r_i(0)|^2> \tag{2}$$

where MSD is the mean square displacement;

Combining Formulas (1) and (2), the following can be obtained:

$$D^* = \frac{MSD}{6t} \rightarrow MSD = 6D^*t \quad (3)$$

The diffusion coefficient of Fe atoms was 1/6 of the slope of the relationship curve between MSD and time t.

Figure 5 summarises the MSD values of Fe atoms along the X, Y, and Z directions at a cooling rate of 2×10^{10} K/s.

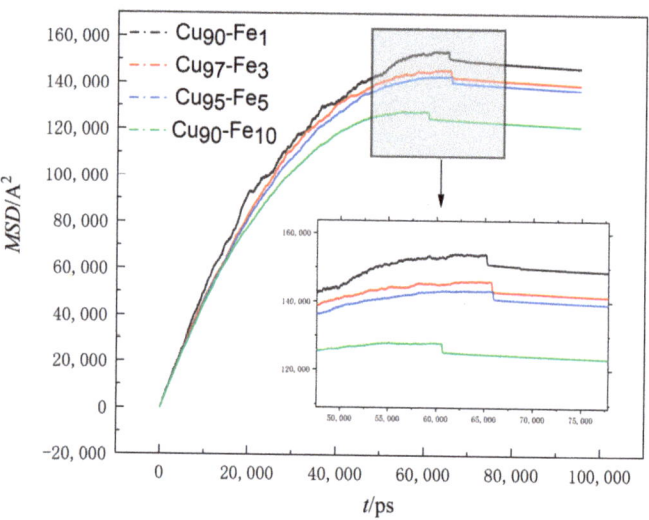

Figure 5. The variation of MSD with time.

The observed decrease in the slope of the MSD (Mean Square Displacement) versus the time curve as time progressed suggests that the diffusion rate of Fe atoms within the system decreased over time [40]. This phenomenon can be attributed to the diminishing temperature of the system as time elapses. The significant subcooling effect caused a reduction in the thermal mobility of the atoms, leading to a decrease in the range of atomic movement. Additionally, from the visual observations, it became apparent that at any given moment, a higher Fe content within the alloy corresponded to a smaller mean square displacement and a reduced slope in the MSD vs. time curve. This further signifies that the diffusion rate of Fe atoms decreased as Fe content increased, and the range of atomic movement became more limited. At 55,000 ps–60,000 ps, the slope of the curve was close to 0, indicating that the diffusion rate of Fe atoms at this stage gradually decreased to 0. However, when analysed in context, non-diffusive local atomic structural rearrangements dominated the structural changes within the system during this period, making this phase an important stage in the formation of large Fe clusters. Here, non-diffusive atomic structure rearrangement refers to the fact that although an atom moves, it is in the bondage of the surrounding atoms; that is, the local structure of the atom and its surroundings is not changed. The aggregation and condensation of small clusters is a specific manifestation of this. At the times of 65,822 ps, 65,762 ps, 65,936 ps, and 60,682 ps, respectively, the curve appeared to be broken, at which time crystallisation took place, followed by minimal changes in the trace, which was mainly due to the low temperature of the system, with the cooling rate being too fast, resulting in a small range of atomic motion.

3.2. Formation Mechanism of Fe Clusters in Cu-Fe Alloys during Rapid Cooling

Figure 6 shows plots of the distribution of Fe atoms for $Cu_{99}Fe_1$, $Cu_{97}Fe_3$, $Cu_{95}Fe_5$, and $Cu_{90}Fe_{10}$ alloys at three distinct temperature points: 2000 K, the phase transition temperature, and 300 K, respectively. The square in Figure 6 represents the simulation box with dimensions of 108.42 × 108.42 × 43.368 Å^3.

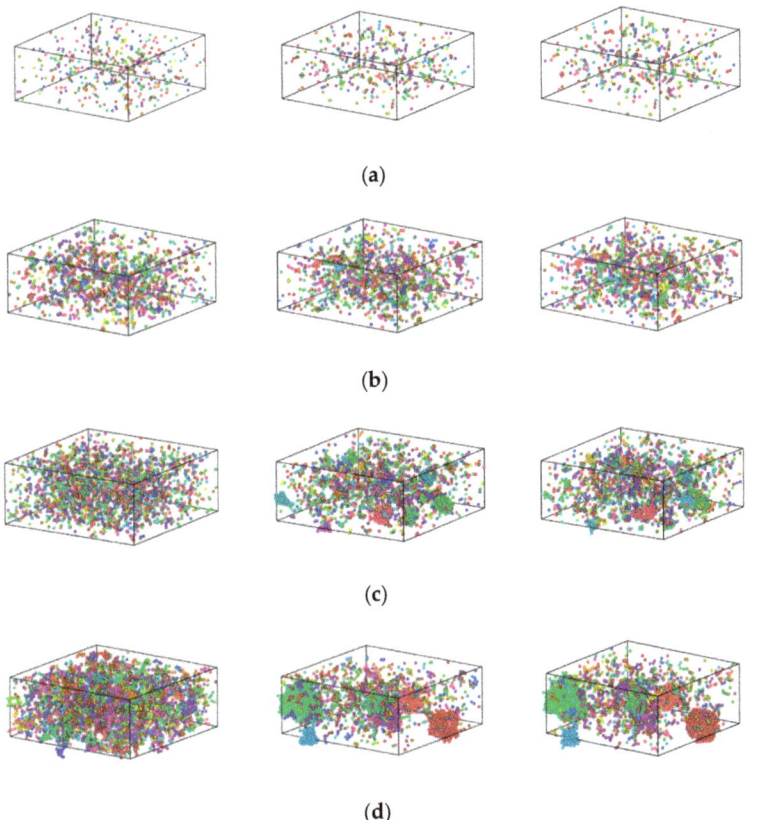

Figure 6. Distribution of Fe atoms in the $Cu_{100-X}Fe_X$ alloy (**a**) $Cu_{99}Fe_1$, (**b**) $Cu_{97}Fe_3$, (**c**) $Cu_{95}Fe_5$, and (**d**) $Cu_{90}Fe_{10}$. Different colors represent clusters of different sizes.

The clusters in the system were determined using cluster analysis in OVITO software (version 3.8.3.). This modifier decomposes the particles into disconnected groups (so-called clusters) based on the selected neighbouring criterion. The neighbouring criterion can be distance-based (cutoff range) or topology-based (bond network); this paper selected distance-based (cutoff range). The cutoff range is determined by selecting the first minimum value of RDF. A cluster is defined as a set of connected particles, each of which is within the (indirect) reach of the other particles in the same cluster. Thus, any two particles from the same cluster are connected by a continuous path consisting of steps that fulfil the selected neighbouring criterion. Coloring particles by cluster gives each identified cluster a unique random colour and colours the particles according to the clusters they belong to. It helps to quickly visualise the results of the clustering algorithm.

$Cu_{99}Fe_1$ alloy did not show the Fe atom aggregation phenomenon with decreasing temperature, and Fe atoms consistently maintained a uniformly dispersed state. Different from the distribution state of Fe atoms in the $Cu_{99}Fe_1$ alloy, the Fe atom aggregation phenomenon was found in the $Cu_{97}Fe_3$ alloy with a decrease in temperature, but the Fe clusters in the $Cu_{97}Fe_3$ alloy were found to be unstable after observation using visualisation software. Here, the maximum number of atoms contained in the clusters was not more than 5. Combined with the phase diagram analysis of Cu-Fe alloys, an observation can be made that when the Fe content of Cu-Fe alloys was lower than 3%, the Fe atoms mainly existed in the form of a Cu-based solid solution, which is consistent with the previous analysis and the results from prior research. When the temperature of $Cu_{95}Fe_5$ and $Cu_{90}Fe_{10}$ alloys was 2000 K, the Fe atoms were uniformly distributed in the system and did not show any structural non-uniformity. With the temperature reduction, the Fe atoms tended to aggregate with one another. Particularly within the temperature interval around 2000 K, which corresponded to the phase transition temperature, a gradual process occurred where Fe atoms transitioned from being dispersed as single atoms to forming small clusters. These small clusters subsequently grew in size, leading to the development of larger clusters with a non-uniform distribution. At the phase transition temperature in K, there was a temperature interval where the size of Fe clusters remained relatively stable, indicating a relatively stable system structure. To facilitate the study of cluster formation and growth, the analysis focused on $Cu_{95}Fe_5$ and $Cu_{90}Fe_{10}$ alloys at a temperature of 300 K when the largest clusters were present. The analysis considered the number of atoms within the largest clusters at different moments, including clusters with more than 5 atoms. This data were used to create a graph that depicts the total number of clusters and their sizes as a function of temperature change. Notably, the term "clusters" here refers to groups of atoms, and the largest cluster consistently remained within the temperature range from the phase transition temperature to 300 K, as illustrated in Figure 7.

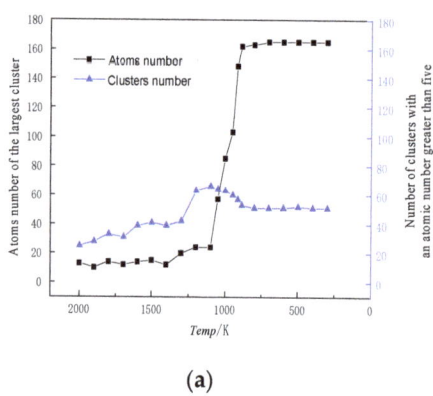
(a)

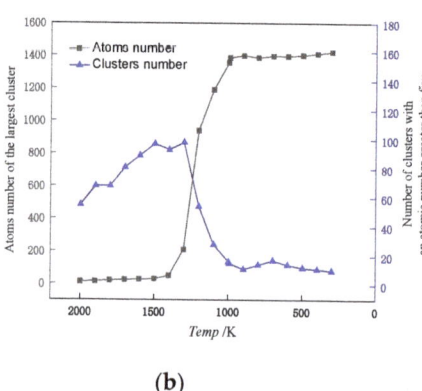
(b)

Figure 7. $Cu_{95}Fe_5$ (**a**), $Cu_{90}Fe_{10}$ (**b**) alloys at 300 K, the maximum cluster contains the number of atoms, the number of atoms greater than 5 total number of clusters versus the temperature change of the graph.

In order to facilitate the investigation into the formation of clusters, the analysis focused on the growth process at a temperature of 300 K when the largest clusters were present, as depicted in Figures 8 and 9. The square in Figures 8 and 9 represents the simulation box with dimensions of $108.42 \times 108.42 \times 43.368$ Å^3.

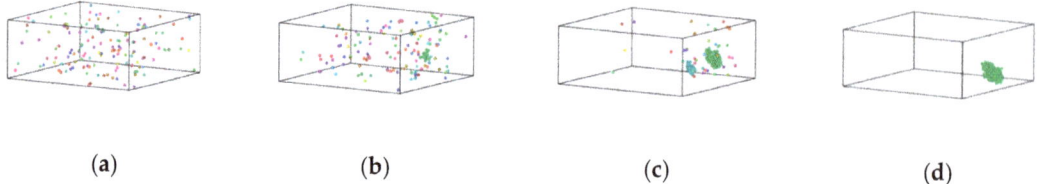

(a)   (b)   (c)   (d)

Figure 8. Distribution of Fe atoms in $Cu_{95}Fe_5$ maximal clusters at different temperatures. (**a**) 2000 K, (**b**) 1050 K, (**c**) 940 K, and (**d**) 300 K.

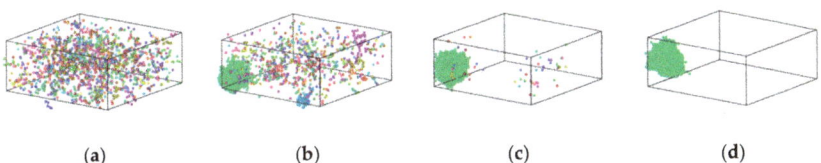

Figure 9. Distribution of Fe atoms in $Cu_{90}Fe_{10}$ maximal clusters at different temperatures. (**a**) 2000 K, (**b**) 1400 K, (**c**) 1000 K, and (**d**) 300 K.

The formation process of Fe clusters in $Cu_{95}Fe_5$ and $Cu_{90}Fe_{10}$ was basically the same, and the largest cluster grew rapidly in a certain temperature range, while the number of clusters decreased rapidly. The precipitation behaviour of Fe atoms could be divided into three phases. In the first phase, occurring between 2000 K and 1050 K for 5% Fe content or between 2000 K and 1400 K for 10% Fe content, the number of atoms within the largest clusters exhibited slight changes with decreasing temperature, while the total count of clusters containing more than 5 atoms significantly increased as the temperature dropped. This behaviour is a result of the interplay between the attractive forces among Fe atoms and the thermal motion of atoms. The mutual attraction of Fe atoms led to their aggregation, but at higher temperatures, intense thermal motion and a higher diffusion rate prevailed, causing an increase in the number of small clusters. The second stage, spanning from 1050 K (for 5% Fe content) to the phase transition temperature or 1400 K (for 10% Fe content) to the phase transition temperature, could be characterised by a rapid increase in the size of the largest clusters. Concurrently, the total number of clusters decreased during this stage. This observation suggests that the swift growth of clusters in this phase primarily arose from the condensation of smaller clusters. This insight was derived from the visualisation of the distribution of Fe atoms using software. For the third stage, as the alloy entered the crystallisation zone and the temperature continued to drop toward −300 °F, the increased cooling rate reduced atomic diffusion rates. As such, this inhibited the substantial growth of larger clusters due to the limited time for atomic diffusion. In summary, the persistent nucleation of Fe clusters throughout the cooling process was driven by Fe-Fe interatomic forces. The size and nucleation rate of iron clusters were influenced by atomic diffusion. Therefore, the formation of clusters at different stages was found to be a result of the combined effects of Fe interatomic forces and atomic diffusion.

4. Conclusions

(1) The Fe-Fe interatomic interaction force was found to be crucial for driving Fe cluster formation. The radial distribution function analysis and coordination number of the alloy during solidification demonstrate that the interaction force between Fe-Fe atoms was considerably stronger than that between other atom pairs. Further, the tendency for Fe atom aggregation increased with a higher content of Fe atoms in the alloy. The mutual attraction and aggregation of Fe Fe is a continuous process that commences when the alloy is in a liquid state.

(2) When the Fe content reached 1%, lowering the alloy temperature did not lead to the aggregation of Fe atoms. Instead, Fe atoms maintained a uniform distribution throughout the matrix, forming a solid solution. The crystal structure remained dominated by FCC-based Cu crystals. At 3% Fe content, a decrease in alloy temperature resulted in Fe atom aggregation. However, the crystal structure, primarily based on BCC, did not undergo significant changes and continued to be dominated by FCC-based Cu crystals. When the Fe content was between 5% and 10%, the Fe atoms formed clusters as the temperature of the alloy decreased.

(3) The higher the iron content, the more apparent the clusters became, and precipitation of Fe atoms occurred in three stages. In the first stage, the increase in the number of iron clusters occurred as a result of the interplay between iron-iron atomic attraction and the thermal motion of atoms. During this phase, the size of the clusters stabilised.

In the second stage, non-diffusible iron atoms underwent rearrangement influenced by the local atomic structure. This stage encompassed non-diffusive rearrangement of atoms, particularly those within condensed and growing small clusters. The third stage could be characterised by the basic stability of cluster size and number following the crystallisation of the alloy.

Author Contributions: Y.L.: Conception, Conceived the research, Discussed and commented on the manuscript, Funding acquisition; X.W.: Conception, Writing—original draft, Wrote the manuscript, Discussed and Commented on the manuscript, Performed the simulation and Analyzed the data. X.G.: Writing—original draft, Analyzed the data, Discussed and commented on the manuscript. Y.J.: Conceived the research, Formal analysis, Writing—original draft. Z.Z.: Wrote the manuscript, Discussed and commented on the manuscript. Z.L.: Writing—original draft, Revise the manuscript, Discussed and commented on the manuscript. H.Z.: Conceived the research, Discussed and commented on the manuscript. All authors have read and agreed to the published version of the manuscript.

Funding: The authors are grateful for the financial support provided by the National Natural Science Foundation of China (Grant Number 51974129) and the Tangshan Technical Innovation Team Training Plan Project (Grant Number 21130207D). Hebei Province Higher Education Science and Technology Research Project (QN2022003); Hebei Province Graduate Innovation Funding Project (CXZZBS2021100).

Institutional Review Board Statement: No applicable.

Informed Consent Statement: Not applicable.

Data Availability Statement: The original contributions presented in the study are included in the article, further inquiries can be directed to the corresponding author.

Conflicts of Interest: The authors declare no conflicts of interest.

References

1. Moon, J.; Choi, Y.; Sasaki, T.; Joo, M.; Shin, H.; Lee, J.S.; Ohkubo, T.; Hono, K.; Baek, S.M.; Kim, H.S. Corrosion-resistant Cu-Fe-based immiscible medium-entropy alloy with tri-layer passivation. *Corros. Sci.* **2021**, *193*, 109888. [CrossRef]
2. Yang, H.Y.; Ma, Z.C.; Lei, C.H.; Meng, L.; Fang, Y.T.; Liu, J.B.; Wang, H.T. High strength and high conductivity Cu alloys: A review. *Sci. China Technol. Sci.* **2020**, *63*, 41–53. [CrossRef]
3. Hauptmann, A. Metals and Alloys. In *Archaeometallurgy—Materials Science Aspects*; Hauptmann, A., Ed.; Springer International Publishing: Cham, Switzerland, 2020; pp. 381–431.
4. Han, Y.; Pan, L.; Zhang, H.; Zeng, Y.; Yin, Z. Effect of lubricant additives of Cu, Fe and bimetallic CuFe nanoparticles on tribological properties. *Wear* **2022**, *508–509*, 204485. [CrossRef]
5. Adam, O.; Jan, V.; Spotz, Z.; Cupera, J.; Pouchly, V. Ultrafine-grained Cu50(FeCo)50 immiscible alloy with excellent thermal stability. *Mater. Charact.* **2021**, *182*, 111532. [CrossRef]
6. Nizinkovskyi, R.; Halle, T.; Krüger, M. Influence of elasticity on the morphology of fcc-Cu precipitates in Fe-Cu alloys. A phase-field study. *J. Nucl. Mater.* **2022**, *566*, 153764. [CrossRef]
7. Huang, S.; Qin, L.; Zhao, J.; Xiang, K.; Luo, S.; Michalik, S.; Mi, J. Revealing atomic structure evolution of an Al-1.5Fe alloy in the liquid state using X-ray total scattering and empirical potential structure refinement. *IOP Conf. Ser. Mater. Sci. Eng.* **2023**, *1274*, 012007. [CrossRef]
8. Tian, Y.Z.; Yang, Y.; Peng, S.Y.; Pang, X.Y.; Li, S.; Jiang, M.; Li, H.X.; Wang, J.W.; Qin, G.W. Managing mechanical and electrical properties of nanostructured Cu-Fe composite by aging treatment. *Mater. Charact.* **2023**, *196*, 112600. [CrossRef]
9. Lu, D.; Osman, M.S.; Khater, M.M.A.; Attia, R.A.M.; Baleanu, D. Analytical and numerical simulations for the kinetics of phase separation in iron (Fe–Cr–X (X=Mo,Cu)) based on ternary alloys. *Phys. A Stat. Mech. Appl.* **2020**, *537*, 122634. [CrossRef]
10. Chatterjee, A.; Popov, D.; Velisavljevic, N.; Misra, A. Phase Transitions of Cu and Fe at Multiscales in an Additively Manufactured Cu-Fe Alloy under High-Pressure. *Nanomaterials* **2022**, *12*, 1514. [CrossRef]
11. Huang, S.; Luo, S.; Qin, L.; Shu, D.; Sun, B.; Lunt, A.J.G.; Korsunsky, A.M.; Mi, J. 3D local atomic structure evolution in a solidifying Al-0.4 Sc dilute alloy melt revealed in operando by synchrotron X-ray total scattering and modelling. *Scr. Mater.* **2022**, *211*, 114484. [CrossRef]
12. Chen, K.X.; Li, Z.X.; Wang, Z.D. Morphological Evolution of Fe-Rich Precipitates in a Cu-2.0Fe Alloy During Isothermal Treatment. *Acta Met. Sin.* **2023**, *59*, 1665–1674.
13. Zhang, H.C.; Wang, X.F.; Li, H.R.; Li, C.Q.; Li, Y.G. Molecular Dynamics Study on the Impact of Cu Clusters at the BCC-Fe Grain Boundary on the Tensile Properties of Crystal. *Metals* **2020**, *10*, 1533. [CrossRef]
14. Nguyen-Trong, D.; Nguyen-Tri, P. Factors affecting the structure, phase transition and crystallization process of AlNi nanoparticles. *J. Alloys Compd.* **2020**, *812*, 152133. [CrossRef]

15. Xu, H.D.; Bao, H.W.; Li, Y.; Bai, H.Z.; Ma, F. Atomic scale insights into the rapid crystallization and precipitation behaviors in FeCu binary alloys. *J. Alloys Compd.* **2021**, *882*, 160725. [CrossRef]
16. Kumar, S. Structural Evolution of Iron–Copper (Fe–Cu) Bimetallic Janus Nanoparticles during Solidification: An Atomistic Investigation. *J. Phys. Chem. C* **2020**, *124*, 1053–1063. [CrossRef]
17. Hirel, P. Atomsk: A tool for manipulating and converting atomic data files. *Comput. Phys. Commun.* **2015**, *197*, 212–219. [CrossRef]
18. Thompson, A.P.; Aktulga, H.M.; Berger, R.; Bolintineanu, D.S.; Brown, W.M.; Crozier, P.S.; in 't Veld, P.J.; Kohlmeyer, A.; Moore, S.G.; Nguyen, T.D.; et al. LAMMPS—A flexible simulation tool for particle-based materials modeling at the atomic, meso, and continuum scales. *Comput. Phys. Commun.* **2022**, *271*, 108171. [CrossRef]
19. Bonny, G.; Pasianot, R.C.; Castin, N.; Malerba, L. Ternary Fe-Cu-Ni many-body potential to model reactor pressure vessel steels: First validation by simulated thermal annealing. *Philos. Mag.* **2009**, *89*, 3531–3546. [CrossRef]
20. Polak, W.Z. Efficiency in identification of internal structure in simulated monoatomic clusters: Comparison between common neighbor analysis and coordination polyhedron method. *Comput. Mater. Sci.* **2022**, *201*, 110882. [CrossRef]
21. Wang, M.J.; Huang, X.X.; Wu, S.P.; Dai, G.X. Molecular dynamics simulations of tensile mechanical properties and microstructures of Al-4.5Cu alloy: The role of temperature and strain rate. *Model. Simul. Mater. Sci. Eng.* **2022**, *30*, 045004. [CrossRef]
22. Mitsui, Y.; Onoue, M.; Kobayashi, R.; Sato, K.; Kuzuhara, S.; Ito, W.; Takahashi, K.; Koyama, K. High Magnetic Field Effects on Cu-precipitation Behavior of Fe-1mass%Cu at 773 K. *ISIJ Int.* **2022**, *62*, 413–417. [CrossRef]
23. Makarenko, K.I.; Konev, S.D.; Dubinin, O.N.; Shishkovsky, I.V. Mechanical characteristics of laser-deposited sandwich structures and quasi-homogeneous alloys of Fe-Cu system. *Mater. Des.* **2022**, *224*, 111313. [CrossRef]
24. Shao, Q.; Guo, J.; Chen, J.; Zhang, Z. Atomic-scale investigation on the structural evolution and deformation behaviors of Cu–Cr nanocrystalline alloys processed by high-pressure torsion. *J. Alloys Compd.* **2020**, *832*, 154994. [CrossRef]
25. Kurz, W.; Rappaz, M.; Trivedi, R. Progress in modelling solidification microstructures in metals and alloys. Part II: Dendrites from 2001 to 2018. *Int. Mater. Rev.* **2021**, *66*, 30–76. [CrossRef]
26. Tang, J.B.; Ahmadi, A.; Alizadeh, A.; Abedinzadeh, R.; Abed, A.M.; Smaisim, G.F.; Hadrawi, S.K.; Nasajpour-Esfahani, N.; Toghraie, D. Investigation of the mechanical properties of different amorphous composites using the molecular dynamics simulation. *J. Mater. Res. Technol.* **2023**, *24*, 1390–1400. [CrossRef]
27. Rawat, S.; Chaturvedi, S. Effect of temperature on the evolution dynamics of voids in dynamic fracture of single crystal iron: A molecular dynamics study. *Philos. Mag.* **2021**, *101*, 657–672. [CrossRef]
28. Rajput, A.; Paul, S.K. Effect of void in deformation and damage mechanism of single crystal copper: A molecular dynamics study. *Model. Simul. Mater. Sci. Eng.* **2021**, *29*, 085013. [CrossRef]
29. Vargas Rubio, K.I.; Medrano Roldán, H.; Reyes Jáquez, D. Simulación mediante dinámica molecular de un nanocluster obtenido de la industria minera. *Acta Univ.* **2022**, *31*, e3010.
30. Celtek, M.; Sengul, S.; Domekeli, U.; Guder, V. Molecular dynamics simulations of glass formation, structural evolution and diffusivity of the Pd-Si alloys during the rapid solidification process. *J. Mol. Liq.* **2023**, *372*, 121163. [CrossRef]
31. Syarif, J.; Gillette, V.; Hussien, H.A.; Badawy, K.; Jisrawi, N. Molecular dynamics simulation of the amorphization and alloying of a mechanically milled Fe-Cu system. *J. Non Cryst. Solids* **2022**, *580*, 121410. [CrossRef]
32. Trong, D.N.; Lu, T.; Tran, T. Study on the Influence of Factors on the Structure and Mechanical Properties of Amorphous Aluminium by Molecular Dynamics Method. *Adv. Mater. Sci. Eng.* **2019**, *2021*, 5564644. [CrossRef]
33. Bochkarev, A.; Lysogorskiy, Y.; Menon, S.; Qamar, M.; Mrovec, M.; Drautz, R. Efficient parametrization of the atomic cluster expansion. *Phys. Rev. Mater.* **2022**, *6*, 013804. [CrossRef]
34. Duan, L.J.; Liu, Y.C.; Duan, J.S. Calculation of radii and atom numbers of different coordination shells in cubic crystals. *Mater. Today Commun.* **2020**, *22*, 100768. [CrossRef]
35. Hansson, P.; Ahadi, A.; Melin, S. Molecular dynamic modelling of the combined influence from strain rate and temperature at tensile loading of nanosized single crystal Cu beams. *Mater. Today Commun.* **2022**, *31*, 103277. [CrossRef]
36. Ranaweera, S.A.; Donnadieu, B.; Henry, W.P.; White, M.G. Effects of electron-donating ability of binding sites on coordination number: The interactions of a cyclic Schiff base with copper ions. *Acta Crystallogr. Sect. C Struct. Chem.* **2023**, *79*, 142–148. [CrossRef]
37. Rogachev, A.S.; Fourmont, A.; Kovalev, D.Y.; Vadchenko, S.G.; Kochetov, N.A.; Shkodich, N.F.; Baras, F.; Politano, O. Mechanical alloying in the Co-Fe-Ni powder mixture: Experimental study and molecular dynamics simulation. *Powder Technol.* **2022**, *399*, 117187. [CrossRef]
38. Dias, M.; Carvalho, P.A.; Gonçalves, A.P.; Alves, E.; Correia, J.B. Hybrid molecular dynamic Monte Carlo simulation and experimental production of a multi-component Cu–Fe–Ni–Mo–W alloy. *Intermetallics* **2023**, *161*, 107960. [CrossRef]
39. Geslin, P.A.; Rodney, D. Microelasticity model of random alloys. Part I: Mean square displacements and stresses. *J. Mech. Phys. Solids* **2021**, *153*, 104479. [CrossRef]
40. Seoane, A.; Farkas, D.; Bai, X.M. Influence of compositional complexity on species diffusion behavior in high-entropy solid-solution alloys. *J. Mater. Res.* **2022**, *37*, 1403–1415. [CrossRef]

Disclaimer/Publisher's Note: The statements, opinions and data contained in all publications are solely those of the individual author(s) and contributor(s) and not of MDPI and/or the editor(s). MDPI and/or the editor(s) disclaim responsibility for any injury to people or property resulting from any ideas, methods, instructions or products referred to in the content.

Article

Study on the Microscopic Mechanism of the Grain Refinement of Al-Ti-B Master Alloy

Lianfeng Yang, Huan Zhang, Xiran Zhao, Bo Liu, Xiumin Chen * and Lei Zhou *

Faculty of Metallurgical and Energy Engineering, Kunming University of Science and Technology, Kunming 650000, China
* Correspondence: chenxiumin@kust.edu.cn (X.C.); zhoulei@kust.edu.cn (L.Z.)

Abstract: In the present work, the structure and properties of Ti_nB_n (n = 2–12) clusters were studied, and the microstructure of a Al-Ti-B system was simulated by molecular dynamics to determine the grain refinement mechanism of an Al-Ti-B master alloy in Al alloy. Based on the density functional theory method, the structural optimization and property calculations of Ti_nB_n (n = 2–12) clusters were carried out. The clusters at the lowest energy levels indicated that the Ti and B atoms were prone to form TiB_2 structures, and the TiB_2 structures tended to be on the surface of the clusters. The $Ti_{10}B_{10}$ cluster was determined to be the most stable structure in the range of n from 2 to 12 by average binding energy and second-order difference energy. The analysis of HOMOs and LUMOs suggested that TiB_2 was the active center in the cluster; the activity of Ti was high, but the activity of B atoms decreased as the cluster size n increased. Meanwhile, the prediction of reaction sites by Fukui function, condensed Fukui function, and condensed dual descriptor identify that Ti atoms were more active than B atoms. Furthermore, TiB_2 structures were found in the Al-Ti-B system simulated by the ab initio molecular dynamics method, and there were Al atoms growing on the Ti atoms in the TiB_2. Based on the above analysis, this study suggests that TiB_2 may be a heterogeneous nucleation center of α-Al. This work helps to further understand the mechanism of Al-Ti-B induced heterogeneous nucleation in Al alloys, which can provide theoretical guidance for related experiments.

Keywords: heterogeneous nucleation; TiB_2; grain refinement; Ti_nB_n clusters

Citation: Yang, L.; Zhang, H.; Zhao, X.; Liu, B.; Chen, X.; Zhou, L. Study on the Microscopic Mechanism of the Grain Refinement of Al-Ti-B Master Alloy. *Metals* **2024**, *14*, 197. https://doi.org/10.3390/met14020197

Academic Editor: Luis Antonio Barrales-Mora

Received: 28 December 2023
Revised: 21 January 2024
Accepted: 4 February 2024
Published: 5 February 2024

Copyright: © 2024 by the authors. Licensee MDPI, Basel, Switzerland. This article is an open access article distributed under the terms and conditions of the Creative Commons Attribution (CC BY) license (https://creativecommons.org/licenses/by/4.0/).

1. Introduction

Al alloys are increasingly being used in the automotive field to meet the demands for lower fuel consumption and emissions. As a result of their widespread use, requirements for performance are also increasing. During the solidification process of alloys, coarse columnar crystals may form, resulting in a large number of defects that affect the quality and performance of the casting. A large number of studies have concentrated on methods of reducing grain size, such as optimizing the casting process of alloys, electromagnetic or ultrasonic vibration, and the addition of grain refiners. The most popular method for microstructure modification is to add refining agents in the preparation process of alloy materials. This method is simple and controllable, low cost, and does not require the assistance of complex equipment. Therefore, grain refinement is an important method to improve the quality of Al alloys. It can increase the nucleation center, reduce the size of a-Al grains, and inhibit the generation of columnar crystals during the cooling crystallization process. The improvement in mechanical properties is largely due to the evolution of the microstructure. Fine grain size can reduce the microporosity and size of second phase particles, reduce the casting defects so as to improve the mechanical properties [1–4].

Al-Ti-B master alloy is one of the most popular grain refiners used in the production of industrial Al alloys; it provides excellent grain refinement performance in continuous and semi-continuous casting of forged alloys. According to reports in the literature, TiB_2 and $TiAl_3$ are the main phases of Al-Ti-B grain refiners [5,6]. Some scholars believe that TiB_2

particles alone do not have a grain refining effect on Al alloys. Only by forming a $TiAl_3$ thin layer on the surface can they activate it to become an effective nucleation centre for α-Al [7–10]. However, Wang's study [11] found that the distance between Al atoms on the Ti-terminated surface of TiB_2 is similar to the distance found in Al_3Ti, so the thin layer of TiB_2 surface observed in the experiment may be strained Al on the Ti-terminated surface. Meanwhile, Al_3Ti particles have high solubility in the melt and can be completely dissolved in a short period of time, so they are unlikely to become a heterogeneous nucleation point of α-Al [12]. Jones [13,14] summarized that TiB_2 introduces multiple heterogeneous nucleation sites, and dendritic α-Al becomes a fine equiaxed grain structure; it can also improve the morphology of eutectic Si from large flake to particles [15–18]. TiB_2 has a good grain refinement effect on α-Al.

Li et al. [19] found experimentally that the addition of TiB_2 in A390 alloy reduced the average grain size of α-Al and eutectic Si, and significantly improved the ductility and mechanical properties of the material. Greer et al. [20] found by model predictions that the heterogeneous nucleus of TiB_2 particles in Al-Ti-B refiners occurs at a lower degree of undercooling. In a recent study, Feng et al. [21] investigated the non-homogeneous nucleus and growth kinetics of Al on a (0001) TiB_2 surface; they concluded that TiB_2 particles can act as nucleating agents for Al, and that α-Al grows directly on the (0001) TiB_2 surface. David [22] compared several possible nucleation mechanisms of TiB_2 terminated with Ti, TiB_2 terminated with B, and TiB_2 covered with $TiAl_3$ by DFT simulation, and obtained the results that TiB_2 terminated with Ti was more stable than with B, and that direct nucleation of α-Al at the Ti-terminated TiB_2 was a more favorable nucleation mechanism than the formation of a $TiAl_3$ thin layer. Liu et al. [23] investigated the effect of Al-Ti-B intermediate alloy, and La on W319 alloy, and found that Al-Ti-B significantly reduced the secondary dendrite arm spacing of the alloy, improved the hardness, and decreased the number of micropores. To a certain extent, the influence of defects produced in the casting process on the properties of the alloy was reduced. Knaislová's group [24] characterised the microstructure of AlSi7Mg0.3 alloys with the addition of Al-Ti-B refiners; compared with the test results, it was found that the grain size of Al was significantly reduced, and the irregular eutectic Si phase was transformed into round or short rod. In studies of Sun et al. [25–29], the number and size of columnar crystals in Al alloys materials significantly decreased with the addition of TiB_2 particles, while the tensile strength and ductility of the materials were improved.

The actual production of alloy materials generally involves a process from liquid to solid, but many phenomena are difficult to observe. Changes of microstructure in the alloy melt during solidification have an important influence on the properties of the material; now computer simulation can be used to understand the nature of microstructure evolution. Simulations are being widely used in the study of alloy materials. By calculating the structure and properties of clusters, it is possible to speculate the changes in the microstructure of the alloy and possible reactions between the components, as well as to understand the geometric structure and interaction of the molecules. At present, cluster simulation is widely discussed in catalysis [30–32], hydrogen storage [33], alloy materials [34], and other fields. In addition, ab initio molecular dynamics simulation is helpful to observe the trajectories of atoms in the alloys, and predict possible results. The electron transfer and bonding of the components can also be analysed by electronic structure calculations. The microstructure of the Al-Ti-B system was analyzed by simulation in this study.

The process of grain refinement in Al-Ti-B grain refiner is still controversial. So far, the study of grain refinement in Al-Ti-B intermediate alloys has been conducted mostly by experimental methods. Observing nucleation behaviour during the early stages of solidification through experimental methods is challenging due to its occurrence at high temperature, microscopic scale, and for very short periods of time; common experimental means do not provide a good indication of changes in the structure of the alloy melt during the cooling process. The evolution of molecules in melts is crucial. Ti_nB_n (n = 2–12) clusters were studied, at the same time, Al-Ti-B system simulation was conducted to understand the possible behavior of Ti and B in this paper. The grain refinement behaviour of Ti and B atoms in Al alloys was investigated by calculating the molecular structure changes and electronic properties of Ti_nB_n clusters and the Al-Ti-B system. This study is significant in exploring the refining mechanism of Al-Ti-B grain refiners.

2. Computational Details

2.1. Calculation of Clusters

Firstly, the primary structures of the Ti_nB_n (n = 2–12) clusters were built and optimized in Materials Studio [35]. The optimization conditions were: DND (double numerical plus d-functions) basis set, and gradient-corrected functionals in the form of generalized gradient approximation (GGA) with the Perdew–Burkee–Ernzerhof (PBE) [36] functional; the energy and gradient convergence accuracy were 10^{-5} Hartree and 0.004 Å; and maximum displacement was 0.3 Å. Secondly, in order to obtain a more comprehensive cluster structure, the lowest energy structure was searched by ab initio molecular dynamics, and the NVT [37] ensemble, the DNP (double numerical plus polarization) basis set, and GGA-PBE exchange correlation function were used; temperature was set to 500 K, the time step was 1 fs and ran for a total of 100 ps. Finally, the cluster structure was reoptimized using Gaussian16 [38], and the energy was calculated on the level of M06L [39] /6-311G (d, p).

The stability of the clusters was calculated using Gaussian16. The Multiwfn program [40–42] was used to calculate the condensed Fukui function and condensed dual descriptor of the $Ti_{10}B_{10}$ cluster. On the basis of the output file of the Multiwfn program, the HOMOs (Highest Occupied Molecular Orbital); LUMOs (Lowest Unoccupied Molecular Orbital) of all clusters; and the Fukui function images were plotted using VMD193 [43] software, and the electronic isosurface of the $Ti_{10}B_{10}$ cluster colored by Fukui function was drawn using GaussView6.0 [44].

2.2. Ab Initio Molecular Dynamics Calculation (AIMD)

Ab initio molecular dynamics simulations of the Al-Ti-B system were carried out using the VASP [45] package and the SCAN [46] Meta-GGA functional (Strongly Constrained and Appropriately Normed Semilocal Density functional) and the NVT ensemble. K mesh was set at $1 \times 1 \times 1$, energy cut was 400 eV, and the convergence standard for energy was 10^{-6} eV/atom. The simulation temperature was 698 K. The simulation time was 10 ps (picosecond) with a time step of 1 fs (femtosecond). The lattice parameter was a = 16.0466 Å; b = 16.08830 Å; c = 15.9796016 Å; α = 90.1209°; β = 90.0634°; γ = 90.2491°, and there were 199 Al; 6 Ti; and 6 B atoms in the supercell.

3. Results and Discussion

3.1. The Ground State Structures of Ti_nB_n (n = 2–12) Clusters

The lowest energy structures of Ti_nB_n (n = 2–12) clusters are shown in Table 1, and all of them have point group symmetry structures of C1, so the symmetry of the clusters is poor. On the whole, Ti atoms are distributed outside the clusters, B atoms are relatively concentrated, and they are gradually surrounded by Ti atoms with the number of atoms increasing. Analysing the structures of the clusters revealed that Ti and B formed TiB_2 structures. As the cluster size increased, the number of TiB_2 structures increased and they were connected by a shared B/Al atom. Therefore, Ti_nB_n (n = 2–12) clusters can be considered to be composed of multiple TiB_2 structures. In addition, atomic arrangement was a major factor influencing growth behaviour [47]. As can be seen from Table 1, TiB_2 tended to be distributed on the surface of the clusters, with the B atoms towards the interior of the clusters, and the Ti atoms towards the surface of the clusters. It has been reported [22] that the Ti termination layer of TiB_2 is more stable to contact with Al melt than the B termination layer of TiB_2.

Table 1. Ground state structures of Ti_nB_n (n = 2–12) clusters.

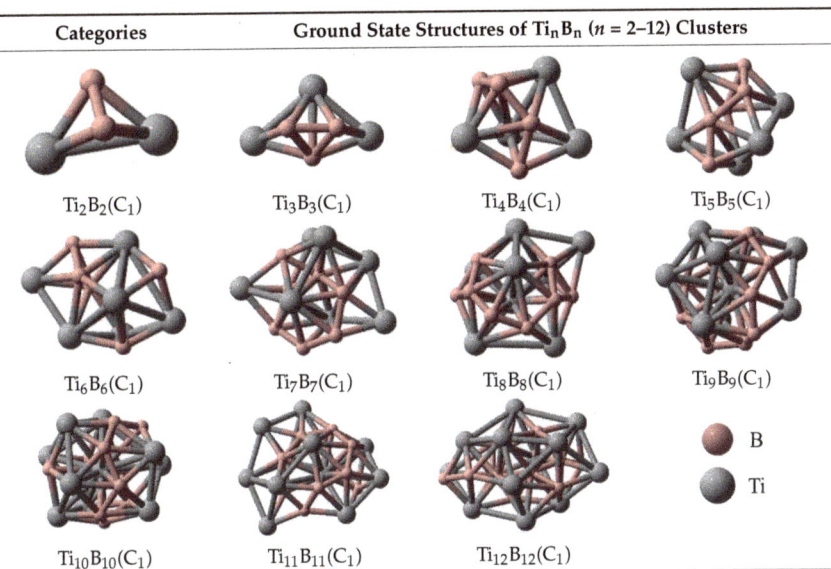

3.2. Stability of Ti_nB_n (n = 2–12) Clusters

The stability of the clusters can be assessed by the average binding energy (Eb) and the second-order difference energy (Δ_2E), which are calculated by the following formulas:

$$Eb = [nE(Ti) + nE(B) - E(Ti_nB_n)]/(2n) \quad (1)$$

$$\Delta_2E = E(Ti_{n+1}B_{n+1}) + E(Ti_{n-1}B_{n-1}) - 2E(Ti_nB_n) \quad (2)$$

E(Ti), E(B), $E(Ti_{n+1}B_{n+1})$, $E(Ti_{n-1}B_{n-1})$ and $E(Ti_nB_n)$ denote the energy of Ti atoms, B atoms, $Ti_{n+1}B_{n+1}$, $Ti_{n-1}B_{n-1}$ and Ti_nB_n clusters, respectively.

If the binding energy is large, it means that the cluster configuration is stable. Figure 1 shows the curve of the average binding energy with the change of n. As can be seen, it gradually increased with the increase of the number of atoms, indicating that the interatomic interaction was enhanced and the clusters became more and more stable. Furthermore, the average binding energy reached a maximum at n = 10, which indicates that the $Ti_{10}B_{10}$ cluster is the most stable.

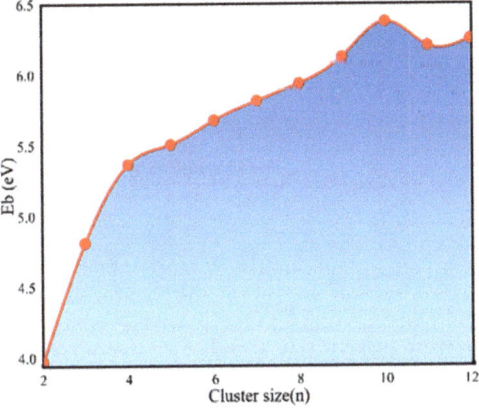

Figure 1. Average binding energy of Ti_nB_n (n = 2–12) clusters.

The second-order difference energy can reflect the stability of the clusters. The larger it is, the better the cluster stability. From Figure 2, it is easy to see when $n \leq 7$, the stability of clusters with an even number of n was higher than that with an odd number. Otherwise, when n = 2, 4 and 10, the second-order difference energy was positive, and much larger than other clusters; it was apparent that the stability of clusters Ti_2B_2, Ti_4B_4, and $Ti_{10}B_{10}$ was higher. In addition, as seen in Figure 2, the maximum value was achieved at n = 10; combined with the average binding energy, it can be considered that the $Ti_{10}B_{10}$ cluster is the most stable structure (n = 2–12).

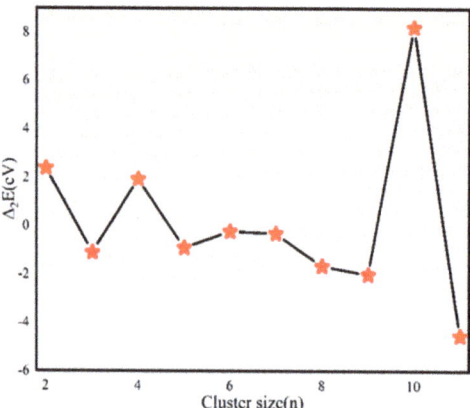

Figure 2. Second-order difference energy of Ti_nB_n (n = 2–12) clusters.

3.3. Prediction of Reaction Sites in Ti_nB_n (n = 2–12) Clusters

The properties of alloys are closely related to the electronic structure, so the HOMOs and LUMOs of Ti_nB_n (n = 2–12) clusters were calculated, as shown in Table 2. They are helpful to analyze the chemical reactivity of molecules. When n = 2, it showed that there were HOMOs and LUMOs on the surface of both Ti and B atoms, and with the gradual increase of n, there was a transfer from B atoms to Ti atoms, which were mainly distributed on the surface of Ti atoms in the TiB_2 structures. This suggested that the TiB_2 structures in the Ti_nB_n clusters were the active region of the cluster, where the activity of Ti atoms was higher than that of B atoms, and the activity of B atoms was gradually decreasing. Zhang et al. [48] reported that there is a strong 3d(Ti)-3p(Al) hybridization between the Ti-terminated interface and α-Al, that is stronger than the Al-B atom bonding; α-Al is more prone to epitaxial grown on the Ti termination surface, whereas the bonding between B-terminated and Al is weak, which cannot induce the continuous growth of Al atoms. Therefore, it is hypothesized that the Ti atom in the TiB_2 structure is the attachment site for α-Al.

Table 2. HOMOs and LUMOs of Ti_nB_n (n = 2–12) clusters.

Categories	Ground State Structures of Ti_nB_n (n = 2–12) Clusters			
HOMO				

Table 2. *Cont.*

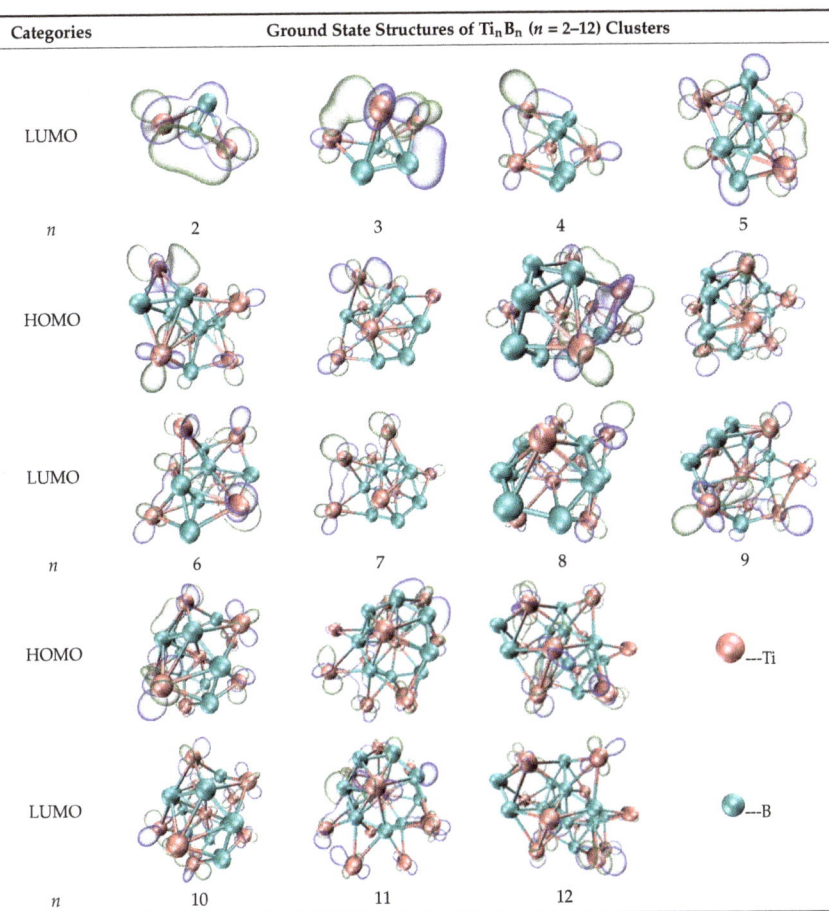

According to the average binding energy and the second-order differential energy, $Ti_{10}B_{10}$ is the most stable structure in Ti_nB_n (n = 2–12). Therefore, the $Ti_{10}B_{10}$ cluster was taken as an example to predict the reaction sites in clusters by means of the Fukui function, condensed Fukui function (CFF), and condensed dual descriptor [49–51]. The electrophilic/nucleophilic reaction formula of the Fukui function [52] are approximated as follows:

$$\text{electrophilic reaction}: f^-(r) = \rho_N(r) - \rho_{N-1}(r) \quad (3)$$

$$\text{nucleophilic reaction}: f^+(r) = \rho_{N+1}(r) - \rho_N(r) \quad (4)$$

The electron density ρ_N, ρ_{N+1} and ρ_{N-1} is represented separately when the system contains N electrons, $N + 1$ electrons, and $N - 1$ electrons. The change of electron density at each position is represented by $f^-(r)$ and $f^+(r)$ due to electron transfer when electrophilic or nucleophilic reaction occurs, respectively.

In Figure 3, (a) and (b) are electron isosurface maps of the Fukui function coloring for electrophilic and nucleophilic reaction predictions, respectively. Blue and red respectively represent positive and negative regions. The darker the blue regions, the larger the function value corresponding to it, and the stronger the reaction activity [53,54]. It is worth noting that the prediction of electrophilic reactions showed that the isosurface of the Fukui function mainly covered the surface of Ti atoms in Figure 3, indicating that the electrophilic reaction is prone to occur on Ti atoms. The isosurface map of the nucleophilic reaction was consistent with the electrophilic reaction, suggesting that the nucleophilic reaction also easily occurs on the Ti atoms. As a consequence, both electrophilic and nucleophilic reactions tend to take place on Ti atoms.

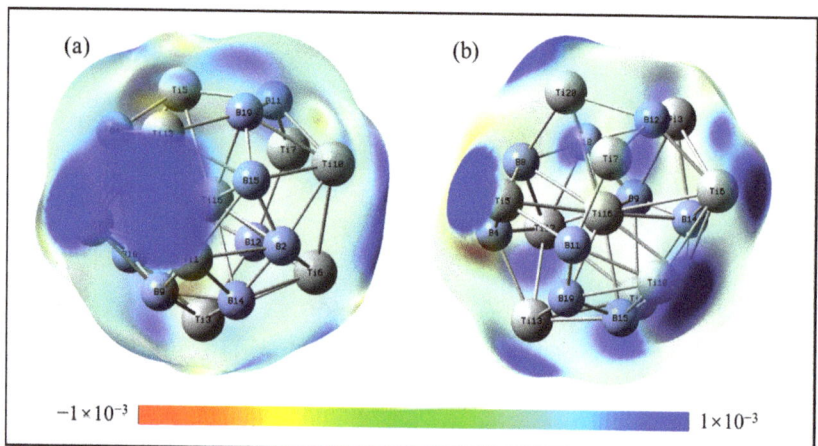

Figure 3. Fukui function isosurface maps of $Ti_{10}B_{10}$ cluster; (**a**) prediction of electrophilic reactions; (**b**) prediction of nucleophilic reactions (isovalue = 0.01 a.u.).

The condensed Fukui function contracts the Fukui function to each atom to obtain their corresponding values to quantify their reactivity [55,56]. The reasonableness of the condensed Fukui function based on Hirschfeld atomic charge calculations for predicting reaction sites has been tested [52]. The values of the condensed Fukui function are listed in Figure 4.

$$\text{electrophilic reaction}: f_A^- = q_{N-1}^A - q_N^A \tag{5}$$

$$\text{nucleophilic reaction}: f_A^+ = q_N^A - q_{N+1}^A \tag{6}$$

where q_{N-1}^A, q_N^A, q_{N+1}^A represent the charges of atom A in the system with $N-1$, N, $N+1$ electrons, and f_A^- and f_A^+ denote the condensed Fukui function corresponding to electrophilic and nucleophilic reactions, respectively.

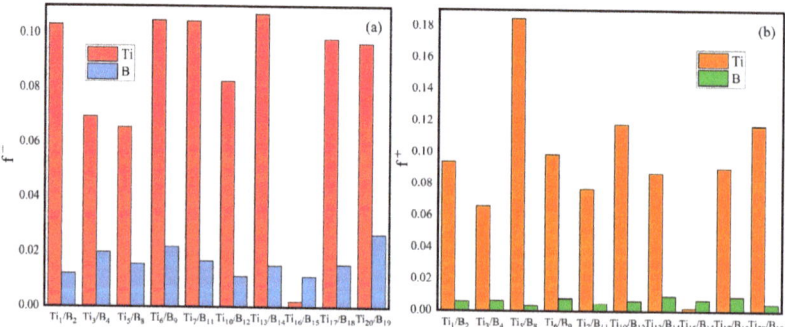

Figure 4. Condensed Fukui functions of $Ti_{10}B_{10}$ cluster; (**a**) f^-, red corresponds to Ti atoms, blue corresponds to B atoms; (**b**) f^+, orange corresponds to Ti atoms, green corresponds to B atoms.

The size of this descriptor can reflect the degree of electrophilicity or nucleophilicity. The condensed Fukui function allows for a more intuitive comparison of the activity of atoms [57]. It can be seen from Figure 4 that in the prediction of electrophilic reaction, the condensed Fukui function value of Ti atoms was significantly greater than that of B atoms, only the value of Ti16 atoms was small, so Ti atoms were vulnerable to electrophilic attack. For the nucleophilic prediction, the condensed Fukui function values for Ti atoms were similarly much larger than those for B atoms (except for Ti_{16} atoms), suggesting that Ti atoms were more likely to attract nucleophilic reagents. The above shows that the reactivity of Ti atoms is higher than that of B atoms; both electrophilic and nucleophilic reactions occur preferentially on Ti atoms. Han [58] et al. also concluded by first principles calculations that the adhesion at the Ti-termination interface is greater than that at the B-termination interface.

The condensed dual descriptor (CDD) is a parameter that defines the nature of the local position in a molecule as electrophilic (positive) or nucleophilic (negative) [59,60]. It is approximated as the second-order derivative of the electron density relative to the number of electrons under a fixed external potential [61]. In this article, the condensed dual descriptor of a $Ti_{10}B_{10}$ cluster was calculated to facilitate comparisons of the electrophilic/nucleophilic reactivity of each atom. The formula was as follows:

$$f_A^{(2)}(r) = f_A^+(r) - f_A^-(r) \qquad (7)$$

The greater the negative value of CDD, the more likely it is to be the electrophilic site; the higher the positive value, the more vulnerable to nuclear attack [62]. The data in Figure 5 show that the Ti atoms had both positive and negative condensed dual descriptors, so it is possible that the Ti atoms were electrophilic or nucleophilic sites. The condensed dual descriptors of the B atoms were all negative, favoring nucleophilic reactions. However, as the maximum and minimum values of the CDD were at the Ti atoms, it can be concluded that the Ti atoms are highly reactive; as mentioned previously, Ti atoms may be the reactive site in clusters.

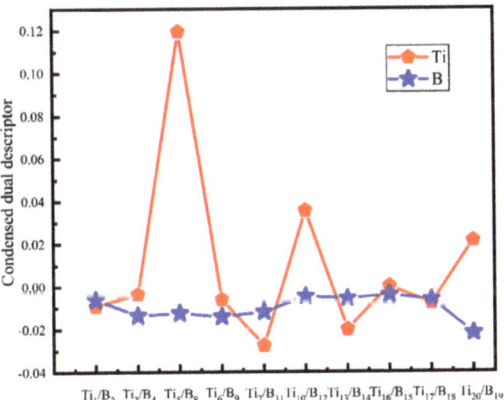

Figure 5. Condensed dual descriptor of $Ti_{10}B_{10}$ cluster.

3.4. Ab Initio Molecular Dynamics Simulation Results

In order to verify the structure and interaction of Ti and B atoms in Al-Ti-B alloys, in this paper, the kinetic behaviour of atoms in the Al-Ti-B system was investigated by ab initio molecular dynamics simulations. As shown in Figure 6, the variation of melt structure in the 698K system was analysed. Figure 6a shows the structure of the Al-Ti-B system before the dynamic simulation, and it can be seen that the three elements were uniformly distributed. After ab initio molecular dynamics simulation, Ti and B atoms formed the triangular structure of TiB_2. Figure 6b shows the TiB_2 structure in the system at 7000 fs, 7005 fs, 7229 fs, and 7230 fs. After ab initio molecular dynamics simulations, the TiB$_2$ structure was observed at 7000 fs, and there was one Al atom bonded to the Ti atom in TiB$_2$. Moreover, at 7005 fs, there were two Al atoms bonded to Ti in TiB$_2$. Then, the epitaxial growth of Al atoms was observed at 7229 fs and 7230 fs. Zhang et al. have reported in the literature [48,63] that Al growth on the Ti-termination surface of the TiB$_2$ layer is then further stacked and extended. In summary, the results of the AIMD simulations are consistent with the cluster results discussed above.

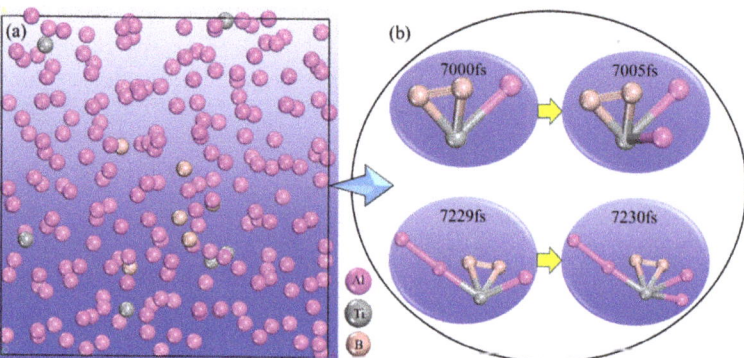

Figure 6. Structure of Al-Ti-B melt; (**a**) structure before dynamic simulation; (**b**) local structures after dynamics simulation.

In order to understand the reasons for the formation of the TiB$_2$ structure in the Al-Ti-B system, the electronic structure of the melt at 7000 fs was analyzed. Figure 7a shows the structure of the Al-Ti-B system. Figure 7b is an electron density difference isosurface map of TiB$_2$; the value of the isosurface was 0.06. According to the electron density, it can be seen that electrons were shared between the B atoms; B–B was a strongly covalent bond. There were shared electrons between Ti and B atoms, but the electrons were biased towards B atoms. The electron density difference section of the TiB$_2$ structure is shown in Figure 7c, and the location of the section is marked with a red dashed line in Figure 7a. The electron transfer between atoms is analyzed by the section diagram. In Figure 7c, blue indicates an increase in electron density, while red indicates a decrease. As can be seen, the charge density of Ti atoms decreased (red), and that of B atoms increased (blue); there were electrons transfers from Ti to B atoms. Therefore, combined with Figure 7b, Ti and B atoms were bonded in the form of ionic covalent bonds. In summary, after first-principles molecular dynamics simulation, Ti and B can generate an in situ TiB$_2$ structure; a heterogeneous nucleation center of Al atoms.

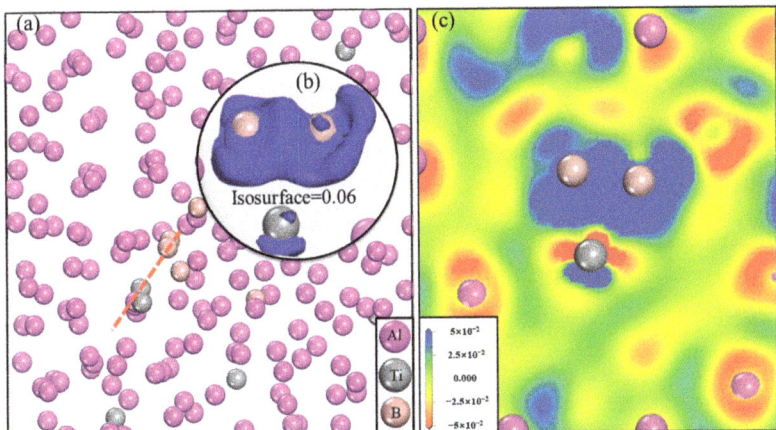

Figure 7. Electron density difference maps; (**a**) Al-Ti-B system; (**b**) electron density difference isosurface; (**c**) electron density difference section.

4. Conclusions

In this paper, Ti_nB_n (n = 2–12) clusters and Al-Ti-B systems were calculated. The possibility of TiB_2 as a heterogeneous nucleation centre of α-Al was discussed by analyzing structural changes and electronic properties. The following are the main conclusions:

(1) TiB_2 triangular structures are present in the cluster structures from n equal to 2 to 12; the amount of TiB_2 increases as the cluster size increases and Ti atoms are oriented towards the surface of clusters. The average binding energy and second-order difference energy of the clusters show that the stability of the clusters increases with the increasing number of atoms, and the most stable structure is obtained at n = 10.

(2) It shows that HOMOs and LUMOs are concentrated on Ti atoms within the TiB_2 structure, indicating that the TiB_2 structure is the more active part of the clusters, and the activity of Ti atoms is higher than that of B atoms. The map of Fukui function shows that the Ti atoms are more reactive than B atoms. The value of the condensed Fukui function and condensed dual descriptor suggests that the value of most Ti atoms is much larger than that of B atoms. Therefore, Ti atoms have higher reactivity.

(3) The AIMD simulation results of the Al-Ti-B system show that Ti and B atoms can form TiB_2 structure in situ, and there are Al atoms adsorbing and growing on the Ti surface in the TiB_2 structure. The electronic structure of TiB_2 was analysed by charge density difference. The results show that B atoms are connected by strong covalent bonds, and Ti and B atoms form ionic covalent bonds.

Author Contributions: L.Y.: Conceptualization, Methodology, Software, Investigation, Writing—original draft; X.C.: Resources, Formal analysis, Software, Writing—review and editing; L.Z.: Resources, Validation, Formal analysis, Software, Writing—review and editing; H.Z.: Formal analysis, Validation; X.Z.: Formal analysis, Visualization; B.L.: Formal analysis, Validation. All authors have read and agreed to the published version of the manuscript.

Funding: This work was supported by the major science and technology special program of Yunnan Province (Grant 202102AB080004).

Data Availability Statement: The original contributions presented in the study are included in the article, further inquiries can be directed to the corresponding authors.

Conflicts of Interest: The authors declare no conflicts of interest.

References

1. Kori, S.; Murty, B.; Chakraborty, M. Development of an efficient grain refiner for Al–7Si alloy and its modification with strontium. *Mater. Sci. Eng. A* **2000**, *283*, 94–104. [CrossRef]
2. Wang, F.; Liu, Z.; Qiu, D.; Taylor, J.A.; Easton, M.A.; Zhang, M.-X. Revisiting the role of peritectics in grain refinement of Al alloys. *Acta Mater.* **2013**, *61*, 360–370. [CrossRef]
3. Quested, T. Understanding mechanisms of grain refinement of aluminium alloys by inoculation. *Mater. Sci. Technol.* **2004**, *20*, 1357–1369. [CrossRef]
4. Fan, Z.; Gao, F.; Wang, Y.; Men, H.; Zhou, L. Effect of solutes on grain refinement. *Prog. Mater. Sci.* **2022**, *123*, 100809. [CrossRef]
5. Ding, W.; Zhao, X.; Chen, T.; Zhang, H.; Liu, X.; Cheng, Y.; Lei, D. Effect of rare earth Y and Al–Ti–B master alloy on the microstructure and mechanical properties of 6063 aluminum alloy. *J. Alloys Compd.* **2020**, *830*, 154685. [CrossRef]
6. Murty, B.; Kori, S.; Chakraborty, M. Grain refinement of aluminium and its alloys by heterogeneous nucleation and alloying. *Int. Mater. Rev.* **2002**, *47*, 3–29. [CrossRef]
7. Fan, Z.; Wang, Y.; Zhang, Y.; Qin, T.; Zhou, X.; Thompson, G.; Pennycook, T.; Hashimoto, T. Grain refining mechanism in the Al/Al–Ti–B system. *Acta Mater.* **2015**, *84*, 292–304. [CrossRef]
8. Huang, B.; Liu, Y.; Zhou, Z.; Cheng, W.; Liu, X. Selective laser melting of 7075 aluminum alloy inoculated by Al–Ti–B: Grain refinement and superior mechanical properties. *Vacuum* **2022**, *200*, 111030. [CrossRef]
9. Greer, A.L. Overview: Application of heterogeneous nucleation in grain-refining of metals. *J. Chem. Phys.* **2016**, *145*, 211704. [CrossRef]
10. Mohanty, P.; Gruzleski, J. Grain refinement mechanisms of hypoeutectic Al Si alloys. *Acta Mater.* **1996**, *44*, 3749–3760. [CrossRef]
11. Wang, J.; Horsfield, A.; Schwingenschlögl, U.; Lee, P.D. Heterogeneous nucleation of solid Al from the melt by TiB 2 and Al 3 Ti: An ab initio molecular dynamics study. *Phys. Rev. B* **2010**, *82*, 184203. [CrossRef]
12. Yu, F.; Liu, Z.; Zhao, R.; Yang, J.; Qiao, J.; Hu, W. Effect of Al–Ti–B master alloy on microstructure and properties of aluminum-air battery anode materials. *J. Mater. Res. Technol.* **2023**, *27*, 4908–4919. [CrossRef]
13. Jones, G.P.; Pearson, J. Factors affecting the grain-refinement of aluminum using titanium and boron additives. *Metall. Trans. B* **1976**, *7*, 223–234. [CrossRef]
14. Sigworth, G.K.; Kuhn, T.A. Grain refinement of aluminum casting alloys. *Int. J. Met.* **2007**, *1*, 31–40. [CrossRef]
15. Han, Y.; Liu, X.; Bian, X. In situ TiB2 particulate reinforced near eutectic Al–Si alloy composites. *Compos. Part A Appl. Sci. Manuf.* **2002**, *33*, 439–444. [CrossRef]
16. Kumar, S.; Chakraborty, M.; Sarma, V.S.; Murty, B.S. Tensile and wear behaviour of in situ Al–7Si/TiB2 particulate composites. *Wear* **2008**, *265*, 134–142. [CrossRef]
17. MAkbari, K.; Baharvandi, H.; Shirvanimoghaddam, K. Tensile and fracture behavior of nano/micro TiB2 particle reinforced casting A356 aluminum alloy composites. *Mater. Des.* **2015**, *66*, 150–161.
18. Dong, B.-X.; Li, Q.; Wang, Z.-F.; Liu, T.-S.; Yang, H.-Y.; Shu, S.-L.; Chen, L.-Y.; Qiu, F.; Jiang, Q.-C.; Zhang, L.-C. Enhancing strength-ductility synergy and mechanisms of Al-based composites by size-tunable in-situ TiB2 particles with specific spatial distribution. *Compos. Part B Eng.* **2021**, *217*, 108912. [CrossRef]
19. Li, P.; Li, Y.; Wu, Y.; Ma, G.; Liu, X. Distribution of TiB2 particles and its effect on the mechanical properties of A390 alloy. *Mater. Sci. Eng. A* **2012**, *546*, 146–152. [CrossRef]
20. Greer, A.; Bunn, A.; Tronche, A.; Evans, P.; Bristow, D. Modelling of inoculation of metallic melts: Application to grain refinement of aluminium by Al–Ti–B. *Acta Mater.* **2000**, *48*, 2823–2835. [CrossRef]
21. Feng, J.; Han, Y.; Han, X.; Wang, X.; Song, S.; Sun, B.; Chen, M.; Liu, P. Atomic insights into heterogeneous nucleation and growth kinetics of Al on TiB2 particles in undercooled Al-5Ti-1B melt. *J. Mater. Sci. Technol.* **2023**, *156*, 72–82. [CrossRef]
22. Wearing, D.; Horsfield, A.P.; Xu, W.; Lee, P.D. Which wets TiB2 inoculant particles: Al or Al3Ti? *J. Alloys Compd.* **2016**, *664*, 460–468. [CrossRef]
23. Liu, X.; Wang, B.; Li, Q.; Wang, J.; Zhang, C.; Xue, C.; Yang, X.; Tian, G.; Liu, X.; Tang, H. Quantifying the effects of grain refiners Al-Ti-B and La on the microstructure and mechanical properties of W319 alloy. *Metals* **2022**, *12*, 627. [CrossRef]
24. Knaislová, A.; Michna, Š.; Hren, I.; Vlach, T.; Michalcová, A.; Novák, P.; Stančeková, D. Microstructural characteristics of Al-Ti-B inoculation wires and their addition to the AlSi7Mg0.3 alloy. *Materials* **2022**, *15*, 7626. [CrossRef] [PubMed]
25. Sun, T.; Wang, H.; Gao, Z.; Wu, Y.; Wang, M.; Jin, X.; Leung, C.L.A.; Lee, P.; Fu, Y.; Wang, H. The role of in-situ nano-TiB2 particles in improving the printability of noncastable 2024Al alloy. *Mater. Res. Lett.* **2022**, *10*, 656–665. [CrossRef]
26. Youssef, Y.; Dashwood, R.; Lee, P. Effect of clustering on particle pushing and solidification behaviour in TiB2 reinforced aluminium PMMCs. *Compos. Part A Appl. Sci. Manuf.* **2005**, *36*, 747–763. [CrossRef]
27. Wang, M.; Chen, D.; Chen, Z.; Wu, Y.; Wang, F.; Ma, N.; Wang, H. Mechanical properties of in-situ TiB2/A356 composites. *Mater. Sci. Eng. A* **2014**, *590*, 246–254. [CrossRef]
28. Xi, L.; Gu, D.; Guo, S.; Wang, R.; Ding, K.; Prashanth, K.G. Grain refinement in laser manufactured Al-based composites with TiB2 ceramic. *J. Mater. Res. Technol.* **2020**, *9*, 2611–2622. [CrossRef]
29. Xiao, Y.; Bian, Z.; Wu, Y.; Ji, G.; Li, Y.; Li, M.; Lian, Q.; Chen, Z.; Addad, A.; Wang, H. Effect of nano-TiB2 particles on the anisotropy in an AlSi10Mg alloy processed by selective laser melting. *J. Alloys Compd.* **2019**, *798*, 644–655. [CrossRef]

30. Guo, X.; Li, X.; Li, Y.; Yang, J.; Wan, X.; Chen, L.; Liu, J.; Liu, X.; Yu, R.; Zheng, L. Molecule template method for precise synthesis of Mo-based alloy clusters and electrocatalytic nitrogen reduction on partially reduced PtMo alloy oxide cluster. *Nano Energy* **2020**, *78*, 105211. [CrossRef]
31. Kurashige, W.; Hayashi, R.; Wakamatsu, K.; Kataoka, Y.; Hossain, S.; Iwase, A.; Kudo, A.; Yamazoe, S.; Negishi, Y. Atomic-level understanding of the effect of heteroatom doping of the cocatalyst on water-splitting activity in AuPd or AuPt alloy cluster-loaded BaLa4Ti4O15. *ACS Appl. Energy Mater.* **2019**, *2*, 4175–4187. [CrossRef]
32. Rodríguez-Kessler, P.; Muñoz-Castro, A.; Alonso-Dávila, P.; Aguilera-Granja, F.; Rodríguez-Domínguez, A. Structural, electronic and catalytic properties of bimetallic PtnAgn (n = 1–7) clusters. *J. Alloys Compd.* **2020**, *845*, 155897. [CrossRef]
33. Huang, H.; Wu, B.; Gao, Q.; Li, P.; Yang, X. Structural, electronic and spectral properties referring to hydrogen storage capacity in binary alloy ScBn (n= 1–12) clusters. *Int. J. Hydrogen Energy* **2017**, *42*, 21086–21095. [CrossRef]
34. Liu, Q.; Cheng, L. Structural evolution and electronic properties of Cu-Zn alloy clusters. *J. Alloys Compd.* **2019**, *771*, 762–768. [CrossRef]
35. Delley, B. An all-electron numerical method for solving the local density functional for polyatomic molecules. *J. Chem. Phys.* **1990**, *92*, 508–517. [CrossRef]
36. Perdew, J.P.; Burke, K.; Ernzerhof, M. Generalized gradient approximation made simple. *Phys. Rev. Lett.* **1996**, *77*, 3865. [CrossRef]
37. Andersen, H.C. Molecular dynamics simulations at constant pressure and/or temperature. *J. Chem. Phys.* **1980**, *72*, 2384–2393. [CrossRef]
38. Frisch, M.; Trucks, G.; Schlegel, H.; Scuseria, G.; Robb, M.; Cheeseman, J.; Scalmani, G.; Barone, V.; Petersson, G.; Nakatsuji, H. *Gaussian 16, Revision A. 03*; Gaussian, Inc.: Wallingford, CT, USA, 2016.
39. Zhao, Y.; Truhlar, D.G. A new local density functional for main-group thermochemistry, transition metal bonding, thermochemical kinetics, and noncovalent interactions. *J. Chem. Phys.* **2006**, *125*, 194101. [CrossRef] [PubMed]
40. Lu, T.; Chen, F. Multiwfn: A multifunctional wavefunction analyzer. *J. Comput. Chem.* **2012**, *33*, 580–592. [CrossRef] [PubMed]
41. Lu, T.; Chen, Q. Realization of Conceptual Density Functional Theory and Information-Theoretic Approach in Multiwfn Program. In *Conceptual Density Functional Theory: Towards a New Chemical Reactivity Theory*; Wiley: Hoboken, NJ, USA, 2022; Volume 2, pp. 631–647.
42. Yang, W.; Mortier, W.J. The use of global and local molecular parameters for the analysis of the gas-phase basicity of amines. *J. Am. Chem. Soc.* **1986**, *108*, 5708–5711. [CrossRef] [PubMed]
43. Humphrey, W.; Dalke, A.; Schulten, K. VMD: Visual molecular dynamics. *J. Mol. Graph.* **1996**, *14*, 33–38. [CrossRef]
44. Dennington, R.; Keith, T.A.; Millam, J.M. *GaussView 6.0. 16*; Semichem Inc.: Shawnee Mission, KS, USA, 2016; pp. 143–150.
45. Hafner, J.; Kresse, G. The vienna ab-initio simulation program VASP: An efficient and versatile tool for studying the structural, dynamic, and electronic properties of materials. In *Properties of Complex Inorganic Solids*; Springer: Berlin/Heidelberg, Germany, 1997; pp. 69–82.
46. Sun, J.; Ruzsinszky, A.; Perdew, J.P. Strongly constrained and appropriately normed semilocal density functional. *Phys. Rev. Lett.* **2015**, *115*, 036402. [CrossRef]
47. Fujinaga, T.; Watanabe, Y.; Shibuta, Y. Nucleation dynamics in Al solidification with Al-Ti refiners by molecular dynamics simulation. *Comput. Mater. Sci.* **2020**, *182*, 109763. [CrossRef]
48. Zhang, H.; Han, Y.; Dai, Y.; Lu, S.; Wang, J.; Zhang, J.; Shu, D.; Sun, B. An ab initio study on the electronic structures of the solid/liquid interface between TiB_2 (0 0 0 1) surface and Al melts. *J. Alloys Compd.* **2014**, *615*, 863–867. [CrossRef]
49. Domingo, L.R.; Ríos-Gutiérrez, M.; Pérez, P. Applications of the conceptual density functional theory indices to organic chemistry reactivity. *Molecules* **2016**, *21*, 748. [CrossRef] [PubMed]
50. De Vleeschouwer, F.; van Speybroeck, V.; Waroquier, M.; Geerlings, P.; de Proft, F. Electrophilicity and nucleophilicity index for radicals. *Org. Lett.* **2007**, *9*, 2721–2724. [CrossRef] [PubMed]
51. Chattaraj, P.K.; Maiti, B.; Sarkar, U. Philicity: A unified treatment of chemical reactivity and selectivity. *J. Phys. Chem. A* **2003**, *107*, 4973–4975. [CrossRef]
52. Oláh, J.; van Alsenoy, C.; Sannigrahi, A. Condensed Fukui functions derived from Stockholder charges: Assessment of their performance as local reactivity descriptors. *J. Phys. Chem. A* **2002**, *106*, 3885–3890. [CrossRef]
53. Martínez-Araya, J.I. Why is the dual descriptor a more accurate local reactivity descriptor than Fukui functions? *J. Math. Chem.* **2015**, *53*, 451–465. [CrossRef]
54. Beck, M.E. Do Fukui function maxima relate to sites of metabolism? A critical case study. *J. Chem. Inf. Model.* **2005**, *45*, 273–282. [CrossRef] [PubMed]
55. Chattaraj, P.K.; Giri, S. Electrophilicity index within a conceptual DFT framework. *Annu. Rep. Sect. C Phys. Chem.* **2009**, *105*, 13–39. [CrossRef]
56. Chattaraj, P.K.; Roy, D.R. Update 1 of: Electrophilicity index. *Chem. Rev.* **2007**, *107*, PR46–PR74. [CrossRef]
57. Fuentealba, P.; Pérez, P. Contreras, On the condensed Fukui function. *J. Chem. Phys.* **2000**, *133*, 2544–2551. [CrossRef]
58. Han, Y.; Dai, Y.; Shu, D.; Wang, J.; Sun, B. First-principles calculations on the stability of Al/Ti B_2 interface. *Appl. Phys. Lett.* **2006**, *89*, 144107. [CrossRef]
59. Edim, M.M.; Enudi, O.C.; Asuquo, B.B.; Louis, H.; Bisong, E.A.; Agwupuye, J.A.; Chioma, A.G.; Odey, J.O.; Joseph, I.; Bassey, F.I. Aromaticity indices, electronic structural properties, and fuzzy atomic space investigations of naphthalene and its aza-derivatives. *Heliyon* **2021**, *7*, e06138. [CrossRef] [PubMed]

60. Cao, J.; Ren, Q.; Chen, F.; Lu, T. Comparative study on the methods for predicting the reactive site of nucleophilic reaction. *Sci. China Chem.* **2015**, *58*, 1845–1852. [CrossRef]
61. Pino-Rios, R.; Inostroza, D.; Cárdenas-Jirón, G.; Tiznado, W. Orbital-weighted dual descriptor for the study of local reactivity of systems with (quasi-) degenerate states. *J. Phys. Chem. A* **2019**, *123*, 10556–10562. [CrossRef]
62. Fievez, T.; Sablon, N.; de Proft, F.; Ayers, P.W.; Geerlings, P. Calculation of Fukui functions without differentiating to the number of electrons. 3. Local Fukui function and dual descriptor. *J. Chem. Theory Comput.* **2008**, *4*, 1065–1072. [CrossRef]
63. Zhang, H.; Han, Y.; Dai, Y.; Wang, J.; Sun, B. An ab initio molecular dynamics study: Liquid-Al/solid-TiB2 interfacial structure during heterogeneous nucleation. *J. Phys. D Appl. Phys.* **2012**, *45*, 455307. [CrossRef]

Disclaimer/Publisher's Note: The statements, opinions and data contained in all publications are solely those of the individual author(s) and contributor(s) and not of MDPI and/or the editor(s). MDPI and/or the editor(s) disclaim responsibility for any injury to people or property resulting from any ideas, methods, instructions or products referred to in the content.

Article

Mechanical and Magnetic Properties of Porous $Ni_{50}Mn_{28}Ga_{22}$ Shape Memory Alloy

Xinyue Li, Kunyu Wang, Yunlong Li, Zhiqiang Wang, Yang Zhao and Jie Zhu *

State Key Laboratory for Advanced Metals and Materials, University of Science and Technology Beijing, 30 Xueyuan Road, Beijing 100083, China; wangky94@hotmail.com (K.W.)
* Correspondence: jiezhu@ustb.edu.cn

Abstract: A porous $Ni_{50}Mn_{28}Ga_{22}$ alloy was produced using powder metallurgy, with NaCl serving as the pore-forming agent. The phase structure, mechanical properties, and magnetic properties of annealed bulk alloys and porous alloys with different pore sizes were analyzed. Vacuum sintering for mixed green billets in a tube furnace was employed, which facilitated the direct evaporation of NaCl, resulting in the formation of porous alloys characterized by a complete sinter neck, uniform pore distribution, and consistent pore size. The study found that porous alloys within this size range exhibit a recoverable shape memory performance of 3.5%, as well as a notable decrease in the critical stress required for martensitic twin shear when compared to that of bulk alloys. Additionally, porous alloys demonstrated a 2% superelastic strain when exposed to 353 K. Notably, under a 1.5 T magnetic field, the porous $Ni_{50}Mn_{28}Ga_{22}$ alloy with a pore size ranging from 20 to 30 μm exhibited a peak saturation magnetization of 62.60 emu/g and a maximum magnetic entropy of 1.93 J/kg·K.

Keywords: porous alloy; shape memory effect; magnetocaloric effect; superelasticity

1. Introduction

Ni–Mn–Ga alloys have gained significant interest in the field of functional materials due to their various mechanical, thermal, and magnetic properties, magnetoresistance, elastic-caloric effect (eCE), and magnetocaloric effect (MCE) [1–7]. The strong superelasticity of the Ni–Mn–Ga alloy is attributed to the stress-induced martensitic transformation. The one-dimensional Ni–Mn–Ga wire produced using melt spinning reportedly exhibits a superelasticity of up to 14% [8]. This superelasticity is attributed to its bamboo-like grain structure, which effectively reduces the obstructive effect of grain boundaries on the martensitic transformation. By reducing the dimensionality, the number of grain boundaries can be decreased, and the constraint of martensitic transformation can be effectively weakened. However, stress-induced martensitic transformation in polycrystalline Ni–Mn–Ga alloy is impeded by grain boundaries, resulting in low macro-superelasticity [9]. To overcome this problem, researchers have decreased the number of alloy grain boundaries by introducing pores into the bulk alloy to reduce the resistance to martensitic transformation [10]. The porous Ni–Mn–Ga alloys prepared using infiltration exhibit a magnetic field-induced strain of 0.12%, surpassing the limit of the bulk alloy. This is due to the reduction in the hindrance to martensitic transformation caused by the addition of pores and subsequent heat treatment, which results in the formation of a large number of coarse grains in the alloy. Consequently, unlike one-dimensional materials, porous materials offer another approach to reducing the dimensionality while also effectively reducing grain boundary constraints.

Ni–Mn–Ga-based alloys exhibit a conventional MCE or eCE in the proximity of both the first-order martensite transformation (MT) and the second-order magnetic transition [11,12]. The MCE in Ni–Mn–Ga alloys has been found to be highly contingent on the alloy's composition, thereby allowing for the tuning of the MT and magnetic transitions to overlap through chemical composition adjustments. This facilitates the achievement of the transformation entropy during paramagnetic austenite and ferromagnetic martensite,

Citation: Li, X.; Wang, K.; Li, Y.; Wang, Z.; Zhao, Y.; Zhu, J. Mechanical and Magnetic Properties of Porous $Ni_{50}Mn_{28}Ga_{22}$ Shape Memory Alloy. *Metals* **2024**, *14*, 291. https://doi.org/10.3390/met14030291

Academic Editors: Andrea Di Schino and Claudio Testani

Received: 17 January 2024
Revised: 6 February 2024
Accepted: 9 February 2024
Published: 29 February 2024

Copyright: © 2024 by the authors. Licensee MDPI, Basel, Switzerland. This article is an open access article distributed under the terms and conditions of the Creative Commons Attribution (CC BY) license (https://creativecommons.org/licenses/by/4.0/).

resulting in an enhancement of the MCE [13]. Furthermore, the tunable transformation temperature characteristic enables the potential design of a magnetic refrigerator with an operating temperature closely approximating room temperature [14–17]. The Ni–Mn–Ga alloy has been designed for use as a temperature sensor or heat accumulator in applications involving higher temperatures and larger temperature ranges, due to large entropy changes under a phase transition. Additionally, the porous structure of the alloy can significantly enhance the heat exchange efficiency between a fluid and a material, owing to its large specific surface area, which can improve the magnetic cooling effect (MCE) characteristics of materials. This paper focuses on the mechanical and MCE properties of porous materials using the cold pressing and sintering method, which differs from the imitation casting method used in a previous study [18–21].

2. Experimental Procedures

The $Ni_{50}Mn_{28}Ga_{22}$ (at.%) alloy ingots were prepared using arc-melting high purity elements of Ni (99.99%), Mn (99.7%), and Ga (99.99%) four times to ensure uniform composition. After the alloy ingot was broken into powder, the alloy powder of 20–30 μm was sieved and mixed with NaCl. To control the pore size of the porous $Ni_{50}Mn_{28}Ga_{22}$ alloy, the 50 vol.% sieved pure NaCl powders were mixed into the Ni–Mn–Ga powder as the space holders, where the NaCl powders were sieved into 20–30 μm, 50–75 μm, 112–200 μm, 200–355 μm, respectively. The mixed powder was obtained using ball milling mixtures consisting of 50 vol.% mechanically ground $Ni_{50}Mn_{28}Ga_{22}$ powder and 50 vol.% NaCl powders of various sizes (>99.5% purity) for 5 h. Then, the mixed powder was placed into a steel mold and a unidirectional pressure of approximately 500 MPa was applied for 1 min. During the sintering process, the crucible containing the green billets was placed into a vacuum furnace and heated to 1023 K for 1 h and then heated to 1373 K and kept for 2 h to make NaCl volatilize from the compact to achieve complete sintering between the powders. It was then cooled to room temperature with a cooling rate of 5 K/min. The remaining NaCl was dissolved by immersing the sample into water for 48 h.

The porous alloys were characterized using an SEM (Carl Zeiss, Oberkochen, Germany) equipped with an EDS system to analyze the microstructure, fracture, and composition. To further examine the grain morphology, the corroded surface was observed using an optical microscope (Carl Zeiss, Oberkochen, Germany). The sample underwent mechanical polishing and etching using Kalling' II solution (100 mL alcohol, 100 mL hydrochloric acid, and 5 g $CuCl_2$) to prepare the porous surface for analysis. The phase analysis of Cu-Kα radiation (λ = 1.5406 Å) at room temperature was conducted using an X-ray diffractometer (XRD, Rigaku, Tokyo, Japan). The martensitic transformation was measured with a heating/cooling rate of 10 K/min using DSC (TA Instruments, New Castle, DE, USA). Isothermal magnetization M(H) curves within a temperature range of 298–353 K were obtained using a vibrating sample magnetometer (VSM) of PPMS (Quantum Design, San Diego, CA, USA) at a heating and cooling rate of 3 K/min, under magnetic fields up to 1.5 T. Specifically, to measure the M(H) curves, the sample was equilibrated at 353 K to achieve a full austenite state before being cooled to the test temperature and maintained for 1 min before starting measurements. Additionally, the magnetic field was applied perpendicular to the direction of green bodies pressing. The magnetocaloric effect (MCE), characterized by magnetic entropy change (ΔS_m), was then calculated from the M(H) curves based on the Maxwell relation.

To investigate the mechanical properties of porous alloys, incremental cyclic compression tests were conducted to assess both their shape memory effect and superelasticity at room temperature and 353 K. Uniaxial loading and unloading experiments were carried out using a universal testing machine (DDL-50), with the axial displacement of the sample being monitored through an extensometer (YUU-25/5). The samples, with a dimension of diameter 3 × 6 mm, were subjected to a temperature of 353 K (A_f + 15 K) for 10 min during the assessment of the superelasticity effect, ensuring a fully austenitic state and complete grain recovery. Additionally, to investigate the dependence of the recoverable shape memory effect on the pre-strain, the samples were immediately unloaded at the set strain value upon reaching the specified strain, under the same strain rate.

3. Results and Discussion

The microstructures of NaCl and $Ni_{50}Mn_{28}Ga_{22}$ alloy powder are depicted in Figure 1a and Figure 1b, respectively. The NaCl powder displays an irregular shape with a size range of 20–30 μm after grinding and screening, with a few particles being smaller than 10 μm. Similarly, the alloy powder exhibits irregularity at a size of 20–30 μm.

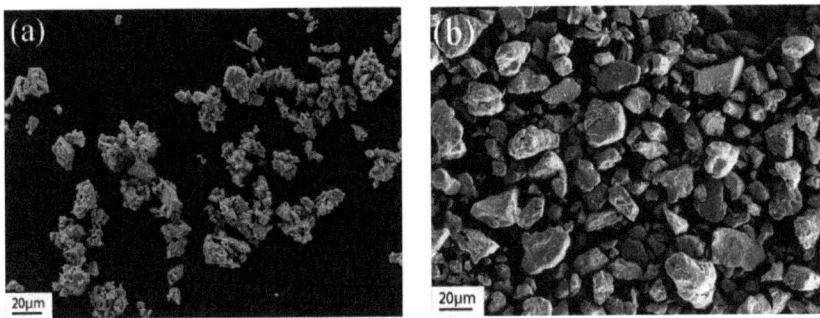

Figure 1. Microstructure of (**a**) NaCl powder and (**b**) $Ni_{50}Mn_{28}Ga_{22}$ alloy powder.

The morphologies of porous alloys with varying pore sizes are presented in Figure 2. The dual pore structure consisted of a regular pore formed using a pore-forming agent and small sintered pores (<10 μm). When NaCl with pore sizes of 20–30 μm (a) and 50–75 μm (b) were added, the porous alloy exhibited an irregular polygonal pore morphology with sharp corners. Conversely, for larger NaCl pore sizes of 112–200 μm (c) and 200–355 μm (d), the pore morphology was relatively regular and the shape was similar to the added NaCl. The magnified view in Figure 2a reveals a complete through-hole structure in the prepared porous alloy, with well-formed sintering necks between the alloy particles, and a uniform distribution and relatively consistent size of pores. The pillar of the porous sample, with a grain size of approximately 25 μm, was composed of multiple martensitic twins arranged in different directions and formed a well-sintered neck. The illustration in the lower right corner of Figure 1 shows a microscopic image of a corroded porous alloy pillar captured using an optical microscope. It can be seen that an obvious lamellar martensite runs through the entire grain and sintering neck, suggesting that coarse martensite twins were achieved through pressing sintering without requiring further heat treatment. Moreover, this structural configuration effectively reduces the number of grain boundaries in the alloy, while the arrangement of martensitic plates perpendicular to the pillar direction facilitates the release of individual grain strain. It can be seen from the enlarged image in Figure 2c that the macropores were caused by the occupation of NaCl particles, but there were still small pores in the matrix. This is caused by the gaps between alloy particles during the green billet pressing process, and most of these pores are closed pores. Similarly, in Figure 2d, the larger pores (red box) in the porous alloy are attributed to the pore-forming agent, whereas the formation of smaller pores (yellow box) occurred between the alloy particles during the sintering process.

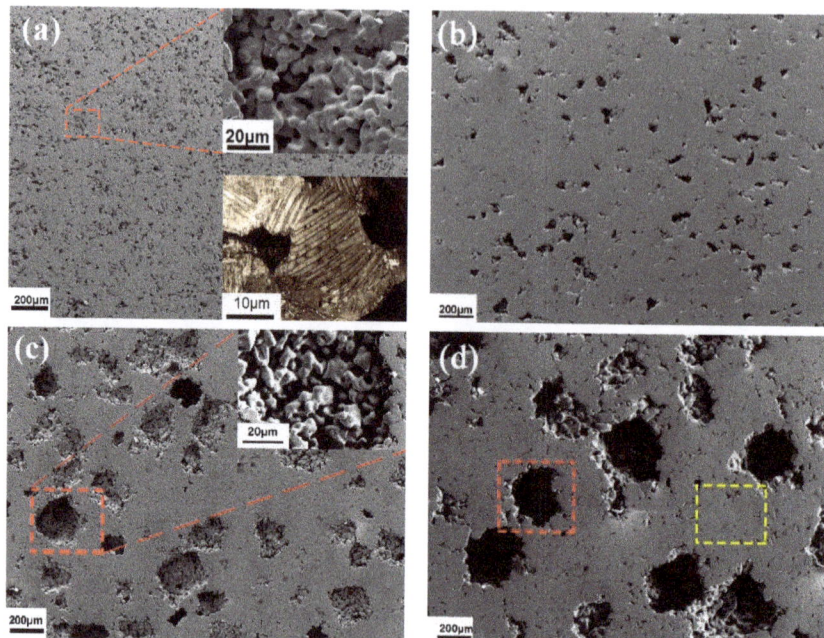

Figure 2. Microstructure of Ni$_{50}$Mn$_{28}$Ga$_{22}$ porous alloy with pore sizes of (**a**) 20–30 μm, (**b**) 50–75 μm, (**c**) 112–200 μm, and (**d**) 200–355 μm.

Table 1 lists the content and porosity of Ni$_{50}$Mn$_{28}$Ga$_{22}$ porous alloy elements obtained using sintering at 1373 K. The actual porosity of the porous alloys was calculated using the following formula:

$$P = (1 - \rho_0/\rho) \times 100\% \tag{1}$$

where P represents the porosity, and ρ_0 and ρ denote the density of porous Ni$_{50}$Mn$_{28}$Ga$_{22}$ alloy and bulk Ni$_{50}$Mn$_{28}$Ga$_{22}$ alloy, respectively [22]. From the table, it can be seen that the composition of porous alloys with different sizes was similar, indicating that the pores did not affect the alloy composition. Compared to the experimental design, the composition of porous alloys with different pore sizes showed a small amount of Mn volatilization, which occurred during the sintering. In addition, the porosity was slightly lower than 50%, which is due to the slight melting of the alloy surface during sintering, causing alloy size shrinkage.

Table 1. Content and porosity of Ni$_{50}$Mn$_{28}$Ga$_{22}$ porous alloys with different pore forming agent size.

Pore Forming Agent Size (μm)	Ni (at.%)	Mn (at.%)	Ga (at.%)	Porosity (vol.%)
20–30	47.10	26.72	26.18	41.24
50–75	50.20	25.34	24.46	43.94
112–200	47.14	25.66	27.20	46.64
200–355	47.38	26.77	25.85	51.22

The DSC curves and room temperature XRD patterns of the Ni$_{50}$Mn$_{28}$Ga$_{22}$ porous alloy with different pore sizes are depicted in Figure 3a and Figure 3b, respectively. The DSC curve reveals a singular endothermic peak during heating and a separate exothermic peak during cooling for samples with various pore sizes, indicating a one-step transformation consistent with the as-cast samples [23]. The phase transition start and finish temperatures (M_s, M_f, A_s, A_f), as well as the Curie point (T_c) of the porous alloy, were determined using the tangent method and are presented in Table 2 for evaluation. The thermal hysteresis ΔT was calculated using $\Delta T = \left(A_s - M_f + A_f - M_s\right)/2$. It is noteworthy that the positions of the endothermic and exothermic peaks and the phase transition temperature across various pore sizes exhibited no significant changes. In addition, as the pore size increased, the height of the phase transition peak decreased and widened, and this change in peak shape was caused by lattice distortion, which was caused by stress [24]. The larger pores cause severe deformation of the alloy powder during the pressing process, resulting in a stress concentration. The transformation of late heat ΔH calculated using the phase transition peak was 10.23, 9.62, 9.62, and 9.48 J/g, in order of pore size from small to large, with no significant difference. This indicates that the pore size does not affect the amount of grains involved in phase transformation in porous alloys. The XRD pattern in Figure 3b illustrates that the $Ni_{50}Mn_{28}Ga_{22}$ porous alloy possesses a non-modulation martensite structure (NM) with a tetragonal arrangement at room temperature. Furthermore, all diffraction peaks can be attributed to the NM martensite phase, implying that the porous alloy retains a uniform composition and does not undergo second-phase precipitation subsequent to sintering [25–28]. The lattice constants and cell volumes of porous alloys with different pore sizes were calculated from XRD curves and are listed in Table 3. It can be seen that the pore size had no significant effect on the lattice constant and cell volume.

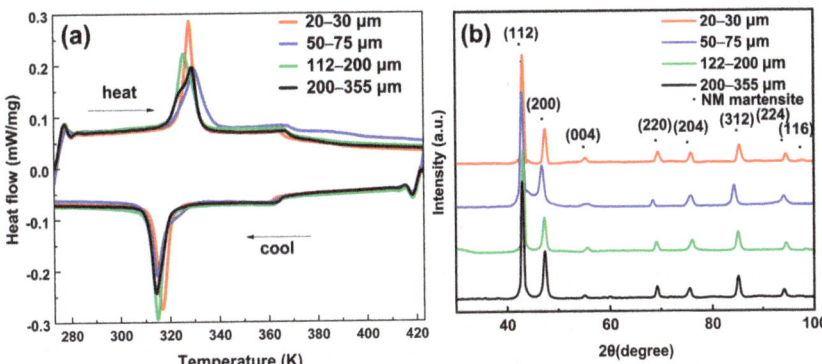

Figure 3. (a) DSC curves and (b) XRD patterns of $Ni_{50}Mn_{28}Ga_{22}$ porous alloy.

Table 2. The phase transformation temperatures (M_s, M_f, A_s, A_f), Curie point (T_c), and hysteresis ΔT of the $Ni_{50}Mn_{28}Ga_{22}$ porous alloy.

Pore-Forming Agent Size (μm)	M_s (K)	M_f (K)	A_s (K)	A_f (K)	T_c (K)	ΔT (K)
20–30	320.93	311.41	321.49	333.16	364.63	14.88
50–75	323.52	307.45	320.37	338.13	365.16	16.08
112–200	320.17	309.14	319.93	333.96	365.04	11.28
200–355	320.74	309.18	318.61	334.29	366.19	13.48

Table 3. Lattice parameters and cell volume of calculated according to the X-ray diffraction data of $Ni_{50}Mn_{28}Ga_{22}$ porous alloys.

Pore-Forming Agent Size (μm)	Lattice Constant (Å)			Cell Volume (Å³)
	a	b	c	
20–30	3.83	3.83	6.66	97.89
50–75	3.87	3.87	6.58	99.06
112–200	3.86	3.86	6.59	98.52
200–355	3.88	3.88	6.57	99.16

3.1. Shape Memory Effect

The mechanical properties of magnetocaloric materials are of crucial importance for their potential applications. The A_f temperature of the samples, as depicted in Figure 3a, was approximately 340 K. Therefore, shape memory effect tests were conducted at room temperature (295 K), wherein the samples were in a completely martensitic state. The comparison between the annealed bulk alloy and porous alloys with different pore sizes (20–30 μm, 112–200 μm, and 200–355 μm) is illustrated in the compression experiment results shown in Figure 4. The sample underwent compression experiments using a cyclic increment method, wherein it was first subjected to a compression recovery of 1%, followed by heating at 473 K for 10 min to achieve shape memory recovery. Subsequently, after the sample cooled down, the next compression process was initiated. Each compression process increased the compression amount by 0.5% compared to the previous one until the sample eventually collapsed.

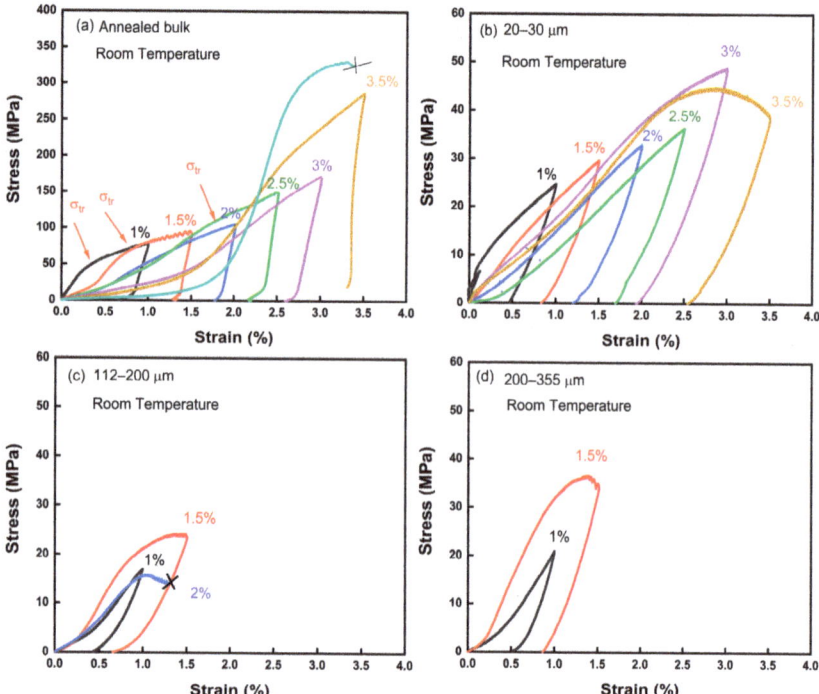

Figure 4. The shape memory effect curves of the $Ni_{50}Mn_{28}Ga_{22}$ (**a**) annealed bulk and porous alloys with different pore sizes of (**b**) 20–30 μm, (**c**) 112–200 μm, and (**d**) 200–355 μm.

In Figure 4a, the annealed bulk alloy exhibits an observable inflection point during the compression process, marked by the red arrow, indicating the de-twinning process of the martensite variant. It is evident from Figure 4b that the porous sample experienced a significant reduction in maximum stress when subjected to the same compression force as the bulk alloy. This reduction can be attributed to the decrease in the equivalent cross-sectional area resulting from the presence of numerous pores. Simultaneously, the abundance of pores led to a decrease in the number of grain boundaries, thereby lowering the resistance of the martensite twin movement. Notably, the 20–30 μm porous alloy demonstrated an enhanced elastic recovery compared to the bulk material. This enhancement can be attributed to the pores serving as a buffer layer during the deformation process, absorbing some strain and augmenting the elastic deformation of the material.

At the same time, the critical stress of martensite variant detwinning gradually decreased with an increase in the compression time. This can be attributed to the path effect resulting from the repeated compression process, which also exists in bulk alloys [5,29–34]. The process leads to the reorientation of dislocations, causing some disadvantageously oriented dislocations to move outside the twin motion path [35]. Consequently, this results in a decreased resistance to twinning motion. Due to the difference in porosity between porous and block structures, when stress is applied to the porous structure, the actual stress borne by the structure is greater than that stress. In addition, if the stress is divided by the relative density of the sample, the actual force acting on the porous alloy pillar node can be obtained. It is worth noting that under a 1% pre-strain, the equivalent stress of porous alloys is $\frac{24.5 \text{ MPa}}{1-41.24\%} = 41.69$ MPa [36], which is significantly lower than that of bulk alloys (76.24 MPa), which also confirms that pores, as elastic layers, store some stress. The increase in twin stress caused by a 3% compression is attributed to work-hardening resulting from the accumulation of a large number of dislocations due to extensive deformation. During a 3.5% compression, the sample experienced a partial pillar fracture. As the pore size increased, the brittleness of samples with pore sizes of 112–200 µm and 200–355 µm also increased, leading to premature fracture (Figure 4c,d). This is because the larger pores reduced the overall structural strength of the alloy, causing significant local stress during the stress process, leading to the early collapse of the specimens.

To better analyze the phase transition process during the sample recovery stage, the recovery rate of each strain component of the annealed bulk and 20–30 µm porous alloy specimen are illustrated in Figure 5. The compression strain during each load–unload–heat recovery cycle comprises four components: elastic recovery strain ε_{el}, superelastic recovery strain ε_{se}, shape memory effect (SME) strain ε_{sme}, and irrecoverable strain ε_{ir}, where the total strain ε_{sum} is the sum of these individual components, as shown in Figure 5a [18] and the following formula:

$$\varepsilon_{sum} = \varepsilon_{el} + \varepsilon_{se} + \varepsilon_{sme} + \varepsilon_{ir} \tag{2}$$

Figure 5b shows the shape memory strain variation with pre-strain of bulk alloys and porous alloys with pore sizes of 20–30 µm. It can be seen that the shape memory strain of the bulk alloys gradually increased with the increase in pre-strain, while the shape memory strain of porous alloys remained constant. This is because the $Ni_{50}Mn_{28}Ga_{22}$ bulk alloy can conduct stress between adjacent grains without loss during deformation. However, due to the addition of pores, stress can only be transmitted along the pillars or nodes, and pores act as elastic layers, absorbing a large amount of strain, resulting in fewer grains in porous alloys where stress reaches the critical stress of martensitic twin reorientation, resulting in lower shape memory strain and higher elastic recovery. It is worth noting from Figure 5c,d that the presence of superelastic strains in both bulk and porous alloys can be attributed to the austenitic phases in the alloys. Notably, the porous alloy exhibited a markedly higher elastic strain than the bulk alloy due to the abundance of pores in the former, effectively serving as a buffer layer. Despite this difference, both types of alloys displayed the same proportion of shape memory strain. However, it is important to highlight that the pores lead to a significant reduction in the irrecoverable strain of the alloy. This phenomenon is attributed to the presence of a large number of dislocations in bulk alloys, which tend to concentrate stress and form cracks at their sites. Conversely, the introduction of pores in porous alloys substantially decreases the number of grain boundaries, making it challenging for stress to concentrate in dislocations, and thereby minimizing the occurrence of grain boundary fractures. Nevertheless, the incorporation of pores also introduces structural drawbacks to the alloy. Owing to the reduced strength of the sintered neck area, the sample is unable to withstand considerable stress, resulting in a lower maximum deformation capacity for the porous alloy compared to the bulk alloy [37].

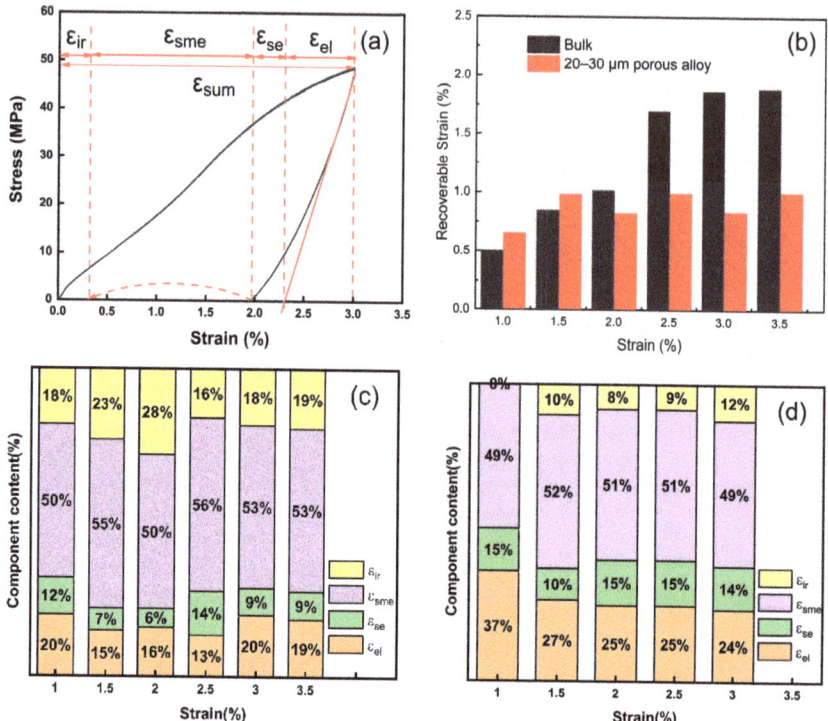

Figure 5. (a) Schematic of strain division during the recovery process. (b) Shape memory strain varies with pre-strain of bulk and porous alloys. Recovery ratio of each strain component curve of (c) $Ni_{50}Mn_{28}Ga_{22}$ bulk alloy, (d) 20–30 μm porous alloy.

To further analyze the fracture process of the alloy, the fracture of the annealed bulk alloy and the 20–30 μm porous alloy were examined. Due to the inherent brittleness of the bulk NiMnGa alloy, most of the fracture surface was an intergranular fracture (Figure 6a), which aligned along the stress direction. There was also a small amount of transgranular fracture. In contrast, Figure 6b reveals a combination of intergranular and transgranular fractures on the porous alloy's fracture surface. Notably, most of the intergranular fractures resulted from weak sintering necks, while well-sintered sintering necks exhibited transgranular fractures. It is obvious that the stress threshold for an intergranular fracture is lower than that for a transgranular fracture. Consequently, stress initially triggers the fracture of weak sintering necks in porous alloys and then concentrates on the grain of the alloy pillar, causing a transgranular fracture. Therefore, enhancing the strength of the sintering neck in porous alloys can significantly enhance their overall strength.

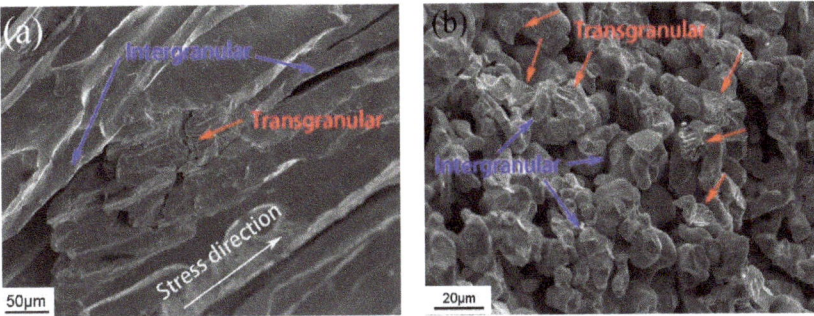

Figure 6. Fracture morphology of (**a**) annealed bulk alloy and (**b**) 20–30 µm porous alloy.

3.2. Superelastic Effect

The specimens of the $Ni_{50}Mn_{28}Ga_{22}$ bulk and porous alloys displayed distinct mechanical properties in the martensitic and austenitic states as a result of the shape memory effect and superelastic effect exhibited in these two states, respectively. Testing the mechanical properties of the samples at various temperatures was therefore imperative. Based on the DSC curve of the $Ni_{50}Mn_{28}Ga_{22}$ alloy, the A_f point was approximately 340 K. Consequently, the samples were heated to 353 K to guarantee the complete transformation of the alloy into the austenitic phase, followed by conducting a superelastic mechanical test on the specimens.

The experimental method was similar to the shape memory testing process, excluding the heating recovery process. For clarity, the stress–strain curve was translated along the X-axis direction. The cyclic stress–strain curve of the bulk alloy is shown in Figure 7a. The bulk alloy exhibited a stress plateau at 42 MPa under a 1.5% pre-strain, while significant work hardening occurred at a 3% pre-strain. Compared with the stress–strain curve at room temperature, the bulk alloy at high temperature exhibited a significant recovery process during the stress release stage, but there was still a large amount of irrecoverable strain, which may be due to part of martensite undergoing reverse transformation during the stress release stage. Due to the accumulation of dislocations caused by a stress concentration, there was still some residual martensite in it [31,38]. As for the $Ni_{50}Mn_{28}Ga_{22}$ alloy in Figure 7b, the recovery stage at high temperatures displayed significant recovery compared to room temperature, indicating that most of the grains were austenite and the sample underwent superelastic strain. Upon reaching a certain stress level, a noticeable inflection point is evident in the 1.5% and 2% compression curves, denoting a stress-induced martensitic transformation. Martensitic transformation occurs in austenite once the deformation of austenite reaches a specific value, resulting in a sudden change in the slope of the curve. During the initial loading stage, there was no significant change in the slope of each compression curve, suggesting elastic deformation. The presence of numerous pores enhances the activation of a dislocation slip or accumulation faults at high stress levels. The porous alloy collapsed at a preset strain of 2.5%, indicating that its ultimate compressive strain was approximately 2% at 353 K, and it could withstand a maximum stress of 100 MPa. The critical stress for martensitic transformation in the sample was approximately 83 MPa, significantly lower than that of polycrystalline bulk alloys (200 MPa) [18].

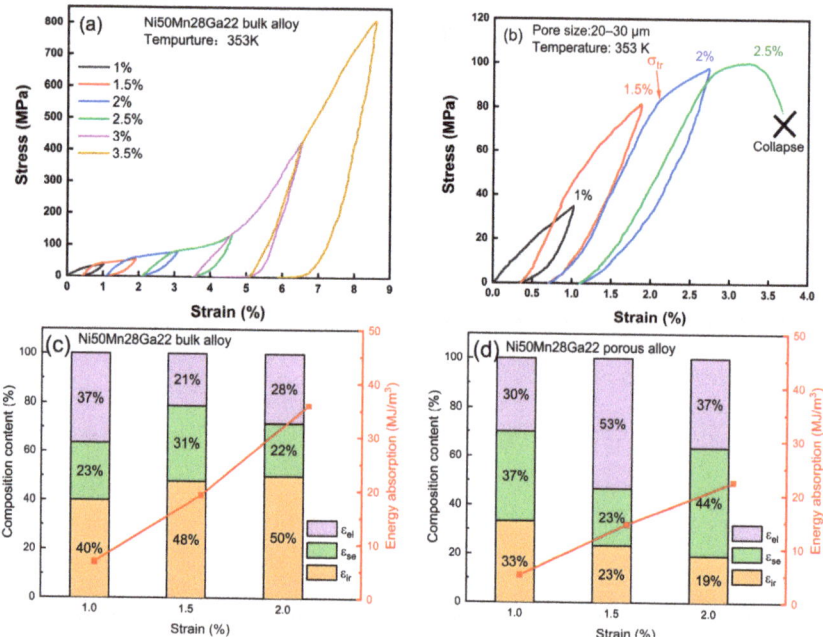

Figure 7. Cyclic compressive stress–strain curve at 353 K of (**a**) $Ni_{50}Mn_{28}Ga_{22}$ bulk alloy and (**b**) $Ni_{50}Mn_{28}Ga_{22}$ porous alloy with a pore size of 20–30 µm. Corresponding recovery ratio of each strain component curve of (**c**) annealed bulk alloy and (**d**) porous alloys.

In order to better compare the stress conditions of bulk and porous alloys, the stress–strain curves were decomposed to obtain Figure 7c,d. Due to the collapse of porous alloys under a 2.5% pre-strain, only the 1–2% pre-strain curve was analyzed. It can be seen that compared with bulk alloys, porous alloys had a significantly lower irreversible deformation and a higher elastic deformation. This is because the addition of pores allows dislocations to only develop along a fixed path, reducing the accumulation of dislocations. Furthermore, a large number of pores act as buffer layers, absorbing some of the strain, leading to an increase in elastic deformation. It is worth noting that during the 1% and 2% pre-strain compression processes, the superelastic strain of porous alloys was significantly higher than that of bulk alloys, indicating once again that the addition of pores is beneficial in reducing the hindrance of grain boundaries to the phase transformation process. The porous structure reduces the number of grain boundaries, lowering the resistance of phase transformation and aligning the critical stress of stress-induced phase transformation of the porous alloy with that of the wire [9]. In addition, the red line in Figure 7c,d represents the energy absorption of the corresponding stress–strain curve. The energy absorption was obtained by calculating the area enclosed by the stress–strain curve and the X-axis in Figure 7a and Figure 7b, respectively. It can be seen that due to the lower critical stress of phase transformation, the energy absorption of porous alloys was significantly lower than that of the bulk alloys. A lower critical stress for phase transformation is advantageous in reducing the energy loss of the sample during cyclic compression.

3.3. Magnetocaloric Effect

The isothermal magnetization curves M(H) were measured to assess the magnetic properties and magnetocaloric effect (MCE) of the $Ni_{50}Mn_{28}Ga_{22}$ porous alloy. Prior to the measurements, the sample was heated to the austenitic state and subsequently cooled to the measurement temperature without the influence of a magnetic field, following the method of zero field cooling (ZFC). After completing the measurement at a specific temperature, the sample was directly cooled down to the next temperature for the continuation of the measurements. The test involved the application of a vertical magnetic field ranging from 0 to 1.5 T, with a temperature interval of 5 K, as shown in Figure 8.

The saturation magnetization of the $Ni_{50}Mn_{28}Ga_{22}$ alloy gradually increased with decreasing temperatures between 298 K and 353 K, as evidenced in Figure 8a,c,e,g, regardless of whether it was a bulk or a porous alloy. The magnetization curves exhibit distinct characteristics, with some belonging to martensite and others to austenite. It is important to note that both the low-temperature martensite and high-temperature austenite phases are ferromagnetic. The introduction of a porous structure into the alloy has a negligible effect on the saturation magnetization. At a magnetic field of 1.5 T, the magnetization of the annealed bulk and porous alloys with different pore size structures was approximately 49.30, 62.60, 61.99, and 59.93 emu/g, respectively. However, the varying pore sizes had a minimal impact on the saturation magnetization of the alloy, and any slight differences observed could be attributed to composition deviations.

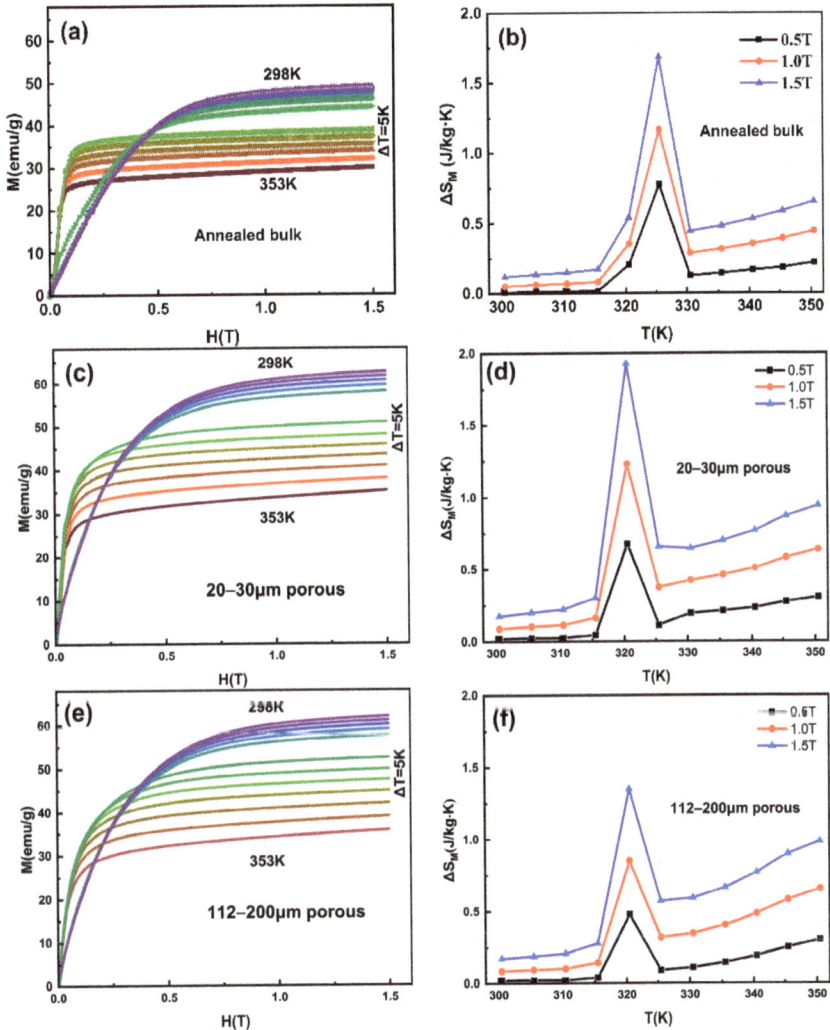

Figure 8. Cont.

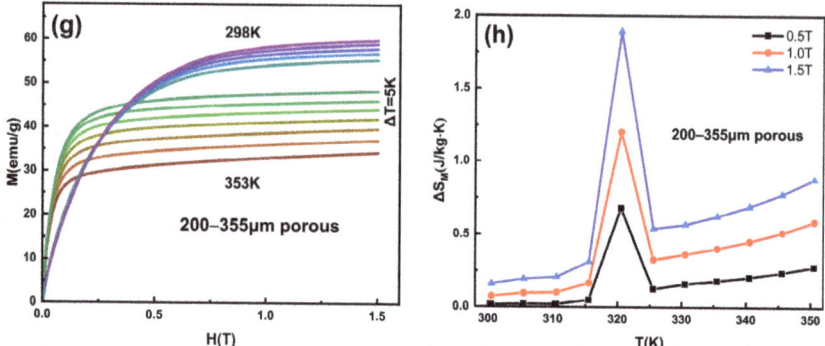

Figure 8. Isothermal magnetization M(H) curves and temperature-dependent magnetic field-induced entropy change ΔS_M in the $Ni_{50}Mn_{28}Ga_{22}$ (**a**,**b**) annealed bulk alloy, and (**c**,**d**) 20–30 μm, (**e**,**f**) 112–200 μm, and (**g**,**h**) 200–355 μm porous alloy.

The magnetic field-induced entropy change ΔS_M can be calculated using the Maxwell relation, where the partial derivative is replaced by finite differences and integration numerically [26,39–45]:

$$\Delta S_M = S_M(T, H_2) - S_M(T, H_1) = \int_{H_1}^{H_2} \left(\frac{\partial M}{\partial T}\right)_H dH \tag{3}$$

The temperature-dependent ΔS_M for annealed bulk and porous alloys with different porous sizes is shown in Figure 8b,d,f,g, respectively. A common characteristic of the samples was that the magnetic entropy change ΔS_M value of the alloy increased gradually with the increase in the magnetic field, and there was a peak in the transformation temperature region. The origin of the large ΔS_M peak in $Ni_{50}Mn_{28}Ga_{22}$ could be attributed to the considerable magnetization jump caused by the phase transformation between martensite and austenite during heating. This is due to the large difference in magnetization between the high temperature austenite phase and the low temperature martensite phase, resulting in the alloy having the largest magnetic entropy change near the transformation temperature. Moreover, the magnetic entropy of alloys with a pore structure in the same magnetic field was larger than that of bulk alloys. Specifically, the magnetic entropy changes in the ΔS_M of bulk and porous alloys with different pore sizes obtained under a 1.5 T magnetic field were 1.68, 1.93, 1.35, and 1.89 J/kg·K, respectively. This indicates that the porous alloy exhibited a better magnetocaloric effect under the same conditions. This may be attributed to the fact that the porous structure reduces the number of grain boundaries, thereby reducing the resistance to domain movement. Additionally, as a magnetic cooling material, the existence of a large number of through-holes can also increase the internal and external heat exchange efficiency of the material, thereby improving the magnetothermal refrigeration efficiency [46,47].

Porous alloys based on Ni–Mn–Ga exhibit various properties and possess an interesting characteristic—the capability to tune the martensitic–austenitic transition over a broad temperature range through adjustments in the Ni, Mn, and Ga content, or the addition of substitute elements [48–50]. Therefore, these alloys with porous structures have low costs, high magnetic entropy changes under relatively low external fields, easy-to-control transition temperatures, and significant surface areas that facilitate effective thermal medium conduction and heat exchange, making them powerful candidates for magnetic refrigerants. However, the current limitations in the application of these alloys stem from their low structural strength and the unclear impact of the pore structure on performance. Adjusting the heat treatment and composition to improve structural strength will further enhance the mechanical and magnetic properties of porous alloys.

4. Conclusions

This paper presents the successful preparation of a porous $Ni_{50}Mn_{28}Ga_{22}$ alloy with a three-dimensional pore structure and approximately 50% porosity through a process involving cold pressing and sintering, using NaCl as a pore-forming agent. The preparation method is simple to implement and offers flexibility in adjusting the pore diameter and the morphology and porosity of the porous alloy. The resulting porous alloy, characterized by a pore size of 20–30 μm, demonstrated the capability to achieve a 3.5% recoverable strain at room temperature and 2% superelastic strain at 353 K. Notably, the introduction of pores in the $Ni_{50}Mn_{28}Ga_{22}$ alloy significantly reduced the critical stress of the phase transformation, as observed at only 80 MPa, in comparison to polycrystalline bulk alloys. This reduction is advantageous for diminishing energy loss during the cycling process. Moreover, the introduction of pores into the alloy led to an improvement in its saturation magnetization and magnetocaloric effect. Specifically, under a 1.5 T magnetic field, the porous $Ni_{50}Mn_{28}Ga_{22}$ alloy with a pore size of 20–30 μm exhibited a maximum saturation magnetization of 62.60 emu/g and a maximum magnetic entropy of 1.93 J/kg·K.

Author Contributions: Conceptualization, X.L. and J.Z.; methodology, K.W.; software, Y.L.; validation, K.W., Y.L. and Z.W.; formal analysis, Z.W.; investigation, Y.Z.; resources, Y.Z.; data curation. X.L.; writing—original draft preparation, X.L.; writing—review and editing, K.W.; visualization, K.W.; supervision, J.Z.; project administration, J.Z.; funding acquisition, J.Z. All authors have read and agreed to the published version of the manuscript.

Funding: This research was funded by State Key Laboratory for Advanced Metals and Materials, grant number [2018Z-26].

Data Availability Statement: The raw data supporting the conclusions of this article will be made available by the authors on request.

Conflicts of Interest: The authors declare no conflict of interest.

References

1. Villa, F.; Morlotti, A.; Fanciulli, C.; Passaretti, F.; Albertini, F.; Villa, E. Anomalous mechanical behavior in NiMnGa alloy sintered through open die pressing method. *Mater. Today Commun.* **2023**, *34*, 105391. [CrossRef]
2. de Souza Silva, F.; Correa, M.A.; Bohn, F.; Bezerra da Silva, R.; dos Passos, T.A.; Torquato, R.A.; de Lima, B.A.G.; Cordeiro, C.H.N.; Oliveira, D.F.d. Effect of Nb doping on NiMnSn Heusler alloys: Mechanical, structural, and magnetic properties modifications. *J. Mater. Res. Technol.* **2023**, *26*, 5167–5176. [CrossRef]
3. Gao, P.; Tian, B.; Xu, J.; Tong, Y.; Chen, F.; Li, L. Investigation on porous NiMnGa alloy and its composite with epoxy resin. *J. Alloys Compd.* **2022**, *892*, 162248. [CrossRef]
4. Gao, P.; Liu, Z.-X.; Tian, B.; Tong, Y.-X.; Chen, F.; Li, L. Microstructure, phase transformation and mechanical properties of NiMnGa particles/Cu composites fabricated by SPS. *Trans. Nonferrous Met. Soc. China* **2022**, *32*, 3291–3300. [CrossRef]
5. Li, D.; Li, Z.; Zhang, X.; Liu, C.; Zhang, G.; Yang, J.; Yang, B.; Yan, H.; Cong, D.; Zhao, X.; et al. Giant Elastocaloric Effect in Ni-Mn-Ga-Based Alloys Boosted by a Large Lattice Volume Change upon the Martensitic Transformation. *ACS Appl. Mater. Interfaces* **2022**, *14*, 1505–1518. [CrossRef] [PubMed]
6. Liu, X.; Bai, J.; Sun, S.; Xu, J.; Jiang, X.; Guan, Z., Gu, J.; Cong, D.; Zhang, Y; Esling, C.; et al. A multielement alloying strategy to tune magnetic and mechanical properties in NiMnTi alloys via Co and B. *J. Appl. Phys.* **2022**, *132*, 095101. [CrossRef]
7. Mishra, A.; Dubey, P.; Chandra, R.; Srivastava, S.K. Observation of large exchange bias field in Mn rich NiMnAl Heusler alloy thin film. *J. Magn. Magn. Mater.* **2021**, *532*, 167964. [CrossRef]
8. Ding, Z.; Liu, D.; Qi, Q.; Zhang, J.; Yao, Y.; Zhang, Y.; Cong, D.; Zhu, J. Multistep superelasticity of Ni-Mn-Ga and Ni-Mn-Ga-Co-Cu microwires under stress-temperature coupling. *Acta Mater.* **2017**, *140*, 326–336. [CrossRef]
9. Hong, S.H.; Park, H.J.; Song, G.; Liaw, P.K.; Kim, K.B. Ultrafine shape memory alloys synthesized using a metastable metallic glass precursor with polymorphic crystallization. *Appl. Mater. Today* **2021**, *22*, 100961. [CrossRef]
10. Lynch, C.S.; Sozinov, A.; Likhachev, A.A.; Ullakko, K. Magnetic and magnetomechanical properties of Ni-Mn-Ga alloys with easy axis and easy plane of magnetization. In *Smart Structures and Materials 2001: Active Materials: Behavior and Mechanics*; SPIE: Bellingham, WA, USA, 2001.
11. Sozinov, A.; Likhachev, A.A.; Lanska, N.; Söderberg, O.; Koho, K.; Ullakko, K.; Lindroos, V.K. Stress-induced variant rearrangement in Ni-Mn-Ga single crystals with nonlayered tetragonal martensitic structure. *J. De Phys. IV* **2004**, *115*, 121–128. [CrossRef]
12. Straka, L.; Heczko, O. Superelastic response of Ni-Mn-Ga martensite in magnetic fields and a simple model. *ITM* **2003**, *39*, 3402–3404. [CrossRef]

13. Huang, L.; Cong, D.Y.; Suo, H.L.; Wang, Y.D. Giant magnetic refrigeration capacity near room temperature in $Ni_{40}Co_{10}Mn_{40}Sn_{10}$ multifunctional alloy. *Appl. Phys. Lett.* **2014**, *104*, 132407. [CrossRef]
14. Chen, F.; Tong, Y.-X.; Tian, B.; Li, L.; Zheng, Y.-F. Martensitic transformation and magnetic properties of Ti-doped NiCoMnSn shape memory alloy. *Rare Met.* **2013**, *33*, 511–515. [CrossRef]
15. Chen, L.; Hu, F.X.; Wang, J.; Bao, L.F.; Zheng, X.Q.; Pan, L.Q.; Yin, J.H.; Sun, J.R.; Shen, B.G. Magnetic entropy change and transport properties in $Ni_{45}Co_5Mn_{36.6}In_{13.4}$ melt-spun ribbons. *J. Alloys Compd.* **2013**, *549*, 170–174. [CrossRef]
16. Dey, S.; Roy, R.K.; Ghosh, M.; Basu Mallick, A.; Mitra, A.; Panda, A.K. Enhancement in magnetocaloric properties of NiMnGa alloy through stoichiometric tuned phase transformation and magneto-thermal transitions. *J. Magn. Magn. Mater.* **2017**, *439*, 305–311. [CrossRef]
17. Hu, F.-X.; Sun, J.-R.; Wu, G.-H.; Shen, B.-G. Magnetic entropy change in $Ni_{50.1}Mn_{20.7}Ga_{29.6}$ single crystal. *J. Appl. Phys.* **2001**, *90*, 5216–5219. [CrossRef]
18. Wang, K.; Hou, R.; Xuan, J.; Li, X.; Zhu, J. Shape memory effect and superelasticity of $Ni_{50}Mn_{30}Ga_{20}$ porous alloy prepared by imitation casting method. *Intermetallics* **2022**, *149*, 107668. [CrossRef]
19. Otaru, A.J. Review on the Acoustical Properties and Characterisation Methods of Sound Absorbing Porous Structures: A Focus on Microcellular Structures Made by a Replication Casting Method. *Met. Mater. Int.* **2019**, *26*, 915–932. [CrossRef]
20. Mohd Razali, R.N.; Abdullah, B.; Ismail, M.H.; Muhamad, N. Characteristic of Modified Geometrical Open-Cell Aluminum Foam by Casting Replication Process. *Mater. Sci. Forum* **2016**, *846*, 37–41. [CrossRef]
21. Boonyongmaneerat, Y.; Chmielus, M.; Dunand, D.C.; Mullner, P. Increasing magnetoplasticity in polycrystalline Ni-Mn-Ga by reducing internal constraints through porosity. *Phys. Rev. Lett.* **2007**, *99*, 247201. [CrossRef] [PubMed]
22. Huang, Y.; Wang, Z.; Zhang, L.; Wei, S. Pore/skeleton structure and compressive strength of porous Mo_3Si-Mo_5Si_3-Mo_5SiB_2 intermetallic compounds prepared by spark plasma sintering and homogenization treatment. *J. Alloys Compd.* **2021**, *856*, 158150. [CrossRef]
23. Laszcz, A.; Hasiak, M.; Kaleta, J. Temperature Dependence of Anisotropy in Ti and Gd Doped NiMnGa-Based Multifunctional Ferromagnetic Shape Memory Alloys. *Materials* **2020**, *13*, 2906. [CrossRef] [PubMed]
24. Hong, S.H.; Kim, J.T.; Park, H.J.; Kim, Y.S.; Suh, J.Y.; Na, Y.S.; Lim, K.R.; Shim, C.H.; Park, J.M.; Kim, K.B. Influence of Zr content on phase formation, transition and mechanical behavior of Ni-Ti-Hf-Zr high temperature shape memory alloys. *J. Alloys Compd.* **2017**, *692*, 77–85. [CrossRef]
25. Wu, Q.; Li, Z.; Chen, F.; Lin, N.Y.; Wang, X.Y.; Zhang, B.Y.; Li, L. Ferromagnetic shape memory alloy $Ni_{52.5}Mn_{22.5}Ga_{25}$: Phase transformation at high pressure and effects on natural gas hydration. *Mater. Lett.* **2023**, *343*, 134371. [CrossRef]
26. Cheng, P.; Zhang, G.; Li, Z.; Yang, B.; Zhang, Z.; Wang, D.; Du, Y. Combining magnetocaloric and elastocaloric effects to achieve a broad refrigeration temperature region in $Ni_{43}Mn_{41}Co_5Sn_{11}$ alloy. *J. Magn. Magn. Mater.* **2022**, *550*, 169082. [CrossRef]
27. Lázpita, P.; Pérez-Checa, A.; Barandiarán, J.M.; Ammerlaan, A.; Zeitler, U.; Chernenko, V. Suppression of martensitic transformation in Ni-Mn-In metamagnetic shape memory alloy under very strong magnetic field. *J. Alloys Compd.* **2021**, *874*, 159814. [CrossRef]
28. Hosoda, H.; Lazarczyk, J.; Sratong-on, P.; Tahara, M.; Chernenko, V. Elaboration of magnetostrain-active NiMnGa particles/polymer layered composites. *Mater. Lett.* **2021**, *289*, 129427. [CrossRef]
29. Chen, P.; Cai, X.; Liu, Y.; Wang, Z.; Jin, M.; Jin, X. Combined effects of grain size and training on fatigue resistance of nanocrystalline NiTi shape memory alloy wires. *Int. J. Fatigue* **2023**, *168*, 107461. [CrossRef]
30. Xuan, J.; Gao, J.; Ding, Z.; Li, X.; Zhu, J. Improved superelasticity and fatigue resistance in nano-precipitate strengthened $Ni_{50}Mn_{23}Ga_{22}Fe_4Cu_1$ microwire. *J. Alloys Compd.* **2021**, *877*, 160296. [CrossRef]
31. Tong, W.; Liang, L.; Xu, J.; Wang, H.J.; Tian, J.; Peng, L.M. Achieving enhanced mechanical, pseudoelastic and elastocaloric properties in Ni-Mn-Ga alloys via Dy micro-alloying and isothermal mechanical cyclic training. *Scr. Mater.* **2022**, *209*, 114393. [CrossRef]
32. Bai, J.; Wang, J.; Shi, S.; Liang, X.; Yang, Y.; Yan, H.; Zhao, X.; Zuo, L.; Zhang, Y.; Esling, C. Theoretical prediction of structural stability, elastic and magnetic properties for Mn_2NiGa alloy. *Mod. Phys. Lett. B* **2021**, *35*, 2150231. [CrossRef]
33. Gao, J.; Ding, Z.; Fu, S.; Wang, K.; Ma, L.; Zhu, J. The magnetization and magnetoresistance of $Ni_{46}Mn_{23}Ga_{22}Co_5Cu_4$ shape memory microwires after mechanical training. *J. Mater. Res. Technol.* **2023**, *23*, 1120–1129. [CrossRef]
34. Chiu, W.-T.; Goto, A.; Tahara, M.; Inamura, T.; Hosoda, H. Investigation of the martensite variant reorientation of the single crystal Ni-Mn-Ga alloy via training processes and a modification with a silicone rubber. *Mater. Chem. Phys.* **2023**, *297*, 127390. [CrossRef]
35. Gaitzsch, U.; Pötschke, M.; Roth, S.; Rellinghaus, B.; Schultz, L. Mechanical training of polycrystalline 7M $Ni_{50}Mn_{30}Ga_{20}$ magnetic shape memory alloy. *Scr. Mater.* **2007**, *57*, 493–495. [CrossRef]
36. Imran, M.; Zhang, X.; Qian, M.; Geng, L. Enhancing the Elastocaloric Cooling Stability of Ni Fe Ga Alloys via Introducing Pores. *Adv. Eng. Mater.* **2020**, *22*, 1901140. [CrossRef]
37. Roth, S.; Gaitzsch, U.; Pötschke, M.; Schultz, L. Magneto-Mechanical Behaviour of Textured Polycrystals of NiMnGa Ferromagnetic Shape Memory Alloys. *Adv. Mater. Res.* **2008**, *52*, 29–34. [CrossRef]
38. Liu, Y.; Zhang, X.; Xing, D.; Shen, H.; Qian, M.; Liu, J.; Chen, D.; Sun, J. Martensite transformation and superelasticity in polycrystalline Ni–Mn–Ga–Fe microwires prepared by melt-extraction technique. *Mater. Sci. Eng. A* **2015**, *636*, 157–163. [CrossRef]

39. Zhou, H.; Wang, D.; Li, Z.; Cong, J.; Yu, Z.; Zhao, S.; Jiang, P.; Cong, D.; Zheng, X.; Qiao, K.; et al. Large enhancement of magnetocaloric effect induced by dual regulation effects of hydrostatic pressure in $Mn_{0.94}Fe_{0.06}NiGe$ compound. *J. Mater. Sci. Technol.* **2022**, *114*, 73–80. [CrossRef]
40. Cheng, P.; Zhou, Z.; Chen, J.; Li, Z.; Yang, B.; Xu, K.; Li, J.; Zhang, Z.; Wang, D.; et al. Combining magnetocaloric and elastocaloric effects in a $Ni_{45}Co_5Mn_{37}In_{13}$ alloy. *J. Mater. Sci. Technol.* **2021**, *94*, 47–52. [CrossRef]
41. Guan, Z.; Jiang, X.; Gu, J.; Bai, J.; Liang, X.; Yan, H.; Zhang, Y.; Esling, C.; Zhao, X.; Zuo, L. Large magnetocaloric effect and excellent mechanical properties near room temperature in Ni-Co-Mn-Ti non-textured polycrystalline alloys. *Appl. Phys. Lett.* **2021**, *119*, 051904. [CrossRef]
42. Bai, J.; Liu, D.; Gu, J.; Jiang, X.; Liang, X.; Guan, Z.; Zhang, Y.; Esling, C.; Zhao, X.; Zuo, L. Excellent mechanical properties and large magnetocaloric effect of spark plasma sintered Ni-Mn-In-Co alloy. *J. Mater. Sci. Technol.* **2021**, *74*, 46–51. [CrossRef]
43. Huang, X.-M.; Zhao, Y.; Yan, H.-L.; Jia, N.; Yang, B.; Li, Z.; Zhang, Y.; Esling, C.; Zhao, X.; Zuo, L. Giant magnetoresistance, magnetostrain and magnetocaloric effects in a Cu-doped<001>-textured $Ni_{45}Co_5Mn_{36}In_{13.2}Cu_{0.8}$ polycrystalline alloy. *J. Alloys Compd.* **2021**, *889*, 161652. [CrossRef]
44. Díaz-García, Á.; Law, J.Y.; Moreno-Ramírez, L.M.; Giri, A.K.; Franco, V. Deconvolution of overlapping first and second order phase transitions in a NiMnIn Heusler alloy using the scaling laws of the magnetocaloric effect. *J. Alloys Compd.* **2021**, *871*, 159621. [CrossRef]
45. Wei, L.; Zhang, X.; Gan, W.; Ding, C.; Liu, C.; Geng, L.; Yan, Y. Large rotating magnetocaloric effects in polycrystalline Ni-Mn-Ga alloys. *J. Alloys Compd.* **2021**, *874*, 159755. [CrossRef]
46. Aydogmus, T.; Bor, S. Superelasticity and compression behavior of porous TiNi alloys produced using Mg spacers. *J. Mech. Behav. Biomed. Mater.* **2012**, *15*, 59–69. [CrossRef] [PubMed]
47. Aydoğmuş, T.; Bor, Ş. Processing of porous TiNi alloys using magnesium as space holder. *J. Alloys Compd.* **2009**, *478*, 705–710. [CrossRef]
48. Golub, V.; L'Vov, V.A.; Salyuk, O.; Barandiaran, J.M.; Chernenko, V.A. Magnetism of nanotwinned martensite in magnetic shape memory alloys. *J. Phys. Condens. Matter* **2020**, *32*, 313001. [CrossRef]
49. Li, Z.; Yang, B.; Zhang, Y.; Esling, C.; Zou, N.; Zhao, X.; Zuo, L. Crystallographic insights into the intermartensitic transformation in Ni–Mn–Ga alloys. *Acta Mater.* **2014**, *74*, 9–17. [CrossRef]
50. Huang, L.; Cong, D.Y.; Wang, Z.L.; Nie, Z.H.; Dong, Y.H.; Zhang, Y.; Ren, Y.; Wang, Y.D. Direct evidence for stress-induced transformation between coexisting multiple martensites in a Ni–Mn–Ga multifunctional alloy. *J. Phys. D Appl. Phys.* **2015**, *48*, 265304. [CrossRef]

Disclaimer/Publisher's Note: The statements, opinions and data contained in all publications are solely those of the individual author(s) and contributor(s) and not of MDPI and/or the editor(s). MDPI and/or the editor(s) disclaim responsibility for any injury to people or property resulting from any ideas, methods, instructions or products referred to in the content.

Article

Application of the Theory of Critical Distance (TCD) to the Breakage of Cardboard Cutting Blades in Al7075 Alloy

Giulia Morettini [1], Luca Landi [1], Luca Burattini [1], Giulia Stornelli [1,*], Gianluca Foffi [2], Andrea Di Schino [1], Filippo Cianetti [1] and Claudio Braccesi [1]

[1] Department of Engineering, University of Perugia, Via G. Duranti 93, 06125 Perugia, Italy; giulia.morettini@unipg.it (G.M.); luca.landi@unipg.it (L.L.); luca.burattini@studenti.unipg.it (L.B.); andrea.dischino@unipg.it (A.D.S.); filippo.cianetti@unipg.it (F.C.); claudio.braccesi@unipg.it (C.B.)
[2] C.M.C. S.p.A., Via C. Marx 13/c, 06012 Città di Castello, Italy; gianluca.foffi@cmcsolutions.com
* Correspondence: giulia.stornelli@unipg.it

Abstract: The study presented in this paper was undertaken in response to two instances of unexpected blade breakage in the cutting blade used in a Carton Wrap machine (CW). Failure of the Al7075 alloy blade occurred at an indentation during typical operational loading conditions. Subsequent metallographic examinations of the fractured samples confirmed that both cases were attributed to fatigue failure. The main objective of this study is to investigate potential causes of fatigue failure in the CW blade using simplified linear elastic static numerical simulations through Finite Element Analysis (FEA). In this research, we employed the well-established Theory of Critical Distance (TCD), and this case study provided a contextualization at an industrial level. Furthermore, the analysis focused on a second key aspect: proposing a new blade geometry aimed at mitigating the identified issues and eliminating possible causes of failure. In this context, the actual stress concentration at the indentation was determined using the TCD with Line Method (LM). The results from the numerical simulations indicated that the new blade geometry significantly reduced stress concentration, resulting in a risk factor reduction of approximately four compared to the original blade design, even under non-optimal operating conditions. Overall, in conjunction with simple linear static FEA, the proposed numerical approach provided substantial support for designers, especially in fault analysis and when comparing different industrial solutions.

Keywords: failure analysis; theory of critical distance (TCD); Al7075-T6 alloy; finite element analysis; linear elastic method

Citation: Morettini, G.; Landi, L.; Burattini, L.; Stornelli, G.; Foffi, G.; Di Schino, A.; Cianetti, F.; Braccesi, C. Application of the Theory of Critical Distance (TCD) to the Breakage of Cardboard Cutting Blades in Al7075 Alloy. *Metals* **2024**, *14*, 301. https://doi.org/10.3390/met14030301

Academic Editor: George A. Pantazopoulos

Received: 29 January 2024
Revised: 23 February 2024
Accepted: 29 February 2024
Published: 3 March 2024

Copyright: © 2024 by the authors. Licensee MDPI, Basel, Switzerland. This article is an open access article distributed under the terms and conditions of the Creative Commons Attribution (CC BY) license (https:// creativecommons.org/licenses/by/ 4.0/).

1. Introduction

In industrial engineering and materials science, the continued pursuit of improving the reliability and efficiency of machinery and components remains a central focus. This is particularly critical in scenarios where unexpected failures can abruptly disrupt or halt production processes. Understanding the root causes of these failures and proposing actionable solutions is paramount to preventing their recurrence.

In this context, the present study delves into the investigation of a recurrent fatigue failure that has plagued carton-cutting production lines. In particular, the failures that have affected the Carton Wrap (CW) cutting blade have consistently manifested in a specific location on the blade where there is a notch. These instances of failure, characterized by fractures originating at notches, present a matter of significant scientific and industrial interest.

If one indeed speaks in a general manner about fatigue testing of a component or entire structure, whether subjected to monotonic cyclic loads or random vibrations, different methodologies are currently employed for the fatigue assessment [1,2]. For instance, the modern unified mechanics theory is used to derive a constitutive model for fatigue life prediction using a three-dimensional computational model [3,4]. Other methodologies, while not a recent development but more industrially established, exploit the frequency domain to rapidly assess the damage caused by the fatigue phenomenon [5,6], even in cases of multiaxial loading [7,8].

To date, there is no universally recognized valid procedure for numerical evaluation through Finite Element Modelling (FEM) in the industrial design of components involving components with notches. Reliance on the designer's expertise is common or on very complex nonlinear plastic analysis. This is due to the well-known limitations of finite element simulation and numerical solutions in accurately resolving stress states at notch apexes, especially when ductile materials are involved. Consequently, scientific literature has introduced various methodologies to address this issue [9]. Many of these methods were initially developed for assessing notches in brittle materials but have evolved to become general techniques for evaluating the structural weakening caused by notches also for ductile materials. These methodologies include criteria based on different approaches: the material's deformation energy storage capacity [10–12]. This approach relies on evaluating the intrinsic characteristics of materials, measuring their ability to store energy during the deformation process to predict potential fracture [13,14]. The cohesive zone technique [15–19] involves analyzing the region near the fracture, focusing on assessing the cohesion between material surfaces. This approach provides a detailed understanding of interactions between fracture surfaces and aids in predicting fracture propagation. The foundational principles of fracture mechanics [20,21] are also used as a prediction methodology. This approach is based on analyzing the mechanical characteristics of the material, focusing on properties that influence fracture propagation. Employing principles of fracture mechanics provides a robust theoretical framework for understanding material behavior under fracture conditions and methods based on the Theory of Critical Distance (TCD) [22–24]. These methodologies are based on the principle that assessing the stress state at the notch is possible by considering a certain distance from its apex [25,26], specifically defined as the critical distance [27–31].

Some of these methodologies can be complex to implement and may not fit well with the stringent time constraints of the industrial context. However, others, such as the TCD, have demonstrated their effectiveness in quickly analyzing notch-related static stress states and fatigue behavior through simple linear elastic finite element simulations, even for highly ductile materials [32,33]. These methodologies rely on defining a length parameter related to notch geometry and material-specific parameters, facilitating straightforward determination of effective stress at the notch, which can be directly compared to material strength characteristics.

In conclusion, looking at the industrial landscape, we can assert that there is an extreme lack of capability for assessing the stress state in notched components. Outdated methodologies, leading to overestimated results, are still being used today due to the absence of validated and applicable alternatives. On the other hand, the scientific community proposes numerous criteria based on different concepts, some of which would align well with the actual industrial needs if their applicability were demonstrated. In response to these challenges, our study embarks on a multifaceted investigation with two primary objectives. Firstly, it aims to assess the possible causes of CW blade failures through fracture surface analysis and by applying the TCD with Line Method (LM), leveraging linear elastic static numerical simulations. Secondly, the study seeks to propose a novel blade geometry designed to mitigate the identified issues and eliminate potential failure causes. We aimed to reduce stress concentrations at critical notches by focusing on blade redesign. This contribution has the potential to enhance the operational reliability of the CW blade, leading to cost savings and improved efficiency for the company, but it also defines a procedure to be used by designers in the design phase.

2. Case Study: Technical Failure Analysis

Packaging machines are used to encase objects of varying sizes and weights. Each box is created within the machine production line and tailored to the specific product to be packaged, starting from a paperboard sheet that is appropriately cut and shaped to give it a three-dimensional form. During the cutting phase, certain grooves are excavated on the paperboard. Figure 1 illustrates the placement of the following elements—marked between brackets ()—within the rotor. A pair of rapidly rotating blades (1), impacting the sheet's surface at high speed, are capable of slicing the material without causing local deformations or damage to the cut edges. The cutting system can be described as a rotor, where the rotating blades are connected to the drive shaft (2), and an electric motor (3) drives the rotation. A gear reducer (4) links both shafts, and an elastic joint (5) is positioned between the electric motor and the gear reducer.

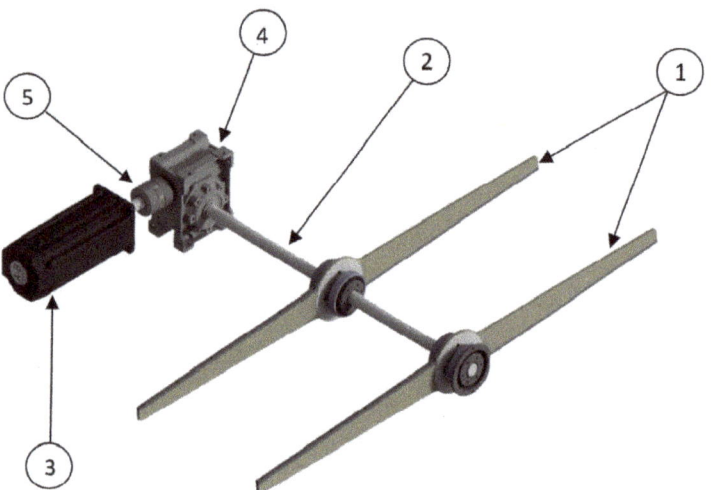

Figure 1. Key components of the die-cutting system.

Incidents of failure occurred during the operational lifespan of the blades, resulting in significant delays in shipments for machine maintenance. Nevertheless, upon comparing the number of broken blades (only two instances occurred after approximately eight million operational cycles) to the number of intact and still operational blades (360), it is conceivable that non-standard breakage conditions may have contributed to these failures. Figure 2 displays the real image of the rolling blade under investigation, visualizing its key dimensions. Watching the blade in Figure 2 through a microscope, it is possible to confirm that what appears to be a perfectly sharp notch possesses instead a small fillet radius due to the tool path during the blade manufacturing process. Therefore, a 0.2 mm radius will be used instead of an ideal sharp edge in subsequent analyses. A larger radius fitting on the notch would not allow the installation of a sufficiently long cutting edge, visible in Figure 2, reducing the maximum depth of the grooves made on the cardboard and making the production of larger boxes impossible.

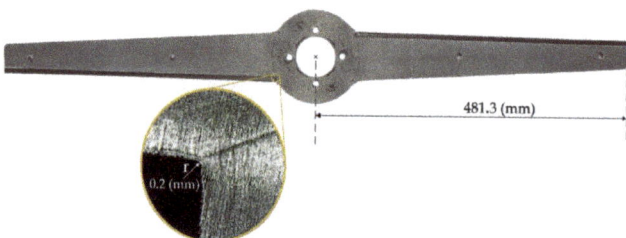

Figure 2. A real model of the rolling blade under investigation with a detail of fillet radius.

2.1. Material Characterization

The cutting system blades are made of Al7075-T6 alloy, and in order to identify the failure causes, the following material and fracture analysis were carried out: (i) microstructural analysis in cross-section close to the fracture, (ii) fracture surface analysis, and (iii) semi-quantitative chemical analysis on the fracture surface and on the cross-section.

The microstructure was examined through optical microscopy (OM—Eclipse LV150 NL, Nikon, Tokyo, Japan) with the aim of identifying possible microstructural defects present in the material. After mechanical polishing and etching, the observations were carried out using Keller reagent for 20 s. The fracture surface analysis was conducted using a stereo-microscope (Nikon SMZ745T) and a Field-Emission Scanning Electron Microscope (FE-SEM—Zeiss, Gemini Supra 25, Jena, Germany) to determine the nature of the fracture phenomena. The chemical analysis was carried out in selected areas using the FE-SEM with an Energy-Dispersive X-ray Spectroscopy (EDS) probe.

The following description was useful in defining the real cause of component failure. In Figure 3, the fracture surface extracted from the component after failure is reported.

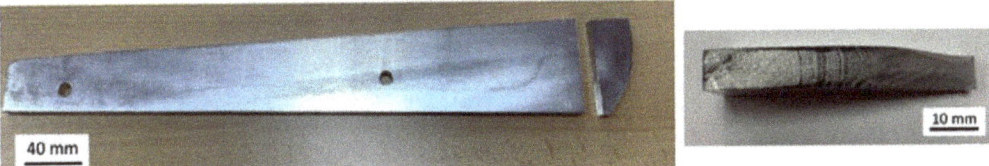

Figure 3. Detail of the extraction of the fracture surface of the cutting system blade.

Therefore, from the complete analysis of the cutting system blades, it emerged that there is a trigger of fracture on the specular part of the component. In Figure 4, it is evident that in the non-fractured portion of the cutting system, there is a crack (a and b) starting from the cutting notch.

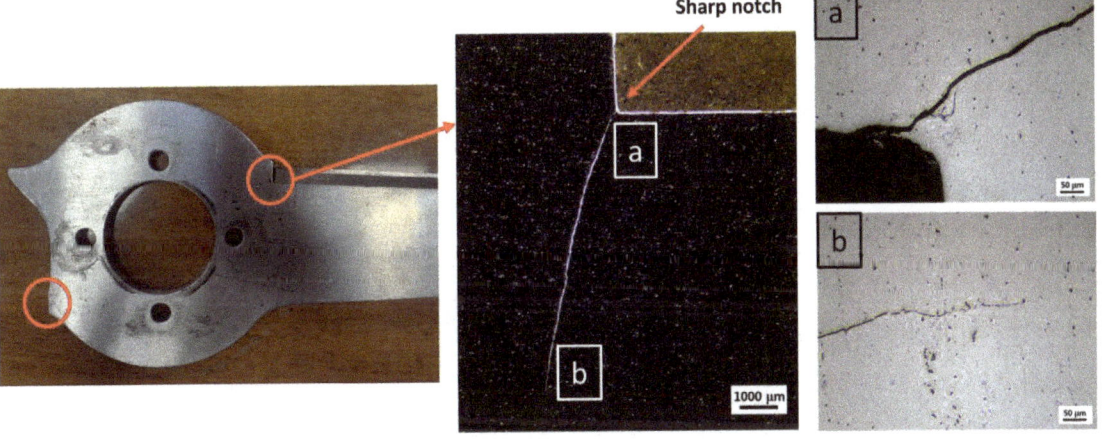

Figure 4. Detail of the sharp notch with evidence of a crack.

From the fracture surface analysis shown in Figure 5, a mix of morphologies distinguishable in 3 different zones is evident, indicated as zone A (starting from the sharp notch of the component), zone B (central zone), and zone C.

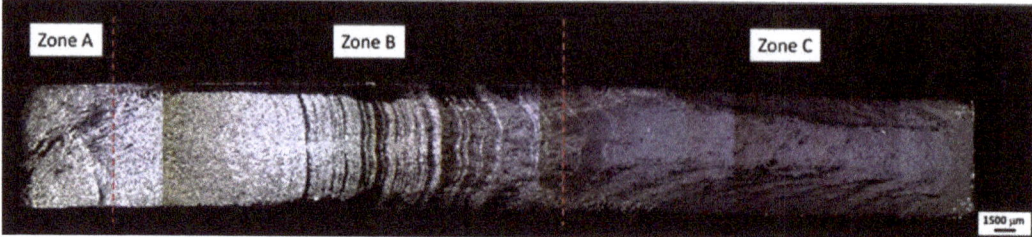

Figure 5. Micrographic analysis of the fracture surface through the stereo-microscope.

From the high magnification micrographic analysis, it emerged that for zone A (Figure 6), the surface is smooth and deformation-free. Beach lines are clearly evident in the central area (Zone B in Figure 7) and for more than half of the fractured surface. Finally, the third zone (zone C in Figure 8) presents a typical morphology of a fracture due to plastic deformation (dimples) on a plane inclined at approximately 45° compared to the rest of the fracture (Figure 9). This morphology is typical of a fracture due to fatigue stress. The fracture trigger section is Zone A, in which the smooth appearance of the surface is attributable to the repeated impacts due to the alternation of traction-compression stresses during the operation of the component. The fracture has propagated to the right (see directionality of the beach lines in Figure 7), and Zone C corresponds to the portion in which the resistant section is insufficient, and the component breaks under the action of static stress, as expected from a fatigue failure.

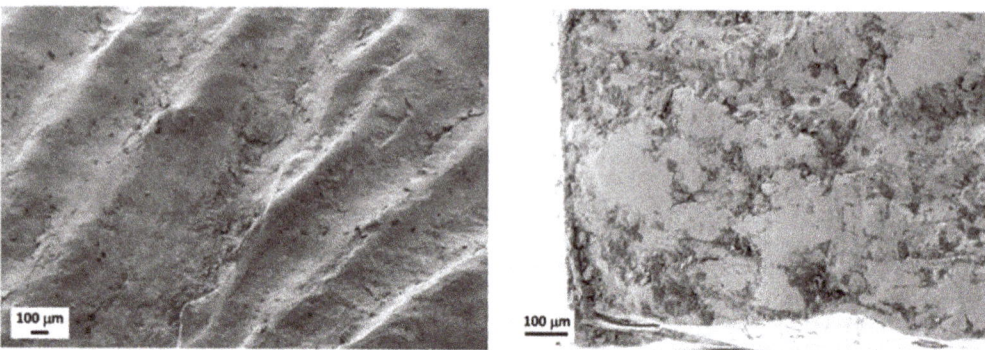

Figure 6. SEM analysis of zone A referred to in Figure 5.

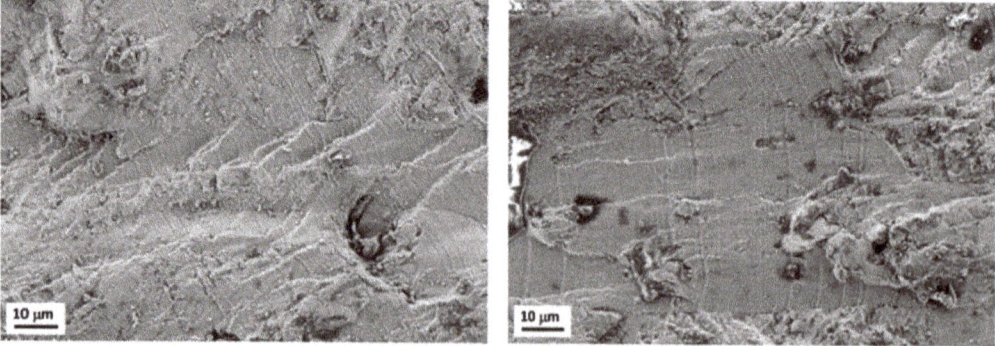

Figure 7. SEM analysis of zone B referred to in Figure 5.

Figure 8. SEM analysis of zone C referred to in Figure 5.

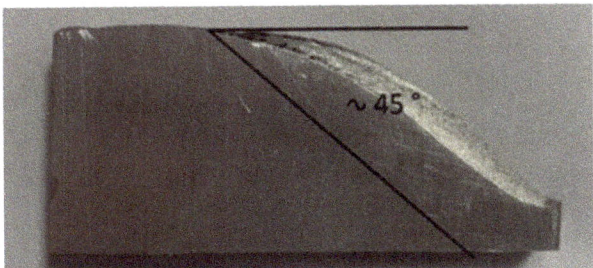

Figure 9. Evidence of the final fracture in zone C which occurred on a plane at 45° with respect to the plane of propagation of the fracture.

Before etching specimens were analyzed using OM for an inclusion state check, the analysis revealed no critical issues in terms of non-metallic inclusion. Following that, specimens were etched, and the micrographic and EDS analysis conducted in cross-section on a sample extracted near the fracture are reported respectively in Figures 10 and 11. The microstructure turned out to be homogeneous and compliant with a tempered Al7075 alloy [34–37], with Zn, Cu, and Mg precipitates homogeneously dispersed. In addition, segregation bands were not revealed.

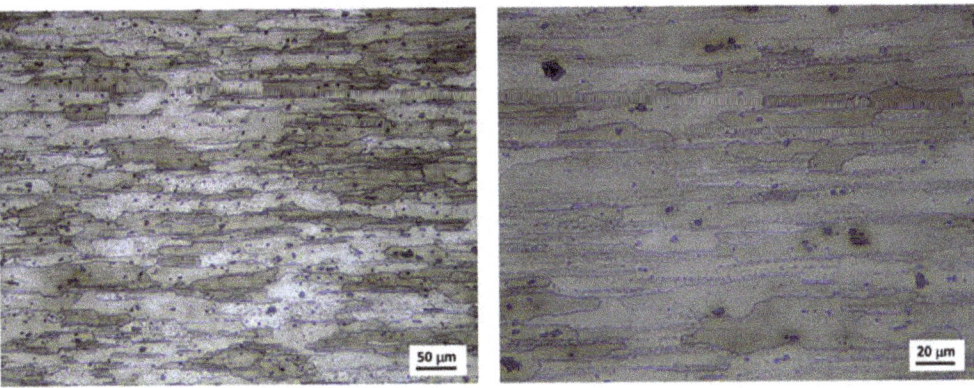

Figure 10. OM micrographic analysis of a cross-section near the fracture.

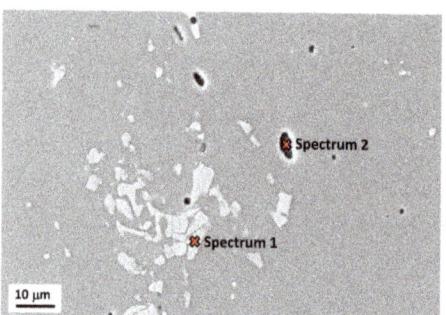

Figure 11. EDS analysis of a cross-section near the fracture.

Finally, from the EDS analysis conducted on the fracture surface (Figure 12), only the presence of Al oxides attributable to the operation of the component emerged, with growth on the surface due to the propagation of the fracture.

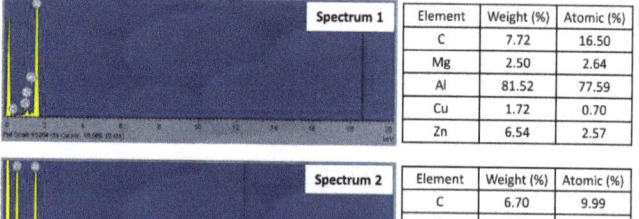

Figure 12. EDS analysis on the fracture surface.

Hence, based on the conducted analysis, it is possible to conclude that the cause of the component failure was fatigue-induced, pinpointing a triggering locus within the notch region.

2.2. Mechanical Characteristics

The mechanical and physical properties obtained from the literature [38–42], which are useful for the description of the material and for the investigations reported in the following paragraphs, are shown in Table 1.

Table 1. Physical and mechanical properties of Al7075-T6 [38–42].

Parameter	Value
Breaking stress (Rm)	560 N/mm^2
Yield stress (Rp0.2)	495 N/mm^2
Elongation (ε)	7%
Density (ρ)	2.81 g/cm^3
Elastic modulus (E)	72,500 N/mm^2
Shear Modulus (G)	26.9 GPa
Shear Strength (τ)	331 MPa
Hardness, Vickers (HV)	175
Poisson's ratio (q)	0.33
Fracture toughness (K_{IC})	50 MPa/$\sqrt{m}$

Table 1. *Cont.*

Parameter	Value
Threshold value of the stress intensity factor ($\Delta k_{TH_{(R=0)}}$)	1.6 MPa/$\sqrt{m}$
Threshold value of the stress intensity factor ($\Delta k_{TH_{(R=-1)}}$)	4.2 MPa/$\sqrt{m}$
Fatigue strength 5×10^8 cycles, fully reversed load ($\Delta \sigma_n'{}_{(R=-1)}$)	318 MPa
Fatigue strength 5×10^8 cycles, repeated load ($\Delta \sigma_n'{}_{(R=0)}$)	229 MPa

3. Analysis of Operational Loads and Constraints

The uncertainty surrounding the force exerted by the blade upon impacting the paperboard has posed a significant challenge in accurately determining the true stress value on the component. By referring to the motor's datasheet, it is possible to confirm that the rated torque, the maximum that can be sustained continuously, amounts to 13.38 Nm. This value can serve as the upper limit for our analysis. It is important to note that, in normal conditions, the motor's torque is distributed between two blades simultaneously engaged in cutting the paperboard. Therefore, it is necessary to halve the aforementioned value to calculate the torque applied to an individual blade. However, any backlash and misalignments in the assembly can lead to brief moments where only one blade encounters the paperboard, causing the entire load to concentrate on it. For this reason, thinking in a safe manner, rated torque was not halved in order to consider this possible critical situation.

A more realistic torque value was obtained through an experimental campaign on the CW. Figure 13 presents the torque curve over time during the cutting test. In the initial segment of the same figure, the curve shows the torque generated by the motor while the blades rotate at a constant speed. As we can see, approximately 10% of the rated torque is necessary to overcome system friction. However, as the blades engage with the paperboard, the torque rapidly escalates to approximately 100% of the rated torque. Therefore, the torque jump, visible in Figure 13, which amounts to 12 Nm, can be considered as that value used by the motor to cut the paperboard (about 90% of the rated torque). The highest peak in Figure 13 has been excluded from consideration because its origin is attributed to the combined impact of two rapid instances involving the blade and cardboard. As discussed earlier, this value is halved to calculate the load for a single blade. Consequently, the torque value for a single blade is established at 6 Nm. We denoted this load condition as the "experimental torque." It is essential to emphasize that this result may vary depending on the type of paperboard used for testing. Nevertheless, these values can be regarded as beneficial for quantitatively assessing loads resembling the operational conditions of the system. To determine the torques at the blades, all the defined values must be multiplied by the transmission ratio, $\tau = 10$. The outcomes are presented in Table 2.

Table 2. Load cases definition.

ID Load Case	Motor Torque (Nm)	Blade Torque (Nm)
Rated case	13.38	133.8
Experimental case	6	60

The blade torques mentioned above can be regarded as impact forces conveniently applied to the tip, and their calculation can be easily performed using the geometric measurement of the arm presented in Figure 2. As a result, the arm of the force was defined as the distance between the tip and the center of gravity of the four bolts used to constrain the blade. Figure 2 illustrates the holes excavated on the blade to allocate bolts and establish the connection with the rolling shaft. In the subsequent sections, detailed information on the force modules applied will be provided in the tables. With a comprehensive understanding of the element's geometry, the applied forces, and the nature of constraints, it becomes possible to construct a blade model in order to investigate the root causes of faults.

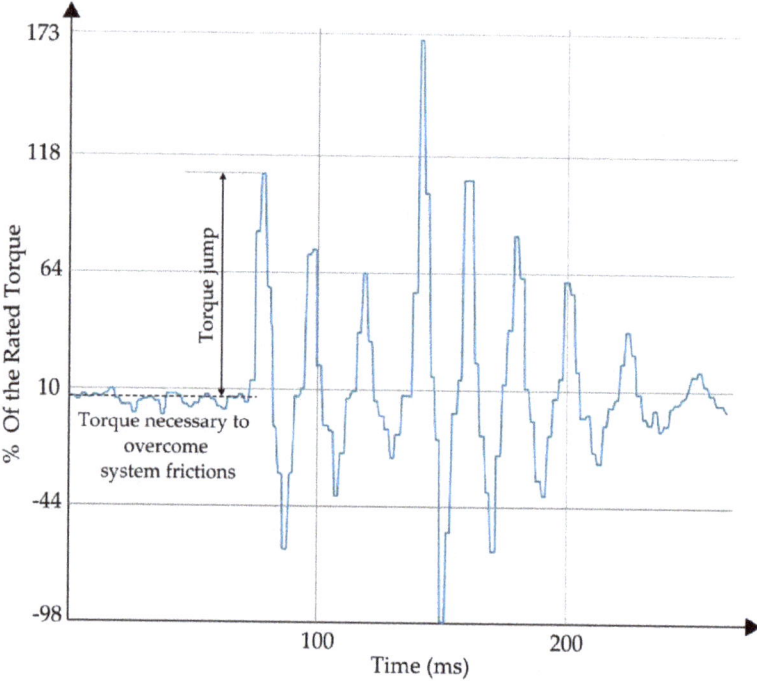

Figure 13. Measure of the experimental load on the blade.

4. Theoretical Background

As can be inferred from Section 2, which describes the geometry under examination, before applying the Finite Element Analysis (FEA) methodology, it is necessary to present the theoretical concepts used to determine the actual stress state in the presence of a notch. It is well known that evaluating the actual stress state in the presence of a notch is analytically challenging. The result obtained from a simple linear-elastic, static, or dynamic analysis performed using finite element software typically does not accurately represent the real stress state at the apex of the notch. For this reason, scientific literature offers various methodologies that, based on energetic and/or punctual geometric concepts, precisely define the stress state of a notched component under static or dynamic loading. In this calculation report, the TCD will be employed, specifically its LM applicative, described in detail below.

4.1. Theory of Critical Distance (TCD)

The origin of TCD can be attributed to the works of Neuber and Peterson [28,29]. In general terms, TCD states that the failure of a notched component can be avoided as long as the effective stress σ_{eff} (calculated based on material-specific parameters) remains below an intrinsic strength value σ_0 (stress related to the applied loading and the material itself), as stated in Equation (1).

$$\sigma_{eff} < \sigma_0 \tag{1}$$

In other words, the entire linear-elastic stress field around the notch apex is represented by an appropriate effective stress σ_{eff}, while the material strength is represented by a suitably designed stress value based on the type of loading or material in use.

The effective stress can be calculated using various methodologies [31], all of which involve a length parameter known as the critical distance L.

In the case of fatigue fracture, the critical distance can be calculated as:

$$L_{fat} = \frac{1}{\pi}\left(\frac{\Delta K_{th}}{\Delta \sigma_0}\right)^2 \quad (2)$$

where ΔK_{th} represents the threshold value of the stress intensity factor and $\Delta \sigma_0$ coincides with the fatigue limit full stress range of the material σ'_n. In this scenario, the application of TCD does not require any calibration.

The TCD method can be formalized using different methodologies [30], one of which is the LM, widely employed in the industrial context due to its applicative simplicity and accuracy. This methodology suggests that the effective stress is evaluated as the average stress over a length of $2 * L_{fat}$ along the bisector of the notch, as described by Equation (3) based on Figure 14.

$$\sigma_{eff\_LM} = \frac{1}{2L_{fat}} \int_0^{2L_{fat}} \sigma(r) \, dr \quad (3)$$

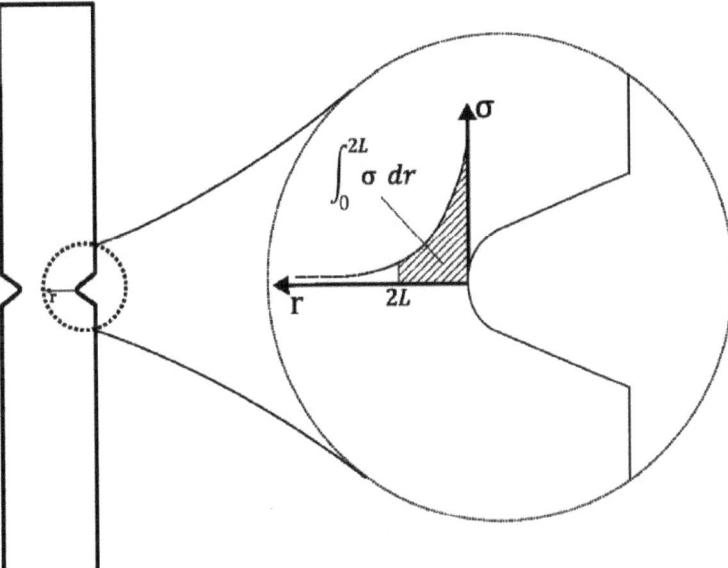

Figure 14. Stress-distance curve from the notch apex and definition of the LM.

In the present real case, the methodology described will be developed with the support of FEA, which allows the replication of the loading conditions and the evaluation of stress distribution around the studied notch. The choice of the stress parameter to consider primarily depends on the type of loading, and in our case, we will refer to the maximum principal stress as a state of plane stress that has been observed.

As mentioned earlier, regardless of the chosen TCD method to apply, an accurate prediction relies on the proper definition of the ΔK_{th} and $\Delta \sigma_0$ parameters, which should ideally be derived from dedicated experimental tests. The parameters, obtained as much as possible from scientific literature and presented in detail in Section 2.2, will be subsequently used for the theoretical application in the specific case of interest presented in the next section.

4.2. Theory Application in the Current Real Case: Definition of the Critical Distance for Fatigue Assessment

For the case analyzed in this report, the definition of L_{fat} in Equation (2) is straightforward. It is possible to easily determine the values of $L_{fat(R=0)} = \frac{1}{\pi}\left(\frac{\Delta k_{TH_{(R=0)}}}{\sigma_n'_{(R=0)}}\right)^2 = 0.0155$ mm and $L_{fat(R=-1)} = \frac{1}{\pi}\left(\frac{\Delta k_{TH_{(R=-1)}}}{\sigma_n'_{(R=-1)}}\right)^2 = 0.0316$, based on the parameter definitions provided in Section 2.2 and consequently the values of $2 L_{fat(R=0)} = 0.031$ mm and $2 L_{fat(R=-1)} = 0.063$ mm. These values will be useful to mediate the stress distribution $\sigma(r)$ along the notch bisector evaluated using FEA, applying the LM methodology expressed in Equation (3).

5. Finite Element Analysis

The company provided the CAD model of the cutting blade, and it has been imported into the Ansys 2023–R2 workbench software for analysis. Given the blade's symmetric shape, only one side of the geometry has been imported into the software. It is important to note that, in order to preserve all the bolt holes, the section line does not align with the symmetrical axis of the component. Furthermore, some lines have been projected onto the surface to define the force application point and establish the notch's stress control path, which is 0.8 mm long, see Figure 15. The model was completed by inserting translational constraints to fix each bolt hole, simulating the bolts' behavior. Using a Cerig command, a rigid region is delineated around each bolt hole by generating constraint equations to correlate nodes on the internal surface of each hole. These slave nodes are rigidly connected to the central master node, which, as specified, is prohibited from translating. Given the blade's flat shape, the use of shell elements during the meshing phase is recommended to provide high result accuracy with low computational cost. Shell 181 element, used in this paper, is suitable for analyzing thin to moderately thick shell structures; it is a well-known four-node element with six degrees of freedom at each node. Three types of mesh sizing were employed [43]: a coarse mesh for discretizing the largest part of the surface and two additional refinement levels applied to enhance result precision around the notch, as shown in Figure 15a,b. Both presented load cases were considered, as indicated in Table 2. It should be noted that the stress values along the control path in the subsequent images are referenced, for simplicity, to the rated case. Figure 16 illustrates the distribution of maximum stress around the notch in a plane stress state hypothesis. Consequently, the maximum principal stress is evaluated along the control path, where 202 nodes are located, using a static structural linear elastic solution. Notably, from Figure 16, it is possible to underline that stress increases as the node approaches the apex of the notch. As previously explained, mathematical equations underlying the FEA linear elastic solution for notches or defects overestimated the stress values at the notch's apex, particularly for non-rounded notches [44]. These findings are subject to the influence of the mesh size employed. Therefore, it is imperative to evaluate the dispersion of the results when the mesh surrounding the control path undergoes variations. The stress pattern along the bisector of the notch, denoted as $\sigma(r)$, exhibits variations as the mesh size is modified, and the attainment of result convergence is realized with a 0.01 mm element size (as employed in our simulations), which strikes a suitable balance between result precision and computational resource utilization [43].

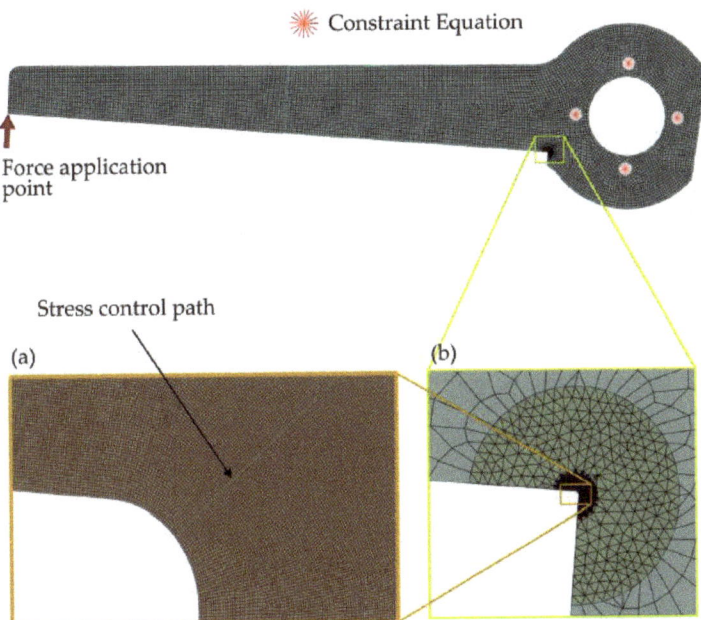

Figure 15. Locations of constraints and forces in the blade model. Detail of the stress control path (**a**) and the mesh sizes around the notch (**b**).

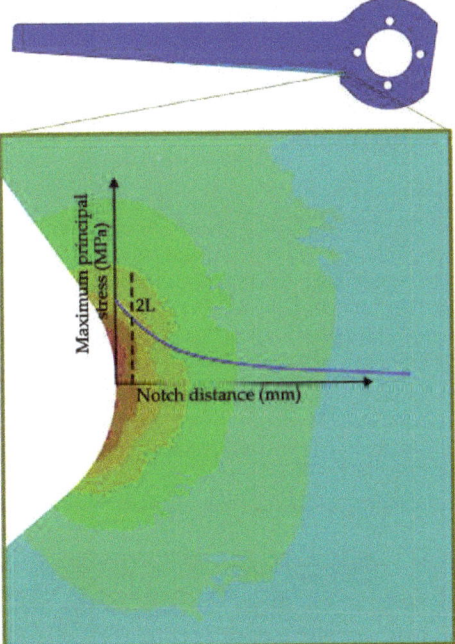

Figure 16. Maximum principal stress shape on notch and definition of the amplitude of the effective stress according to the Line Method.

Table 3 reports the effective maximum stress σeff_LM calculated through Equation (3) using LM for both cases of load.

Table 3. Maximum principal stress on notch according to LM in each load case.

ID Load Case	Force Applied on the Tip (N)	Effective Maximum Stress σeff_LM (R = 0) (MPa)	Effective Maximum Stress σeff_LM (R = −1) (MPa)
Rated case	278	129	107.4
Experimental case	125	58	48.3

From the material data reported in Table 1, the maximum stress value for the fatigue strength limit depends on the typology of the load applied. Its value is 159 MPa for R = −1 and 229 MPa for R = 0; these values conventionally guarantee 5×10^6 load cycles before the fatigue break [38]. Since it is not possible to define with certainty the type of load applied to the blade, it is necessary to carry out a combined comparison of the two fatigue limits (R = 0 and R = −1) with the corresponding σeff_LM derived from the maximum principal stress analysis. Thus, the following results reported in Table 4 can be highlighted.

Table 4. Fatigue resistance of the rolling blade under different load conditions.

Load Case	σeff_LM (Mpa)	Experimental Case Fatigue Limit (Mpa)	Result	σeff_LM (Mpa)	Rated Case Fatigue Limit (Mpa)	Result
R = 0	58	229	No failure	129	159	No failure
R = −1	48.3	159	No failure	107.4	229	No failure

Based on the previous considerations shown in Table 4, it can be concluded that the rolling blade is not susceptible to fatigue failure, which explains why the majority of them did not break. It is evident, therefore, that under standard operating conditions, the blade does not experience failure. The root cause of the breakage remains unidentified. Therefore, further investigation is required to identify any potential phenomena that may alter the loads or constraint conditions.

6. Possible Cause of Failure

The connection of the blade system to the shaft is supported by four bolts, as shown in Figure 17. In order to investigate a possible cause of blade breakage attributable to incorrect loading of the component, the conformation of each bolt hole was studied. Figure 17 shows in detail the images of the four holes from which it is clear that the holes called 1, 2, and 3 present an oval shape with evident plastic deformation in the vertical or horizontal direction (see the white arrows). Hole 4 presents a little plastic deformation localized on the entire perimeter and does not present any geometric alteration. Therefore, the evident ovalization and plastic deformation along the perimeter of holes 1, 2, and 3 is indicative of a possible misalignment of the coupling attributable to an incorrect grip of the bolts on the component. The hypothesis that the component was coupled only by bolt 4 is therefore relevant, thus resulting in an incorrect distribution of loads.

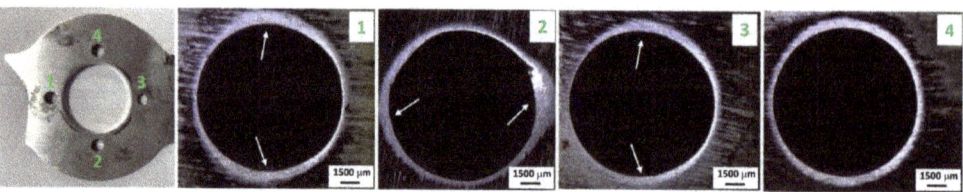

Figure 17. Detail of the bolt holes.

6.1. Finite Element Analysis in the Hypothesised Constrain Condition

In this paragraph, an unconventional working condition was investigated. Maintaining the same load cases as presented in Table 2 and utilizing the identical mesh size depicted in Figure 18, modifications are introduced to the number of holes where translational constraints are applied. This adjustment allows the emulation of a specific scenario in which one or more bolts are unscrewed, enabling assessing the resulting stress changes along the control path. It is worth noting that the control path referred to in the subsequent analysis pertains to the force application for the rated case, though the same procedure is equally applicable to the other load case (the results are reported for both conditions). The FEA of the rolling blade, considering observations made regarding bolt holes, is suitably adjusted. With reference to Figure 15, illustrating the use of four bolts to connect the blade to the rolling shaft, constraint equations are now applied exclusively to hole 4. However, it is crucial to recognize that merely restraining the translational movements of a single bolt does not guarantee the complete stabilization of the model. Consequently, an additional constraint is necessary to fix rotations. Because the rolling blade retains the capability to rotate about the bolt axis, a node constraint between the shaft surface and the blade central hole is used to achieve equilibrium, as illustrated in Figure 18.

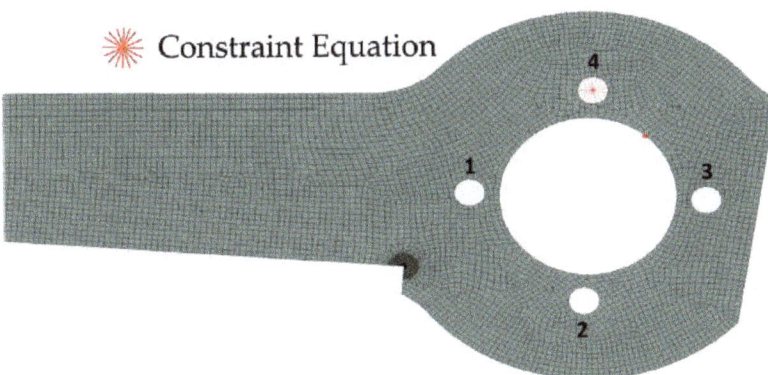

Figure 18. Modified constraint conditions to emulate a single socket of hole number 4.

6.2. Results and Discussions

Table 5 shows that stress significantly escalates in the stress control path for both load cases when only one bolt is fastened.

Table 5. Maximum principal stress on notch for a single bolt fastened according to LM.

ID Load Case	Force Applied on the Tip (N)	Effective Maximum Stress σeff_LM (R = 0) (MPa)	Effective Maximum Stress σeff_LM (R = −1) (MPa)
Rated case	278	223.8	176.1
Experimental case	125	100.6	79.2

In the experimental load case, the stress on the notch increases, decreasing safety margins, without, however, reaching the fatigue limit (159 MPa for $R = -1$ or 229 MPa for $R = 0$). However, this load condition does not appear to be more plausible; in fact, due to the misalignments induced by unscrewing the bolts, applying the torque to a single blade will be more plausible.

If we consider, therefore, the rated load condition case (highlighted in red in Table 5), the fatigue stress limit was exceeded under the hypothesis of fully reversed load condition ($R = -1$), and it dangerously approaches the fatigue limit for ($R = 0$).

In conclusion, it can be affirmed that the examined constraint condition can reasonably induce fatigue failure in the rolling blade.

7. New Proposed Geometry

The conventional approach used to alleviate stress on the notch involves replacing the sharp edge with a rounded one. To maximize stress reduction on the notch, the maximum possible radius is employed, minimizing connection issues between the rolling blade and the cutting edge and minimizing the size reduction of the produced boxes as much as possible. For these reasons, the proposed geometric model implements a 10 mm radius for the notch under investigation, see Figure 19. As a result of this modification, an increased area exposed to high-stress values due to the presence of a larger notch is obtained; however, as a beneficial effect, the magnitude of the maximum stress should be reduced.

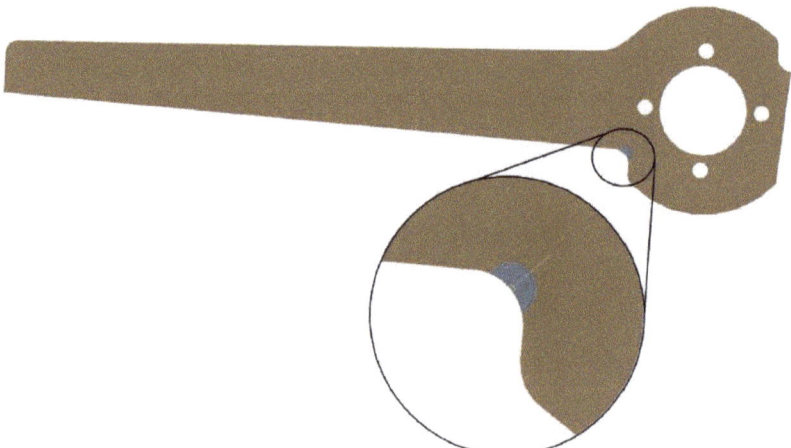

Figure 19. New blade geometry proposed.

7.1. New Geometry Analysis

The new geometry is analyzed using the same loads as analyzed in paragraph 3 and incorporating the constraints described in paragraph 6. The mesh is generated in a manner similar to the one proposed previously, with the objective of enabling an accurate comparison of the model's results.

The size of the mesh along the stress control path has been refined; even in this case, the result's stability is verified by varying the mesh size around the notch. Also, this analysis confirms the good accuracy of the results using 0.01 mm elements.

Table 6 demonstrates that in no instance does fatigue failure occur in the proposed geometry. Indeed, the fatigue limit is not reached under each of the assessed load and constraint conditions. Furthermore, when examining the values of σ_{eff_LM} for the new geometry, a substantial safety margin against fatigue failure is achieved.

Table 6. Maximum principal stress on the modified geometry notch according to LM.

ID Load Case	Force Applied on the Tip (N)	Effective Maximum Stress σ_{eff_LM} (R = 0) one Bolt Fastened (MPa)	Effective Maximum Stress σ_{eff_LM} (R = −1) one Bolt Fastened (MPa)
Rated case	278	66.7	64.7
Experimental case	125	30	29.1

7.2. Comparison of the Obtained Results

Both the stress distributions along the path, with reference to the sharp notch and the rounded one, can be visually compared when the force of the rated case is applied to the rolling blade, as depicted in Figure 20.

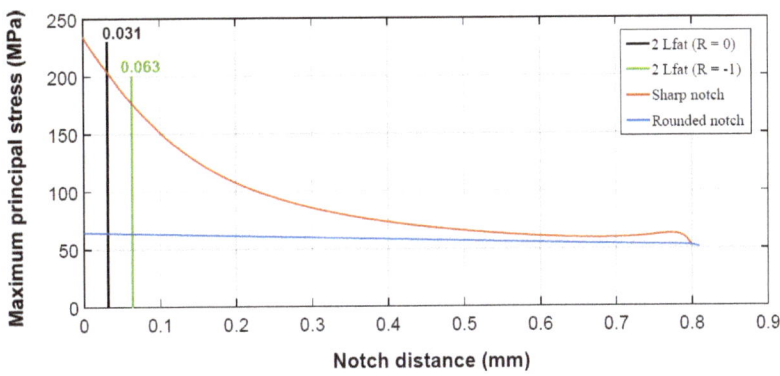

Figure 20. Stress distributions along the path with reference to the sharp and rounded notch. According to Equation (2), 2 L_{fat} represents the double value of the critical length of the TCD method for the cases R = 0 and R = −1.

Based on Figure 20, a substantial decrease in stress levels around the notch is evident. This effect becomes more pronounced when observing the stress differential in proximity to the notch apex in relation to the proposed blade geometry. The black vertical line, as well as the green line, establishes the boundary with which the effective maximum stress σ_{eff_LM} (MPa) was computed. In fact, all node stresses located on the left side of the corresponding line must be used as per Equation (3). A stress reduction factor can be calculated from the ratio of the effective maximum stresses of the two cases reported. This factor serves as a valuable parameter for assessing how much the new blade geometry can abate stress on the notch. Table 7 summarizes the effective maximum stresses calculated using LM for various load and constraint conditions.

Table 7. Fatigue behavior of the rolling blade under different load conditions.

Constrain Condition	ID Load Case	Effective Maximum Stress σ_{eff_LM} (MPa)				Stress Reduction Factor	
		Sharped Notch		Rounded Notch			
		R = 0	R = −1	R = 0	R = −1	R = 0	R = −1
Four bolts fastened	Experimental case	58	48.3	-	-	-	-
	Rated case	129	107.4	-	-		
One bolt fastened	Experimental case	100.6	79.2	30	29.1	3.4	2.7
	Rated case	223.8	176.1	66.7	64.7		

8. Conclusions

This study comprehensively analyzed factors contributing to the failure of some paperboard cutting blades. Metallographic investigation revealed that fatigue load conditions caused the breakage. Only through the implementation of TCD via FEA was it possible to identify the potential failure causes, with bolt loosening identified as a triggering factor.

Additionally, a geometric modification involving a rounded notch design was analyzed thanks to the proposed method. Numerical simulations employing the TCD Line method significantly reduced the stress concentration compared to the sharp-notch design.

The potential of the methodology proposed lies in being able to quantify this reduction. The rounded notch design exhibited a maximum reduction factor of around 3, enhancing the overall blade resistance without compromising operational performance.

In conclusion, this real case study aims to fill the literature gap by demonstrating the method's validity on real case studies and emphasizing the potential for industrial designers and scientific researchers to utilize numerical methodologies like the TCD method for predicting fatigue failure in notched components.

Author Contributions: Conceptualization, G.M., L.L. and L.B.; Formal analysis, G.M., L.L. and L.B.; Investigation, G.M., L.L., L.B. and G.S.; Methodology, G.M.; Software, L.B.; Validation, G.M., L.B. and G.S.; Visuali-zation, G.F.; Writing—original draft, G.M., L.B. and G.S.; Writing—review & editing, G.M., L.L., L.B., G.S., G.F., A.D.S., F.C. and C.B.; Supervision, L.L., A.D.S., F.C. and C.B. All authors have read and agreed to the published version of the manuscript.

Funding: This research received no external funding.

Institutional Review Board Statement: Not applicable.

Informed Consent Statement: Not applicable.

Data Availability Statement: The data presented in this study are available on request from the corresponding author due to restrictions.

Conflicts of Interest: Author Gianluca Foffi was employed by the company C.M.C. S.p.A. The remaining authors declare that the research was conducted in the absence of any commercial or financial relationships that could be construed as a potential conflict of interest.

References

1. Lee, H.W.; Basaran, C. A Review of Damage, Void Evolution, and Fatigue Life Prediction Models. *Metals* **2021**, *11*, 609. [CrossRef]
2. Lee, H.W.; Basaran, C. Predicting High Cycle Fatigue Life with Unified Mechanics Theory. *Mech. Mater.* **2022**, *164*, 104116. [CrossRef]
3. Basaran, C. *Introduction to Unified Mechanics Theory with Applications*; Springer International Publishing: Cham, Germany, 2021; ISBN 978-3-030-57771-1.
4. Canale, G.; Lepore, M.; Bagherifard, S.; Guagliano, M.; Maligno, A. An Experimental Validation of Unified Mechanics Theory for Predicting Stainless Steel Low and High Cycle Fatigue Damage Initiation. *Forces Mech.* **2023**, *10*, 100162. [CrossRef]
5. Lee, H.W.; Djukic, M.B.; Basaran, C. Modeling Fatigue Life and Hydrogen Embrittlement of Bcc Steel with Unified Mechanics Theory. *Int. J. Hydrog. Energy* **2023**, *48*, 20773–20803. [CrossRef]
6. Lee, H.W.; Basaran, C.; Egner, H.; Lipski, A.; Piotrowski, M.; Mroziński, S.; Bin Jamal, M.N.; Lakshmana Rao, C. Modeling Ultrasonic Vibration Fatigue with Unified Mechanics Theory. *Int. J. Solids Struct.* **2022**, *236–237*, 111313. [CrossRef]
7. Morettini, G.; Braccesi, C.; Cianetti, F.; Razavi, N.; Solberg, K.; Capponi, L. Collection of Experimental Data for Multiaxial Fatigue Criteria Verification. *Fatigue Fract. Eng. Mater. Struct.* **2020**, *43*, 162–174. [CrossRef]
8. Braccesi, C.; Morettini, G.; Cianetti, F.; Palmieri, M. Development of a New Simple Energy Method for Life Prediction in Multiaxial Fatigue. *Int. J. Fatigue* **2018**, *112*, 1–8. [CrossRef]
9. Berto, F.; Lazzarin, P. Recent Developments in Brittle and Quasi-Brittle Failure Assessment of Engineering Materials by Means of Local Approaches. *Mater. Sci. Eng. R Rep.* **2014**, *75*, 1–48. [CrossRef]
10. Foti, P.; Razavi, N.; Berto, F. Fracture Assessment of U-Notched PMMA under Mixed Mode I/II Loading Conditions by Means of Local Approaches. *Procedia Struct. Integr.* **2021**, *33*, 482–490. [CrossRef]
11. Sánchez, M.; Cicero, S.; Arrieta, S.; Torabi, A.R. Fracture Load Prediction of Non-Linear Structural Steels through Calibration of the ASED Criterion. *Metals* **2023**, *13*, 1211. [CrossRef]
12. Milone, A.; Foti, P.; Filippi, S.; Landolfo, R.; Berto, F. Evaluation of the Influence of Mean Stress on the Fatigue Behavior of Notched and Smooth Medium Carbon Steel Components through an Energetic Local Approach. *Fatigue Fract. Eng. Mater. Struct.* **2023**, *46*, 4315–4332. [CrossRef]
13. Foti, P.; Razavi, N.; Marsavina, L.; Berto, F. Volume Free Strain Energy Density Method for Applications to Blunt V-Notches. *Procedia Struct. Integr.* **2020**, *28*, 734–742. [CrossRef]
14. Ding, K.; Yang, Y.; Wang, Z.; Zhang, T.; Guo, W. Relationship between Local Strain Energy Density and Fatigue Life of Riveted Al-Li Alloy Plate. *Theor. Appl. Fract. Mech.* **2023**, *125*, 103672. [CrossRef]
15. Dugdale, D.S. Yielding of Steel Sheets Containing Slits. *J. Mech. Phys. Solids* **1960**, *8*, 100–104. [CrossRef]
16. Barenblatt, G.I. The Formation of Equilibrium Cracks during Brittle Fracture. General Ideas and Hypotheses. Axially-Symmetric Cracks. *J. Appl. Math. Mech.* **1959**, *23*, 622–636. [CrossRef]
17. Hillerborg, A.; Modéer, M.; Petersson, P.-E. Analysis of Crack Formation and Crack Growth in Concrete by Means of Fracture Mechanics and Finite Elements. *Cem. Concr. Res.* **1976**, *6*, 773–781. [CrossRef]
18. Gómez, F.J.; Elices, M.; Valiente, A. Cracking in PMMA Containing U-shaped Notches. *Fatigue Fract. Eng. Mater. Struct.* **2000**, *23*, 795–803. [CrossRef]

19. Jahanshahi, S.; Chakherlou, T.N.; Rostampoureh, A.; Aalami, M.R. Evaluating the Validity of the Cohesive Zone Model in Mixed Mode I + III Fracture of Al-Alloy 2024-T3 Adhesive Joints Using DBM-DCB Tests. *Int. J. Fract.* **2023**, *240*, 143–165. [CrossRef]
20. Ritchie, R.O.; Knott, J.F.; Rice, J.R. On the Relationship between Critical Tensile Stress and Fracture Toughness in Mild Steel. *J. Mech. Phys. Solids* **1973**, *21*, 395–410. [CrossRef]
21. Liu, Y.; Mahadevan, S. Threshold Stress Intensity Factor and Crack Growth Rate Prediction under Mixed-Mode Loading. *Eng. Fract. Mech.* **2007**, *74*, 332–345. [CrossRef]
22. Taylor, D. *The Theory of Critical Distances: A New Perspective in Fracture Mechanics*; Elsevier: Amsterdam, The Netherlands; Boston, MA, USA, 2007.
23. Susmel, L.; Taylor, D. On the Use of the Theory of Critical Distances to Predict Static Failures in Ductile Metallic Materials Containing Different Geometrical Features. *Eng. Fract. Mech.* **2008**, *75*, 4410–4421. [CrossRef]
24. Susmel, L.; Taylor, D. The Theory of Critical Distances to Predict Static Strength of Notched Brittle Components Subjected to Mixed-Mode Loading. *Eng. Fract. Mech.* **2008**, *75*, 534–550. [CrossRef]
25. Taylor, D.; Merlo, M.; Pegley, R.; Cavatorta, M.P. The Effect of Stress Concentrations on the Fracture Strength of Polymethylmethacrylate. *Mater. Sci. Eng. A* **2004**, *382*, 288–294. [CrossRef]
26. Cicero, S.; Arroyo, B.; Álvarez, J.A.; González, P.; Flores, L.A. Development of a Theory of Critical Distances Based Methodology for Environmentally Assisted Cracking Analyses. In *Advances in Accelerated Testing and Predictive Methods in Creep, Fatigue, and Environmental Cracking*; ASTM International: West Conshohocken, PA, USA, 2023; pp. 277–299.
27. Madrazo, V.; Cicero, S.; Carrascal, I.A. On the Point Method and the Line Method Notch Effect Predictions in Al7075-T651. *Eng. Fract. Mech.* **2012**, *79*, 363–379. [CrossRef]
28. Neuber, H. *Theory of Notch Stresses: Principles for Exact Calculation of Strength with Reference to Structural Form and Material*, 2nd ed.; Springer: Berlin/Heidelberg, Germany, 1958; Volume 292.
29. Peterson, R.E. Notch-Sensitivity. In *Metal Fatigue*; Sines, G., Waisman, J.L., Eds.; McGraw Hill: New York, NY, USA, 1959; Volume 12, pp. 293–306.
30. Taylor, D. The Theory of Critical Distances. *Eng. Fract. Mech.* **2008**, *75*, 1696–1705. [CrossRef]
31. Taylor, D. Geometrical Effects in Fatigue: A Unifying Theoretical Model. *Int. J. Fatigue* **1999**, *21*, 413–420. [CrossRef]
32. Kinloch, A.J.; Shaw, S.J.; Hunston, D.L. Crack Propagation in Rubber-Toughened Epoxy. In Proceedings of the International Conference on Yield, Deformation and Fracture, Buffalo, NY, USA, 29 March–1 April 1982; Cambridge Plastics and Rubber Institute: London, UK, 1982; pp. 1–6.
33. Morettini, G.; Razavi, S.M.J.; Staffa, A.; Palmieri, M.; Berto, F.; Cianetti, F.; Braccesi, C. On the Combined Use of Averaged Strain Energy Density Criteria (ASED) and Equivalent Material Concept (ECC) for the Fracture Load Prediction of Additively Manufactured PLA v-Notched Specimens. *Procedia Struct. Integr.* **2023**, *47*, 296–309. [CrossRef]
34. Li, G.; Jadhav, S.D.; Martín, A.; Montero-Sistiaga, M.L.; Soete, J.; Sebastian, M.S.; Cepeda-Jiménez, C.M.; Vanmeensel, K. Investigation of Solidification and Precipitation Behavior of Si-Modified 7075 Aluminum Alloy Fabricated by Laser-Based Powder Bed Fusion. *Metall. Mater. Trans. A* **2021**, *52*, 194–210. [CrossRef]
35. Goswami, R.; Lynch, S.; Holroyd, N.J.H.; Knight, S.P.; Holtz, R.L. Evolution of Grain Boundary Precipitates in Al 7075 Upon Aging and Correlation with Stress Corrosion Cracking Behavior. *Metall. Mater. Trans. A* **2013**, *44*, 1268–1278. [CrossRef]
36. Di Schino, A.; Montanari, R.; Testani, C.; Varone, A. Dislocation Breakaway Damping in AA7050 Alloy. *Metals* **2020**, *10*, 1682. [CrossRef]
37. Angella, G.; Di Schino, A.; Donnini, R.; Richetta, M.; Testani, C.; Varone, A. AA7050 Al Alloy Hot-Forging Process for Improved Fracture Toughness Properties. *Metals* **2019**, *9*, 64. [CrossRef]
38. Torabi, A.R.; Campagnolo, A.; Berto, F. Mixed Mode I/II Crack Initiation from U-Notches in Al 7075-T6 Thin Plates by Large-Scale Yielding Regime. *Theor. Appl. Fract. Mech.* **2016**, *86*, 284–291. [CrossRef]
39. Tokaji, K.; Ogawa, T.; Kameyama, Y. The Effects of Stress Ratio on the Growth Behaviour of Small Fatigue Cracks in an Aluminium Alloy 7075-T6. *Fatigue Fract. Eng. Mater. Struct.* **1990**, *13*, 411–421. [CrossRef]
40. Smith, C.R. *S-N Characteristics of Notched Specimens*; NASA: Washington, DC, USA, 1966.
41. Akio, O.; Keiichiro, T.; Hidenobu, M. Fatigue Crack Initiation and Growth under Mixed Mode Loading in Aluminum Alloys 2017-T3 and 7075-T6. *Eng. Fract. Mech.* **1987**, *28*, 721–732.
42. Santus, C.; Taylor, D.; Benedetti, M. Experimental Determination and Sensitivity Analysis of the Fatigue Critical Distance Obtained with Rounded V-Notched Specimens. *Int. J. Fatigue* **2018**, *113*, 113–125. [CrossRef]
43. Foti, P.; Berto, F. Evaluation of the Effect of the TIG-Dressing Technique on Welded Joints through the Strain Energy Density Method. *Procedia Struct. Integr.* **2020**, *25*, 201–208. [CrossRef]
44. Torabi, A.R.; Alaei, M. Mixed-Mode Ductile Failure Analysis of V-Notched Al 7075-T6 Thin Sheets. *Eng. Fract. Mech.* **2015**, *150*, 70–95. [CrossRef]

Disclaimer/Publisher's Note: The statements, opinions and data contained in all publications are solely those of the individual author(s) and contributor(s) and not of MDPI and/or the editor(s). MDPI and/or the editor(s) disclaim responsibility for any injury to people or property resulting from any ideas, methods, instructions or products referred to in the content.

Article

Effect of Deformation Degree on Microstructure and Properties of Ni-Based Alloy Forgings

Ruifeng Dong [1,*], Jian Li [1], Zishuai Chen [2], Wei Zhang [1] and Xing Zhou [1]

1. College of Materials Science and Engineering, Inner Mongolia University of Technology, Hohhot 010051, China
2. College of Materials Science and Chemical Engineering, Harbin University of Science and Technology, Harbin 150040, China
* Correspondence: drfcsp@imut.edu.cn

Abstract: The primary objective of this paper is to investigate the influence of deformation degree on the microstructure and properties of a Ni-based superalloy. An upsetting experiment was conducted using a free-forging hammer to achieve a deformation degree ranging from 60% to 80%. The impact of the forging deformation degree on the hardness and high-temperature erosion performance was evaluated using the Rockwell hardness tester (HRC) and high-temperature erosion tester, respectively. The experimental results indicate that as the deformation degree increased, the hardness of the forged material progressively increased while the rate of high-temperature erosion gradually decreased. In order to comprehensively study the mechanism behind the variations in forging performance, optical microscopy (OM), scanning electron microscopy (SEM), electron backscatter diffraction (EBSD), and transmission electron microscopy (TEM) were employed. The findings reveal that as the deformation degree increased, the presence of small-angle grain boundaries and an increase in grain boundary area contributed to enhanced hardness in the alloy forgings. Furthermore, it was discovered that grain boundaries with twin orientation promoted dynamic recrystallization during deformation, specifically through a discontinuous dynamic recrystallization mechanism. Additionally, the precipitated γ' phase in the alloy exhibited particle sizes ranging from 40 to 100 nm. This particle size range resulted in a higher critical shear stress value and a more pronounced strengthening effect on the alloy.

Keywords: deformation degree; erosion rate; low angle grain boundary; precipitated phase

Citation: Dong, R.; Li, J.; Chen, Z.; Zhang, W.; Zhou, X. Effect of Deformation Degree on Microstructure and Properties of Ni-Based Alloy Forgings. *Metals* **2024**, *14*, 340. https://doi.org/10.3390/met14030340

Academic Editors: Andrea Di Schino, Boris B. Straumal and Claudio Testani

Received: 14 August 2023
Revised: 18 September 2023
Accepted: 26 September 2023
Published: 15 March 2024

Copyright: © 2024 by the authors. Licensee MDPI, Basel, Switzerland. This article is an open access article distributed under the terms and conditions of the Creative Commons Attribution (CC BY) license (https://creativecommons.org/licenses/by/4.0/).

1. Introduction

Superalloys are alloys that exhibit exceptional resistance to deformation and maintain their functionality under high external forces and temperatures exceeding 600 °C. The performance of fighter aircraft heavily relies on their cutting-edge aeroengines, which predominantly utilize superalloys. Consequently, research and development on superalloys have become globally significant strategic topics. The United Kingdom, the United States, and the former Soviet Union were the initial pioneers in superalloy production. The United Kingdom, as the forefather of superalloys, has established two distinct systems: cast superalloys and wrought superalloys. The primary utilization is focused on nickel-based wrought superalloys. Simultaneously, the United States played a creative role by introducing powder superalloys, incorporating elements such as cobalt, molybdenum, tungsten, and others into nickel-based alloys to produce various grades of superalloys with different service temperature capabilities. Currently, the United States boasts the widest variety and highest quality of superalloys worldwide. Following suit, China embarked on high-temperature alloy research and development [1–4].

In comparison to international efforts, China's research and development in high-temperature alloys started relatively later. The first high-temperature alloy brand was a replica of the Soviet Union's GH3030. Subsequently, several successful high-temperature alloys were developed, including GH3044, GH4033, and GH2036 [5]. Iron-based superalloys fail to meet the demands of aerospace development due to their lower service temperature. With limited cobalt reserves in the natural environment, the development of cobalt-based superalloys gained prominence. As a result, Chinese researchers shifted their focus to the creation of nickel-based superalloys, with particular attention given to GH4738, a widely utilized nickel-based superalloy [6].

This research investigates the influence of deformation degree on the microstructure and characteristics of nickel-based superalloys, specifically focusing on GH4738 alloy, a phase precipitation hardening type of deformed nickel-based superalloy. Below 760 °C, this alloy exhibits excellent tensile strength and durability, while below 870 °C, it demonstrates outstanding oxidation resistance. Because of its strong crack propagation resistance and corrosion resistance, GH4738 alloy is widely used in turbine disk components of aero engines [7–10]. However, due to the high-volume percentage of the ′ phase (approximately 23.5%), achieving microstructure uniformity during the forging process becomes challenging. The uniform distribution of grains and dispersed ′ phase in forgings is crucial for ensuring superior mechanical properties in wrought superalloys [11,12]. Moreover, GH4738 alloy contains a significant amount of alloy elements, leading to substantial solid solution and precipitation strengthening effects, which further complicates subsequent processing [13]. Thus, reasonable deformation process parameters are essential for achieving a homogeneous microstructure in alloy forgings [14]. Therefore, optimizing the hot deformation process of this alloy holds great importance in controlling grain size, ′ phase size, and distribution after deformation. These optimizations can enhance the alloy's stability, extend its useful life, and improve its overall mechanical properties, considering its high service temperature and creep qualities [15].

Apart from deformation-related investigations, it is also crucial to examine high-temperature erosion wear, which refers to the material loss, fracture, or displacement caused by repetitive impacts of solid particles on a solid surface. Dust ingestion during engine operation accelerates wear and shortens the engine's lifespan. Hence, researching the performance of high-temperature erosion wear is vital for increasing engine longevity [16].

Through quantitative description, Li Wang et al. [17] constructed a thermal processing map to determine the ideal deformation process parameters based on microstructure evolution during thermal deformation. In a systematic investigation of the GH4738 alloy's dynamic recrystallization microstructure distribution under hot processing factors, Liu Hui [18] obtained specific results. Additionally, several studies have been conducted on the rheological mechanical behavior and microstructure evolution of the GH4738 alloy during hot deformation [19–21]. However, all of the aforementioned researchers utilized the Gleeble-1500 thermal simulation testing apparatus, which resulted in a slight disparity between the experimental conditions and actual production. To study the creep behavior of a Ni-based solid solution-reinforced NiMoCr alloy under different hot rolling and cold rolling process conditions, T. Kvackaj et al. [22] discovered that the failure process involves fracture nucleation and crack propagation, both of which are strongly influenced by grain size. It was found that the fine recrystallized structure exhibits significantly lower creep resistance compared to the coarse grain size. Therefore, this study focuses on investigating the hot working process scheme of the GH4738 alloy turbine disk using the free forging process under identical deformation temperatures and varying degrees of deformation. Hot deformation tests were performed on GH4738 alloy bars with different levels of deformation to examine grain refinement behavior and the mechanism of recrystallized grain nucleation post-deformation. Analyzing the alloy's strengthening mechanism after deformation will facilitate the design and optimization of deformation process parameters for GH4738 alloy bars, based on both experimental and theoretical approaches.

2. Experimental Material and Methods

Vacuum consumable remelting (VAR) and vacuum induction melting (VIM) were employed to create the GH4738 alloy that was used in this study. The 200 mm billet was produced after billet opening and homogenization treatment. The chemical make-up of the nickel-based superalloy GH4738 is shown in Table 1.

Table 1. Chemical composition table of GH4738 superalloy (mass %).

Element	C	Cr	Co	Mo	Al	Ti	Fe	Ni
Mass %	0.035	19.41	13.22	4.30	1.35	2.98	1.00	bal

As shown in Figure 1, four small rod blanks, each with a diameter of 45 mm, were extracted at intervals of 200 mm to ensure consistency in the initial tissue state. The rod billet was then heated in a stepwise manner to a temperature of 1140 °C in the heating furnace. Immediately after heating, the billets were taken out and subjected to repeated upsetting using a 1 t free forging hammer. This process transformed the four rod billets into disk forgings at a height of 80 cm from the material. The undercutting rates employed were 5×10^{-2} s^{-1}, 6×10^{-2} s^{-1}, 7×10^{-2} s^{-1}, and 8×10^{-2} s^{-1}. Subsequently, the forged disk forgings were placed in designated positions for cooling, ensuring that they retained their shape without any deformation. The forged disks exhibited no signs of deformation cracks. The specific details of the thermal deformation process are outlined in Table 2.

Table 2. Actual thermal processing parameters of the samples.

Scheme	Deformation Temperature (°C)	Billet Size (mm)	Finished Size (mm)	Deformation Degree (%)
1	1140	φ45 × 110	φ69 × 42	62
2			φ76 × 35	68
3			φ82 × 30	72
4			φ92 × 23	80

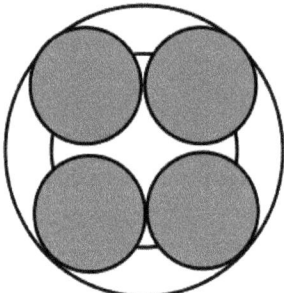

Figure 1. Schematic diagram of the material selection location.

To obtain samples of 10 × 10 × 10 mm dimensions, disc forgings with varying degrees of deformation were cut along the 1/4 diameter direction. Electrolytic polishing, electrolytic corrosion, and polishing were necessary for metallographic observation. Electrolytic removal of surface scratches resulted in a clearer metallographic image. An etching agent of 3 g CuCl$_2$ + 20 mL HCl + 30 mL C$_2$H$_5$OH was used, while the voltage and time for electropolishing, as well as the polishing solution, were both at 20% HCl and 80% CH$_3$OH. SEM and EBSD were utilized for observations. The electrolytic corrosion solution consisted of 170 mL H$_3$PO$_4$ + 10 mL H$_2$SO$_4$ + 15 g Cr$_2$O$_3$ at a voltage of 3–5 V and time range of 5–10 s. The microstructure of the four groups of samples was examined using Zeiss metallographic and scanning electron microscopes (Oberkochen, Germany) to study the size, size distribution, and size variation of precipitated phases and grains under varying degrees of deformation. The nucleation mechanism of recrystallized grains, substructure changes, and γ′ phase evolution during hot deformation of the alloy were studied using EBSD and a transmission electron microscope.

The HR-150A Rockwell hardness tester (Lanzhou Zhongke Kaihua Technology Development Co. China, Lanzhou, China) was utilized to examine the hardness variations of the four sample groups. The process involved testing five samples from each group, eliminating the highest and lowest values, and calculating the average as a representation of the sample's hardness. For the high-temperature erosion wear test, the ASTM G76 high-temperature erosion wear test machine was employed. A 50 mm by 50 mm by 10 mm sample was used for this test. The experimental process entailed the following details: The abrasive temperature was set at 750 °C, with an objective temperature of 750 °C. A total of 2 kg of abrasive was used, and the erosion speed was maintained at 88 m/s. The erosion angle was set to 90°, and the erosion time was 20 s. Under these conditions, the morphology of the GH4738 alloy, subjected to various degrees of distortion due to high-temperature wear, was examined, along with the influence of hardness on high-temperature wear.

To prepare material samples for transmission electron microscopy (TEM) investigation, a square sheet measuring $8 \times 8 \times 2$ mm was thinned using metallographic sandpaper until it reached a thickness of under 65 mm. The electrolyte used for the preparation of these sheet samples in TEM was a mixed solution consisting of 10% $HClO_4$ and 90% C_2H_5OH. The preparation was carried out in a current environment ranging from -40 °C to -35 °C, with a voltage of 50 V and a current intensity of 40 mA to 45 mA.

3. Results and Discussion

3.1. Effect of Deformation Degree on Microstructure of Ni-Based Alloy

The microstructure of the GH4738 alloy under various degrees of deformation is shown in Figure 2. The figure illustrates how, following the upsetting deformation experiment, the sample's grain size steadily reduced as the degree of distortion increased and there was no mixed crystal. The image also shows that the samples with the four deformation degrees had two sizes of grains, with the smaller grains around the bigger grains. This event shows that recrystallization is essentially finished. At the same time, when the degree of deformation is very low, the deformation is too slight, and the stored energy is insufficient to cause recrystallization; thus, the change in grain size is not immediately apparent. The grains are refined following recrystallization when the deformation exceeds the critical deformation, and the bigger the deformation is, the finer the grains become. Because the nucleation rate grows quickly and the stored energy driving nucleation and growth increases constantly as the deformation size increases, refining occurs as the G/N ratio decreases.

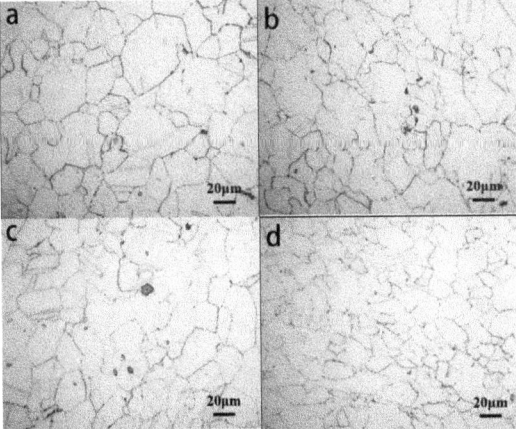

Figure 2. Microstructure of forgings of the GH4738 alloy with different degrees of deformation: (**a**) 62%; (**b**) 68%; (**c**) 72%; (**d**) 80%.

The precipitated γ′ phase diagram of GH4738 alloy forgings at various degrees of deformation is shown in Figure 3. The size of the γ′ phase in forgings steadily reduced with the increasing deformation amount, as indicated in the figure, although the amount of γ′ phase precipitation did not significantly increase. The grain border area was expanded as the grain size was gradually fine-tuned. During the deformation process, this effect encouraged the diffusion of solute atoms. As a result, the size of the γ′ phase in the sample was finer and more uniform, and it had a tendency to be spherical. The foundation for the dispersion distribution of the γ′ phase is laid by the uniform distribution of solute atoms in the matrix [23,24].

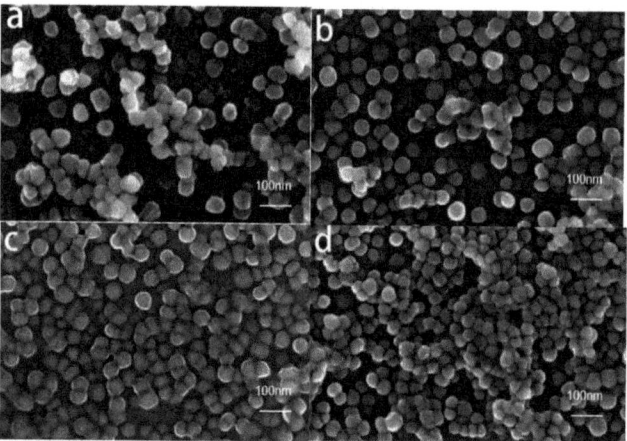

Figure 3. γ′ phase in GH4738 alloy forgings at different deformation levels: (**a**) 62%; (**b**) 68%; (**c**) 72%; (**d**) 80%.

3.2. Effect of Deformation Degree on Properties of Ni-Based Alloy

The average grain size, γ′ phase particle size, and hardness of samples with varying degrees of deformation are compared in Figure 4. The Image J software 2.3.0 (National Institutes of Health, Bethesda, MD, USA) was used to measure the diameter of each grain in the optical microscopy (OM) images, enabling the determination of the average grain size. Measurements were taken from three different angles and typical orientations for each grain, and the average value was derived from these three sets of data. The grains in the images were categorized, and the distribution percentage for each grade was determined based on the ASTM grade list and the measured values. Thus, the average value of grains within the same grade was obtained. The average grain size of the material was calculated by summing the average value and the proportion of grains in each grade. The "phase" amount in the image, referred to as "phase size," was determined using the Image Pro Plus software 6.0 (Media Cybernetics, Rockville, MD, USA) by calculating the percentage of the pixel area occupied by the brightest "phase" in the image. Each test group was counted three times, and the average value was calculated. According to the diagram, significant levels of deformation can result in the formation of fine and uniform grains as well as the γ′ phase. Furthermore, it leads to an increase in hardness from 38.9 HRC to 44 HRC. Grain refinement serves to provide a strengthening effect through the presence of fine grains. Additionally, under substantial degrees of deformation, complete recrystallization may occur, resulting in the formation of a strong material [25,26].

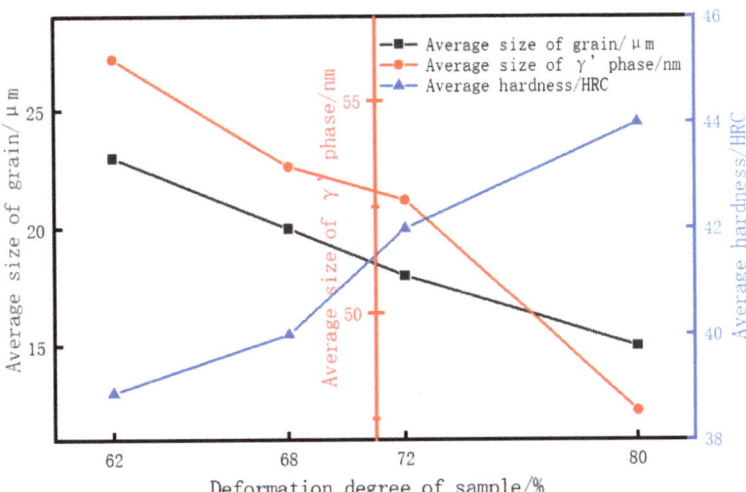

Figure 4. Comparison of average grain size, γ′ phase particle size, and hardness of samples with different deformation degrees.

Figure 5 presents the morphology of high-temperature erosion observed using various methods. The image reveals an increased number of craters and a slight cutting effect, indicating the prevalence of impact deformation mechanisms. Figure 6 illustrates the impact of deformation degree on the hardness and high-temperature wear rate of GH4738 alloy. It can be observed that as the degree of deformation increased, so did the hardness. Calculations demonstrate that the hardness at 80% deformation was 14.3% higher than that at 62% deformation. At a deformation degree of 62%, the sample surface exhibited low strength, making it vulnerable to erosion and wear. Consequently, the erosion pits displayed a significant amount of cutting, which limits plastic deformation. Moreover, through computation and comparison, it was determined that the sample with an 80% deformation degree exhibited a reduced high-temperature erosion rate, which decreased by 10.3% compared to the sample with a 62% deformation degree. Furthermore, it is worth noting that the harder the alloy is, the faster it tends to corrode at high temperatures. The mechanism of partial dislocation cutting the small size phase of the GH4738 alloy under various processes undergoes changes during high-temperature erosion at 750 °C. This leads to a decrease in intracrystalline strength to some extent but an increase in plasticity at high temperature. However, the grain boundary strengthening mechanism provides better resistance against high-temperature erosion. When the content of the phase is close to or equal to the grain boundary fraction, grain boundary strengthening becomes the dominant mechanism in enhancing resistance to high-temperature erosion.

3.3. The Nucleation Mechanism of Recrystallized Grains in the Alloy and the Evolution of Substructure Inside the Grains

Researchers can utilize Electron Backscatter Diffraction (EBSD) to investigate the mechanism of dynamic recrystallization in alloys by analyzing orientation, microstrain, and dislocation density. Dislocation density provides insights into the presence of substructures within a specific area and can be determined by comparing the cumulative orientation angle along a designated line. In a study conducted by Azarbarmas et al. [27], the EBSD technique was employed to analyze the dynamic recrystallization behavior of In718 alloy during thermal deformation. By examining the cumulative orientation angle within the deformed grains, it was observed that most subgranular boundaries were still forming at lower strains, indicating the absence of continuous dynamic recrystallization (CDRX). However, as dynamic recrystallization (DRX) progressed under higher strains, strain-free grains emerged, leading to a gradual decrease in dislocations and a reduction in the occurrence of small-angle grain boundaries. At this stage, medium to high angular dislocation boundaries were observed near the original grain boundaries, signifying the influence of successive progressive subgrain rotations and promoting CDRX at higher strains. Hence, the primary deformation mechanism identified for the In718 alloy at lower strains is the dynamic discontinuous recrystallization (DDRX) mechanism. However, as the strain increases, the prevalence of the CDRX mechanism becomes more pronounced.

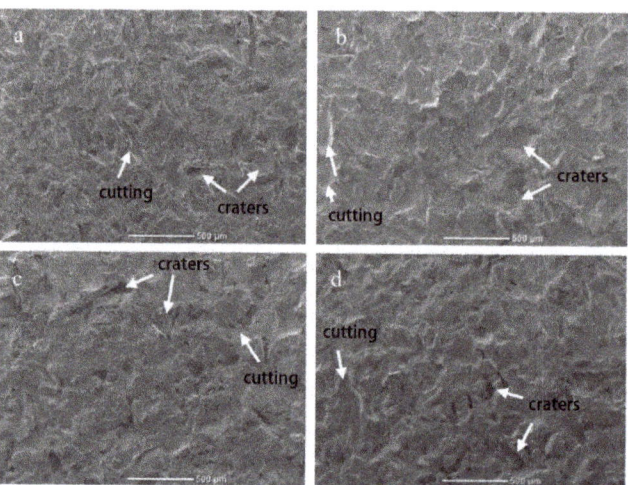

Figure 5. High-temperature erosion morphology of specimens with different degrees of deformation: (**a**) 62%; (**b**) 68%; (**c**) 72%; (**d**) 80%.

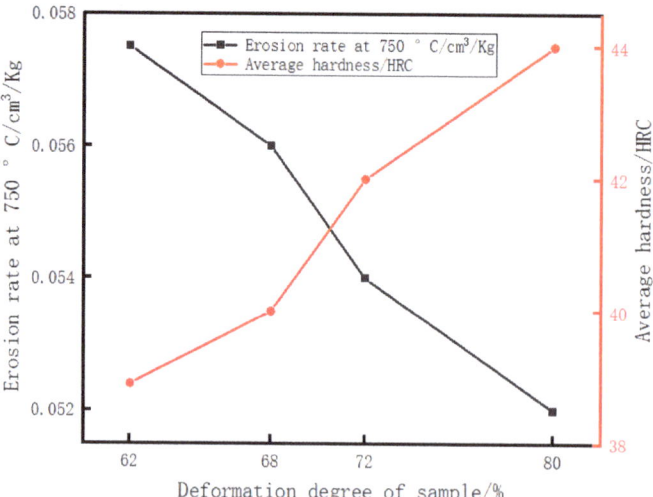

Figure 6. Effect of deformation degree on hardness and high temperature wear rate of the GH4738 alloy, cm^3/Kg.

The distribution diagram in Figure 7 shows the grain boundaries of the GH4738 alloy with different phase differences, while the EBSD diagram displays the alloy's microstructure. It can be observed from Figure 7a that the forging had a consistently small grain size. The presence of carbides inhibited the migration of original grain boundaries, leading to a significant increase in the nucleation rate of recrystallized grains. Recrystallization in this alloy was primarily achieved through dynamic recrystallization (DDRX). Figure 7b reveals that large angle grain boundaries accounted for 43% of the deformation in GH4738 alloy forgings, while small-angle grain boundaries accounted for 56%. There were relatively few medium-angle grain boundaries, and their characteristics were similar. Figure 7c illustrates the distribution of various phase differences (ranging from 0° to 2–5°) at low-angle grain boundaries in GH4738 alloy. When the proportion of 2° phase differences in these boundaries reaches around 80%, it indicates the presence of subgrain boundaries and dislocation substructures in the alloy. During the hot processing of highly deformed alloys, a large number of dislocations interact with grain boundaries, resulting in dislocation entanglement and stacking, which contributes to the development of substructures within the alloy [28]. In metals with low stacking fault energy, complete dislocations tend to transform into partial dislocations with lower energy, thereby promoting the formation of twin structures and improving overall performance.

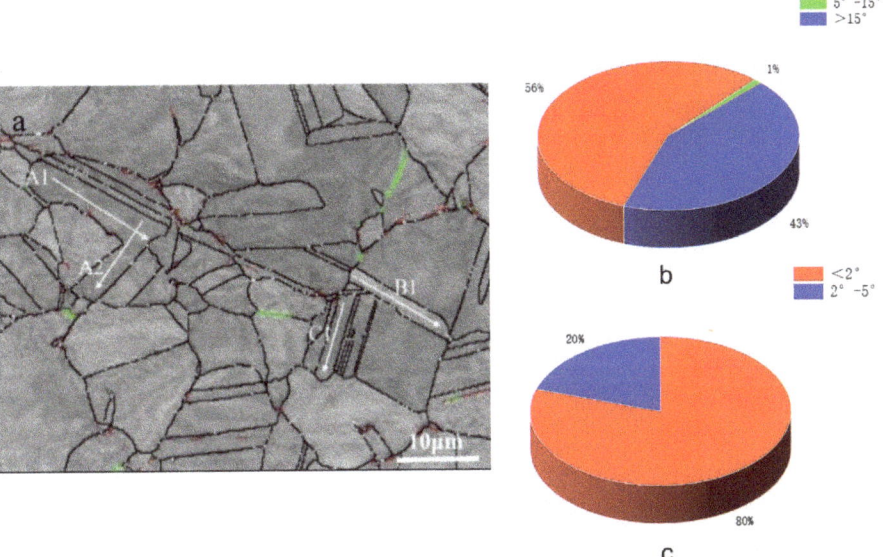

Figure 7. EBSD diagram and distribution pattern of grain boundaries with different phase differences for GH4738 alloy. (**a**) EBSD diagram of forgings with a deformation of 80%; (**b**) the distribution of grain boundaries with orientation differences in different states of the alloy; (**c**) the distribution of <2° and 2–5° orientation difference grain boundaries in different states of the alloy.

The distribution of the orientation angle along Figure 7a along straight lines is illustrated in Figure 8. According to the analysis of cumulative orientation angles in figures, the highest cumulative orientation angle of deformed grains in GH4738 alloy forgings was only 5.8°. During the heating process prior to forging, the second phase in the alloy and the majority of primary carbides have completely returned to the matrix [29,30]. Consequently, there was minimal presence of precipitated phases impeding the grain boundary during deformation. The primary mechanism for the nucleation of recrystallized grains was the initial bending of the grain boundary, while the major nucleation process was the discontinuous dynamic recrystallization mechanism. The granular $M_{23}C_6$ carbides were dispersed and precipitated on the recrystallized grain boundaries and twin boundaries during the cooling process following deformation. This formed the basis for alloy strengthening through the pinning of recrystallized grain boundaries and twin boundaries.

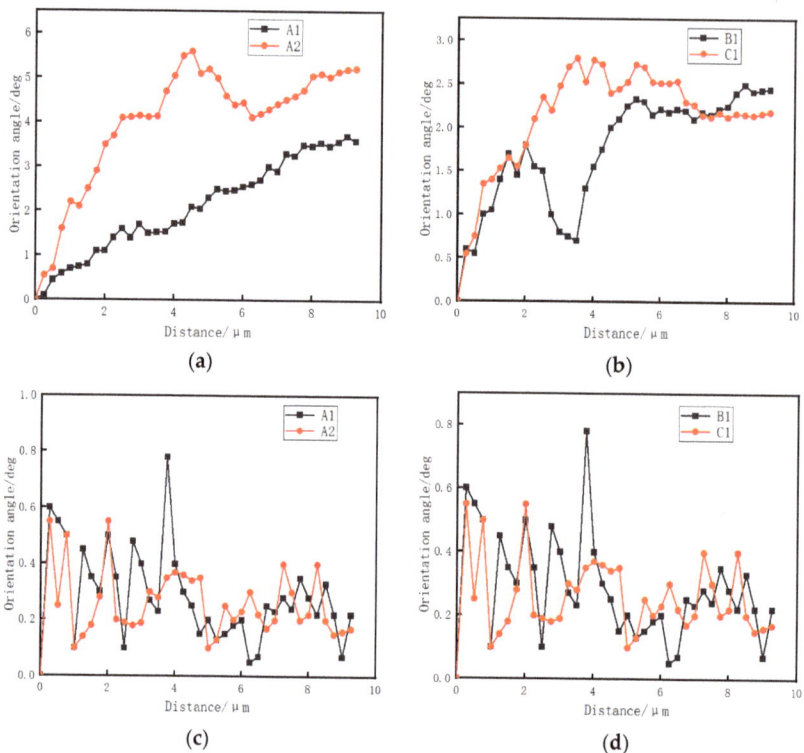

Figure 8. Distribution of orientation angles along the direction of the straight-line segment in Figure 6. ((**a**,**b**) Point to origin, (**c**,**d**) point to point).

The nickel-based superalloy layer exhibited a low fault energy, which enhanced the likelihood of twin formation during deformation. The dynamic and static recrystallization processes lead to the creation of different types of stepped grain boundaries due to the transition from regular crystal to the low refractive index crystal plane. This transition reduces energy and widens the grain boundary [31]. Twin formation reduces the grain boundary energy of recrystallized grains, thereby promoting dynamic recrystallization. In the heat deformation process of alloy GH4738, primary Σ3 twin boundaries with a 60° orientation and Σ9 twin boundaries with a 28.9° structure were observed. The "coincidence site lattice" rule [32] states that the interaction of two Σ3 twin boundaries can generate a Σ9 twin boundary, the interaction of a Σ3 and Σ9 twin boundary can produce a Σ3 twin boundary, and a Σ27 twin boundary can be formed. Higher-order twin boundaries between neighboring twins may appear at triple junctions. These twin boundaries contribute to the material's hardening effect and lower the average free energy of dislocations.

The pinning effect of the strengthening phase on the grain boundary is determined by the interfacial energy of the grain boundary when it interacts with the strengthening phase, as described in Zener's principle [33]. In nickel-based superalloys, the interfacial energy of high-angle grain boundaries is typically 0.69 J/m². In contrast, coherent twin barriers have a much lower interface energy of only 0.03 J/m². Consequently, the pinning effect of the strengthening phase on the twin boundary is relatively weaker. Figure 9 shows that the forgings contained additional subgrain boundaries and dislocation substructures. Twin boundaries are usually absent in recrystallized grains of smaller sizes but are more likely to appear in larger grains, suggesting that they form along with expanding recrystallized grains. The creation of twin boundaries is commonly observed at three-fork grain boundaries in forgings, as it reduces the system's interface energy. The majority of the twin boundaries in alloy forgings are primary three twin boundaries, with a few high-order Σ9 twin boundaries and scarce high-order Σ27 twin boundaries. The development of twin boundaries is also influenced by the degree of deformation, which promotes the nucleation of recrystallized grains. The most notable consequence of twin boundaries is the expansion of the grain boundary region, leading to enhanced alloy strength [34].

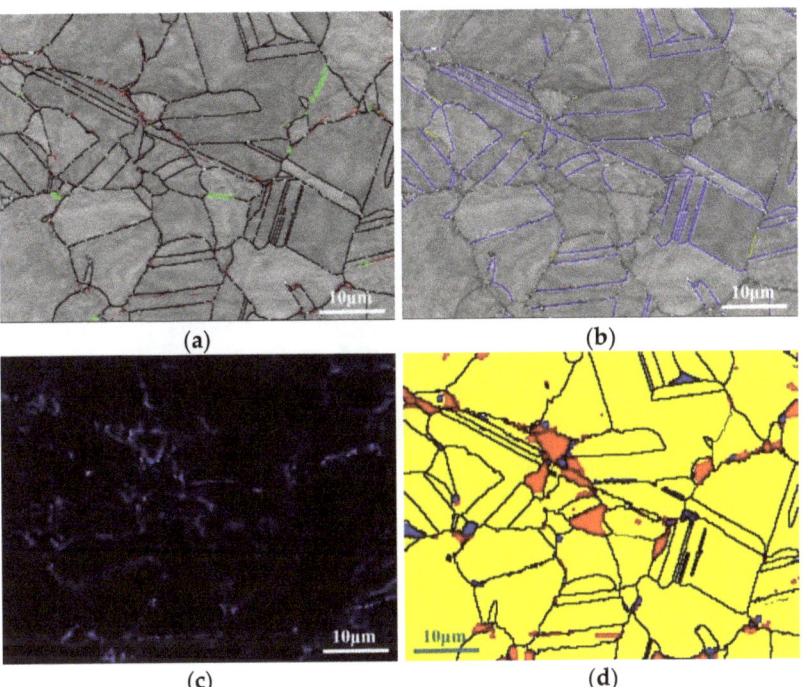

Figure 9. EBSD diagram of GH4738 alloy forgings with 80% deformation. (**a**) Angle diagram of orientation difference between grains (>15°, 5~15°, and 2~5° black line, green line, and red line are used in turn); (**b**) Σ3, Σ9 twin boundary (indicated by blue line and yellow line in turn); (**c**) dislocation density diagram between grains (increased from blue to white in turn); (**d**) distribution of recrystallized grains (indicated by blue, yellow, and red areas for complete recrystallization, substructure, and incomplete recrystallization in turn).

3.4. γ′ Phase Evolution in Nickel-Based Superalloys

The alloy was enriched with approximately 4.5 wt% of Al and Ti elements, which combined to form the γ′ phase. The γ′ phase adopted a face-centered cubic ordered structure and could only exist in the matrix due to its compatibility with the matrix. The primary component of the γ′ phase is Ni3(Al, Ti), and the content and dissolution temperature of the γ′ phase vary based on the Al and Ti content. In this study, the dissolution temperature of the Al + Ti element was approximately 1040 °C [35,36]. The main method of strengthening the alloy was through the precipitation strengthening of the γ′ phase. The performance of the alloy is heavily influenced by the quantity, distribution, and size of the γ′ phase within it. In the research on strengthening GH4720Li alloy [37–39], two types of coupling cutting relationships were observed between the γ′ phase and dislocations: strong coupling dislocation cutting and weak coupling cutting, depending on whether the particle size of the γ′ phase was greater than or equal to 40 nm. It has been found that the optimum particle size for the γ′ phase in the best precipitation strengthening state is 40 nm. Within this range, the critical shear stress value ranges from 70% to 100% of the peak value, indicating a significant strengthening effect. Even in the range of 30–100 nm, the particle size of the γ′ phase still contributes to a notable strengthening impact. Although there may be differences in composition and shape, the essential shear stress and particle size discussed above are also applicable to the γ′ phase in the GH4738 alloy, as it differs from the γ′ phase in the GH4720Li alloy primarily in terms of quantity.

The TEM image of the GH4738 alloy forgings in Figure 10 reveals a dislocation spacing of approximately 40 nm. The most effective strengthening occurred when the initial dislocation cut into the γ′ phase, while the subsequent dislocation simply sheared off the γ′ phase. This observation further confirms that 40 nm is the optimal particle size for the γ′ phase, and this size significantly influences the precipitation strengthening of the alloy. Consequently, as the level of deformation increased, the size of the γ′ phase particles and the spacing between them decreased, eventually converging toward 40 nm. This reduction in particle size and spacing enhanced the performance of the alloy.

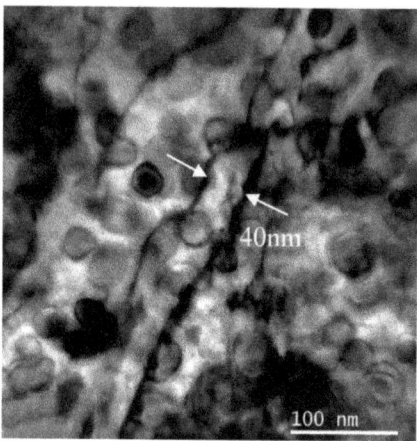

Figure 10. TEM image of GH4738 alloy forgings with an 80% deformation degree.

4. Conclusions

(1) The grain size of nickel-based superalloys and the particle size of the "phase" steadily shrunk as the degree of deformation increased, and the hardness gradually rose. At 80% deformation, the hardness was 14.3% greater than it was at 62% distortion.

(2) Chiseling and a limited amount of cutting were the major characteristics of high-temperature erosion at 750 °C, which was mostly dependent on the impact deformation mechanism. The sample with an 80% deformation degree had a lower high-temperature erosion rate, which was decreased by 10.3% when compared to the sample with a 62% deformation degree. Furthermore, the harder the alloy was, the faster it corroded at high temperatures.

(3) The method of grain recrystallization used in Ni-based alloy forgings is known as discontinuous dynamic recrystallization, and it is characterized by a high dislocation density at the recrystallized grain boundaries. As a result, the sample's microstructure frequently had numerous substructures and twin borders. This structure effectively refined the grains and increased the grain boundary area, which increased the alloy's functionality even more.

(4) As the degree of deformation increased during the hot working of the Ni-based alloy, the particle size and dislocation spacing of the phase decreased. According to studies, the 'phase's critical shear stress value is high, and its particle size ranges from 40 to 100 nm, which has a favorable strengthening impact on the alloy.

Author Contributions: Data curation, Z.C. and W.Z.; Writing—original draft, J.L.; Writing—review & editing, R.D.; Supervision, X.Z. All authors have read and agreed to the published version of the manuscript.

Funding: This research was funded by the Inner Mongolia Autonomous Region Science and Technology Plan Project, grant number 2021GG0266, and Inner Mongolia University of Technology 2022 Science and Technology Innovation Fund for Undergraduates, grant number 88.

Data Availability Statement: Data is contained within the article.

Conflicts of Interest: The authors declare no conflict of interest.

References

1. Qian, Y.-H.; Li, H.-K. *Superalloy*; Metallurgical Industry Press: Beijing, China, 2022; pp. 1–99.
2. Wang, H.-Y.; An, Y.-Q.; Li, C.-Y. Research Progress of Ni-based Superalloys. *Mater. Rep.* **2011**, *25*, 482–486.
3. Qu, J.-L.; Yi, C.-S.; Chen, J.-W. Research progress of precipitated phase in GH4720Li superalloy. *J. Mater. Eng.* **2020**, *48*, 73–83.
4. Du, J.; Zhao, G.; Deng, Q.; Lv, X.; Zhang, B. Development of Wrought Superalloy in China. *J. Aeronaut. Mater.* **2016**, *36*, 27–39.
5. Ma, J.; Li, W.; Zhang, X.; Kou, H.; Shao, J.; Geng, P.; Deng, Y.; Fang, D. Tensile properties and temperature-dependent yield strength prediction of GH4033 wrought superalloy. *Mater. Sci. Eng. A* **2016**, *676*, 165–172. [CrossRef]
6. Guo, J.-T. *Materials Science and Engineering for Superalloys*; Science Press: Beijing, China, 2010.
7. Shi, C.-X.; Zhong, Z.-Y. Development and innovation of superalloy in China. *Acta Metall. Sin.* **2010**, *46*, 1281–1288. [CrossRef]
8. Liu, Y.; Zhang, Y.-A.; Wang, W.; Li, D.-S.; Ma, J.-Y. Influence of rare earth Y on microstructure and high temperature oxidation behavior of Ni-Fe-Co-Cu alloy. *Chin. J. Rare Met.* **2020**, *44*, 9–15.
9. Velikanova, N.-P.; Protasova, N.-A.; Velikanov, P.-G.; Kiselev, A.S.; Salih, S.I.S. Influence of operating time on the strength reliability of the gas generator turbine disk of the aircraft drive for the GPA. *J. Phys. Conf. Ser.* **2021**, *1891*, 012044. [CrossRef]
10. Li, C.-M.; Tan, Y.-B.; Zhao, F. Modification of Flow Stress Curve and Processing Maps of Inconel 718 Superalloy. *Chin. J. Rare Met.* **2020**, *44*, 585–596.
11. Liu, W.-C.; Xiao, F.-R.; Yao, M.; Chen, Z.-L.; Jiang, Z.-Q.; Wang, S.-G. Relationship between the lattice constant of γ phase and the content of ö phase, γ″ and γ′ phases in inconel 718. *Scr. Mater.* **1997**, *37*, 59–64. [CrossRef]
12. Yoo, B.; Im, H.-J.; Seol, J.-B.; Choi, P.-P. On the microstructural evolution and partitioning behavior of L12-structured γ′-based Co-Ti-W alloys upon Cr and Al alloying. *Intermetallics* **2019**, *104*, 97–102. [CrossRef]
13. Su, X.; Lv, X.-D. Thermal deformation behavior of as-cast and as-forged GH4738 alloy. *Heat Treat. Met.* **2021**, *46*, 46–52.
14. Li, H.-Y.; Dong, J.-X.; Li, L.-H. Evolution of microstructure and hot deformation behavior of GH4738 alloy during homogenization. *Trans. Mater. Heat Treat.* **2017**, *38*, 61–69.
15. Goodfellow, A.-J.; Galindo-Nava, E.-I.; Christofidou, K.-A.; Jones, N.G.; Boyer, C.D.; Martin, T.L.; Bagot, P.A.J.; Hardy, M.C.; Stone, H.J. The effect of phase chemistry on the extent of strengthening mechanisms in model Ni-Cr-Al-Ti-Mo based superalloys. *Acta Mater.* **2018**, *153*, 290–302. [CrossRef]
16. Dai, C.-W.; Ding, W.-F.; Xu, J.-H.; Fu, Y.-C.; Yu, T.-Y. Influence of Grain Wear on Material Removal Behavior during Grinding Nickel-based Superalloy with a Single Diamond Grain. *Int. J. Mach. Tools Manuf.* **2017**, *113*, 49–58. [CrossRef]
17. Li, W.; Yang, G.; Lei, T.; Yin, S.-B.; Wang, L. Hot Deformation Behavior of GH738 for A-USC Turbine Blades. *J. Iron Steel Res.* **2015**, *22*, 1043–1048.
18. Liu, H.; Cai, X.-Y. Effects of Hot Working Parameters on Dynamic Recrystallization Behaviors of GH738 Alloy. *J. Iron Steel Res.* **2014**, *26*, 46–50.
19. Jiang, H.; Li, Y.-J.; Liu, Q.-Y.; Dong, J.-X. Effect of Final Forging Temperature on Hot Deformation Behavior of GH4738 Superalloy. *Rare Met. Mater. Eng.* **2021**, *50*, 2552–2556.
20. Wang, L.-Y.; Wang, J.-G.; Liu, D.; Wang, H.-P.; Xu, W.-S.; Zhao, X.-D.; Tong, J.; Qin, W.-D. Characterization and verification on hot working process windows for GH4738 alloy. *Forg. Stamp. Technol.* **2021**, *46*, 207–215.
21. Wang, W.-W.; Yi, Y.-P.; Li, P.-C.; Huang, S.-Q. Simulation on Forging Process for Superalloy Waspaloy Turbine Disc Using Composite Sheathed Technology. *Hot Work. Technol.* **2011**, *40*, 22–25.
22. Kvackaj, T.; Zrnik, J.; Vrchovinsky, V. Influence of plastic deformation on creep behaviour of NiMoCr alloy. *High Temp. Mater. Process.* **2003**, *22*, 57–62. [CrossRef]
23. Wang, H.-P.; Liu, D.; Shi, Y.-Z.; Wang, J.; Yang, Y.; Wang, L.; Qin, W. Matrix-Diffusion-Controlled Coarsening of the γ′ Phase in Waspaloy. *Met. Mater. Int.* **2019**, *25*, 1410–1419. [CrossRef]
24. Yao, Z.-H.; Dong, J.-X.; Zhang, M.-C.; Hong, C.-M. Influence of solution and stabilization heat treatment on carbide and gamma prime for super alloy GH738. *Trans. Mater. Heat Treat.* **2013**, *34*, 43–49.
25. Jackson, M.P.; Reed, R.C. Heat treatment of UDIMET 720Li: The effect of microstructure on properties. *Mater. Sci. Eng. A* **1999**, *259*, 85–97. [CrossRef]
26. Decker, R. The evolution of wrought age-hardenable superalloys. *JOM* **2006**, *58*, 32–36. [CrossRef]
27. Azarbarmas, M.; Aghaie-Khafri, M.; Cabrera, J.M.; Calvo, J. Dynamic recrystallization mechanisms and twining evolution during hot deformation of Inconel 718. *Mater. Sci. Eng. A* **2016**, *678*, 137–152. [CrossRef]
28. Drapier, J.; Coutsouradis, D.; Habraken, L. Measurements of stacking-fault energies in CoNi and CoNi-Cr alloys. *Acta Metall.* **1967**, *15*, 673–675. [CrossRef]
29. Wan, Z.-P. Study on High Temperature Deformation Behavior and Microstructure and Properties Control of GH4720LI Nickel-Based Alloy. Ph.D. Thesis, Harbin Institute of Technology, Harbin, China, 2019.

30. Xiang, S.; Ju, Q.; Liu, G.-Q. Dynamic recrystallization mechanism of 15Cr-25Ni-Fe base superalloy. *J. Jiangsu Univ.* **2010**, *31*, 665–669.
31. Shi, C.-Y.; Zhang, M.-C.; Guo, J. Recrystallization Mechanism of Typical Ni-based Superalloys. *Rare Met. Mater. Eng.* **2023**, *52*, 63–73.
32. Chen, X.-F.; Yao, Z.-H.; Dong, J.-X.; Shen, H.-W.; Wang, Y. The effect of stress on primary MC carbides degeneration of Waspaloy during long term thermal exposure. *J. Alloys Compd.* **2018**, *735*, 928–937. [CrossRef]
33. Zener, C.; Hollomon, J.H. Effect of Strain Rate Upon Plastic Flow of Steel. *J. Appl. Phys.* **1944**, *15*, 22–32. [CrossRef]
34. Chen, Z.-S. Effect of Hot Working Process on Microstructure and Properties of GH4738 Superalloy. Master's Thesis, Inner Mongolia University of Technology, Hohhot, China, 2021.
35. Chen, Z.-Q.; Tai, Q.-A.; Zhao, X.-D.; Wang, J.-Y.; Li, C.-Y. Precipitation and dissolution on carbide for superalloy GH738. *J. Iron Steel Res.* **2013**, *25*, 37–45.
36. Liu, X.-B.; Wang, T.; Chen, S.; Wei, K.; Fu, S.-H.; Li, Z.; Wan, Z.-P. Effect of Al and Ti elements on γ' phase dissolution law of GH4738 alloy during heat treatment. *Hot Work. Technol.* **2022**, *51*, 93–96.
37. Chen, Z.S.; Dong, R.F.; Li, J.N.; Deng, X.T.; Zhang, W.; Ren, X.L. Reasons for the stable existence of gamma' phase and strengthening mechanism of GH4720Li nickel-based superalloy. *Mater. Res. Experss* **2021**, *8*, 056508.
38. Li, J.-N. Effect of Hot Working Technology on Microstructure and Properties of GH4720Li Superalloy and Research on Its Mechanism. Master's Thesis, Inner Mongolia University of Technology, Hohhot, China, 2020.
39. Zhang, W.; Li, J.-N.; Dong, R.-F.; Chen, Z.-S.; Li, J.; Zhou, X.; Wang, Q.-Z.; Qu, J.-L. Effect of heat treatment process parameters on the microstructure and properties of GH4720Li superalloy. *Mater. Res. Experss* **2023**, *10*, 016514.

Disclaimer/Publisher's Note: The statements, opinions and data contained in all publications are solely those of the individual author(s) and contributor(s) and not of MDPI and/or the editor(s). MDPI and/or the editor(s) disclaim responsibility for any injury to people or property resulting from any ideas, methods, instructions or products referred to in the content.

Article

Improved High-Temperature Stability and Hydrogen Penetration through a Pd/Ta Composite Membrane with a TaTiNbZr Intermediate Layer

Haoxin Sun [1], Bo Liu [1,*] and Guo Pu [2,*]

[1] Key Laboratory of Radiation Physics and Technology of Ministry of Education, Institute of Nuclear Science and Technology, Sichuan University, Chengdu 610064, China; sunhaoxin1024@163.com
[2] School of National Defense Science and Technology, Southwest University of Science and Technology, Mianyang 621010, China
* Correspondence: liubo2009720@sohu.com (B.L.); puguo0605@163.com (G.P.)

Abstract: In the hydrogen separation membrane, a dense TaTiNbZr amorphous layer was prepared between Pd and Ta to form a Pd/TaTiNbZr/Ta membrane system to prevent the reaction between Pd and Ta at high temperatures. The structural and chemical stability of the Pd/TaTiNbZr/Ta film system at high temperatures were investigated by annealing at 600 °C for 24 h. The high-temperature hydrogen permeation properties of the Pd/TaTiNbZr/Ta film systems were investigated by hydrogen permeation experiments at 600 °C after heat treatment for 6 h. The TaTiNbZr layer was significantly hydrogen-permeable. With the increase in the thickness of the barrier layer, the hydrogen permeability of Pd/TaTiNbZr/Ta decreased, but its hydrogen permeation flux was smaller than that of the highest value of Pd/Ta when it reached the steady state. The presence of the TaTiNbZr layer effectively blocks the interdiffusion between Pd and Ta to form $TaPd_3$, improving the sustained working ability of the Pd/TaTiNbZr/Ta membrane system. The results show that TaTiNbZr is a candidate material for the intermediate layer to improve the high-temperature stability of metal-composite hydrogen separation membranes.

Keywords: hydrogen permeation; barrier layer; high-entropy alloy

1. Introduction

Against the backdrop of today's growing energy challenges, hydrogen energy is attracting attention as a clean and efficient form of energy. There are still many challenges to realizing the widespread use of hydrogen energy in the energy transition, including technical issues in the production, purification, storage, transportation and utilization of hydrogen [1,2]. Among them, hydrogen purification has been receiving a lot of attention from researchers. Hydrogen separation membranes are key components of membrane reactors for the separation of hydrogen and its isotopes, e.g., for the recovery of deuterium and tritium from fusion reactors and for the steam reforming of natural gas. In previous studies [3–6], researchers have proposed solid alloy membrane separation technology to improve the recovery of hydrogen isotopes. Alloy film separation of hydrogen is characterized by high hydrogen selectivity, high hydrogen permeability, and good chemical stability [7–9]. Therefore, the development of new alloy membranes for the extraction and purification of hydrogen-permeable isotopes is a hot topic in current research.

Thick Pd layers reduce the rate of hydrogen penetration and have a high material cost [10]. Therefore, the development of new ultrathin films with good mechanical properties is the main goal of many researchers in this field. This goal is usually achieved by incorporating a thin Pd layer on the surface of the alloy [3–6,11–15]. Group V transition metals theoretically have a higher permeability to hydrogen atoms than Pd, a feature that has been experimentally demonstrated in a number of systems [4]. Cooney et al. showed

Citation: Sun, H.; Liu, B.; Pu, G. Improved High-Temperature Stability and Hydrogen Penetration through a Pd/Ta Composite Membrane with a TaTiNbZr Intermediate Layer. *Coatings* **2024**, *14*, 370. https://doi.org/10.3390/coatings14030370

Academic Editors: Andrea Di Schino and Claudio Testani

Received: 27 February 2024
Revised: 18 March 2024
Accepted: 19 March 2024
Published: 20 March 2024

Copyright: © 2024 by the authors. Licensee MDPI, Basel, Switzerland. This article is an open access article distributed under the terms and conditions of the Creative Commons Attribution (CC BY) license (https://creativecommons.org/licenses/by/4.0/).

that the high permeability of group V transition metals is attributed to their more open body-centered cubic (BCC) structure with greater hydrogen solubility and faster diffusion rates [4]. In addition, the cost of these metals is much lower than that of pure Pd. Due to the high H_2 permeability and low cost of these BCC metals, thicker films can be used as support materials. When Pd/Ta multilayers were subjected to hydrogen permeation experiments at 500 °C, the decrease in the hydrogen permeation ability of Pd/Ta was mainly due to the formation of compounds by the interdiffusion of Pd and Ta, leading to a loss of the catalytic ability of Pd [4–6,14]. Therefore, inserting a diffusion barrier layer between Pd and Ta is a strategy that can prevent the formation of PdTa alloys. Nozaki et al. [16] chose to use hafnium nitride (HfN) as a barrier layer and developed a Pd/HfN/Ta membrane system, and after annealing at different temperatures, the results showed that the HfN layer exhibited significant hydrogen permeability, with the shortcoming that the hydrogen permeability of HfN was smaller than that of Pd and Ta. Nozaki et al. [14] analyzed the effect of the thickness of the interlayer HfN on the hydrogen permeability and showed that HfN samples with a thickness of 50 nm had a better blocking effect than those of 20 nm. However, for temperatures above 600 °C, HfN cannot effectively block Ta-Pd interdiffusion. In interconnect structures, a diffusion barrier with high thermal stability is very important to inhibit Ta-Pd interdiffusion. Because of the microstructural defects of these conventional barrier materials, higher requirements are placed on suitable barrier layers. A large numbers of reports [17–19] revealed that multi-principal component alloys with amorphous structures are outstanding diffusion barriers, and that amorphous high-entropy alloys (HEAs) have excellent properties as diffusion barriers. Based on the above research, a Pd/TaTiNbZr/Ta multi-layer was constructed to suppress the interdiffusion between the Pd and Ta membrane in the current work. According to previous studies and theoretical calculations, The group V transition metals (Ta, Ti, and Nb) have better hydrogen permeability than Pd [20]. Jayalakshmi et al. [15] investigated that a Ni-Nb-Zr-Ta alloy with a low Zr content demonstrated a larger dilatation of the amorphous structure upon hydrogen charging. Dock-Young Lee et al. [21] also found that Zr has a large atomic size, and the introduction of Zr in the amorphous matrix increases the free volume, which is conducive to the diffusion of hydrogen. Therefore, a TaTiNbZr high-entropy alloy was prepared as a barrier layer to inhibit the interdiffusion of Pd and Ta. The TaTiNbZr amorphous alloy has the characteristics of complicated compositions, hysteretic diffusion, and serious lattice distortion. In addition, TaTiNbZr amorphous alloys with free grain boundaries can be used as a diffusion barrier layer [22]. Nevertheless, there is the question as to whether the TaTiNbZr intermediate layer could inhibit the hydrogen permeability on account of its distinct sluggish diffusion effect for the solute or solvent atoms, even in high-temperature conditions.

In this study, Pd/TaTiNbZr/Ta multilayers were prepared using the magnetron sputtering technique. The hydrogen permeability, structure, and chemical stability of the Pd/TaTiNbZr/Ta films were investigated at 600 °C. The purpose of this study is to test whether the high-entropy alloy TaTiNbZr barrier layer can penetrate hydrogen, as well as the effect of the high-entropy alloy TaTiNbZr barrier layer on the high-temperature stability of Pd films. The Pd/TaTiNbZr/Ta samples with different barrier thicknesses (50 nm, 100 nm, 150 nm) were heat-treated under vacuum at 600 °C, and the hydrogen absorption of Pd/TaTiNbZr/Ta with barrier thickness of 50 nm and 100 nm was measured at 600 °C. The changes in the morphology and surface state of the alloy after heat treatment were also studied.

2. Experimental Details

2.1. Specimen Preparation

For the study of structural and chemical stability, Pd/TaTiNbZr/Ta multilayer thin film samples were prepared by magnetron sputtering on a single-crystal silicon substrate (Si). For the study of hydrogen permeability, Pd/TaTiNbZr/Ta multilayer films were prepared using a 300 μm Ta sheet as a substrate. The substrate was put into acetone and anhydrous ethanol, respectively, and cleaned in an ultrasonic cleaning machine for 10 min. After the ultrasonic cleaning was completed, the substrate was repeatedly rinsed with deionized water and dried for use. At room temperature, Pd/TaTiNbZr/Ta films were deposited on the substrate surface by magnetron sputtering. Before the deposition experiment, the vacuum chamber pressure was controlled below 4.0×10^{-4} Pa and then stabilized at 0.4 Pa through 38 sccm pure argon gas. The TaTiNbZr target was prepared by powder metallurgy, and the target raw materials were 99.99% pure titanium (Ti), zirconium (Zr), niobium (Nb), and tantalum (Ta) powder. The sputtering powers of the Pd, Ta, and TaTiNbZr targets were all 200 W, and deposition distance was 10 cm. The load bias was −60 V. The thickness of the TaTiNbZr barrier layer was regulated by controlling the deposition time, and three kinds of samples were prepared: #1: Pd/TaTiNbZr (50 nm)/Ta, #2: Pd/TaTiNbZr (100 nm)/Ta, and #3: Pd/TaTiNbZr (150 nm)/Ta.

2.2. Annealing and Hydrogen Permeation Test

The deposited multilayer film was selected for an annealing experiment. The annealing experiment was carried out in a vacuum tube furnace instrument. The annealing vacuum was 4×10^{-4} Pa, the heating rate was 10 °C/min, the annealing temperature was 600 °C, and the holding time was set to 24 h. There are three reasons for using 600 °C for the experiment. First, hydrogen embrittlement tends to occur at temperatures below 350 °C, and this temperature is high enough to effectively avoid the effects of hydrogen embrittlement [23,24]. The second is that the interdiffusion between Ta and Pd becomes very obvious above 400 °C, and choosing 600 °C will make this diffusion easier to observe [25,26]. The third is that at 350 °C–600 °C, Pd can catalyze the cracking of carbon and hydrogen isotopes in alkanes, which can be applied to the extraction of deuterium and tritium from the treatment of exhaust gas in fusion reactors [27,28].

The hydrogen permeation experiment was carried out on a plasma gas-driven hydrogen permeation device independently designed and developed by the Advanced Nuclear Energy Laboratory of Sichuan University. The hydrogen permeation flux was obtained by detecting the changes in H^+ ion flow in the downstream chamber by means of a quadrupole mass spectrometer. The hydrogen penetration chamber is divided into an upper chamber and a lower chamber. The background vacuum degree of the upper chamber is pumped to $\sim 10^{-4}$ Pa, and the background vacuum degree of the lower chamber is kept at 10^{-5} Pa~10^{-6} Pa. Then, after the sample was kept at 600 °C for 6 h, hydrogen was injected into the upper chamber so that the hydrogen pressure in the upper chamber was maintained at 90 kPa; the temperature during hydrogen penetration was 873 K and the inner diameter of the sample film was 20.5 mm.

2.3. Micro-Structure Characterization

The phase structure of the Pd/TaTiNbZr/Ta multilayer films was analyzed using a small-angle grazing-incidence X-ray diffractometer (GIXRD, Bruker D8 Advanced, Billerica, MA, USA). The analysis conditions were as follows: the Cu target Kα ray wavelength was 0.15408 nm, the scanning speed was 4°/min, and the step size was 0.02°. Scanning electron microscopy (SEM, FEI Inspect F50, Hillsboro, OR, USA) was used to observe the surface morphology and cross-section morphology of the Pd/TaTiNbZr/Ta multilayer films, and an energy-dispersive spectrometer (EDS, FEI Thermo Scientific EDAX EDS, Waltham, MA, USA) was used to detect the distribution and content of elements in the multilayer films.

3. Results and Discussions

3.1. Influence of Annealing Temperature on Microstructure of Multilayer Films

Figure 1 shows the GIXRD patterns of samples with three different barrier layer thicknesses before and after annealing. It was observed that after annealing at 600 °C for 24 h, the diffraction peaks of Pd and Ta of the three multilayer film samples were basically consistent with the deposited state and had the same crystal structure. Moreover, no TaPd$_3$ diffraction peak was found in the GIXRD pattern of the annealed samples, indicating that TaTiNbZr barrier layers can effectively prevent the formation of TaPd$_3$ compounds under this annealing condition. The deposited TaTiNbZr multi-principal component showed a wide diffraction peak with a span of about 10° at 37°, and no obvious peak was observed, indicating that the deposited TiTaNbZr multi-principal component alloy film had an amorphous structure. Previous studies have shown that amorphous barrier layers have better blocking properties [13]. Furthermore, the thickness of the barrier layer has a significant impact on the hydrogen permeability of the material [14,16], and the three kinds of TaTiNbZr barrier layers can prevent the formation of TaPd$_3$. Therefore, Pd/TaTiNbZr (50 nm)/Ta is mainly used as the research object in the subsequent study.

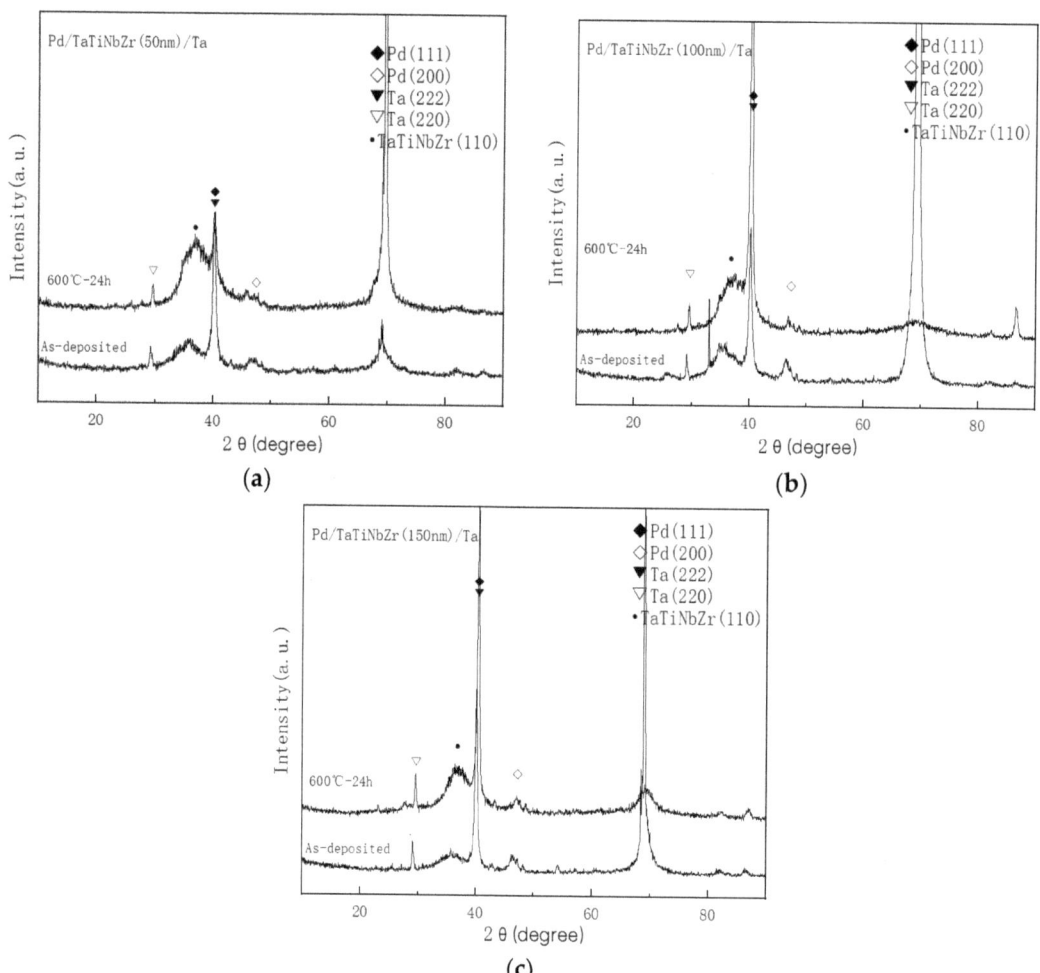

Figure 1. (a–c) XRD patterns of Pd/TaTiNbZr (50 nm)/Ta, Pd/TaTiNbZr (100 nm)/Ta, Pd/TaTiNbZr (150 nm)/Ta before and after annealing at 600 °C for 24 h.

Figure 2 shows the SEM images between annealed Pd/TaTiNbZr (50 nm)/Ta multilayers and deposited Pd/TaTiNbZr (50 nm)/Ta multilayers. It can be observed from Figure 2a that the surface of the film system is covered by a dense Pd layer without exposing the underlying barrier layer. After annealing 24 h at a temperature of 600 °C, the surface Pd layer has a flocculent structure (Figure 2b,c), the agglomeration of Pd film occurs after annealing, and there is no membrane separation phenomenon. Nozaki et al. [14,16] observed that

Pd/Ta after high-temperature treatment became porous in structure, and Daniel et al. [4] found that one of the reasons for the failure of a Pd/Ta multilayer film was that the Pd layer would be stratified after annealing, which reduced the effective contact area between Pd and hydrogen. Neither of the above changes appeared in our experiment. Through the flocculent structure of the Pd film in Figure 2d,e, it can be seen that the TaTiNbZr barrier layer surface below is still intact, and no Pd diffusion into the barrier layer is observed.

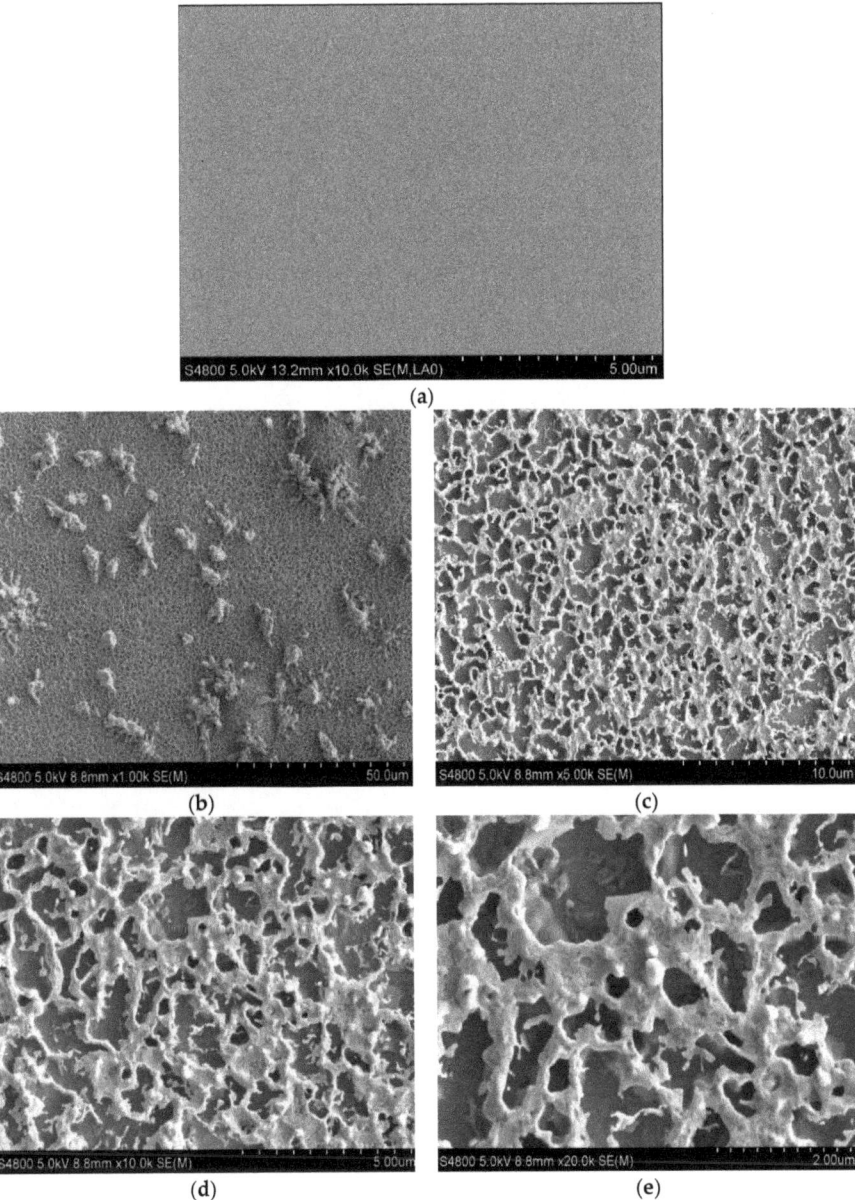

Figure 2. (**a**) The sedimentary surface morphology of Pd/TaTiNbZr (50 nm)/Ta sample. (**b**–**e**) The surface morphology of Pd/TaTiNbZr (50 nm)/Ta sample after annealing at 600 °C for 24 h.

The deposited and annealed Pd/TaTiNbZr (50 nm)/Ta samples were analyzed by sectional scanning and energy spectrum analysis, as shown in Figure 3. It can be clearly observed from Figure 3a that the sample has a multi-layer structure. After annealing, the multi-layer structure remains intact, the interlayer interface is clear, and the structure is stable. The Pd/NbTiCo/Pd prepared by Xiao et al. formed a boundary layer between the Pd layer and Nb30Ti35Co35 substrate after annealing at high temperatures, which contributes to the fast declination of hydrogen flux from the beginning of the permeation process [11,29]. This indicates that after annealing at 600 °C for 24 h, the TaTiNbZr barrier layer with a thickness of 50 nm can effectively prevent the interdiffusion of Pd and Ta. Xiao et al. [29] indicated that HfN could only improve the permeability stability in the temperature range of 500–550 °C, and could not effectively prevent the diffusion of Pd at 600 °C. The distribution of various elements in Figure 3e shows the same distribution as the structure of the multilayer film, demonstrating a homogeneous distribution of elements. The EDS spectrum in Figure 3f scanned the distribution of Pd elements; only the surface had a high Pd content, and there was no Pd element inside the sample. In the reported Pd/NbTiCo/Pd film, Pd diffused significantly into the substrate after annealing at 600 °C [11,29]. This shows that TaTiNbZr barrier layers can effectively prevent Pd diffusion.

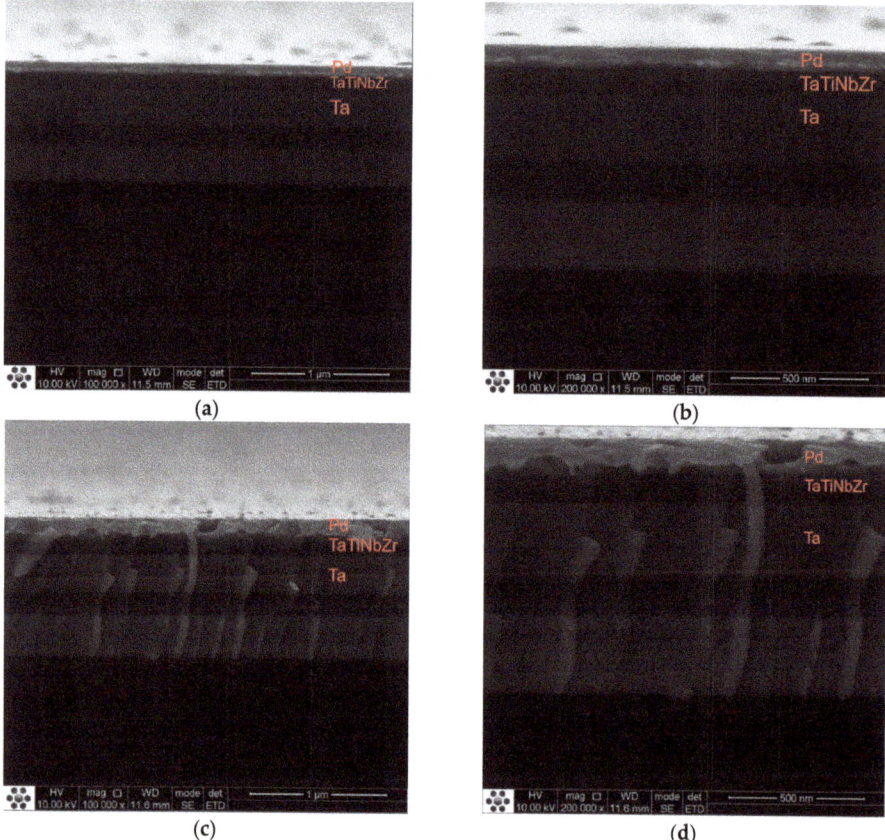

Figure 3. *Cont.*

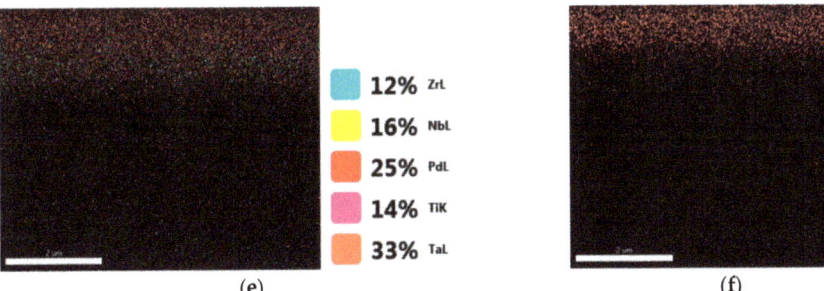

(e) **(f)**

Figure 3. (**a**,**b**) Depositional cross-section morphology of Pd/TaTiNbZr (50 nm)/Ta sample. (**c**,**d**) Cross-section morphology of Pd/TaTiNbZr (50 nm)/Ta sample after annealing at 600 °C for 24 h. (**e**) EDS surface scanning of Pd/TaTiNbZr (50 nm)/Ta samples after annealing at 600 °C for 24 h. (**f**) Pd distribution of Pd/TaTiNbZr (50 nm)/Ta sample scanned on EDS surface after annealing at 600 °C for 24 h.

The microstructural changes of the Pd/TaTiNbZr (50 nm)/Ta multilayers were mainly investigated after annealing at 600 °C for 24 h. The Pd on the surface after annealing changed from providing dense and uniform coverage on the sample surface in the as-deposited state to agglomeration on the sample surface. From the cross-section SEM, it can be seen that the TaTiNbZr high-entropy alloy and the supporting material, Ta, are dense and uniform, and the multilayer structure of the sample remains intact, with a clear interlayer interface and a stable structure. From previous studies on high-entropy alloys, it can be seen that new phases appear in XRD when the high-entropy alloys fail; for example, when FeCoBTiNb or NbMoTaW/TiVCr is used as a Cu/Si barrier layer, the Cu_3Si phase will appear when the barrier layer fails [30,31]. In this study, after annealing at 600 °C for 24 h, the amorphous structure of the high-entropy alloy is stabilized without the appearance of new phases, which is consistent with the phase of the as-deposited state.

3.2. Hydrogen Permeation through Pd/TaTiNbZr/Ta and Pd/Ta

In order to investigate the permeation behavior of hydrogen in Pd/TaTiNbZr/Ta multilayer films, gas-driven permeation behavior experiments of hydrogen through the multilayer films were conducted. The hydrogen-permeable samples were Pd/Ta, Pd/TaTiNbZr (50 nm)/Ta, and Pd/TaTiNbZr (100 nm)/Ta. A quadrupole mass spectrometer was used to detect changes in H^+ ion flow. The hydrogen permeable flux at 600 °C was obtained by the following transformations:

The output data of the quadrupole mass spectrometer are the ion current (I), which can be converted into gas pressure by the following formula:

$$K = I \cdot P^{-1} \tag{1}$$

where P is the lower chamber pressure in mbar and K is the sensitivity of the quadrupole mass spectrometer; in this work, K = 9.66 × 10^{-4} (A·mBAR^{-1}). The vacuum pumping speed of the lower chamber is S = 0.7 ($m^3 \cdot S^{-1}$). The penetration rate, Q (Pa·$m^3 \cdot S^{-1}$), can be calculated by the following formula:

$$Q = (P - P_{base}) \cdot S \tag{2}$$

where P_{base} is the pressure before penetration in the lower chamber, and the conversion of the permeability rate (Q) into permeability flux (F) (molecule·s^{-1}) can be expressed as follows [32]:

$$1 \text{ Pa} \cdot m^3 \cdot s^{-1} = 10 \text{ cc} \cdot s^{-1} = 10^{-5} \text{ } m^3 \cdot s^{-1}/24.5 \text{ Lmol}^{-1} \approx 4.08 \text{ mol} \cdot s^{-1} \tag{3}$$

The total permeation flux (F) is obtained, and the planar diaphragm permeation flux (J) (mol·s^{-1}·m^{-2}) passes through

$$J = F/A \qquad (4)$$

where A (m^{-2}) is the effective penetration area of the diaphragm.

At the beginning of the hydrogen permeation experiment, hydrogen molecules are first adsorbed on the Pd surface. It can be seen from Figure 4 that Pd/Ta has a crystal structure, so H atoms can rapidly diffuse to the lower surface along the grain boundary. When hydrogen atoms dissolve in the metal material and diffuse into the near-surface region, they recombine with another near-surface hydrogen atom to form hydrogen molecules and escape from the lower surface [4–6]. The diffusion of H atoms in Pd/TaTiNbZr/Ta is a similar process. The difference is that when H atoms diffuse along the Pd grain boundary towards the barrier layer, due to the lattice distortion of the amorphous TaTiNbZr HEA, the diffusion distance increases, resulting in a longer diffusion time for H$_2$ to reach a steady state in Pd/TaTiNbZr/Ta [18,31,33,34]. The permeable flux of the three film samples is shown in Figure 5. From Figure 5, it can be seen that Pd/Ta shows excellent hydrogen permeation performance at the beginning of hydrogen permeation and can reach a steady state very quickly, and after keeping the steady state for a period of time, the hydrogen permeation performance decreases sharply. It can be seen from Figure 4 that this is because Pd and Ta form TaPd$_3$ at a high temperature of 600 °C, and Pd loses its catalytic performance. Pd/TaTiNbZr(50 nm)/Ta and Pd/TaTiNbZr(100 nm)/Ta reached a steady state after a long time after the initial introduction of hydrogen, but the permeability flux at steady state was lower than Pd/Ta, because the addition of a barrier layer affected the rate of hydrogen penetration. It can be seen from the GIXRD pattern in Figure 4 that Pd/TaTiNbZr(50 nm)/Ta does not form TaPd$_3$ after the hydrogen penetration experiment at a high temperature, indicating that the TaTiNbZr barrier layer can effectively prevent the mutual diffusion of Pd and Ta. The addition of the barrier layer will affect the hydrogen permeation flux, but effectively extend the continuous working ability of the multilayer film. When the experimental temperature and upstream pressure are kept constant, the thickness of the barrier layer is the main reason for affecting the hydrogen permeation flux, but the thickness of the barrier layer does not affect the continuous working ability of the sample.

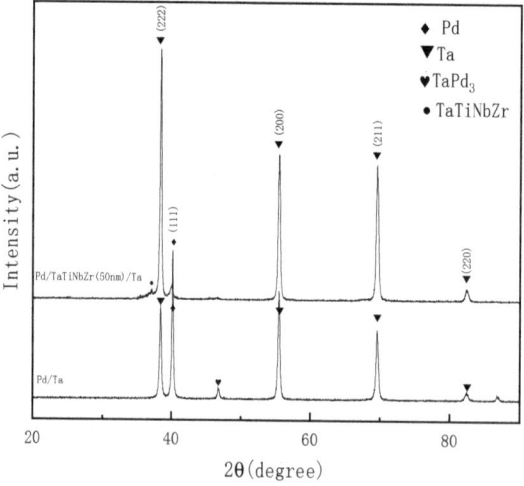

Figure 4. XRD spectra of Pd/Ta and Pd/TaTiNbZr(50 nm)/Ta after hydrogen penetration.

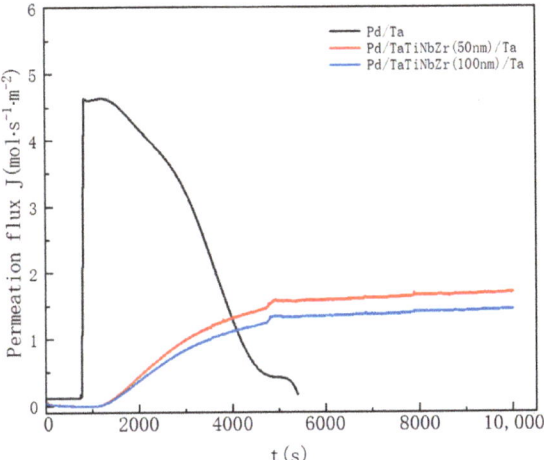

Figure 5. Time evolution curve of gas-driven hydrogen permeation flux of Pd/Ta, Pd/TaTiNbZr (50 nm)/Ta, and Pd/TaTiNbZr (100 nm)/Ta at 600 °C.

In this study, the effect of adding a barrier layer and the thickness of the barrier layer on hydrogen permeability was discussed. It is well known that the experimental results of hydrogen permeation are subject to changes in pressure, temperature, and other parameters. The rate of hydrogen permeation usually increases as the pressure increases. This is because high pressure increases the frequency of collisions between gas molecules, thereby facilitating the rate at which hydrogen molecules pass through the permeable layer. On the other hand, the rate of hydrogen permeation also increases as the temperature increases because the increase in temperature increases the average kinetic energy of the gas molecules, thus giving them a higher penetration capacity [35].

4. Conclusions

1. After the addition of an amorphous TaTiNbZr barrier layer, Pd/TaTiNbZr/Ta can maintain structural and morphological stability before and after annealing.
2. The high-temperature hydrogen permeation experiment shows that the barrier layer can effectively increase the working time of the multilayer film. Pd/TaTiNbZr(50 nm)/Ta has better hydrogen permeation performance than Pd/TaTiNbZr(100 nm)/Ta, indicating that the thickness of the barrier layer is a key factor affecting the hydrogen permeation ability.
3. A layer of TaTiNbZr was inserted between a Ta substrate and a Pd film as a barrier layer. No $TaPd_3$ was produced after the hydrogen infiltration experiment at 600 °C, indicating that TaTiNbZr can effectively block the mutual diffusion of Ta and Pd.

Author Contributions: Conceptualization, H.S.; methodology, H.S. and G.P.; software, H.S. and G.P.; validation, H.S. and G.P.; formal analysis, H.S. and G.P.; investigation, H.S.; resources, H.S.; data curation, H.S.; writing—original draft preparation, H.S.; writing—review and editing, H.S., B.L. and G.P.; visualization, H.S. and B.L.; supervision, H.S. and B.L.; project administration, B.L.; funding acquisition, G.P. All authors have read and agreed to the published version of the manuscript.

Funding: This research received no external funding.

Institutional Review Board Statement: Not applicable.

Informed Consent Statement: Not applicable.

Data Availability Statement: Data are contained within the article.

Conflicts of Interest: The authors declare no conflict of interest.

References

1. Hassanpouryouzband, A.; Wilkinson, M.; Haszeldine, R.S. Hydrogen energy futures–foraging or farming? *Chem. Soc. Rev.* **2024**, *53*, 2258–2263. [CrossRef] [PubMed]
2. Hassanpouryouzband, A.; Joonaki, E.; Edlmann, K.; Haszeldine, R.S. Offshore geological storage of hydrogen: Is this our best option to achieve net-zero? *ACS Energy Lett.* **2021**, *6*, 2181–2186. [CrossRef]
3. Alimov, V.N.; Hatano, Y.; Busnyuk, A.O.; Livshits, D.A.; Notkin, M.E.; Livshits, A.I. Hydrogen permeation through the Pd-Nb-Pd composite membrane: Surface effects and thermal degradation. *Int. J. Hydrogen Energy* **2011**, *36*, 7737–7746. [CrossRef]
4. Cooney, D.A.; Way, J.D.; Wolden, C.A. A comparison of the performance and stability of Pd/BCC metal composite membranes for hydrogen purification. *Int. J. Hydrogen Energy* **2014**, *39*, 19009–19017. [CrossRef]
5. Jo, Y.S.; Lee, C.H.; Kong, S.Y.; Lee, K.Y.; Yoon, C.W.; Nam, S.W.; Han, J. Characterization of a Pd/Ta composite membrane and its application to a large scale high-purity hydrogen separation from mixed gas. *Sep. Purif. Technol.* **2018**, *200*, 221–229. [CrossRef]
6. Park, Y.; Kwak, Y.; Yu, S.; Badakhsh, A.; Lee, Y.-J.; Jeong, H.; Kim, Y.; Sohn, H.; Nam, S.W.; Yoon, C.W.; et al. Degradation mechanism of a Pd/Ta composite membrane: Catalytic surface fouling with inter-diffusion. *J. Alloys Compd.* **2021**, *854*, 157196. [CrossRef]
7. Zhang, Y.; Lu, J.; Ikehara, T.; Maeda, R.; Nishimura, C. Characterization and permeation of microfabricated palladium membrane. *Mater. Trans.* **2006**, *47*, 255–258. [CrossRef]
8. Zhang, Y.; Maeda, R.; Komaki, M.; Nishimura, C. Hydrogen permeation and diffusion of metallic composite membranes. *J. Membr. Sci.* **2006**, *269*, 60–65. [CrossRef]
9. Zhang, Y.; Gwak, J.; Murakoshi, Y.; Ikehara, T.; Maeda, R.; Nishimura, C. Hydrogen permeation characteristics of thin Pd membrane prepared by microfabrication technology. *J. Membr. Sci.* **2006**, *277*, 203–209. [CrossRef]
10. Demange, D.; Glugla, M.; Günther, K.; Le, T.L.; Simon, K.H.; Wagner, R.; Welte, S. Tritium processing tests for the validation of upgraded PERMCAT mechanical design. *Fusion Sci. Technol.* **2008**, *54*, 14–17. [CrossRef]
11. Liang, X.; Li, X.; Nagaumi, H.; Guo, J.; Gallucci, F.; van Sint Annaland, M.; Liu, D. Degradation of Pd/Nb30Ti35Co35/Pd hydrogen permeable membrane: A numerical description. *J. Membr. Sci.* **2020**, *601*, 117922. [CrossRef]
12. Wong, T.; Yu, Z.; Suzuki, K.; Gibson, M.; Ishikawa, K.; Aoki, K. Effect of Annealing on the Hydrogen Permeation and Mechanical Behaviour of Nb-Ni-Zr Alloy Membranes. *Mater. Sci. Forum* **2010**, *654–656*, 2839–2842. [CrossRef]
13. Chin, H.-S.; Suh, J.-Y.; Park, K.-W.; Lee, W.; Fleury, E. Hydrogen permeability of glass-forming Ni-Nb-Zr-Ta crystalline membranes. *Met. Mater. Int.* **2011**, *17*, 541–545. [CrossRef]
14. Nozaki, T.; Hatano, Y. Hydrogen permeation through a Pd/Ta composite membrane with a HfN intermediate layer. *Int. J. Hydrogen Energy* **2013**, *38*, 11983–11987. [CrossRef]
15. Jayalakshmi, S.; Choi, Y.; Kim, Y.; Fleury, E. Hydrogenation properties of Ni-Nb-Zr-Ta amorphous ribbons. *Intermetallics* **2010**, *18*, 1988–1993. [CrossRef]
16. Nozaki, T.; Hatano, Y.; Yamakawa, E.; Hachikawa, A.; Ichinose, K. Improvement of high temperature stability of Pd coating on Ta by HfN intermediate layer. *Int. J. Hydrogen Energy* **2010**, *35*, 12454–12460. [CrossRef]
17. Liang, S.C.; Tsai, D.-C.; Chang, Z.-C.; Lin, T.-N.; Shiao, M.-H.; Shieu, F.-S. Thermally Stable TiVCrZrHf Nitride Films as Diffusion Barriers in Copper Metallization. *Electrochem. Solid State Lett.* **2012**, *15*, H5–H8. [CrossRef]
18. Tsai, M.H.; Yeh, J.W.; Gan, J.Y. Diffusion barrier properties of AlMoNbSiTaTiVZr high-entropy alloy layer between copper and silicon. *Thin Solid Film.* **2008**, *516*, 5527–5530. [CrossRef]
19. Chang, S.Y.; Li, C.-E.; Huang, Y.-C.; Hsu, H.-F.; Yeh, J.-W.; Lin, S.-J. Structural and thermodynamic factors of suppressed interdiffusion kinetics in multi-component high-entropy materials. *Sci. Rep.* **2014**, *4*, 4162. [CrossRef]
20. Steward, S. *Review of Hydrogen Isotope Permeability through Materials*; Lawrence Livermore National Lab. (LLNL): Livermore, CA, USA, 1983.
21. Lee, D.-Y. Hydrogen Permeation Properties of Pd-Coated Ni-Nb-Ti-Zr Amorphous Alloys. *Met. Mater. Int.* **2008**, *14*, 545–548. [CrossRef]
22. Li, J.; Pu, G.; Sun, H.; Du, X.; Lin, L.; Ren, D.; Zhang, K.; Liu, B. Effect of He-ions irradiation on the microstructure and mechanical properties of TiTaNbZr refractory medium-entropy alloy film. *Vacuum* **2023**, *217*, 112545. [CrossRef]
23. Paglieri, S.N.; Wermer, J.R.; Buxbaum, R.E.; Ciocco, M.V.; Howard, B.H.; Morreale, B.D. Development of membranes for hydrogen separation: Pd coated V-10Pd. *Energy Mater.* **2008**, *3*, 169–176. [CrossRef]
24. Nambu, T.; Shimizu, K.; Matsumoto, Y.; Rong, R.; Watanabe, N.; Yukawa, H.; Morinaga, M.; Yasuda, I. Enhanced hydrogen embrittlement of Pd-coated niobium metal membrane detected by in situ small punch test under hydrogen permeation. *J. Alloys Compd.* **2007**, *446*, 588–592. [CrossRef]
25. Peachey, N.M.; Snow, R.C.; Dye, R.C. Composite Pd/Ta metal membranes for hydrogen separation. *J. Membr. Sci.* **1996**, *111*, 123–133. [CrossRef]
26. Edlund, D.J.; McCarthy, J. The relationship between intermetallic diffusion and flux decline in composite-metal membranes-implications for achieving long membrane lifetime. *J. Membr. Sci.* **1995**, *107*, 147–153. [CrossRef]
27. Konishi, S.; Glugla, M.; Hayashi, T. Fuel cycle design for ITER and its extrapolation to DEMO. *Fusion Eng. Des.* **2008**, *83*, 954–958. [CrossRef]
28. Glugla, M.; Dörr, L.; Lässer, R.; Murdoch, D.; Yoshida, H. Recovery of tritium from different sources by the ITER Tokamak exhaust processing system. *Fusion Eng. Des.* **2002**, *61–62*, 569–574. [CrossRef]

29. Liang, X.; Li, X.; Chen, R.; Nagaumi, H.; Guo, J.; Liu, D. Enhancement of hydrogen permeation stability at high temperatures for Pd/Nb30Ti35Co35/Pd composite membranes by HfN intermediate layer. *J. Membr. Sci.* **2022**, *643*, 120062. [CrossRef]
30. Li, P.F.; Ma, Y.; Ma, H.; Ta, S.; Yang, Z.; Han, X.; Kai, M.; Chen, J.; Cao, Z. Enhanced diffusion barrier property of nanolayered NbMoTaW/TiVCr high entropy alloy for copper metallization. *J. Alloys Compd.* **2022**, *895*, 162574. [CrossRef]
31. Fang, J.-S.; Yang, L.-C.; Lee, Y.-C. Low resistivity Fe-Co-B-Ti-Nb amorphous thin film as a copper barrier. *J. Alloys Compd.* **2014**, *586*, S348–S352. [CrossRef]
32. Liu, Q. Experimental Study on Hydrogen Permeation Behavior in Niobium Driven by Gas and Plasma. Master's Thesis, Xihua University, Chengdu, China, March 2023.
33. Jiang, C.; Li, R.; Wang, X.; Shang, H.; Zhang, Y.; Liaw, P.K. Diffusion Barrier Performance of AlCrTaTiZr/AlCrTaTiZr-N High-Entropy Alloy Films for Cu/Si Connect System. *Entropy* **2020**, *22*, 234. [CrossRef] [PubMed]
34. Peng, Y.; Wang, H.; Li, Q.; Wang, L.; Zhang, W.; Zhang, L.; Guo, S.; Liu, Y.; Liu, S.; Ma, Q. Effect of Mo on interdifussion behaviors and interfacial characteristics in multicomponent diffusion couple of FeCoCrNi high entropy alloys and diamond. *Mater. Des.* **2022**, *215*, 110522. [CrossRef]
35. Jing, W.-N.; Liu, J.; Guo, H.; Wang, S.; Bi, H.; Chen, B.; Chen, J.; Wang, H.; Wei, J.; Ye, Z.; et al. Gas- and plasma-driven hydrogen permeation behavior of stagnant eutectic-solid GaInSn/Fe double-layer structure. *Chin. Phys. B* **2023**, *32*, 045201. [CrossRef]

Disclaimer/Publisher's Note: The statements, opinions and data contained in all publications are solely those of the individual author(s) and contributor(s) and not of MDPI and/or the editor(s). MDPI and/or the editor(s) disclaim responsibility for any injury to people or property resulting from any ideas, methods, instructions or products referred to in the content.

Article

Nickel Nanoparticles: Insights into Sintering Dynamics

Lucia Bajtošová, Barbora Kihoulou, Rostislav Králík, Jan Hanuš and Miroslav Cieslar *

Faculty of Mathematics and Physics, Charles University, Ke Karlovu 3, 12116 Prague, Czech Republic; 96473937@o365.cuni.cz (L.B.); krivska.barbora@seznam.cz (B.K.); rkralik96@gmail.com (R.K.); jan.hanus@gmail.com (J.H.)
* Correspondence: miroslav.cieslar@mff.cuni.cz

Abstract: The sintering dynamics of nickel nanoparticles (Ni NPs) were investigated through a comprehensive approach that included in situ transmission electron microscopy annealing and molecular dynamics simulations. This study systematically examines the transformation behaviors of Ni NP agglomerates over a temperature spectrum from room temperature to 850 °C. Experimental observations, supported by molecular dynamics simulations, revealed the essential influence of rotational and translational motions of particles, especially at lower temperatures, on sintering outcomes. The effect of the orientation of particles on the sintering process was confirmed, with initial configurations markedly determining sintering efficiency and dynamics. Calculated activation energies from this investigation follow those reported in the literature, confirming surface diffusion as the predominant mechanism driving the sintering of Ni NPs.

Keywords: nickel nanoparticles; sintering dynamics; molecular dynamics simulations; in situ TEM annealing; surface diffusion; particle orientation

1. Introduction

The characteristics of nanomaterials fundamentally differ from those of their bulk counterparts due to the pronounced influence of their surface areas. This difference is primarily attributed to their substantial surface-to-volume ratio, which ensures that the surface properties significantly shape the overall behavior of nanomaterials. The decrease in all sizes to tens of nanometers leads to the formation of monocrystalline particles with unique properties. Several metals, such as Au, Ag, Co, Ni, Cu, and Pd, or alloys, such as Fe-Cu, Co-Cu, Fe-Ag, or Ni-Ag, have been investigated in nanoparticle form and found use in a variety of applications [1–4]. Whether it is the magnetic properties of Fe-Co-Ni particles [5], the luminescence properties of ZnO nanoparticles [6], catalytic properties of Au [7], Pd [8], or CdS NPs [9], the size of the particles plays a crucial role.

A broad range of applications was found for nickel nanoparticles (Ni NPs), which can be used in electrocatalysis, photocatalysis, or biomedicine [10,11]. These unique properties of NPs as catalysts derive from the large percentage of coordinatively unsaturated atoms located at the surface, edges, and corners of the NPs compared to the total number of atoms [12]. The unsaturated atoms have the highest catalytic activity because they tend to increase their coordination number by coordinating with their surroundings. Furthermore, potential applications in magnetic sensors [13], memory devices [14,15], and conducting materials [16] have been examined. For the particles below a critical size, a single magnetic domain is energetically more stable, which results in their supermagnetic properties [17,18].

The synthesis of Ni NPs is achievable through various methods categorized into physical, chemical, and biological techniques, each with distinct advantages [19]. Chemical synthesis includes processes such as chemical reduction [20], electrochemical techniques [21], and thermal decomposition [22]. Biological synthesis uses biological sources like plants, bacteria, fungi, and other microorganisms for the synthesis [23]. Physical approaches, including mechanical milling and arc discharge [24], offer robust solutions for achieving precise control over particle size and morphology through top-down strategies. Notably, gas aggregation stands out for producing highly uniform and pure NiNPs with controlled size and shape [25].

Consolidation of individual nanoparticles occurs through the elevation of temperature, pressure, or application of electric fields [26]. This consolidation of individual monocrystalline particles to a dense microstructure causes a deterioration of the desired properties originating from the nano-size of the particles. When sintering nickel nanoparticles, the process differs significantly from that in bulk recrystallization [27] due to the large surface-to-volume ratio of nanoparticles. Studies of the sintering of Ni gas nanoparticles have been used to calculate diffusion coefficients [28]. However, there is a lack of studies that allow for the direct observation of the behavior of agglomerates of Ni nanocrystals at elevated temperatures.

The literature reveals a few key findings in molecular dynamics (MD) simulations for nickel nanoparticle sintering. Studies have focused on the effects of particle size and temperature on the melting and sintering processes, providing insights into the size-dependent melting phenomena and surface melting of nickel nanoparticles [29]. Additionally, research has explored the sintering mechanisms, particularly looking into nanoscale materials, which can offer a broader understanding applicable to nickel nanoparticles [30]. Another critical area of investigation involves using specific interatomic potentials, like Morse parameters and EAM functions, to accurately simulate the behavior of nickel nanoparticles and their interactions with other materials [31]. These studies collectively contribute to a deeper understanding of the sintering behavior of nickel nanoparticles at the atomic level. Crucially, for a comprehensive understanding, these simulations should be rigorously compared with experimental data to validate the accuracy and relevance of the theoretical models. Most molecular dynamics studies have predominantly concentrated on the sintering of two particles, not examining the effects that arise from the interaction of multiple particles.

In this work, pure Ni nanoparticles were prepared using magnetron sputtering. In situ annealing in the transmission electron microscope is used to monitor the coalescence of these particles in temperature ranges from room temperature to 800 °C. Experimental observations are complemented with MD simulations.

2. Materials and Methods

Pure Ni nanoparticles were prepared by DC magnetron sputtering at the following deposition conditions: pressure of Ar 13 Pa, current of 500 mA, and voltage of 335 V. The nanoparticles were sputtered from a 1.5 mm thick Ni target with a purity of at least 99%. The maximum amount of impurities is presented in Table 1. Details of the used geometry and procedure can be found in [32].

Table 1. Composition of the sputtering target.

Ni	Fe	Mn	Si	Cu	S	C
99.0 min	0.40 max	0.35 max	0.35 max	0.25 max	0.01 max	0.02 max

The samples for TEM were prepared by mixing the particles with methanol and dropping the solution on silicon dioxide ultrathin-film TEM window grids. The particles were characterized on a Jeol 2200FS electron microscope operated at 200 keV using conventional, scanning, and high-resolution transmission electron microscopy (TEM, STEM, and HRTEM, respectively). Bright-field (BF), high-angle annular dark-field (HAADF), and secondary electron (SE) detectors were used in the scanning mode. The annealing of particles was performed in situ up to 900 °C with the same microscope using a Gatan double-tilt heating holder. The calibration of the holder was performed using the phase transformation temperature of alpha to beta titanium at high temperatures and phase transformations in the Al-Sn system at low temperatures. The statistical evaluation of the number and size of nanoparticles was performed with ImageJ 1.45 software [33].

A pair and a pile of randomly oriented Ni nanoparticles with a radius of 3.5 nm for the molecular dynamics simulation were created in Atomsk [34]. The LAMMPS software

23 June 2022 version [35] with MEAM potential [36] was used for MD simulations. The particles were held at constant temperatures with a Nose–Hoover thermostat [37] for one ns, using a one fs time step. Simulated atomic configurations were visualized and analyzed using Ovito 3.2.1. [38].

3. Results
3.1. Experimental TEM Observations
3.1.1. Characterization of Initial State

In Figure 1, the TEM images reveal an agglomeration of nickel particles, predominantly of a spherical shape. A detailed view in Figure 1b shows the atomic distances within some of these particles. Fourier transformation analysis of these images confirms the presence of fcc nickel, evidenced by the matching interatomic distances to nickel 111 and 200 planes. The fcc structure of nickel is further confirmed by the selected area diffraction pattern (Figure 1c). No signs of surface oxidation were detected.

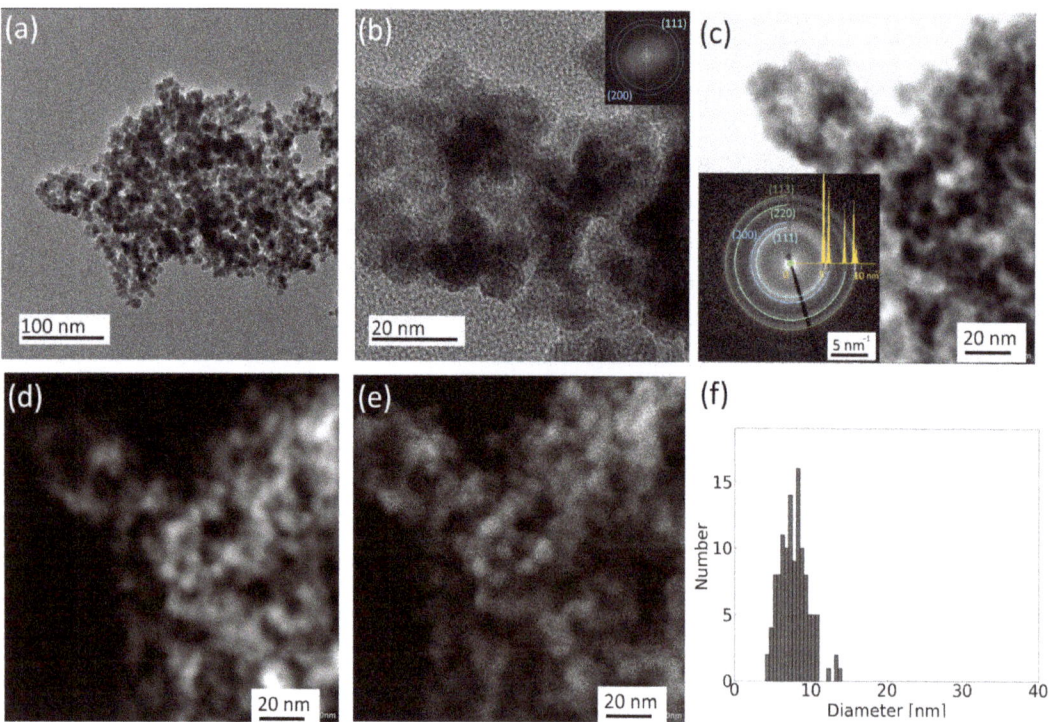

Figure 1. TEM images of the initial state of an agglomerate of Ni nanoparticles—(**a**) TEM BF showing a representative agglomerate at room temperature, (**b**) detailed HRSTEM view with a confirmation of fcc Ni structure by Fourier transformation image in the inset, (**c**) STEM BF—diffraction contrast with selected area diffraction pattern (SAED) overlaid by fcc Ni diffraction pattern, (**d**) STEM HAADF—mass thickness contrast, (**e**) SE—surface contrast of the agglomerate detail, and (**f**) histogram of particles size distribution.

Images in the HAADF (Figure 1d) and SE (Figure 1e) detectors provide additional information about the thickness and surface morphology. The total number of particles of the agglomerate exceeding 2000 can be roughly estimated from these images. Regarding the size distribution of these nanoparticles, as depicted in Figure 1f, their diameters range from 4 to 14 nm, with an average value of 8 nm and a standard deviation of 2 nm.

3.1.2. In Situ TEM Annealing

The particles were annealed up to 100 °C, 300 °C, and 500 °C, then in 50 °C steps up to 850 °C. Each temperature was held for 5 min for the stabilization. The annealing process of the isolated nanoparticle agglomerate is shown in Figure 2. Initially, up to a temperature of 300 °C, the nanoparticles exhibit minimal changes in shape. However, upon reaching 500 °C, the particles begin to sinter, coalescing into larger entities that are two to three times their original size while approximately retaining the overall form of the agglomerate. The particles undergo progressive reshaping as temperatures increase between 500 °C and 800 °C. The process culminates at 850 °C, where the nanoparticles merge to form a single large particle.

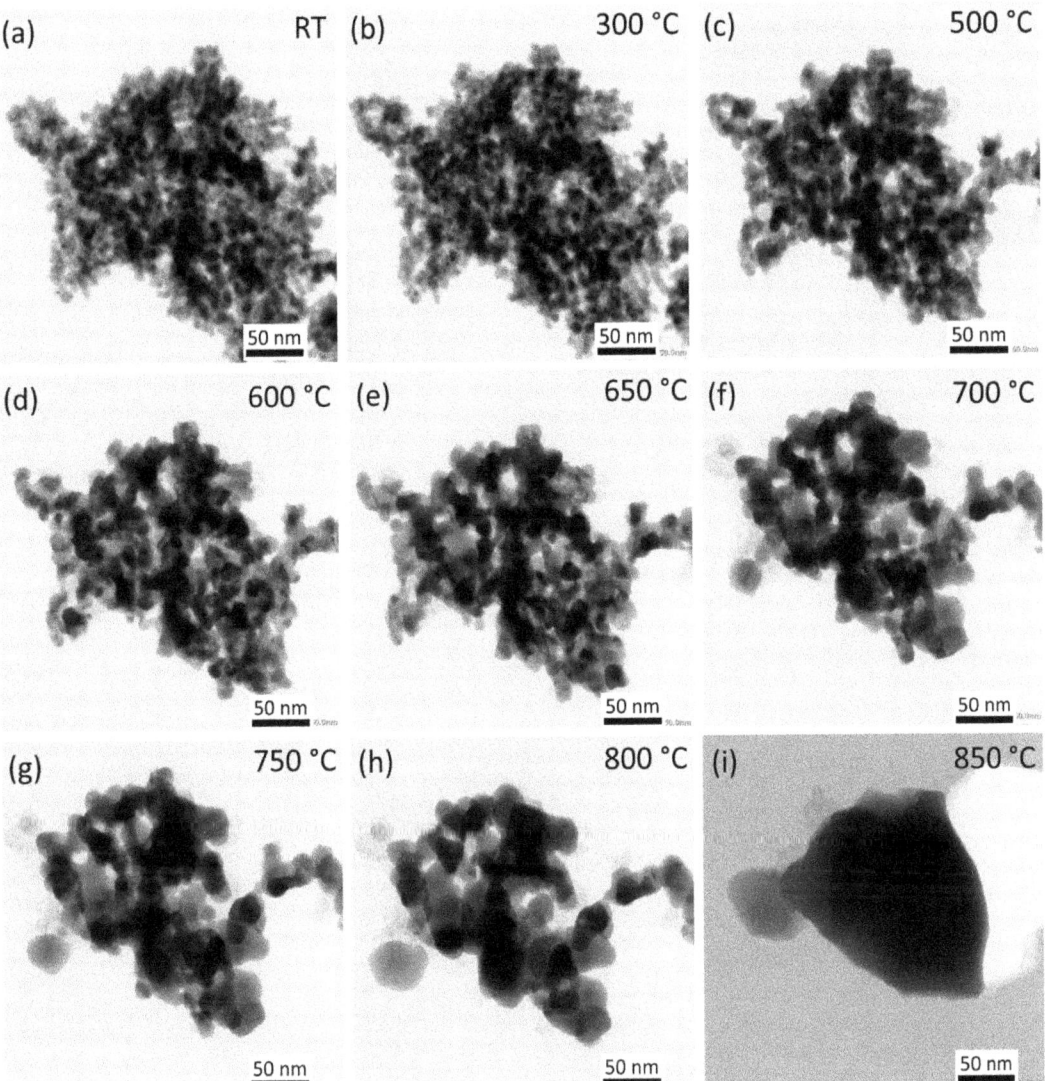

Figure 2. In situ annealing of the nanoparticle agglomerate. STEM BF images in the temperature range of 25–850 °C: (**a**,**b**) initial stages of sintering, (**c**–**h**) coalescence of smaller particles, and (**i**) coalescence of the whole agglomerate into a single particle.

Figure 3 presents the changes in the size distribution of nanoparticles within an agglomerate across different temperatures. These statistical distributions were derived by measuring diameters for 120 distinct particles at room temperature (RT) and 300 °C. As temperatures escalate, there is a noticeable reduction in the total number of particles, and the distributions at 500 °C and 700 °C measured on 80 and 70 particles, respectively. The data reveal a significant increase in the mean diameter of the nanoparticles: from 8 nm at room temperature, it enlarges to 12 nm at 500 °C and expands further to 19 nm at 700 °C.

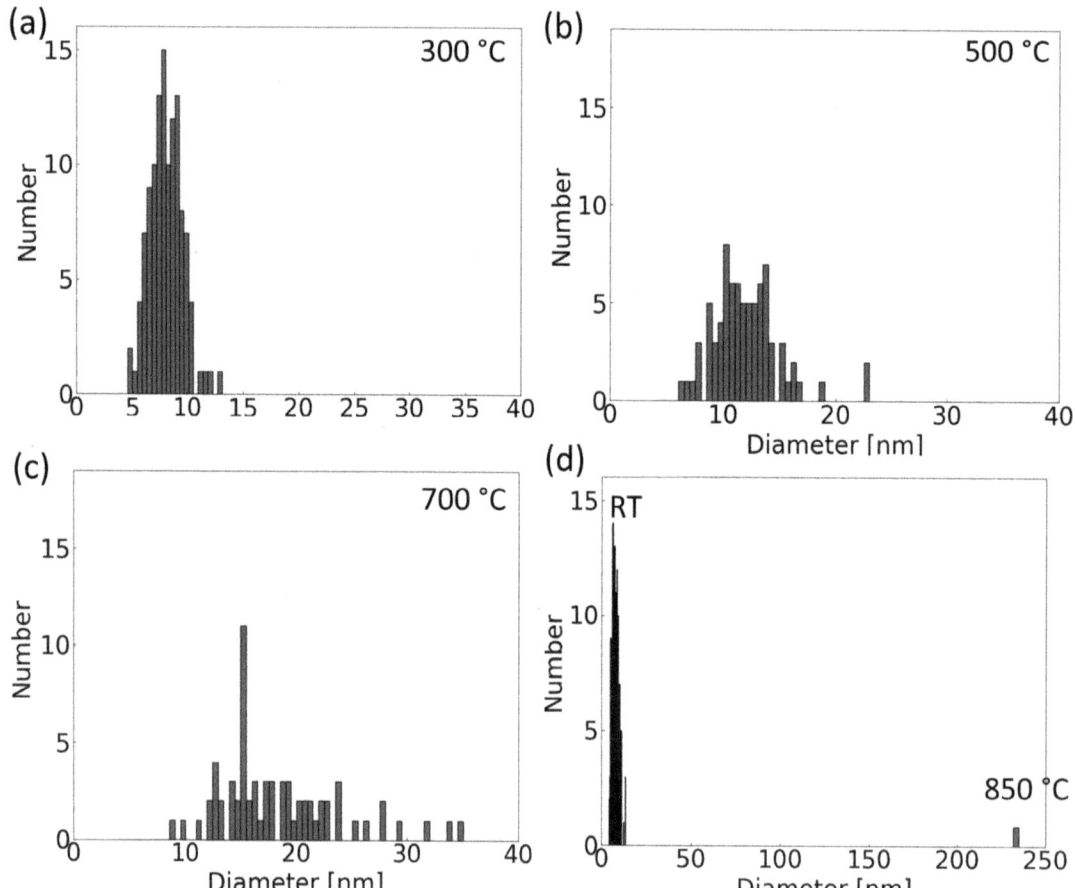

Figure 3. Histograms of the size distribution of particles in the agglomerate in Figure 2 at chosen temperatures: (**a**) 300 °C, (**b**) 500 °C, (**c**) 700 °C, and (**d**) 850 °C compared to the room temperature. The total number of particles decreases with increasing temperature.

The characteristic sintering time (τ) of nickel particles is defined as the time required for the length of the sintered neck between two particles to reach 83% of the initial radius of the particles. The formula of characteristic sintering time as a function of the radius of the primary particles can be derived from macroscopic continuum theories of sintering via surface diffusion [28,39]:

$$\tau = k_B \times T/(25 \times D \times \gamma \times a^4) \times R^4, \tag{1}$$

where R represents the radius of the primary particles, D denotes the diffusion constant, γ signifies the surface free energy, and a corresponds to the diameter of the atoms. The model incorporates the temperature-dependent diffusion coefficient, D, which follows the Arrhenius relationship $D = D_0 \times \exp(-E_a/k_B T)$, where D_0 is the pre-exponential factor, E_a is the activation energy, k_B is the Boltzmann constant, and T is the sintering temperature in Kelvin.

Additional information from HAADF detectors (Figure 4a–d) allows for the distinction of thickness contrast from the diffraction contrast visible in BF. Compared to images of the surface from the SE detector (Figure 4e–h), all the dark parts corresponding to the holes in the SE images align well with dark parts in the HAADF detector corresponding to a reduced thickness. A conclusion that can be drawn from these observations is that the surface inhomogeneities are primarily attributable to the perspective of the agglomerates, as seen from above. The lack of significant cavities on the lower side suggests the adhesion of the particles to the SiO_2 membrane. Furthermore, while the HAADF detector images suggest that particles appear as individual entities, the necking process—signifying the onset of sintering among these particles—is distinctly observed through the SE detector.

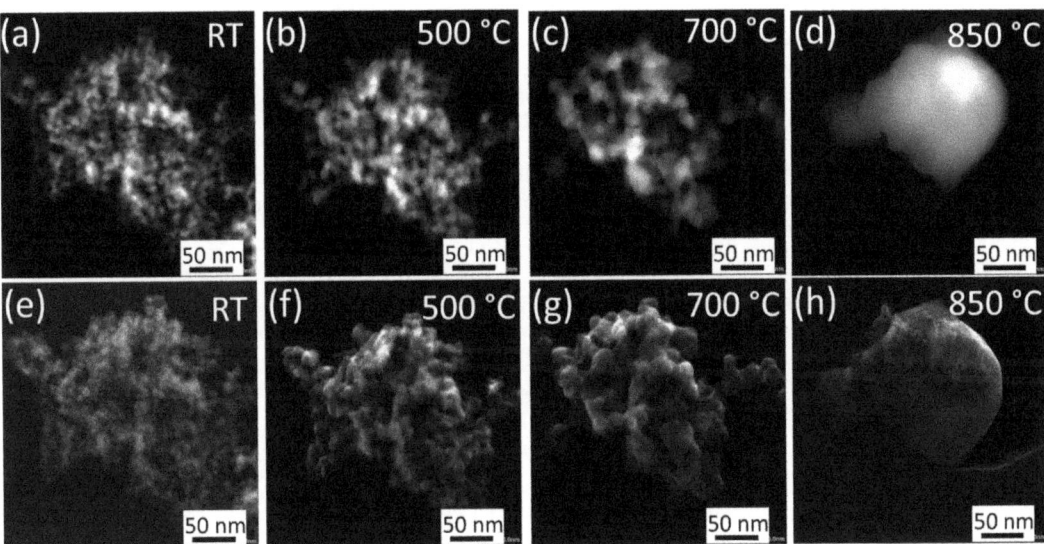

Figure 4. Annealing of the nanoparticle agglomerate at temperatures with the most prominent shape changes, comparison of mass thickness contrast in STEM HAADF images (**a–d**), and surface morphology contrast in SE detector images (**e–h**).

The information from the HAADF, BF, and SE detectors and size distribution for temperatures of 300, 500, 700, and 850 °C (Figure 4) allow for estimating the surface area change. Implementing the modified phenomenological model of Koch and Friedlander, described by Tsyganov et al. [28], enables a quantitative analysis of this dependence. Initially, the volume of the two individual particles at 850 °C and the volume of the aggregate of particles at 800 °C were computed. In the case of non-spherical particles, the volume estimation employed a spherical model using the average radii obtained from measurements. With volume conservation validated between these temperatures, this principle of volume consistency was extrapolated across all examined temperatures. The particle count was estimated, adhering to their empirically measured distribution, enabling the subsequent surface area calculation, which was predicted using the spherical particle assumption. Area changes as a function of temperature are plotted in Figure 5a.

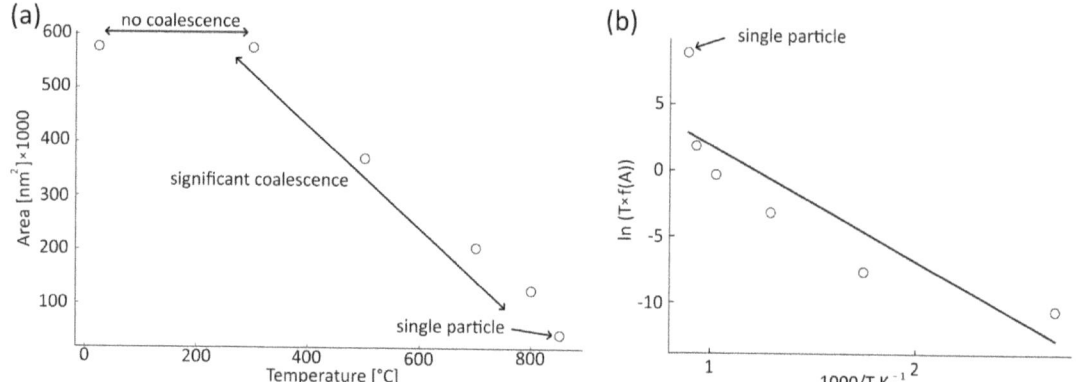

Figure 5. Analysis of the surface area of the annealed agglomerate: (**a**) change in the estimated surface as a function of temperature, the ranges where no coalescence, coalescence of smaller particles, and coalescence of the whole agglomerate are indicated in the plot; (**b**) Arrhenius plot according to Equation (2). The solid line represents the linear regression of the measured data (empty circles). The initial and final areas (points marked as a single particle) were not used in the fit.

Employing Equation (1), combined with the principle of conserved volume, and integration of the Koch–Friedlander equation as well as diffusion coefficient $D = D_0 \times \exp(-E_a/k_B T)$, the formulation of Equation (2) follows (detailed derivation in [28]):

$$T \times f(A) = C \times \exp(-E_a/k_B T), \quad (2)$$

where T is temperature, E_a is the activation energy, k_B is the Boltzmann constant, C is a constant associated with sintering conditions, and $f(A)$ is a function solely dependent on the agglomerate surface area before sintering (A_0), during sintering (A), and in the fully sintered state (A_s). If we denote A_s/A_0 as α, A_s/A as β, and A_0/A as γ, then

$$f(A) = f(\alpha, \beta, \gamma) = \ln\left|\frac{1-\alpha}{1-\beta}\right| - \alpha(\gamma - 1) - \frac{1}{2}\alpha^2(\gamma^2 - 1) - \frac{1}{3}\alpha^3(\gamma^3 - 1) \quad (3)$$

By plotting the logarithm of $T \cdot f(A)$ as a function of inverse temperature (Figure 5b), the activation energy was determined as 0.75 eV/atom for a temperature range of 300–800 °C. The value is much smaller than volume diffusion, comparable to some values of the grain boundary and surface diffusion [40,41]. At 850 °C, when the nanoparticles merge into a larger single particle, there is a noticeable deviation from the linear fit in Figure 5b. This indicates a change in the sintering mechanism.

3.1.3. Initial Stages of Annealing

The details of the initial phases of sintering under 450 °C are depicted in Figure 6. Figure 6a–d capture the merging of particles of various sizes, forming larger aggregates, specifically a 7 nm, and a 6 nm particle with a larger 10 nm particle at a temperature of 300 °C. The increase in temperature to 400 °C facilitates the merging of slightly larger particles between 13 and 15 nm, as shown in Figure 6e–h. Both particle size and temperature significantly influence the initial stages of particle coalescence, forming larger, more consolidated structures.

Beyond the primary coalescence process, subtle positional adjustments among individual particles within the agglomerate are noticeable (Figure 7). These adjustments illustrate a tendency towards reducing inter-particle distances, contributing to the densification and overall compaction of the agglomerate.

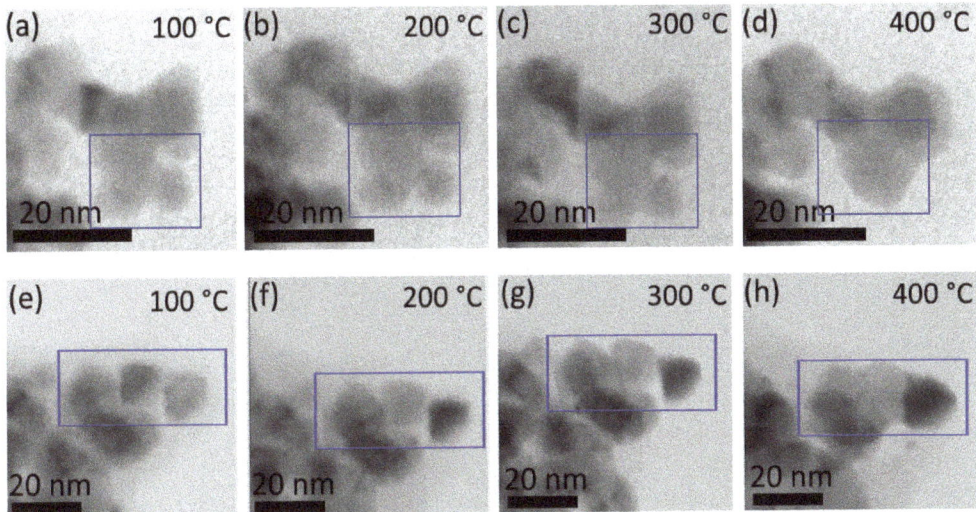

Figure 6. Annealing of different agglomerates of Ni nanoparticles at a temperature range of 100–400 °C; detail of initial stages of sintering. STEM BF image sequences of two different areas: (**a**–**d**) first area and (**e**–**h**) the second area. Blue rectangles mark the coalescing particles.

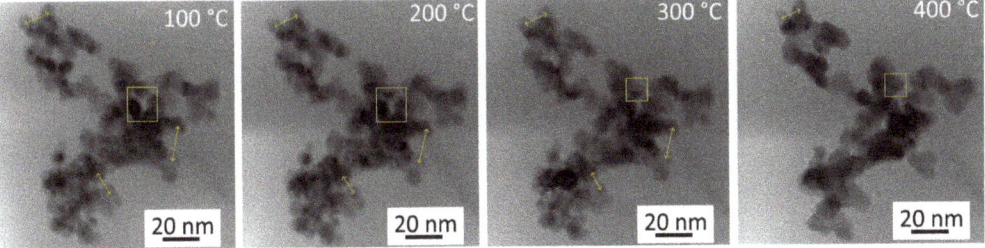

Figure 7. Initial stages of annealing of a smaller agglomerate of Ni nanoparticles; STEM BF images at different temperatures. Yellow arrows and rectangles mark position changes between individual particles.

3.2. Characterization by Molecular Dynamics Simulations

3.2.1. Annealing Simulations of a Four-Particle Agglomerate: Particle Motion Analysis

The molecular dynamics simulation of the sintering process of the four nickel nanoparticles with the initial configuration shown in Figure 8 was run for two different temperatures (273 and 1100 K).

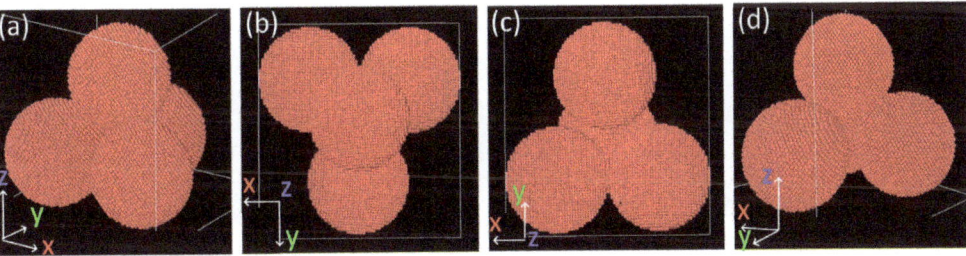

Figure 8. Ovito visualization of the initial configuration of a pile of nanoparticles used in MD simulations: (**a**) front, (**b**) up, (**c**) down, and (**d**) back views.

The dynamics of a four-particle agglomerate over a 1 ns timescale at room temperature is illustrated in Figure 9. The whole body motion of all the particles can be observed to make contact between non-touching particles, causing a decrease in their distance (Figure 9e,f) and contact (Figure 9g). The small rotations between crystalline lattices of individual particles accompany this translational motion. Their behavior is similar to that observed at lower temperatures in the in situ TEM experiment. The contact area between two particles expands minimally, extending only a few interatomic distances. The translational and small rotational motion of individual particles in the agglomerate results from minimizing the surface energy and the boundary energy between two lattices.

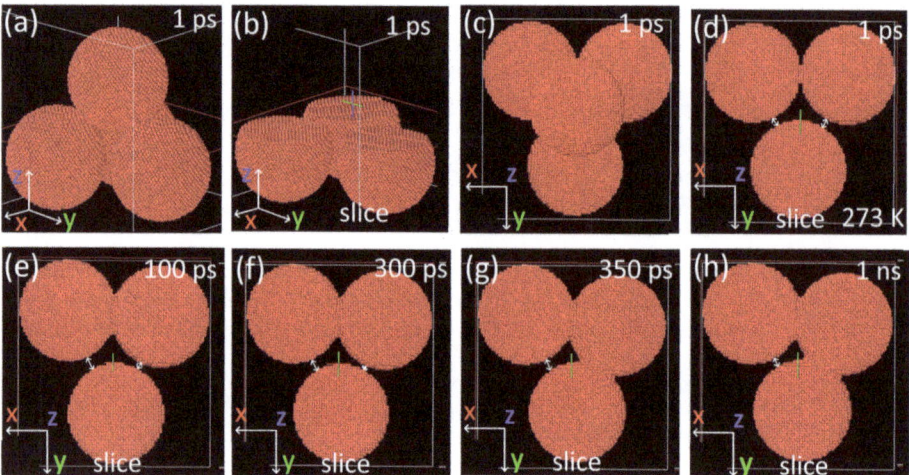

Figure 9. Ovito visualization of results of MD simulations of the nanoparticles annealing at 273 K. (**a**–**c**) A 1 ps of simulation: (**a**) front view, (**b**) how the pile was sliced for a better view, (**c**) top view, and (**d**–**h**) top view sliced at different timesteps. White arrows mark the distance between individual particles.

Aside from the whole body motion at a high temperature of 1100 K (Figure 10), the traveling of individual atoms from the surface of the particles through the interspace between two particles can be detected. By combining these mechanisms, particles meet and coalesce into a single mass (Figure 10d). The initial contact and neck formation between initially non-touching particles occur in a shorter time (100 ps) than during the room temperature annealing (350 ps). This is a result of particle motion and rotations caused by attractive van der Waals forces [42] and energy minimization by lattice rotations between all four particles in the pile. After the contact, the neck growth continues, and the sintering time obeys predictions based on diffusion mechanisms.

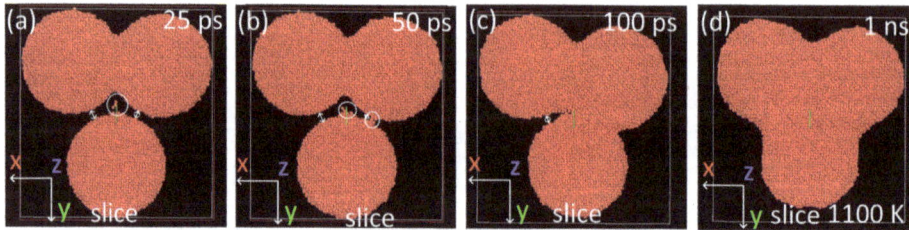

Figure 10. Ovito visualization of results of annealing of nanoparticles at a temperature of 1100 K. Top view sliced at different timesteps (25 ps–1 ns): (**a**) 25 ps, (**b**) 50 ps, (**c**) 100 ps, and (**d**) 1 ns. White arrows mark the distance between individual particles, and white circles mark the motion of individual atoms through the void.

3.2.2. Motion of Individual Atoms

A selection of atoms from the vicinity of the neck of two joined particles was made at one ns of simulation (Figure 11d,h). Images of this particle selection at lower simulation times (Figure 11a–h) indicate the original position of atoms contributing to the necking. Most of these atoms are close to the contact area (the distance is lower than four lattice parameters—Figure 11a). Surface atoms are the only ones traveling more considerable distances (Figure 11f). Thus, the maximum final displacement of inner atoms is approximately 1.4 nm, while the final displacement of surface atoms can reach 4 nm.

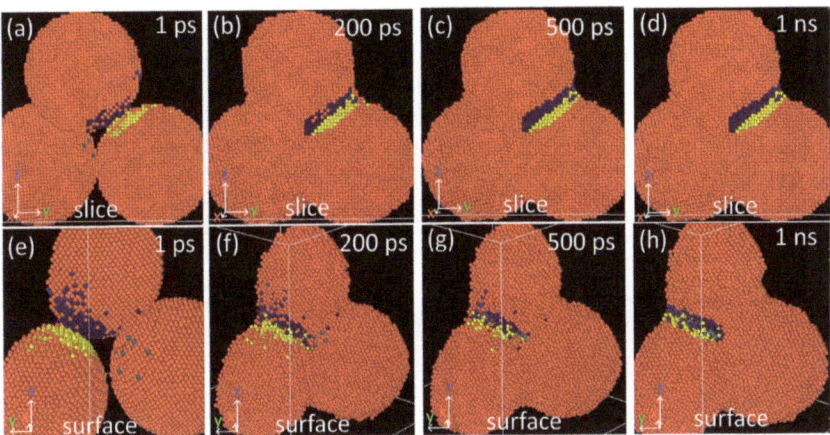

Figure 11. Detail of necking of nanoparticles annealing at 1100 K at different time steps. The selection is colored according to the particle index: yellow, blue, and green atoms correspond to the selection in each particle; red atoms are outside of the selection. (**a**–**d**) Front view, a cross-section through the interior of particles, and (**e**–**h**) back view of the surface.

The motion of inner atoms unaffected by necking at 1100 K is visualized in Figure 12. A truncated conical volume was selected at the beginning of the simulation time, cutting through both particle surfaces (Figure 12a–d). The images of the selection at 1 ns show that only surface atoms shift and the inner ones remain stationary.

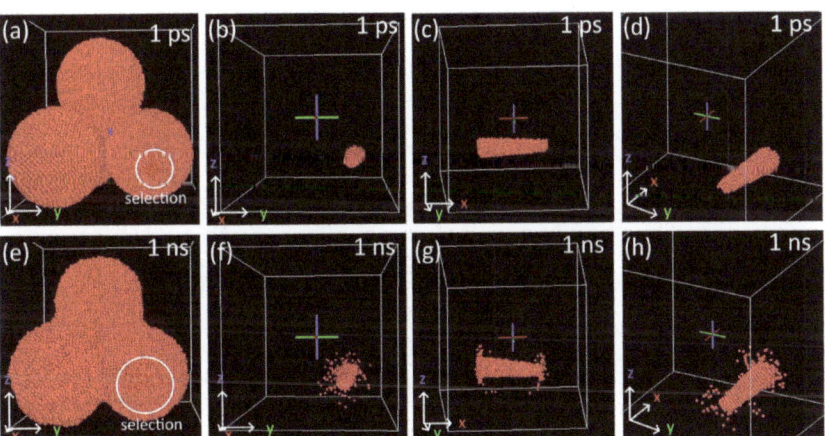

Figure 12. Selection of atoms cutting through the nanoparticle. The selection was made manually at 1 ps of simulation: (**a**–**d**) front and side views at RT; (**e**–**h**) the corresponding views at 1 ns. The simulation was made at 1100 K.

3.2.3. Diffusion Coefficients

The diffusion coefficient was calculated from the sintering of two particles (Figure 12) with a radius of 3.5 nm at 1100 K from the initial slope of the atomic mean-square displacement, MSD, using the Einstein relation:

$$D = \frac{1}{2dN} \lim_{t \to \infty} \frac{d}{dt} \langle \sum_{i=1}^{N} (r_i(t) - r_i(t=0))^2 \rangle \quad (4)$$

where N is the number of atoms in a particle, d is the dimensionality, and r_i is the position of each atom. The calculated values for the same and different orientations of particles are 3.4×10^{-12} m^2/s and 1.1×10^{-11} m^2/s, respectively. The values match well with the diffusion coefficient for 1100 K that Rahbar et al. [40] obtained from calculations with different interatomic potentials. The increase in the total diffusion coefficient in misoriented particles is related to the formation of an additional non-crystalline area at the interface. The images of the nanoparticles with the same and different orientations at three simulation timesteps (Figure 13) confirm a lower sintering time for differently oriented particles. This phenomenon aligns with observations made in simulations of Al and Au nanoparticles [43,44], where a slower sintering evolution of particles with initially larger areas was reported.

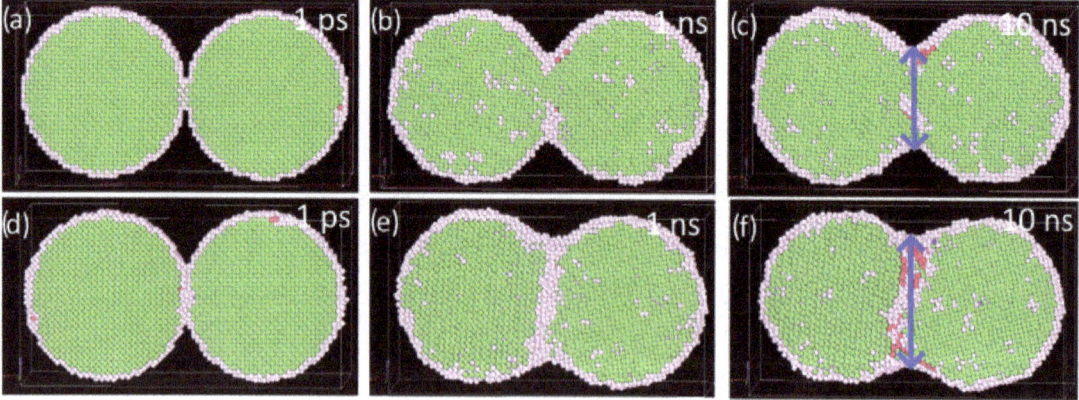

Figure 13. Sintering of two particles at 1100 K for two different particle orientations: (**a–c**) same initial orientation of particles; (**d–f**) differently oriented nanoparticles ([100], [010], [001], and [1–10], [001], [110] directions parallel to x, y, and z axes). The atoms are colored according to the common neighbor analysis: green—fcc, red—hcp, and gray—not defined lattice structure. The final length of the sintered particles is marked by a blue arrow.

The diffusion coefficient value, 7.4×10^{-12} m^2/s, of individual particles in the nanoparticle pile with the same initial orientation (Figure 13) was computed at 1100 K. This is between the values of diffusion coefficients of two nanoparticles at the same temperature. The value indicates variations in the area of the disoriented lattice at the particle boundary due to mutual rotations of a many-body system.

4. Discussion

The diffusion in MD simulations occurs mainly through the diffusion of surface atoms. The mechanism matches the mechanism predicted in the experimental part, where the calculated activation energy corresponds to surface and grain boundary diffusion values. For nanoparticles, boundaries can be excluded or considered as the surfaces of two particles in contact. This result is expected and corresponds well to other experimental and simulation findings of nanoparticle sintering [28,43,44].

The TEM observations indicate a noticeable change in particle shapes at given temperatures. The images were captured within a few seconds, a time range resulting from the time required to acquire an STEM image. Consequently, this timeframe restricted the ability to detect processes that occurred faster. Each temperature was held for several minutes, allowing slower processes to come into play. When taking the values of gamma and D as constants from the established literature values, the characteristic sintering time as a function of temperature for different values of particle radii can be computed from Equation (1). For the sintering of particles with a diameter under 10 nm, a characteristic sintering time of 100 s is expected at room temperature, while it decreases to several seconds at 300 °C. The estimated characteristic sintering time ranges from seconds to hundreds of seconds for particles with diameters under 20 nm and temperatures over 500 °C. Sintering time estimates obtained from molecular dynamics simulations [29,40] cannot be directly applied to match the experimental results due to the significantly smaller particle sizes and higher temperatures involved in the simulations.

This estimation fits the temperature scale when the sintering of individual particles was observed with a microscope. The coalescence of smaller particles to ones with larger diameters was observed for the temperatures when characteristic sintering times were close to seconds. However, after the quick transformation after each temperature step, the newly formed structures remained unchanged for the ten-minute step, where they were held at a constant temperature. The necking of particles might have occurred. However, it did not result in coalescence, even after a longer time. Monte Carlo simulations [45] show that the increase in the sintering time is linked to the presence of faceted elongated particles. A more isotropic equilibrium shape is formed more slowly than expected for spherical particles because crystal growth on the facets requires the nucleation of new islands.

The influence of particle orientation on the sintering rate was shown in MD simulations, where the lattice mismatch between two particles resulted in lower sintering times. Similar results were reported in MD simulation studies of the sintering of Au and Al nanoparticles [43,44]. Many random orientations exist in our experiments using nanoparticle agglomerates without a preferred initial-contact-area orientation. Consequently, this influence should neutralize across the sample and not significantly impact the results.

Within the context of molecular dynamics simulations, nanoparticles exhibit rotational and translational motions when they are part of agglomerates. Small rotations of particles in MD simulations have been reported by Zhu and Averback [46] on two-particle simulations. They observed an elastic deformation during the early sintering stages, after which a relative rotation between the two particles begins until a low minimum energy grain boundary is achieved. The rotation effect was also observed in experimental Au nanoparticles [47] and numerical Au/Cu nanoparticles [48]. Generally, it is considered to be a process of boundary elimination.

However, the motion in our simulation appears to be unrelated to the alignment of lattice planes, but contributes to accelerating surface minimization. This fact could be the result of complications caused by the presence of several particles. Notably, this effect has not been accounted for in the calculated diffusion coefficients or the subsequently estimated time rates, thus representing an overlooked factor that speeds up the sintering process.

5. Conclusions

In situ TEM observations were used to study the coalescence and sintering of fcc Ni nanoparticles with an initial mean diameter of 8 nm from room temperature to 850 °C. The analysis of the images allowed the measurement of the particle size from which the activation energy of diffusion was determined for temperatures up to 800 °C. The activation energy of diffusion calculated during the direct in situ annealing observation aligns with the existing literature, reinforcing the dominance of surface diffusion as the primary mechanism driving these sintering phenomena through previously unexplored temperature ranges. The phenomena observed in the experimental findings, including rotational and translational movements during sintering at lower temperatures and sintering via surface diffusion, align with molecular dynamics simulations, validating the use of MD simulations for calculations in the temperature range of RT—800 °C. Between 800 and 850 °C, well under the melting temperature predicted for nanoparticles by molecular dynamics simulations, the coalescence of a nanoparticle agglomerate to one single particle was observed, indicating a change in the sintering mechanism. Direct experimental observations of fcc Ni nanoparticle sintering offer valuable data for validating the findings via various MD simulations.

Author Contributions: The following specifies the individual contributions of each author to this research article: conceptualization, methodology, and writing—original draft preparation, L.B.; data curation and validation, R.K.; visualization and investigation, B.K.; resources, software, and writing—review and editing, J.H.; supervision, project administration, and funding acquisition, M.C. All authors have read and agreed to the published version of the manuscript.

Funding: This research was funded by the Czech Science Foundation, grant number 22-22572S.

Data Availability Statement: The raw data supporting the conclusions of this article will be made available by the authors on request.

Conflicts of Interest: The authors declare no conflicts of interest.

References

1. Dreaden, E.C.; Alkilany, A.M.; Huang, X.; Murphy, C.J.; El-Sayed, M.A. The golden age: Gold nanoparticles for biomedicine. *Chem. Soc. Rev.* **2012**, *41*, 2740–2779. [CrossRef]
2. Kuang, X.; Wang, Z.; Luo, Z.; He, Z.; Liang, L.; Gao, Q.; Li, Y.; Xia, K.; Xie, Z.; Chang, R.; et al. Ag nanoparticles enhance immune checkpoint blockade efficacy by promotion of immune surveillance in melanoma. *J. Colloid Interface Sci.* **2022**, *616*, 189–200. [CrossRef]
3. Jadhav, P.; Khalid, Z.B.; Krishnan, S.; Bhuyar, P.; Zularisam, A.W.; Razak, A.S.A.; Nasrullah, M. Application of iron-cobalt-copper (Fe-Co-Cu) trimetallic nanoparticles on anaerobic digestion (AD) for biogas production. *Biomass Convers. Bioref.* **2024**, *14*, 7591–7601. [CrossRef]
4. Kumar, K.H.; Venkatesh, N.; Bhowmik, H.; Kuila, A. Metallic Nanoparticle: A Review. *Biomed. J. Sci. Technol. Res.* **2018**, *4*, 3765–3775.
5. Yang, C.J.; Kim, K.S.; Wu, J. Isolated Fe–Co–Ni nanoparticles in a random arrangement and their magnetic properties. *J. Appl. Phys.* **2001**, *90*, 5741–5746. [CrossRef]
6. Wang, X.; Zhao, F.; Xie, P.; Deng, S.; Xu, N.; Wang, H. Surface emission characteristics of ZnO nanoparticles. *Chem. Phys. Lett.* **2006**, *423*, 361–365. [CrossRef]
7. Faramarzi, M.A.; Forootanfar, H. Biosynthesis and characterization of gold nanoparticles produced by laccase from Paraconiothyrium variable. *Colloids Surf. B Biointerfaces* **2011**, *87*, 23–27. [CrossRef] [PubMed]
8. Das, S.K.; Parandhaman, T.; Pentela, N.; Islam, A.K.M.; Mandal, A.B.; Mukherjee, M. Understanding the Biosynthesis and Catalytic Activity of Pd, Pt, and Ag Nanoparticles in Hydrogenation and Suzuki Coupling Reactions at the Nano–Bio Interface. *J. Phys. Chem. C* **2014**, *118*, 24623–24632. [CrossRef]
9. Lakowicz, J.R.; Gryczynski, I.; Gryczynski, Z.; Murphy, C.J. Luminescence Spectral Properties of CdS Nanoparticles. *J. Phys. Chem. B* **1999**, *103*, 7613–7620. [CrossRef]
10. Seok, S.; Choi, M.; Lee, Y.; Jang, D.; Shin, Y.; Kim, Y.-H.; Jo, C.; Park, S. Ni Nanoparticles on Ni Core/N-Doped Carbon Shell Heterostructures for Electrocatalytic Oxygen Evolution. *ACS Appl. Nano Mater.* **2021**, *4*, 9418–9429. [CrossRef]
11. Xie, Z.; Zhang, T.; Zhao, Z. Ni Nanoparticles Grown on SiO_2 Supports Using a Carbon Interlayer Sacrificial Strategy for Chemoselective Hydrogenation of Nitrobenzene and m-Cresol. *ACS Appl. Nano Mater.* **2021**, *4*, 9353–9360. [CrossRef]
12. Navalón, S.; Garcia, H. Nanoparticles for Catalysis. *Nanomaterials* **2016**, *6*, 123. [CrossRef] [PubMed]
13. Wang, Z.K.; Kuok, M.H.; Ng, S.C.; Lockwood, D.J.; Cottam, M.G.; Nielsch, K.; Wehrspohn, R.B.; Gösele, U. Spin-Wave Quantization in Ferromagnetic Nickel Nanowires. *Phys. Rev. Lett.* **2002**, *89*, 027201. [CrossRef] [PubMed]
14. Zheng, W.T.; Sun, C.Q. Electronic process of nitriding: Mechanism and applications. *Prog. Solid State Chem.* **2006**, *34*, 1–20. [CrossRef]
15. Yoon, D.H.; Kim, S.J.; Jung, J.; Lima, H.S.; Kim, H.J. Low-voltage driving solution-processed nickel oxide based unipolar resistive switching memory with Ni nanoparticles. *J. Mater. Chem.* **2012**, *22*, 20117–20124. [CrossRef]
16. Sze, J.Y.; Tay, B.K.; Pakes, C.I.; Jamieson, D.N.; Prawer, S. Conducting Ni nanoparticles in an ion-modified polymer. *J. Appl. Phys.* **2005**, *98*, 066101. [CrossRef]
17. Fonseca, F.C.; Goya, G.F.; Jardim, R.F.; Muccillo, R.; Carreño, N.L.V.; Longo, E.; Leite, E.R. Superparamagnetism and magnetic properties of Ni nanoparticles embedded in SiO_2. *Phys. Rev. B* **2002**, *66*, 104406. [CrossRef]
18. Li, Q.; Kartikowati, C.W.; Horie, S.; Ogi, T.; Iwaki, T.; Okuyama, K. Correlation between particle size/domain structure and magnetic properties of highly crystalline Fe_3O_4 nanoparticles. *Sci. Rep.* **2017**, *7*, 9894. [CrossRef]
19. Hassan, M.R.; Yasmin, F.; Noor, F.K.; Rahman, M.S.; Uddin, M.S.; Bhowmik, S. Synthesis and Applications of Nickel Nanoparticles (NiNPs)—Comprehensive Review. *JUC* **2023**, *19*, 9–37. [CrossRef]

20. Su, F.; Qiu, X.; Liang, F.; Tanaka, M.; Qu, T.; Yao, Y.; Ma, W.; Yang, B.; Dai, Y.; Hayashi, K.; et al. Preparation of Nickel Nanoparticles by Direct Current Arc Discharge Method and Their Catalytic Application in Hybrid Na-Air Battery. *Nanomaterials* **2018**, *8*, 684. [CrossRef]
21. Pandey, A.; Manivannan, R. A Study on Synthesis of Nickel Nanoparticles Using Chemical Reduction Technique. *Recent Pat. Nanomed.* **2015**, *5*, 33–37. [CrossRef]
22. Ramos, R.; Valdez, B.; Nedev, N.; Curiel, M.; Perez, O.; Salvador, J. Electric discharge synthesis of nickel nanoparticles with virtual instrument control. *Instrum. Sci. Technol.* **2021**, *49*, 499–508. [CrossRef]
23. Mourdikoudis, S.; Simeonidis, K.; Vilalta-Clemente, A.; Tuna, F.; Tsiaoussis, I.; Angelakeris, M.; Dendrinou-Samara, C.; Kalogirou, O. Controlling the crystal structure of Ni nanoparticles by the use of alkylamines. *J. Magn. Magn. Mater.* **2009**, *321*, 2723–2728. [CrossRef]
24. Ahghari, M.R.; Soltaninejad, V.; Maleki, A. Synthesis of nickel nanoparticles by a green and convenient method as a magnetic mirror with antibacterial activities. *Sci. Rep.* **2020**, *10*, 12627. [CrossRef] [PubMed]
25. Maicas, M.; Sanz, M.; Cui, H.; Aroca, C.; Sánchez, P. Magnetic properties and morphology of Ni nanoparticles synthesized in gas phase. *J. Magn. Magn. Mater.* **2010**, *322*, 3485–3489. [CrossRef]
26. Kang, S.-J.L. *Sintering: Densification, Grain Growth, and Microstructure*; Elsevier: Oxford, UK, 2005; pp. 3–7.
27. Jahani, N.; Reihanian, M.; Gheisari, K. Kinetics of recrystallization and microstructure distribution during isothermal annealing of cold rolled nickel. *Mater. Res. Express* **2019**, *6*, 096504. [CrossRef]
28. Tsyganov, S.; Kästner, J.; Rellinghaus, B.; Kauffeldt, T.; Westerhoff, F.; Wolf, D. Analysis of Ni nanoparticle gas phase sintering. *Phys. Rev. B* **2007**, *75*, 045421. [CrossRef]
29. Song, P.; Wen, D. Surface melting and sintering of metallic nanoparticles. *J. Nanosci. Nanotechnol.* **2010**, *10*, 8010–8017. [CrossRef] [PubMed]
30. Zhang, Y.; Zhang, J. Sintering phenomena and mechanical strength of nickel-based materials in direct metal laser sintering process—A molecular dynamics study. *J. Mater. Res.* **2016**, *31*, 2233–2243. [CrossRef]
31. Safina, L.; Baimova, J.; Mulyukov, R. Nickel nanoparticles inside carbon nanostructures: Atomistic simulation. *Mech. Adv. Mater. Mod. Process.* **2019**, *5*, 2. [CrossRef]
32. Hanuš, J.; Vaidulych, M.; Kylián, O.; Choukourov, A.; Kousal, J.; Khalakhan, I.; Cieslar, M.; Solař, P.; Biederman, H.J. Fabrication of Ni@Ti core–shell nanoparticles by modified gas aggregation source. *Phys. D Appl. Phys.* **2017**, *50*, 475307. [CrossRef]
33. Schneider, C.A.; Rasband, W.S.; Eliceiri, K.W. NIH Image to ImageJ: 25 years of image analysis. *Nat. Methods* **2012**, *9*, 671–675. [CrossRef]
34. Hirel, P. Atomsk: A tool for manipulating and converting atomic data files. *Comput. Phys. Comm.* **2015**, *197*, 212–219. [CrossRef]
35. Thompson, A.P.; Aktulga, H.M.; Berger, R.; Bolintineanu, D.S.; Brown, W.M.; Crozier, P.S.; in 't Veld, P.J.; Kohlmeyer, A.; Moore, S.G.; Nguyen, T.D.; et al. LAMMPS—A flexible simulation tool for particle-based materials modeling at the atomic, meso, and continuum scales. *Comp. Phys. Comm.* **2022**, *271*, 10817. [CrossRef]
36. Kavousi, S.; Novak, B.R.; Baskes, M.I.; Zaeem, M.A.; Moldovan, D. Modified embedded-atom method potential for high-temperature crystal-melt properties of Ti–Ni alloys and its application to phase field simulation of solidification. *Model. Simul. Mater. Sci. Eng.* **2019**, *28*, 015006. [CrossRef]
37. Shinoda, W.; Shiga, M.; Mikami, M. Rapid estimation of elastic constants by molecular dynamics simulation under constant stress. *Phys. Rev. B* **2004**, *69*, 134103. [CrossRef]
38. Stukowski, A. Visualization and analysis of atomistic simulation data with OVITO—The Open Visualization Tool. *Model. Simul. Mater. Sci. Eng.* **2010**, *18*, 015012. [CrossRef]
39. Nichols, F.A. Coalescence of two spheres by surface diffusion. *J. Appl. Phys.* **1966**, *37*, 2805. [CrossRef]
40. Rahbar, H. Sintering rate of nickel nanoparticles by molecular dynamics. *J. Phys. Chem.* **2023**, *127*, 6802–6812. [CrossRef]
41. Kelchner, C.L.; Plimpton, S.J.; Hamilton, J.C. Dislocation nucleation and defect structure during surface indentation. *Phys. Rev. B* **1998**, *58*, 11085. [CrossRef]
42. Shrestha, S.; Wang, B.; Dutta, P. Nanoparticle processing: Understanding and controlling aggregation. *Adv. Colloid Interface Sci.* **2020**, *279*, 102162. [CrossRef]
43. Arcidiacono, S.; Bieri, N.R.; Poulikakos, D.; Grigoropoulos, C.P. On the coalescence of gold nanoparticles. *Int. J. Multiph. Flow* **2004**, *30*, 979–994. [CrossRef]
44. Raut, J.S.; Bhagat, R.B.; Fichthorn, K.A. Sintering of aluminum nanoparticles: A molecular dynamics study. *Nanostruct. Mater.* **1998**, *10*, 837–851. [CrossRef]
45. Combe, N.; Jensen, P.; Pimpinelli, A. Changing shapes in the nanoworld. *Phys. Rev. Lett.* **2000**, *85*, 110. [CrossRef]
46. Zhu, H.; Averback, R.S. Sintering of Nanoparticle Powders: Simulations and Experiments. *Mater. Manuf. Process.* **1996**, *11*, 905–923. [CrossRef]
47. Iijima, S.; Ajayan, P.M. Substrate and size effects on the coalescence of small particles. *J. Appl. Phys.* **1991**, *70*, 5138–5140. [CrossRef]
48. Zeng, P.; Zajac, S.; Clapp, P.C.; Rifkin, J.A. Nanoparticle sintering simulations. *Mater. Sci. Eng. A* **1998**, *252*, 301–306. [CrossRef]

Disclaimer/Publisher's Note: The statements, opinions and data contained in all publications are solely those of the individual author(s) and contributor(s) and not of MDPI and/or the editor(s). MDPI and/or the editor(s) disclaim responsibility for any injury to people or property resulting from any ideas, methods, instructions or products referred to in the content.

MDPI
St. Alban-Anlage 66
4052 Basel
Switzerland
www.mdpi.com

MDPI Books Editorial Office
E-mail: books@mdpi.com
www.mdpi.com/books

Disclaimer/Publisher's Note: The statements, opinions and data contained in all publications are solely those of the individual author(s) and contributor(s) and not of MDPI and/or the editor(s). MDPI and/or the editor(s) disclaim responsibility for any injury to people or property resulting from any ideas, methods, instructions or products referred to in the content.

www.ingramcontent.com/pod-product-compliance
Lightning Source LLC
LaVergne TN
LVHW070234100526
838202LV00015B/2129